GASEOUS DIELECTRICS X

Gaseous Dielectrics X

Edited by

Loucas G. Christophorou
Academy of Athens
Athens, Greece

and

James K. Olthoff
National Institute of Standards and Technology
Gaithersburg, Maryland

and

Panayota Vassiliou
National Technical University of Athens
Athens, Greece

Springer

Proceedings of the Tenth International Symposium on Gaseous Dielectrics,
held March 29–April 2, 2004, at the Astir Palace Vouliagmeni Hotel in Athens, Greece

ISBN 978-1-4613-4745-3 ISBN 978-1-4419-8979-6 (eBook)
DOI 10.1007/978-1-4419-8979-6

Originally published by Springer Science+Business Media, Inc in 2004
Softcover reprint of the hardcover 1st edition 2004

9 8 7 6 5 4 3 2 1

springeronline.com

PREFACE

The Tenth International Symposium on Gaseous Dielectrics was held at the Astir Palace Vouliagmeni Hotel, Athens, Greece, March 29-April 2, 2004. The symposium continued the interdisciplinary character and comprehensive approach of the preceding nine symposia.

Gaseous Dielectrics X is a detailed record of the symposium proceedings. It covers recent advances and developments in a wide range of basic, applied, and industrial areas of gaseous dielectrics. It is hoped that *Gaseous Dielectrics X* will aid future research and development in, and encourage wider industrial use of, gaseous dielectrics.

The Organizing Committee of the Tenth International Symposium on Gaseous Dielectrics consisted of L. G. Christophorou (Chairman, Greece), J. K. Olthoff (co-Chairman, USA), A. Bulinski (Canada), A. H. Cookson (USA), C. T. Dervos (Greece), J. de Urquijo (Mexico), J. Blackman (USA), O. Farish (UK), M. E. Fréchette (Canada), I. Gillimberti (Italy), A. Garscadden (USA), A. Gleizes (France), H. Hama (Japan), T. Kawamura (Japan), E. Marode (France), I. W. McAllister (Denmark), H. Morrison (Canada), A. H. Mufti (Saudi Arabia), L. Niemeyer (Switzerland), W. Pfeiffer (Germany), Y. Qiu (China), I. Sauers (USA), M. Schmidt (Germany), H.-H. Schramm (Germany), L. van der Zel (USA), S. Yanabu (Japan), Y. Wang (USA), and J. W. Wetzer (The Netherlands). The Local Arrangements Committee consisted of J. N. Avaritsiotis, P. Vassiliou, C. T. Dervos of The National Technical University of Athens, C. A. Stassinopoulos of the Aristotelian University of Thessaloniki, and D. Stavropoulos of the Public Power Corporation of Greece. The contribution of each member of these committees, the work of the session chairmen, the interest of the participants, and the advice of many colleagues are gratefully acknowledged. I am especially indebted to Professors C. T. Dervos and P. Vassiliou of the National Technical University of Athens for their excellent collaboration and hard work that made this Symposium such a successful meeting, and to Dr. A. H. Cookson of NIST for organizing the Session on Alternative Technologies and to Dr. H. H. Schramm of Siemens for organizing the Session on Gas-Insulated Equipment.

The Tenth International Symposium on Gaseous Dielectrics was hosted by the Academy of Athens (Greece), the National Technical University of Athens (Greece), and the National Institute of Standards and Technology (USA). It was sponsored by the Academy of Athens (Greece); the National Technical University of Athens (Greece); the Public Power Corporation of Greece; ABB Switzerland Ltd.; Siemens PTD, High Voltage Division (Germany); the European Office of Aerospace Research and Development of USAF; Tokyo Electric Power Company (Japan); the Kansai Electric Power Co. (Japan); Chubu Electric Power Co. (Japan); TMT&D Corporation (Japan); and Japan AE Power Systems Corporation (Japan). The symposium was technically sponsored by the IEEE Dielectrics and Electrical Insulation Society. The support of the host institutions and the financial assistance of the sponsors are acknowledged with gratitude.

Loucas G. Christophorou, Symposium Chairman
Athens, Greece

CONTENTS

SECTION 1: BASIC PHYSICS OF GASEOUS DIELECTRICS

SECTION 2: GAS-DISCHARGE BEHAVIOR: BASIC MECHANISMS

SECTION 3: PARTIAL DISCHARGES: MECHANISMS AND APPLICATIONS

SECTION 4: GASEOUS DIELECTRICS AS PLASMA PROCESSING GASES

SECTION 5: HIGH PRESSURE GAS DIELECTRICS / MIXTURES

SECTION 6: GAS MIXTURES AS ALTERNATIVES TO SF_6

SECTION 7: SF_6 INSULATED APPARATUS: DIAGNOSTICS

SECTION 8: GAS-INSULATED TRANSMISSION LINES

SECTION 9: GAS-INSULATED EQUIPMENT

SECTION 10: ALTERNATIVE TECHNOLOGIES

SECTION 11: INDUSTRIAL OUTLOOK: REDUCTION IN SF_6 USE AND EMISSIONS

APPENDIX A : BANQUET TALK

GASEOUS DIELECTRICS X

SECTION 1: BASIC PHYSICS OF GASEOUS DIELECTRICS

TOTAL IONIZATION CROSS SECTIONS OF MOLECULES BY ELECTRON IMPACT*

Yong-Ki Kim†

Summary: The theoretical basis for the binary-encounter-Bethe (BEB) model is outlined, and the model is compared to available experiments and other theories for samples of stable molecules, radicals, and positive ions. Possible directions for improving the BEB model are also presented.

1. INTRODUCTION

Electron-impact ionization cross sections of atoms and molecules are widely used in many applications such as the modeling of discharge phenomena, plasma processing of materials, fusion plasma modeling, and design of accelerators. Although quantum mechanics was invented almost 80 years ago, we still have not found a reliable theory that can predict ionization cross sections of atoms and molecules of practical interest. During the last decade, theorists finally succeeded in solving the problem of electron-hydrogen atom scattering.[1–3] However, there is no clean *ab initio* theory for ionization cross sections of molecules, even for small, simple molecules such as H_2. The basic difficulty in developing such a theory is in describing the continuum states of the incident as well as ejected electrons before and after the collision. Hence, most theories for molecular ionization use various levels of approximations and often require empirical parameters, which further limit the applicability of such theories.

The classical theory by Gryzinski[4] does not require empirical parameters, but it works well only on single-orbital molecules such as H_2. Other models, such as the DM formalism,[5] contain empirical parameters the choice of which is not

---

* Work supported in part by the U.S. Department of Energy, Office of Fusion Sciences.

† National Institute of Standards and Technology, Gaithersburg, MD 20899-8422, U.S.A.

Gaseous Dielectrics X.
Edited by Christophorou *et al.*, Springer, New York, 2004.

always transparent when a user wants to apply the theory to a new class of molecules.

In this article, I will summarize a theory, called the Binary-Encounter-Bethe (BEB) model,[6] which (a) does not contain empirical parameters, (b) offers a simple equation and recipe for the ionization cross section of individual molecular orbitals, and (c) provides reliable cross sections at low as well as high incident electron energies. The BEB model has been verified to produce reliable cross sections for dozens of stable, neutral, large and small molecules (e.g., H_2, SF_6), radicals, and singly-charged positive ions.

The theoretical basis for the BEB model is outlined in Sec. 2, the BEB cross sections for ionization of sample molecules, radicals, and ions are compared to existing experiments and other semi-empirical theories in Sec. 3, and unsolved issues and future prospects are listed in Sec. 4.

2. THEORETICAL BASIS FOR THE BEB MODEL

Ionizing collisions can be qualitatively divided into two categories; soft and hard collisions. Soft collisions involve small momentum transfers from the incident electron to target bound electrons, and primarily generate slow ejected electrons. Hard collisions involve large momentum transfers and mostly generate fast ejected electrons. This division, however, is only qualitative and hence arbitrary. Roughly two-thirds of total ionization cross sections come from the ejection of slow electrons.[7]

Most collision theories based on quantum mechanics address both types of collisions, a typical one being the Born approximation. An exception is the Mott cross section, which is the exact solution for an idealized problem, namely, the collision of two *free* electrons. The Mott cross section does not include soft collisions arising from the electric dipole ($E1$) interaction between the incident and target electron, while the Born approximation does. The $E1$ interaction occurs only for bound electrons.

There have been many attempts in the past by theorists to combine the Mott cross section with the Born cross section because these two theories are the simplest ones based on quantum mechanics. The main hurdle in this approach is the fact that it is *quantitatively* difficult to isolate the part in the Born cross section that is also included in the Mott cross section—i.e., to avoid duplication. This is the primary reason that existing theories of this type had to introduce empirical parameters.

The BEB model overcame this difficulty by requiring the asymptotic forms (high incident electron energy) of the ionization cross section itself and the matching stopping power to satisfy the correct forms predicted by the Born approximation.[6] This requirement produced an equation for differential ionization cross section, i.e., the energy distribution of ejected electrons as a function of the incident electron energy, which we labeled as the binary-encounter-dipole (BED) model. The total ionization cross section is obtained by integrating the differential ionization cross section over the ejected electron energy.

The BED model uses optical oscillator strengths (OOSs) for individual orbitals as input. The OOSs for individual orbitals, however, are not easy to calculate or measure, except for simple targets such as H, He, and H_2. To overcome this difficulty, a simplified form of OOS was adopted, and a versatile expression was derived for the total ionization cross section of individual molecular orbitals. This expression is the BEB model, and provides the ionization cross section as a function of the incident electron energy T for each molecular orbital. The BEB model uses the orbital binding energy B, orbital kinetic energy U, and the orbital electron occupation number N:

$$\sigma_{BEB}(t) = \frac{S}{t+(u+1)/n}\left[\frac{\ln t}{2}\left(1-\frac{1}{t^2}\right)+1-\frac{1}{t}-\frac{\ln t}{t+1}\right], \qquad (1)$$

where $t=T/B$, $S=4\pi a_0^2 N(R/B)^2$, $u=U/B$, and n=principal quantum number of the dominant atomic orbital if $n > 2$. For orbitals in K and L shells, n is set to unity. The total ionization cross sections of a molecule is the sum of Eq. (1) for all occupied orbitals of a molecule. The orbital parameters B, U, and N are readily available from popular molecular wave function codes such as Gaussian and GAMESS.

The first logarithmic term in Eq. (1) comes from the soft collision part of the Born approximation, the rest in the square brackets comes from the Mott cross section. However, the denominator, $t+(u+1)/n=[T+(U+B)/n]/B$, is not based on rigorous quantum mechanics. It comes from the binary-encounter theory with the argument that the *effective* energy difference between the incident and target electrons is the incident energy T plus the potential energy $B+U$ of the target electron.[8] The kinetic energy becomes high for orbitals with many radial function nodes, and we found it necessary to reduce it for radial functions with many nodes. This is the reason that we introduced the division by n in Eq. (1) to avoid making the cross section unrealistically small for valence orbitals with many nodes.[9,10] Note that $t+(u+1)/n$ approaches t as the incident energy T increases. The Born approximation is valid at high t, and it has only t in the denominator.

For many stable molecules, Eq. (1) has produced total ionization cross sections in excellent agreement both in magnitude (15% or better at the peak) and shape from threshold to several keV in T. In addition to its simplicity and freedom from adjustable parameters, the BEB model:

(a) Uses orbital constants specific to a molecule rather than using simple additivity rules based on atoms in a molecule
(b) Has the correct asymptotic behavior ($\ln t/t$) predicted by the Born approximation, unlike theories based on classical physics
(c) Applies to stable molecules and radicals of all sizes

(d) Produces reliable ionization cross sections for singly-charged positive ions by simply replacing n in Eq. (1) by $2n$
(e) Can be applied to atoms essentially in the same form.[11,12]

On the other hand, the BEB model (and its parent BED model) cannot account for:
(f) Neutral dissociation of the target molecule
(g) Ionization of negative ions
(h) Ionization of two or more electrons from the same molecular orbital
(i) Indirect ionization such as ionization resulting from the excitation of inner-shell electrons (this is known as *excitation-autoionization*)
(j) Resonances commonly found near ionization thresholds.

Since the BEB model assumes that an ion is produced whenever the energy transferred to the target exceeds an orbital binding energy, the BEB model overestimates an ionization cross section if neutral dissociation is a significant fraction of the events that follow such an energy transfer. Conversely, if a significant contribution comes from excitation-autoionization, then the BEB model underestimates the ionization cross section.

3. COMPARISONS TO EXPERIMENTS AND OTHER THEORIES

We compare BEB cross sections with available experimental data for N_2, CF_4, SF_6, CH_3, N_2^+ and SF_5 in Figs. 1–6. It is clear that the BEB model can produce reliable ionization cross sections for small as well as large, stable molecules. It can also produce reliable cross sections for hydrocarbon radicals and singly-charged positive ions. BEB cross sections of about 90 molecules are available from a public NIST web site.[13]

However, the BEB cross sections calculated from Eq. (1) do not agree well with the experimental data on radicals containing fluorine, CF_x, SF_x, and NF_x, although the BEB cross sections for the stable molecules CF_4 and SF_6 agree well with experiments.

Huo has modified the BED model by replacing the first logarithmic term in Eq. (1), which represents the $E1$ interaction, with the Born cross section.[19–21] Her model is called a *simplified version of the improved* BED, or siBED model. She replaced the $E1$ interaction term with her version of the full Born cross section, which contains two adjustable parameters and a generic OOS more elaborate than the one used in the BEB model. Although the thorny issue of avoiding the duplication of hard collisions contained both in the Born and Mott cross sections has not been addressed directly by Huo, presumably it is solved by setting two adjustable parameters in the model to reproduce well known cross sections of other molecules. The siBED cross sections are in excellent agreement with the experimental data on CF_x, NF_x, and SF_x.

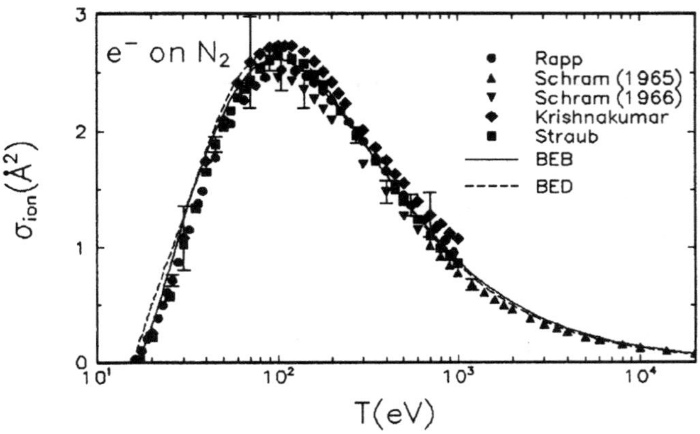

Figure 1. Total ionization cross section of N_2 (from Fig. 2 of Ref. 14). Solid symbols are experimental data. Optical oscillator strengths needed for the BED model was deduced from photoionization experiments.

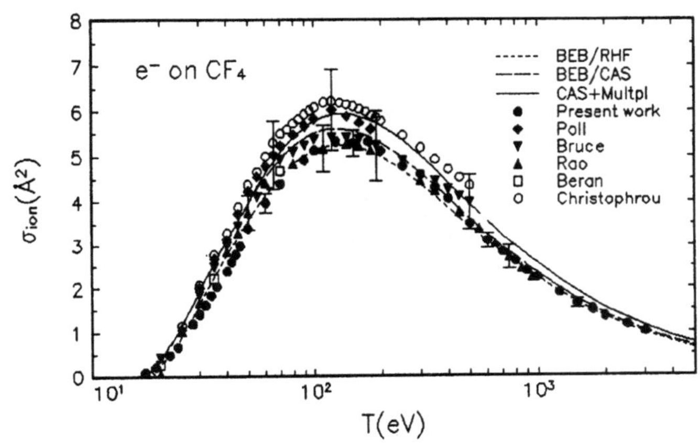

Figure 2. Total ionization cross section of CF_4 (from Fig. 4 of Ref. 15). RHF=restricted Hartree-Fock (uncorrelated) wave function, CAS=complete active space (correlated) wave function, Multipl=includes multiple ionization, solid circle=experiment by Nishimura (Ref. 15), open circle=recommended cross section by Christophorou and Olthoff, other symbols are experimental data.

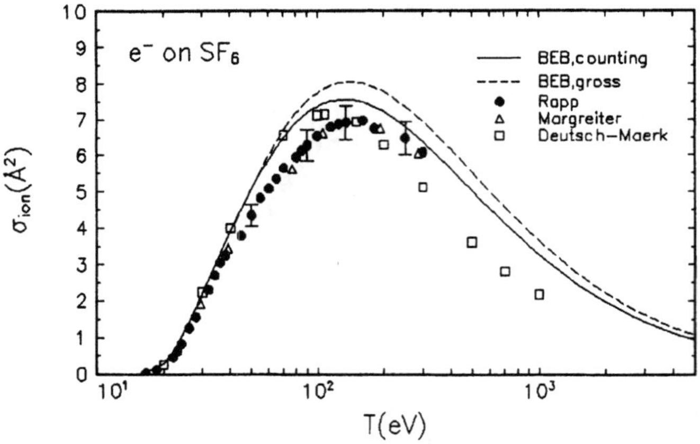

Figure 3. Total ionization cross section of SF_6 (adapted from Fig. 1 of Ref. 16). Rapp=experimental absolute cross section, Margreiter=relative cross section normalized to Rapp, Deutsch-Märk=DM formalism (Ref. 5).

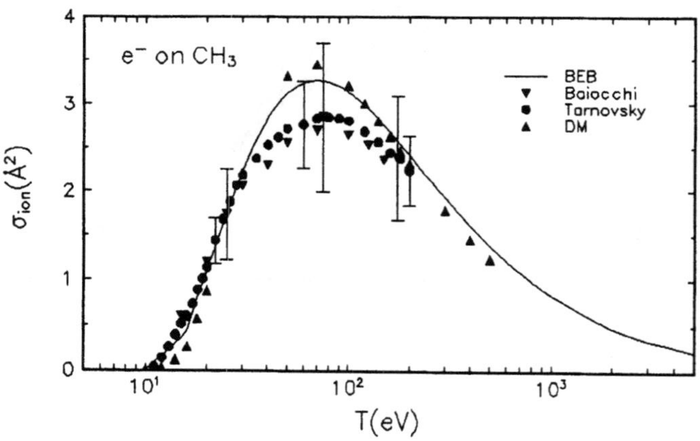

Figure 4. Total ionization cross section of CH_3 (from Fig. 5 of Ref. 17). DM=DM formalism (Ref. 5), other symbols are experimental data.

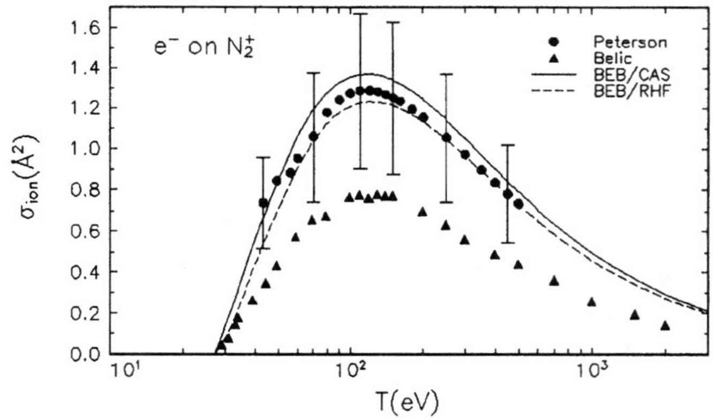

Figure 5. Total ionization cross section of N_2^+ (from Fig. 2 of Ref. 18). Symbols are experimental data. See caption of Fig. 2 for CAS, RHF.

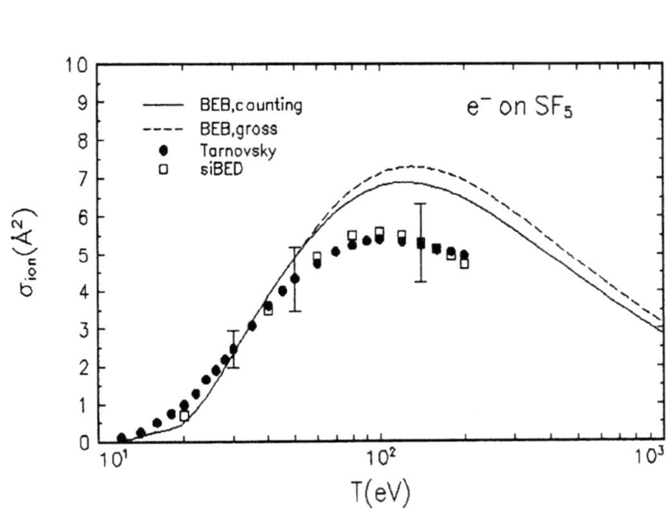

Figure 6. Total ionization cross section of SF_5 (adapted from Fig. 2 of Ref. 16). Tarnovsky=experiment, siBED=new model by Huo (Ref. 21).

Neutral dissociation amounts to about 20% of the total ionization cross section in CF_4.[22] We found that the difference between the experimental ionization cross section measured by Nishimura[15] and the BEB cross section[15] closely follows the shape of the measured cross section for the neutral dissociation,[15] strongly indicating that about one-half of the neutral dissociation comes from energy transfers exceeding the ionization energy of CF_4. The other half presumably resulted from energy transfers less than the ionization energy. The magnitudes of neutral dissociation in CF_x, NF_x, and SF_x are unknown.

4. CONCLUSIONS AND FUTURE OUTLOOK

The BEB model has demonstrated its versatility in providing reliable total ionization cross sections for a wide range of molecules, large and small, including radicals and positive ions. The formula and procedure to use for the model are simple enough that anyone who has access to popular quantum chemistry codes can produce the necessary molecular orbital constants and calculate the desired ionization cross sections. For many molecules, ionization cross sections are readily available from the NIST web site,[13] which is updated as new BEB cross sections become available. The web site also offers instantaneous on-line calculations of total ionization cross sections of the molecules included in the web site by simply typing in the incident electron energy.

Huo has offered an improved version of the BED (siBED) model.[19–21] Her siBED model is not as simple to use as the original BED/BEB model, and it has been tested mostly on fluorine-containing radicals, after adjusting two empirical parameters in the model to known cross sections of other stable molecules. It is desirable to establish the versatility of the model and the universality of the empirical parameters in the model by applying it further to a wider range of molecules.

As for the present BEB model, we have begun to explore the possibility of accounting for the excitation-autoionization (EA) using a new scaling method, which turned out to be very successful in atoms. For electric dipole ($E1$) and spin allowed excitations in atoms, it was found that the plane-wave Born (PWB) excitation cross sections were converted to accurate cross sections by the following scaling:

$$\sigma_{BE}(T) = \frac{T}{T+B+E}\sigma_{PWB}(T), \qquad (2)$$

where E=excitation energy. This scaling is called the BE scaling,[23] and it has been verified for discrete excitations of many neutral atoms, hydrogen through thallium. For singly-charged positive ions of atoms, a similar scaling converted Coulomb-Born (CB) excitation cross sections into accurate cross sections:

$$\sigma_E(T) = \frac{T}{T+E}\sigma_{CB}(T). \quad (3)$$

This scaling is called the E scaling.[24] The BE and E scaling methods provide easy and efficient ways to calculate accurate excitation cross sections not only for the excitations to bound excited states, but also to the core-excited, autoionizing states, which are extremely difficult to calculate with conventional theories such as the R-matrix and close-coupling methods. We have used these two scaling methods to reveal that almost one-half of the total ionization cross sections of Al, Ga, and In come from excitation-autoionization.[11]

For almost all neutral atoms, the BEB model alone was insufficient to account for total ionization cross sections because of the contributions from EA, which is not included in the BEB model. The fact that the BEB model has been very successful on so many molecules indicates that EA may not play a significant role in molecules. Core-excited states in molecules may preferably lead to neutral dissociation. This is a worthwhile subject to pursue.

On the other hand, comparisons of BEB ionization cross sections of CS and S_2 with available experiments[25] strongly suggest that the contribution from EA is significant. A similar situation also exists for OH. Our preliminary study indicates that the E scaling may convert PWB cross sections to reliable ones for the excitation of neutral molecules by electron impact. If the applicability of the E scaling—or a variation of it—to neutral molecules is verified, then not only we can get accurate cross sections for excitations to bound excited states, but also we can accurately predict the contribution to total ionization cross sections through excitations of core electrons to auoionizing states.

5. ACKNOWLEDGEMENTS

The author is grateful to Dr. W. M Huo for providing her results on SF_x prior to publication, and to Prof. M. A. Ali and Dr. K. K. Irikura for generating ionization cross sections of many molecules and enlightening discussions of molecular properties.

6. REFERENCES

1. I. Bray and A. T. Stelbovics, Adv. At. Mol. Phys. **35**, 290 (1995).
2. M. Baertschy, T. N. Resigno, W. A. Issacs, X. Li, and C. W. McCurdy, Phys. Rev. A **63**, 022712 (2001).
3. F. Robischeaux, M. S. Pindzola, and D. R. Plante, Phys. Rev. A **55**, 3573 (1997).
4. M. Gryzinski, Phys. Rev. **138**, A305 (1965); **138**, A322 (1965); **138**, A336 (1965).
5. H. Deutsch, K. Becker, S. Matt, T. D. Märk, Int. J. Mass Spectrom. **197**, 37 (2000).
6. Y.-K. Kim and M. E. Rudd, Phys. Rev. A **50**, 3954 (1994).
7. Y.-K. Kim, Radiat. Res. **64**, 205 (1975).
8. L. Vriens, in *Case Studies in Atomic Physics*, Vol. 1, edited by E. W. McDaniel and M. R. C. McDowell (North Holland, Amsterdam, 1969), p. 335.

9. W. Hwang, Y.-K. Kim, and M. E. Rudd, J. Chem. Phys. **104**, 2956 (1996).
10. M. A. Ali, K. K. Irikura, and Y.-K. Kim, Int. J. Mass Spectrom. **201**, 187 (2000).
11. Y.-K. Kim and P. M. Stone, Phys. Rev. A **64**, 052707 (2001).
12. Y.-K. Kim and J. P. Desclaux, Phys. Rev. A **66**, 042708 (2002).
13. http://physics.nist.gov/ionxsec.
14. W. Hwang, Y.-K. Kim, and M. E. Rudd, J. Chem. Phys. **104**, 2956 (1996).
15. H. Nishimura, W. M. Huo, M. A. Ali, and Y.-K. Kim, J. Chem. Phys. **110**, 3811 (1999).
16. M. A. Ali, K.K. Irikura, and Y.-K. Kim, Int. J. Mass Spectrom. **201**, 187 (2000).
17. Y.-K. Kim and K. K. Irikura, in ***Atomic and Molecular Data and Their Applications***, edited by K. A. Berrington and K. L. Bell (American Inst. of Physics, 2000), p 220.
18. Y.-K. Kim, K. K. Irikura, and M. A. Ali, J. Res. NIST **105**, 285 (2000).
19. W. M. Huo, Phys. Rev. A **64**, 042719 (2001).
20. W. M. Huo, V. Tarnovsky, and K. H. Becker, Chem. Phys. Lett. **358**, 328 (2002).
21. W. M. Huo, V. Tarnovsky, and K. H. Becker, Int. J. Mass Spectrom. in print (2004).
22. S. Motlag and J. H. Moore, J. Chem. Phys. **109**, 432 (1998).
23. Y.-K. Kim, Phys. Rev. A **64**, 032713 (2001).
24. Y.-K. Kim, Phys. Rev. A **65**, 022705 (2002).
25. Y.-K. Kim, W. Hwang, N. M. Weinberger, M. A. Ali, and M. E. Rudd, J. Chem. Phys. **106**, 1026 (1997).

SF_6 DIELECTRIC MOLECULES: ELECTRON SCATTERING DYNAMICS AND POSSIBLE APPLICATIONS

Casten Makochekanwa, Mineo Kimura* and Osamu Sueoka

Abstract: Electron scattering dynamics have been investigated experimentally for the energy range of 0.8-1000eV, using an absorption type time-of-flight apparatus. The elastic process has been also studied theoretically using the continuum multiple scattering method. The present cross section results are found to reasonably agree with the previous similar results by other groups in the energy region we overlap. Peaks have been observed at 2.5, 7.0, 12.0 and ~30 eV and attributed to the a_{1g}, t_{1g}, t_{2g} and e_g resonances studied by Dehmer *et al.*

1. INTRIDUCTION

SF_6, sulfur hexafluoride, is a man-made gas, and now is one of the most widely and extensively used gases for commercial and research applications.[1] Therefore, its various physical and chemical properties, and dynamical behavior under various application conditions have been broadly studied. However, it is also now known as one of the most serious greenhouse gases since it is an efficient absorber of infrared radiation at wavelength near 10.5 μm, and is solidly stable in chemical and photolytic reactions and hence, stays permanently once released in air. Hence, concerns over its environmental impact have newly acquired interests for more comprehensive study of this gas.

Earlier studies on the dynamical aspects of this molecule include, total cross section measurements by electron and positron impact by Dababneh *et al.*,[2] electron impact by Kennerly *et al.*, [3] Kasperski *et al.*,[4] and electron attachment studies by Spence and Schulz[5]. Phelps and Van Brunt[6] have compiled a lot of available data on electron impact, and published a large set of cross section data for a wide range of energy from 1 meV to 1 keV.

* Corresponding author: Graduate School of Science and Engineering, Yamaguchi University, Tokiwadai 2-16-1, Ube, Yamaguchi 755-8611, Japan.

E-mail: mineo@yamaguchi-u.ac.jp TEL: (81) 836 85 9806, FAX: (81) 836 85 9801.

Gaseous Dielectrics X.
Edited by Christophorou *et al.*, Springer, New York, 2004.

2. EXPERIMENTAL AND THEORETICAL PROCEDURES

2.1 Experimental

The apparatus used for the total cross section (TCS) measurements investigated here is a linear transmission type time-of-flight (TOF) with a 600 mm path length. Details about the apparatus and procedure can be found in our earlier report.[7] Only a few characteristics are highlighted here. A ^{22}Na radioisotope with an activity of 90 μ Ci was used for the beam source, with the slow electron beams, of energy width of around 1 eV (FWHM), being produced as secondary electrons resulting from multiple scattering from the tungsten moderator.

In this experiment a magnetic field is used for beam transportation. Hence, a forward scattering correction is necessary for the measured TCS values. Details about the simulation method for this correction have been published elsewhere.[7] The correction for these TCSs for electron scattering was carried out using the differential cross sections of Srivastava *et al.*[9] and Sakae *et al.*[10]

The TCS values, Q_t, are given as

$$Q_t = (-1/nl)\ \ln(I_g/I_v),$$

where n and l are gas density in the collision cell and the effective length of the collision cell. I_g and I_v are the beam intensities in the gas run and the vacuum run, respectively. The independence of TCS values on the cell pressure was ensured for electron collision. As shown in Figure 1(a), no systematic variation showing the pressure dependence of TCS is observed.

2.2. Theoretical

The theoretical approach employed is the continuum multiple-scattering (CMS) method, which is a simple but efficient model for treating electron scattering from polyatomic molecules.[11] The scattering part of the method is based on the static-exchange-polarization potential model within the fixed-nuclei approximation. The Schrödinger equation in each region[11] is solved numerically under separate boundary conditions, and by matching the wave functions and their derivatives from each region, we can determine the total wave functions of the scattered electron and hence, the scattering matrix. The scattering cross section can be easily determined by a conventional procedure.

3. RESULTS AND DISCUSSION

The total cross section (TCS) and EMS elastic scattering cross section (CMS ECS) results are shown in Fig. 1(b). As observed in our TCS and ECS data, Figure 1(b), the energy region 2 – 30 eV, with the TCS showing a series of resonance peaks at 2.5, 7.0, 12.0 and the broad one at ~ 30 eV with cross sections of 24.8±1.3, 30.8±1.7, 31.2±1.7 and 29.4±1.6 × 10^{-16} cm^2, respectively. The CMS ECS result almost shows the same energy dependence as that of the TCS curve, i.e. showing the same four resonance peaks, albeit slightly shifted to higher energies of 3, 7, 17 and the change of slope at ~ 40 eV. These structures correspond to the a_{1g}, t_{1g}, t_{2g} and e_g orbital assignments carried out by Dehmer *et al.*[12] in their study of eigenphase sums, where they found fingerprints of these resonances at 2.1, 7.2, 12.7 and 27.0 eV, i.e. in good agreement with our results.

It is possibly worth noting here that in the multichannel contributions to the peaks at 2.5, 7.0 and 12.0 eV, the total vibration excitation channel seems to be the dominant inelastic channel, with the conspicuous coincidence of its peaks at 7.0 and 12.0 eV, Figure 2(b), with those observed in the TCS. At these peaks, whose TCS values have been quoted above, the sum of the CMS ECS and the total vibration excitation amounts to ~ 24 × 10^{-16} cm^2, ~ 30 × 10^{-16} cm^2 and ~ 31 × 10^{-16} cm^2, respectively. Though these figures are nearly equal to our TCS values at these energies, no discrepancies are arising, and besides, the next significant inelastic channel is dissociative attachment, which however is only maximally ~ 0.05 × 10^{-16} cm^2 at this energy range. Outside this 2 – 30 eV energy region of resonances, (i) the *low energy region* below 2 eV shows TCSs that decrease to show a minimum at 1.2 eV before rising again, in accordance with the prediction by Dehmer *et al.*[12] who explained the trend as due to an a_{1g} resonance giving rise to sharply increasing cross sections near zero eV. This trend is better seen in the results by the other groups, shown in Figure 2(a), whose lower energies are down to a few meV. That the electron attachment cross section tend towards the TCS in magnitude at this energy range just reinforces the already established property of these molecules that has made them useful as a dielectric gas, i.e. helping to reduce the number of free electrons in a dielectric material by electron attachment at low energies. (ii) The *higher energy region* above 30 eV shows that though ionization and neutral dissociation processes become the dominant inelastic channels, with ionization dominating above 40 eV, their contributions to the TCS are still very minimal compared to that due to elastic scattering.

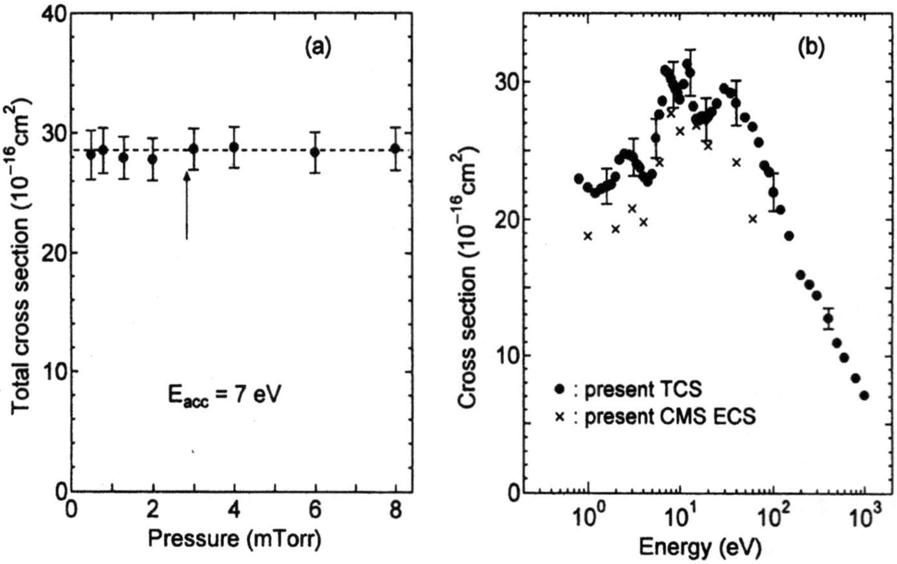

Figure 1. Electron scattering (a) TCSs for SF_6 molecules against collision gas pressure. The beam intensity attenuation (I_0/I_g) of 3 used in our TCS measurements for acceleration energy E_{acc} = 7 eV, is shown by the arrow; (b) present TCS (•) and CMS ECS (×) results for SF_6. Error bars show total uncertainties, consisting of statistical and errors due to gas density and estimation of collision cell effective length.

For instance, ionization peaks at about 100 eV with a contribution of only 29% to the TCS. This is a desirable property for this gas for the gaseous dielectric application, i.e. this minimal ionization means the gas will not produce many electrons by impact ionisation, and hence keeping the number of free electrons low.

Good qualitative agreement is observed between our data and all earlier four data sets shown in Figure 2(a), i.e. in regards to the energies of the resonances, monotonic increase below 1.2 eV and decrease above 70 eV. However, quantitative differences, i.e. being lower than the other three below 1.2 eV, greater than all at 2 – 4 eV and less than the other three above 100 eV, are also observed. At the 7.0, 12.0 and ~ 30 eV resonance peaks, our data is respectively ~3.5% greater, nearly equal to and ~5% greater than that of Kennerly *et al.*,[3] whilst less than or equal to the other three data sets.

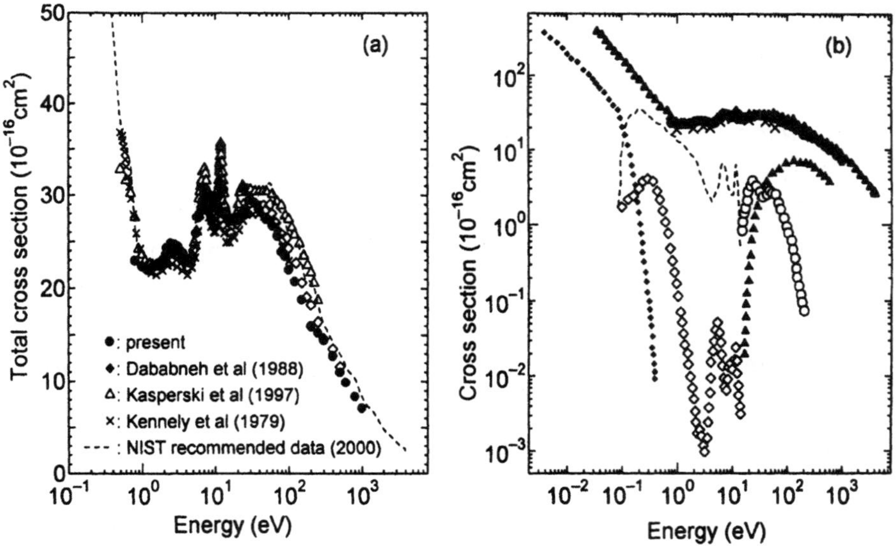

Figure 2. SF_6 Electron scattering (a) present (•); Dababneh *et al.*[2] (♦); Kasperski *et al.*[4] (Δ); Kennerly *et al.*[3] (×); NIST recommended[13] (- - -) TCS results; (b) present TCS (•) and CMS ECS (×), NIST[13] recommended TCS (Δ), total vibration (- - -), attachment (♦), total ionisation (▲), dissociative attachment (◊), and neutral dissociation (○).

Summary

Electron TCSs and ECSs for SF_6 have been investigated experimentally and theoretically over the energy range of 0.8-1000 eV. Peaks have been observed at 2.5, 7.0, 12.0 and ~30 eV and attributed to the a_{1g}, t_{1g}, t_{2g} and e_g resonances observed and assigned by Dehmer *et al.* These results agree well with earlier partial and total cross sections. The phenomena of electron attachment cross sections tending toward the TCS in magnitude below 1.2 eV, and the low contribution of the ionization channel to the TCS at all energies above its threshold, i.e. ideal properties for a gaseous dielectric, have been confirmed.

Acknowledgement
The work was supported in part by the Ministry of Education, Sport, Culture and Technology, Japan Society for Promotion of Science, and Cooperative grant from National Instutute for Fusion Science.

References

1. L. G. Christophorou, J. K. Olthoff, and D. S. Green, Gases for Electrical Insulation and Arc Interruption: Possible Present and Future Alternatives to SF_6, *National Institute of Standards and Technology (NIST) Technical Note 1425*, 1997, p. 1 – 44,
2. M. S. Dabaneh, Y. –F. Hsieh, W. E. kauppila, C. K. Kwan, Steven J. Smith, T. S. Stein and M. N. Uddin, Total-cross-section measurements for positron and electron scattering by O_2, CH_4 and SF_6, *Phys. Rev. A* **38**(3), 1207-1216 (1988).
3. R. E. Kennerly, R. A. Bohman and M. McMillan, The total absolute electron scattering cross sections for SF_6 for incident electron energies between 0.5 and 100 eV including resonance structure, *J. Chem. Phys.* **70**(4), 2039-2041 (1979).
4. G. Kasperski, P. Mozejko and C. Szmykowski, Electron scattering on C_6F_6 and SF_6 molecules, *Z. Phys. D* **24**, 187-191 (1997).
5. D. Spence and G. J. Schulz, Temperature dependence of electron attachment at low energies for polyatomic molecules, *J. Chem. Phys.* **58**(5), 1800-1803 (1973).
6. A. V. Phelps and R. J. Van Brunt, Electron-transport, ionisation, attachment, and dissociation coefficients in SF_6 and its mixtures, *J Appl. Phys*. **64**(9), 4269-4277 (1988).
7. O. Sueoka, S. Mori and A. Hamada, Total cross sections for positrons and electrons colliding with molecules: I. SiH_4 and CF_4, *J. Phys. B* **27**, 1453-1465 (1994).
8. A. Hamada and O. Sueoka, Total cross sections for positrons and electrons colliding with molecules: II. HCl, *J. Phys. B* **27**, 5055-5064 (1994).
9. S. K. Srivastava, S. Trajmar, A. Chutjian, and W. Williams, *J. Chem. Phys.* **64**, 2767-2771 (1976).
10. T. Sakae, S. Sumiyoshi, E. Mukarami, Y. Matsumoto, K. Ishibashi, and A. Katase, Scattering of electrons by CH_4, CF_4 and SF_6 in the 75-700 eV range, *J. Phys. B* **22**, 1385-1394 (1989).
11. M. Kimura and H. Sato, *Comments At. Mol. Phys*. **26**, 333-345 (1991).
12. J. L. Dehmer, J. Siegel, and D. Dill, Shape resonances in *e*-SF_6 scattering, *J. Chem. Phys.* **69**, 5205-5206 (1978).
13. L. G. Chrisophorou and J. K. Olthoff, Electron interactions with SF_6, *National Institute of Standards and Technology (NIST) Reference Data* (last updated June 13, 2000); http://www.eeel.nist.gov/811/refdata/sf6/cross.htm

ELECTRON ATTACHMENT TO SULPHUR HEXAFLUORIDE AND OTHER FLUORINE CONTAINING MOLECULES INVESTIGATED USING ELECTRON SWARM TECHNIQUES

Chris A. Mayhew*

1. INTRODUCTION

1.1 The need to study electron attachment processes

The collision of electrons with molecules is one of the most fundamental of physical processes. Many mechanisms can take place, including inelastic and elastic scattering, ionization, and capture. These mechanisms are being used in a wide and expanding variety of industrial applications. A thorough understanding of each of these processes provides information which could, for example, be used to optimize electron and ion densities in plasmas to enhance critical reactions involved in many technologies and which will be essential to the development of the next generation of plasma processing plants. Such information is considered to be a priority in a report from a panel on "Database Needs for Modeling and Simulation of Plasma Processing", for the Board of Physics and Astronomy, National Research Council, USA.[1] Our ability to control electron interactions therefore provides exciting new opportunities that can be exploited by both the research and technological communities.

In a recent report on "Electron-Driven Processes: Scientific Challenges and Technological Opportunities"[2] the importance of studies of electron-molecule processes has been further highlighted with reference to not only the physical

* Molecular Physics Group, School of Physics and Astronomy, University of Birmingham, Edgbaston, Birmingham, B15 2TT, UK

Gaseous Dielectrics X.
Edited by Christophorou *et al.*, Springer, New York, 2004.

sciences, e.g. radiation chemistry, plasma processing of materials for microelectronic devices, planetary atmospheres and lighting applications, but also in areas of the life sciences. Recent studies have shown that the interaction of low energy (i.e. subionization) electrons with living tissue can be highly destructive,[3] the process being electron attachment which leads to new reactive species. The investigation of the effects of dissociative electron attachment with biologically relevant molecules is rapidly becoming an active area of research.[4-7]

By its very nature, the study of electron capture processes is thus multidisciplinary, fundamental, and is also of practical value. Electron attachment data need to be incorporated into models used to characterize gaseous discharges and industrial plasmas. Furthermore, the creation of reactive intermediate species, via dissociative electron attachment routes, initiate and drive chemical and physical changes in many diverse environments ranging from interstellar molecular clouds to technological plasmas, through to living tissue.

1.2 Sulphur hexafluoride and related molecules

Of all the electron attaching molecules investigated to date, SF_6 is the most studied. This is mainly because of its application as an insulating dielectric, and its use in technological plasmas as a fluorine atom source for the etching of semiconductors.[8-13] A thorough understanding of the electron attaching properties of SF_6 is important for modeling and characterizing such plasmas.

Comparisons of electron attachment mechanisms to related molecules and derivatives of SF_6 provide interesting insights into the physical and chemical molecular properties of electron attachment, improve our understanding of this fundamental process, are informative and are of general interest. In this article we report electron swarm measurements to a series of homologous molecules - SF_6, SeF_6, and TeF_6, and to two derivatives of SF_6, namely SF_5CF_3 and SF_5Cl.

1.3 Cyclic perfluorocarbons

Under the Kyoto Protocol of 1997, SF_6 is one of the greenhouse gases to be regulated. Alternative molecules to replace SF_6 are being proposed, including perfluorocarbons. Here we present non-thermal electron-swarm data over a mean electron energy range of approximately 0.05 – 1.8 eV for electron attachment to various cyclic perfluorocarbons; perfluoro-cyclohexane (c-C_6F_{12}), perfluoro-methylcyclohexane (CF_3-c-C_6F_{11}), perfluoro-ethyl-cyclohexane (C_2F_5-c-C_6F_{11}), perfluoro-1,2-dimethyl-cyclohexane (1,2-$(CF_3)_2$-c-C_6F_{10}), perfluoro-1,3-dimethylcyclohexane (1,3-$(CF_3)_2$-c-C_6F_{10}), perfluoro-1,4-dimethylcyclohexane (1,4-$(CF_3)_2$-c-C_6F_{10}), and perfluoro-1,3,5-trimethyl-cyclohexane (1,3,5-$(CF_3)_3$-c-C_6F_9). Studies such as the one presented here, which investigate electron attachment to molecules of different structures and isomeric forms, help to reveal the molecular properties which influence electron attachment processes.

2. A BRIEF NOTE ON EXPERIMENTAL TECHNIQUES

The fundamental importance of electron attachment processes, together with desirable practical applications of these processes, has led to many measurements of electron attachment cross sections and rate constants of molecules. Several experimental methods have been developed for such measurements, including the flowing afterglow-Langmuir probe,[14] threshold photoionisation,[15] electron beam (coupled with molecular beam[16] or gas collision chamber[17]) and the electron swarm techniques.[18-19] For thermal electron energy distributions the flowing afterglow technique has been successfully used to obtain capture rate constants and to identify anion products. For electron energies above the thermal value, the electron beam and electron swarm techniques have been extensively used.

Beam studies investigate electron attachment processes using nearly mono-energetic electrons over an energy range of typically 0.1 – 10 eV. The capture occurs under collision free conditions. Relative attachment cross-sections and competing decay channels as a function of electron energy are obtained. The technique is used to identify stable, long-lived anions ($\tau > 10^{-6}$ s).

Swarm measurements provide absolute data averaged over a broad (and usually non-thermal) electron energy distribution. The attachment takes place in a high pressure environment so that collisional relaxation of the excess internal energy in the anions produced from electron attachment may take place. Swarm measurements provide data which can be used to place the relative capture cross sections determined from electron beam studies onto an absolute scale.

2.1. The electron swarm technique

The electron swarm technique has resulted in a large body of data on electron attachment processes, most notably from the work by Christophorou and colleagues.[18-20] The technique relies on the production of pulses of electrons, thereby ensuring that they are temporally distinct from the anion products, in a drift chamber containing a non-electron attaching buffer gas of number density N. Under the influence of an applied electric field E, the electrons are drawn through the gas towards a collector plate, where the intensity of the electron pulses can be measured. Within the drift region, electrons usually attain a non-thermal equilibrium energy distribution, $f(\varepsilon)$, determined by a dynamic balance between the kinetic energy gained from the electric field and energy loss through multiple collisions with the molecules of the buffer gas. Thus $f(\varepsilon)$ primarily depends on the nature of the buffer gas and the ratio E/N. This distribution has been well characterized for a number of gases, e.g. He, Ar and N_2. Trace amounts of an electron attaching gas mixed into the buffer gas results in the removal of electrons, without changing the electron energy distribution. The electron pulse intensity is subsequently reduced at the detector. This reduction in intensity is used in swarm experiments to obtain absolute values of the density reduced electron attachment rate coefficient, from which, with knowledge of the mean electron drift velocities, the electron attachment rate constant as a function of mean electron energy can be determined. Electron attachment cross sections, $\sigma(\varepsilon)$, as a function of electron energy, ε, can in principle be determined by deconvoluting the swarm data, because $f(\varepsilon)$ is known as a function of E/N.

2.2 The Electron Swarm Mass Spectrometer

We have developed the electron swarm technique by attaching a mass spectrometer to a drift tube used for swarm studies. A 70 μm orifice in the collector permits product anions to pass into a region of the instrument where they are focussed and mass analysed. This instrument, called an Electron Swarm Mass Spectrometer (ESMS) therefore has the capability of identifying anion products resulting from electron attachment inside the drift tube. This provides valuable additional information on electron capture processes, allowing the mechanisms involved to be elucidated. A similar instrument has been developed by the group of Grimsrud et al., which has been used most notably to investigate electron attachment to $POCl_3$ and $PSCl_3$.[21-22]

The ESMS we have used for our studies of electron attachment processes has been described in detail in the literature.[23] Recently, we have modified the drift chamber of our instrument for the following reasons. In our original design, electrons are produced via ionization of the buffer gas by electrons emitted from a cylindrical 11 mCi ^{63}Ni β-ray source. To minimize electron attachment between the source and drift tube, two buffer gas flows are required; one through the ionization source and the other (in the opposite direction) through the drift tube. This has the potential of leading to flows of gases of different chemical composition, which in turn leads to concentration gradients. These are difficult to predict and/or control. Furthermore, the electric field close to the source region can not be uniform, leading to uncertainties in the electron energy distribution around this region. The new design of drift tube uses a planar ^{63}Ni radioactive source, and forms an integral part of the drift tube. This allows the use of a single gas stream to fill all parts of the chamber, eliminating any inhomogeneity in gas concentration and uncertainties in the reaction length. The varying electric field gradients between the electron source and electrical gate are also eliminated. Figure 1 presents a schematic diagram of the new drift tube.

As in our initial design,[23] metal ring electrodes inside a glass envelope form the drift tube. The rings are made of aluminium, and are coated with molybdenum to reduce charging effects. They are physically separated from each other by 1 mm ceramic spacers, but are electrically connected via a chain of 10 MΩ resistors. The drift tube is 9.7 cm long, measured from the radioactive source to the Faraday plate. The Faraday plate is electrically isolated from the drift tube, but is normally held at ground potential. The electron gate is used to convert the constant stream of electrons produced in the source region into a chain of pulses. The gate is based on the Bradbury-Nielsen design, consisting of two interdigitated wire arrays. One array is held at a potential of 9 V above the voltage of the drift tube at the position of the gate, V_d, and the other at 9 V below V_d. When the gate is closed, electrons experience a large electric field deflecting them sideways, stopping them from entering the remaining section of the drift tube. To open the gate, the potentials on the arrays are simultaneously switched to V_d for typically 1 ms at a frequency of 25 Hz. The Faraday plate collects the electrons and anions that pass through the drift tube. The electron current pulse is converted to a voltage pulse by a current to voltage converter. The signal is amplified and passed to a gated integrator to measure amplitude.

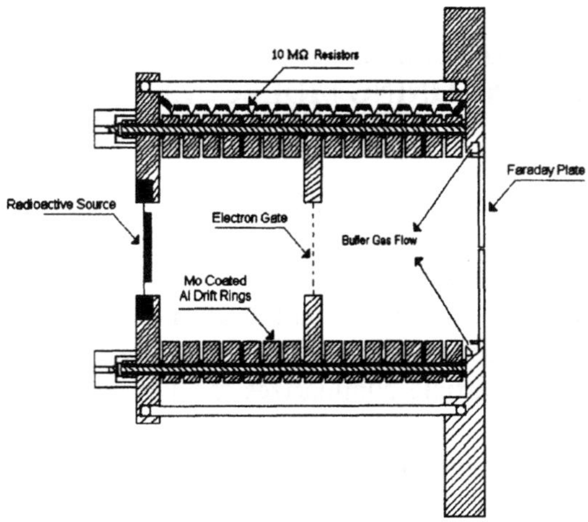

Figure 1. A schematic representation of the new drift tube used in our electron swarm mass spectrometer. The diagram represents a cross section of the cylindrically symmetric drift tube.

The determination of the density normalized electron attachment coefficient, α, is readily deduced in the following straightforward manner. Let the intensity of electrons at $x = 0$, the start of the drift tube, be I_0. Between the source and the gate, separated by a distance l_1, diffusion and attachment of electrons occur, so that the electron intensity at the gate will be $I(l_1)$ such that $I(l_1) = I_0 e^{-(n\alpha+\beta)l_1}$, where β is the diffusive loss coefficient (defined as the probability of diffusive loss of electrons per unit length) and n is the number density of the electron attaching molecule. A fraction of $I(l_1)$, γ, is pulsed into the next section of the drift tube. The initial intensity of the pulse in this section is $A(l_1)$, where $A(l_1) = \gamma I(l_1) = \gamma I_0 e^{-(n\alpha+\beta)l_1}$. As this electron pulse propagates down the drift tube further diffusional and attachment loss of electrons will occur, so that the amplitude of the pulse at the Faraday plate will be $A(l_2)$, where

$A(l_2) = A(l_1)e^{-(n\alpha+\beta)(l_2-l_1)} = \gamma I_0 e^{-\beta l_2} e^{-n\alpha l_2}$, where l_2 represents the total length of the drift tube. α can then be readily determined from $\alpha = -\ln\left\{ A(l_2) \Big/ \gamma I_0 e^{-\beta l_2} \right\} \Big/ n l_2$ which is independent of the position of the gate in the drift tube. A measurement of $\ln A(l_2)$ as a function of n for a fixed E/N is made. Typically 5-8 concentrations (including $n = 0$) are used. The maximum concentration is chosen to produce an attenuation of the pulse amplitude of about 90%. A linear least-squares fit of the data, with all points weighted equally gives $\alpha(E/N)$. By multiplying the density normalized electron attachment coefficients by the appropriate mean electron drift velocities, the electron attachment rate constants, $k_a(E/N)$, are obtained.

3. ASSOCIATIVE AND DISSOCIATIVE ELECTRON ATTACHMENT BY FLUORINE CONTAINING MOLECULES

3.1 SF_6 and related molecules; SeF_6, TeF_6, SF_5CF_3 and SF_5Cl

Studies of low energy electron attachment to SF_6 and the related molecules, SeF_6, TeF_6, and SF_5CF_3 under non-thermal electron-swarm conditions for mean electron energies from 0.05 eV through to about 2 eV have been previously reported by us.[23-24] This article summarizes the key data on these molecules together with those obtained for SF_5Cl, so that useful comparisons can be made.

SF_6 attaches electrons much more rapidly than either SeF_6 or TeF_6. For all three molecules, attachment is dominated by the capture of near zero-energy electrons. The ratios $k_a(SF_6) : k_a(SeF_6) : k_a(TeF_6) \approx 3000 : 10 : 1$ are found not to vary with E/N. Through extrapolation of the results to a mean electron energy of 0.038 eV, the estimated thermal (300 K) electron attachment rate constants are $k_{th}(SF_6) \approx (2.5 \pm 0.3) \times 10^{-7}$ cm^3 s^{-1}, $k_{th}(SeF_6) \approx (8.0 \pm 1.2) \times 10^{-10}$ cm^3 s^{-1}, and $k_{th}(TeF_6) \approx (8.2 \pm 1.1) \times 10^{-11}$ cm^3 s^{-1}. In each case the dominant anion product is XF_6^- (X = S, Se, Te), accompanied by XF_5^-. No other anion products directly arising from electron attachment to XF_6 are observed over the mean electron energy range investigated.

SF_5CF_3 also attaches electrons most rapidly for near-zero electron energy, with the estimated thermal electron attachment rate constant being approximately 25% of that for SF_6. Over the mean electron energy range investigated one anion product is observed, namely SF_5^-, i.e. the parent anion $(SF_5CF_3^-)^*$ is not sufficiently long-lived to be detected.

The electron attachment rate constants measured for SF_5Cl also decrease with increasing mean electron energy, with the estimated thermal value being significantly less than the expected s-wave capture value. Thus although the electron attachment cross section for SF_5Cl peaks at zero electron energy, its value is smaller than that for SF_6. Deconvolution of our rate constant data provides an estimate of the cross section at zero electron energy to be approximately 1×10^{-15} cm^2, which is two orders of magnitude smaller than that determined for SF_6 ($\sigma(\varepsilon = 0) \sim 1 \times 10^{-13}$ cm^2).

In the case of SF_6, we noted that replacement of an F atom by Cl leads to a large decrease in the density normalised electron attachment coefficients (and related quantities). In contrast, we have observed that in going from $CHClF_2$ to $CHCl_2F$, a large increase in the density normalised electron attachment coefficients results.[25] In the case of CHF_3, we were unable to detect any attachment. Thus, $\alpha(SF_6) > \alpha(SF_5Cl)$ but $\alpha(CHF_3) < \alpha(CHClF_2) < \alpha(CHCl_2F)$.

SF_5Cl is observed to undergo predominantly dissociative electron attachment, but unlike SF_5CF_3 (which also extensively fragments upon electron attachment through one channel leading to the product anion SF_5^-), $(SF_5Cl^-)^*$ fragments through a number of channels, leading to the product anions F_2^-, SF_4^-, and SF_5^-, the branching ratios of which are dependent on E/N. The parent anion SF_5Cl^- is also an observed product resulting from associative attachment. Anion reactions with SF_5Cl made it difficult to accurately assign branching ratios. Nevertheless, we are able to make the following statements. Firstly, the branching ratios smoothly transfer over from the N_2 to the Ar data - indicating that the electron attachment process is independent of the buffer gas. Secondly, as the mean electron energy tends to zero, the branching ratios associated with the SF_4^- and F_2^- product anions extrapolate to zero percent. Thirdly, two product anions remain as the mean electron energy tends to zero; SF_5Cl^- (~10%) and SF_5^- (~90%). These two product anions, with similar branching ratios, are the only ones observed in a thermal measurement described below.

3.2 Thermal electron attachment rate measurements of SF_6, SF_5CF_3 and SF_5Cl

Most swarm studies have been performed in buffer gases within which the electron energy distribution is dependent on E/N. However, by using CO_2 as the buffer gas an electron swarm with a thermal energy distribution over a large range of E/N values can be created. Measurement of the thermal attachment rate constant, k_{th}, is then directly possible. Figure 2 shows k_{th}(300 K) measured for SF_6 over a large range in E/N. For comparison, the rate constants measured by Hunter et al. in an N_2 buffer gas over the same E/N range are shown.[20] The constancy of the electron attachment rate constant for SF_6 measured using a CO_2 buffer gas throughout the E/N range shown in figure 2 serves to illustrate that the electron energy distribution in the swarm environment remains thermal. This is confirmed further from the value we obtained for $k_{th}(SF_6)$.

The swarm data provides a thermal (300 K) electron attachment rate constant for SF_6 of $k_{th}(SF_6) = (2.38 \pm 0.15) \times 10^{-7}$ cm^3 s^{-1}, which agrees well with the value determined by Hunter et al. through extrapolation,[20] $(2.3 \pm 0.1) \times 10^{-7}$ cm^3 s^{-1}.

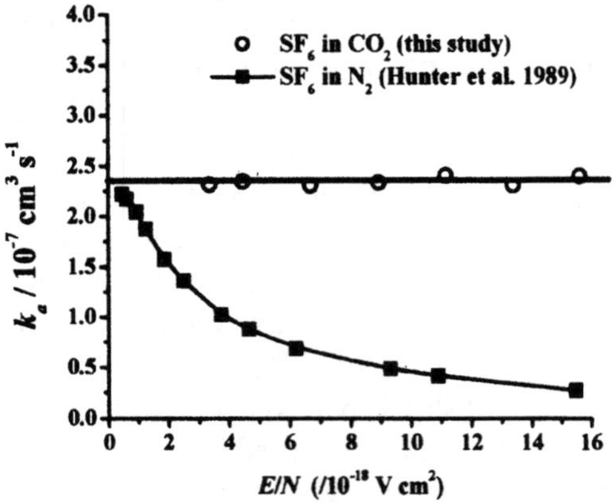

Figure 2. The dependence of the electron attachment rate constants on E/N for SF_6 determined in a swarm environment using N_2 (from the work of Hunter et al.[20]) and CO_2 buffer gases.

Similar sets of measurements to those taken for SF_6 in a CO_2 buffer gas in our swarm apparatus have been obtained for SF_5CF_3 and SF_5Cl. These provide the following 300 K k_{th} values; $k_{th}(SF_5CF_3) = (8.0 \pm 0.3) \times 10^{-8}$ cm^3 s^{-1} and $k_{th}(SF_5Cl) = (2.0 \pm 0.3) \times 10^{-8}$ cm^3 s^{-1}. Both of these values are below the expected s-wave capture value. The value we have obtained for SF_5CF_3 agrees well with a FALP determination by Miller et al., who obtained $k_{th}(SF_5CF_3) = (8.6 \pm 2.2) \times 10^{-8}$ cm^3 s^{-1}.[26] The product anions observed for all three molecules are in good agreement with those found in the non-thermal measurements.

3.3 The cyclic perfluorocarbons (PFCS), c-C_6F_{12}, CF_3-c-C_6F_{11} C_2F_5-c-C_6F_{11}, 1,2-$(CF_3)_2$-c-C_6F_{10} , 1,3-$(CF_3)_2$-c-C_6F_{10} , 1,4-$(CF_3)_2$-c-C_6F_{10} and 1,3,5-$(CF_3)_3$-c-C_6F_9

Of the above closely related cyclic PFCs, electron attachment using swarm techniques have only been studied for CF_3-c-C_6F_{11} and 1,3-$(CF_3)_2$-c-C_6F_{10}.[27-28]

c-C_6F_{12} is the only cyclic PFC of this study for which the maximum in the electron attachment rate constant occurs at zero mean electron energy. This is illustrated in figure 3. The electron attachment rate constants as a function of mean electron energy, $\bar{\varepsilon}$, for the other cyclic PFCs are shown in figure 4. The data shown in both figures were determined in an atmospheric pressure buffer gas of either N_2, for $\bar{\varepsilon} < 0.5$ eV, or Ar, for $\bar{\varepsilon} > 0.5$ eV. Nevertheless, the data shown for CF_3-c-C_6F_{11} is in good agreement with the study by Christodoulides et al.[27] For 1,3-$(CF_3)_2$-c-C_6F_{10} the attachment rate constants we have obtained as a function of $\bar{\varepsilon}$ are approximately 30% higher than those obtained in the earlier study.[28] The major uncertainties in our measurements relate to difficulties in obtaining accurate values of attaching gas concentrations.

For all the cyclic PFCs investigated in this study, the dominant product anion resulting from electron attachment across the mean electron energy range studied is the parent c-$C_nF_{2n}^-$ (n = 6-9), i.e. dissociative electron attachment is found not to be important within our swarm environment.

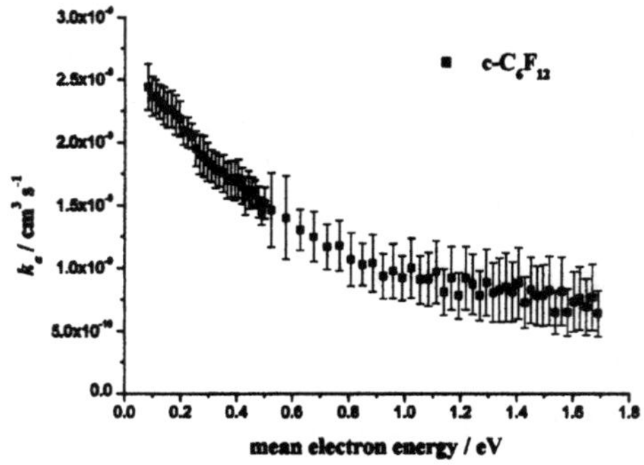

Figure 3. Electron attachment rate constants, k_a, as a function of mean electron energy for c-C_6F_{12}.

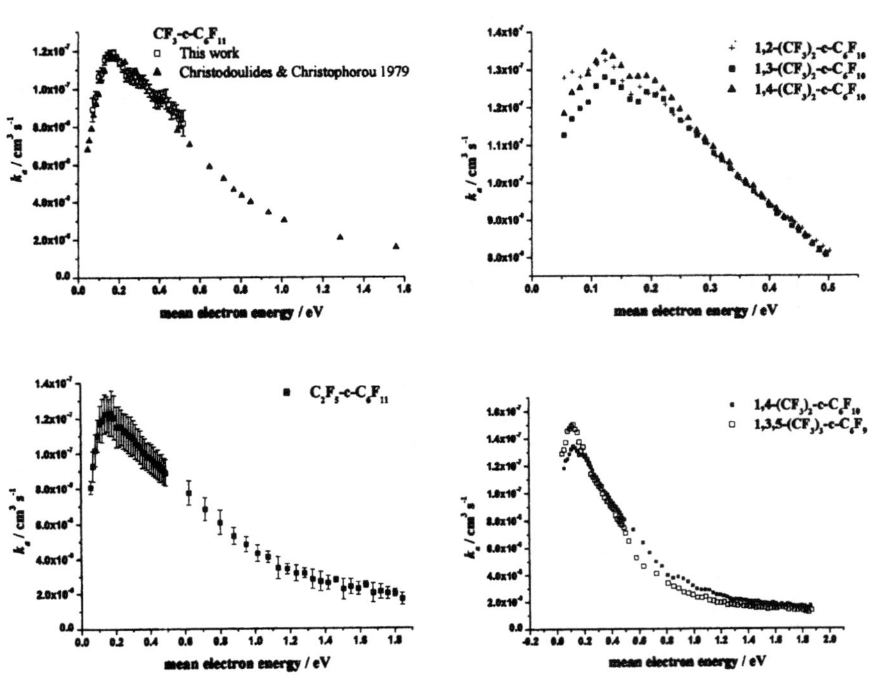

Figure 4. Electron attachment rate constants, k_a, as a function of mean electron energy for CF_3-c-C_6F_{11} C_2F_5-c-C_6F_{11}, 1,2-$(CF_3)_2$-c-C_6F_{10} , 1,3-$(CF_3)_2$-c-C_6F_{10}, 1,4-$(CF_3)_2$-c-C_6F_{10} and 1,3,5-$(CF_3)_3$-c-C_6F_9.

Acknowledgements

I wish to thank the support of the UK funding agency EPSRC, and G. K. Jarvis and A. D. J. Critchley for their help in obtaining some of the data presented here.

References

1. *Database Needs for Modelling and Simulation of Plasma Processing*, National Research Council, National Academy Press, Washington DC (1996).
2. *Electron-Driven Processes: Scientific Challenges and Technological Opportunities*, National Research Council, National Academy Press, Washington DC (2000).
3. B. Boudaïffa, P. Cloutier, D. Hunting, M. A. Huels, and L. Sanche, *Science* **287**, 1658-1660 (2000).
4. M. A. Huels, L. Sanche, I. Hahndorf, and E. Illenberger, *J. Chem. Phys.* **108**, 1309-1312 (1998).
5. H. Abdoul-Carime, M. A. Huels, L. Sanche, F. Brüning, and E. Illenberger, *J. Chem. Phys.* **113**, 2517-2521 (2000).
6. H. Abdoul-Carime, M. A. Huels, L. Sanche, and E. Illenberger, *J. Am. Chem. Soc.* **123**, 5354-5355 (2001).
7. G. Hanel, B. Gstir, S. Denifl, P. Scheier, M. Probst, B. Farizon, E. Illenberger, and T. D. Märk, *Phys. Rev. Lett.* **90**, 188104-1-188104-4 (2003).
8. C L Lugez, M E Jacox, R A King, and H F Schaefer III, *J. Chem. Phys.* **108**, 9639-9650 (1998).
9. E R Fisher, B L Kickel, and P B Armentrout, *J. Chem. Phys.* **97**, 4859-4870 (1992).
10. H.-X. Wan, J. H. Moore, J. K. Olthoff, and R. J. Van Brunt, *Plasma Chem. Plasma Processes* **13**, 1-16 (1993).
11. I Sauers, L. G. Christophorou, and S. M Spyrou, *Plasma Chem. Plasma Processes* **13**, 17-35 (1993).
12. V Tarnovsky, H Deutsch, K E Martus, and K Becker, *J. Chem. Phys.* **109**, 6596-6600 (1998).
13. L G Christophorou, and J K Olthoff (Eds.) *Proceedings of the VIII International Symposium on Gaseous Dielectrics* (Plenum Press, New York, 1998).
14. D. Smith and P. Španěl, *Ad. At. Mol. And Opt. Physics* **32**, 307-343 (1994).
15. J. M. Ajello and A. Chutjian, *J. Chem. Phys.* **71**, 1079-1087 (1979).
16. M. Fenzlaff, R. Gerhard, and E. Illenberger, *J. Chem. Phys.* **88**, 149-155 (1988).
17. L. E. Kline, D. K. Davies, C. L. Chen, and P. J. Chantry, J. Appl. Phys. **50**, 6789-6796 (1979).
18. L. G. Christophorou, E. L. Chaney, and A. A. Christodoulides, *Chem. Phys. Letts.* **3(6)**, 363-366 (1969).
19. L. G. Christophorou, D. L. McCorkle, and A. A. Christodoulides, *Electron-Molecule Interactions and Their Applications* (Academic Press, New York, 1984)
20. S. R. Hunter, J. G. Carter, and L. G. Christophorou, *J. Chem. Phys.* **90**, 4879-4891 (1989).
21. D. H. Williamson, C. A Mayhew, W. B Knighton, and E. P. Grimsrud, *J. Chem. Phys.* **113**, 11036-11043 (2000).
22. W. B Knighton, T. M. Miller, E. P. Grimsrud, and A. A. Viggiano, *J. Chem. Phys.* **120**, 211-216 (2004).
23. G. K. Jarvis, R. A. Kennedy, and C. A. Mayhew, *Int. J. Mass Spectrom. Ion Processes* **205**, 253-270 (2001).
24. R. A. Kennedy and C. A. Mayhew, *Int. J. Mass Spectrom. Ion Processes* **206**, vii-x (2001).
25. G. K. Jarvis, C. A. Mayhew, L. Singleton, and S. Spyrou, *Int. J. Mass Spectrom. Ion Processes* **164**, 207- 223 (1997) 207.
26. T. M. Miller, S. T. Arnold, A. A. Viggiano, and W. B. Knighton, *J. Chem. Phys.* **116** 6021-6027 (2002).
27. A. A. Christodoulides and L. G. Christophorou, *Chem. Phys. Lett.* **61(3)**, 533-557 (1979).
28. R. Y. Pai, L. G. Christophorou, A. A. Christodoulides, *J.Chem. Phys.* **70**, 1169-1176 (1979).

LOW ENERGY ELECTRON ATTACHMENT BY HALOALKANES

I.Szamrej[1], M.Foryś[1], N.L. Asfandiarov[2]

ABSTRACT

Electron attachment process in three halopropanes, $CH_3CH_2CH_2Br$, $CH_3CHBrCH_3$ and $CH_3CHBrCF_3$, have been investigated using an electron swarm and electron capture negative ion mass spectrometry techniques. All compounds attach electrons in dissociative process, and the main product is bromine ion. The rate constants for thermal electron attachment are equal to $1.1 \cdot 10^{-11}$, $1.4 \cdot 10^{-12}$ and $4.1 \cdot 10^{-10}$ $cm^3 molec.^{-1} s^{-1}$ for $CH_3CH_2CH_2Br$, $CH_3CHBrCH_3$ and $CH_3CHBrCF_3$, respectively.

INTRODUCTION

Low energy electron attachment studies provide fundamental information on electron-molecule interaction, which is of practical importance for many applications, among them gaseous electronics.

If electrons are in thermal equilibrium with the environmental molecules, the rate constant strongly depends on the shape and position of the resonance. The measured rate constants differ by many orders of magnitude, from 10^{-7} to less than 10^{-13} $cm^3 molec.^{-1} s^{-1}$.

Some years ago we have started the systematic investigation of the electron capture process by haloalkanes to get a clearer picture of how the effectiveness and the mechanism of the electron attachment depend on the structure of the attaching molecule. This is probably the largest and most thoroughly investigated homologous group of compounds.

For now, there exist two approaches to systematize the available rate constants for electron attachment process. Christophorou[1] have demonstrated that the experimental thermal rate constants for a large number of halocarbons depend on the energy position of the electron attaching state as measured by an electron transmition spectroscopy, Szamrej et al.[2,3] have concentrated on the link between the structure of the molecule and its ability to accept thermal electrons.

---

[1] Chemistry Department, University of Podlasie, Siedlce, Poland
[2] Institute of Physics of Molecules and Crystals, Russian Academy of Science, Ufa, Russia

Gaseous Dielectrics X.
Edited by Christophorou *et al.*, Springer, New York, 2004.

RESULTS AND DISCUSSION

The new data presented in this paper on the kinetics of the thermal electron capture process by bromo- and bromofluoropropanes are the continuation of our effort to find the connection between the structure of the molecule and its ability to accept electrons.

The two parameters of the attachment process have been measured: rate constants at thermal electron energies and dissociation products at energies up to 10 eV.

For the investigations two experimental techniques have been used. An electron swarm method with ionization chamber[2,4] and carbon dioxide as a carrier gas has been applied for the measurements of the electron capture rate constants. An electron capture negative ion mass spectrometry has been used for the identification of the attachment products and the energy at which they are produced.[5]

The experimental set up for the electron swarm experiment was fully described earlier. The electron swarm is produced by an α-particle in the chamber with two parallel electrodes. The swarm moves under influence of the uniform electric field to the collecting electrode traversing a distance ca. 2 cm. The output pulse goes through preamplifier to the digital memory of the oscilloscope and then the data are transferred to computer memory. From the shape of the output pulse the rate of electron disappearance while reacting with electron-attaching molecules is calculated.

The details of the electron capture negative ion mass spectrometry technique have been describe previously [5]. Negative ion mass spectra has been obtained using a modified MI-1201 mass spectrometer. The experimental conditions were as follows: accelerating voltage 4 kV, electron trap current ~ 1 μA, FWHM of electron energy distribution $\Delta E_{1/2} = 0.4$ eV, E_{el} varies in the range 0 – 12 eV.

We have investigated three halopropane molecules: $CH_3CH_2CH_2Br$, $CH_3CHBrCH_3$ and $CH_3CHBrCF_3$.

Bromopropanes were from Aldrich Chemical Company with stated purity of 99% for 1- and >98% for 2-bromopropane while bromofluoropropane was synthesized in the Institute of Organic Chemistry of Polish Academy of Science with the purity > 98%. For the kinetic measurements all the compounds were carefully purified both from higher and lower volatile admixtures. Carbon dioxide used as the carrier gas in the swarm experiment was also purified using freeze-pump-thaw technique. The experiment has been repeated for different concentrations of halocarbons as well as different pressures (0.5 – 2 atmospheres) of CO_2. The rate of the attachment process depended only on halocarbons concentration, so the results for the rate constants were averaged over all measurements. The example experimental results are shown in Fig.1.

The obtained thermal rate constants are presented in Table 1. As to our best knowledge they are the first in the literature.

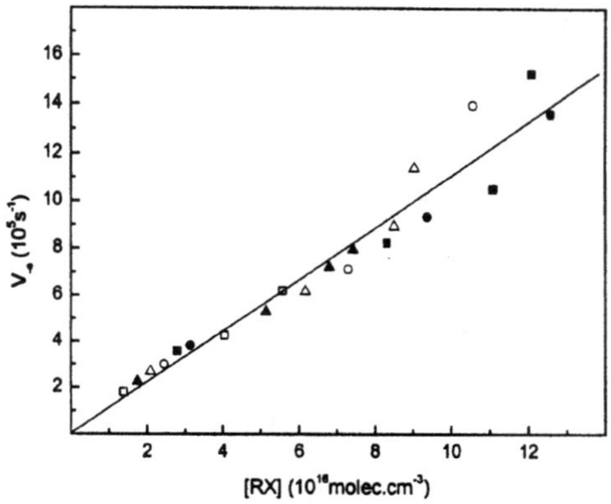

Fig.1 The rate of electron disappearance from the swarm as a function of [$CH_3CH_2CH_2Br$]: [CO_2] [10^{19} molec.cm^{-3}] ● - 2.97, ■ - 2.64, ○ - 2.31, △ - 1.98, ▲ - 1.65, □ - 1.32.

Table 1. Thermal electron capture rate constants for halopropanes.

molecule	k (cm^{3}molec.$^{-1}$s^{-1})
$CH_3CH_2CH_2Br$	$1.1 \cdot 10^{-11}$
$CH_3CHBrCH_3$	$1.4 \cdot 10^{-12}$
$CH_3CHBrCF_3$	$4.1 \cdot 10^{-10}$

As it is seen from the table changing the position of bromine atom from 1 to 2 decreases the rate constant for about one order of magnitude while changing one of CH_3 group into CF_3 increases it more than by two orders.

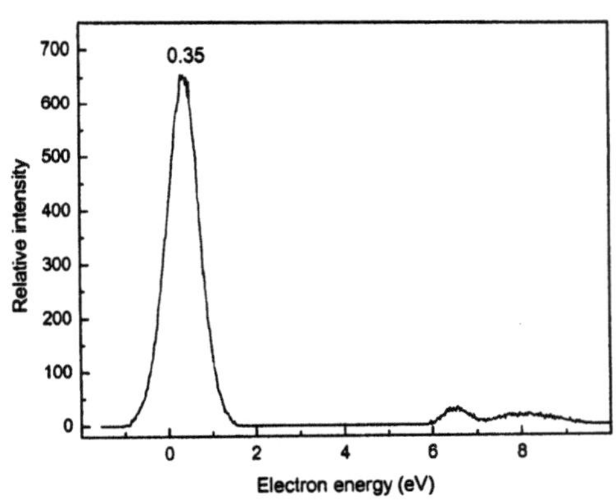

Fig.2 Curve of Br^- ion effective yield for 1-bromopropane.

The difference in the rate constants between 1- and 2-dibromopropane can be caused by some steric factor. Bromine atom in position 2 is screened by two methyl radicals so it seems more difficult for electron to be captured then in the case of 1-bromopropane.

The kinetic results are fully compatible with electron capture negative ion mass spectra for these compounds. In all three cases the dominant negative ion is Br^-. The curves of its effective yields are presented in Figures 2-4.

The changes in the position of the maximum of negative ion resonance are clearly visible. The higher rate constant of the electron attachment process the lower position of the maximum for the negative ion formation.

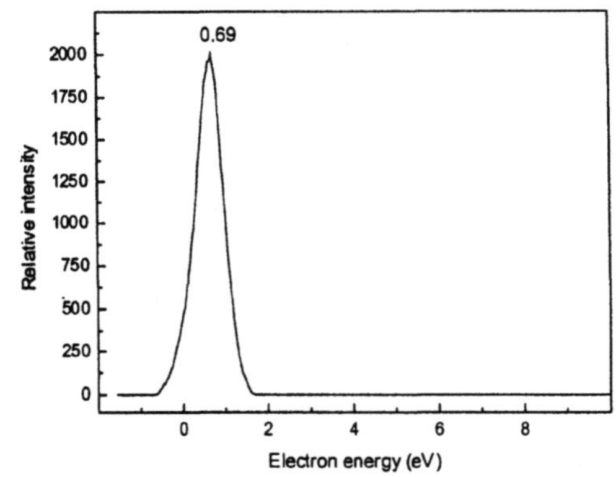

Fig.3 Curve of Br^- ion effective yield for 2-bromopropane.

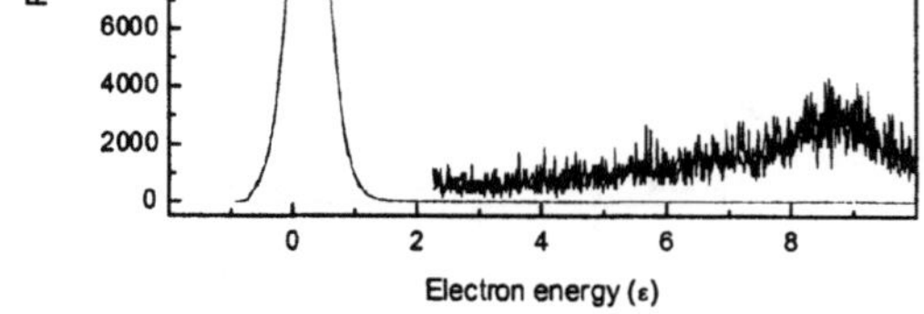

Fig.4 Curve of Br^- ion effective yield for 2-bromo-1,1,1-fluoropropane.

ACKNOWLEDGMENTS

This work was supported in part by the Polish State Committee for Scientific Research under grant 3T09A01018 and NATO grant JSTC.RCLG.979570.

References

1. L.G.Christophorou, The dependence of the thermal electron attachment rate constant in gases and liquids on the energy position of the electron attaching state, Z.Phys.Chem., **1996**, 195, 195,
2. A. Rosa, W. Barszczewska, M. Foryś, I. Szamrej, Electron capture by haloethanes in a carbon dioxide buffer gas, Int. J. Mass Spectrom., **205**, 85 (2001)
3. I.Szamrej, The dependence of electron capture rate constants on electronic polarizability, Gaseous Dielectrics VIII, Kluwer Academic/Plenum Publishers, N.N., p.63 (1998)
4. W.Barszczewska, J.Kopyra, J.Wnorowska, I.Szamrej, Low energy electron attachment by chloroalkanes, J.Phys.Chem. A, **107**, 11427 (2003)
5. V.I. Khvostenko, Negative ion mass spectrometry in organic chemistry, Nauka, Moscow, (1981)

ELECTRON ATTACHMENT TO BIOLOGICALLY RELEVANT MOLECULES: TOWARDS THE INITIAL MOLECULAR STEPS IN RADIATION DAMAGE

Sascha Gohlke, Hassan Abdoul-Carime and Eugen Illenberger

1. INTRODUCTION

The interaction of low energy electrons with biomolecules is directly related to the old problem of radiation damage. After the discovery of X-rays, radioactivity and nuclear fission it became soon obvious that the exposure of living beings to high energy radiation (particles and photons) can result in fatal effects for the concerned individual. The variety of such effects is subsumed under the term *radiation damage*. It includes damage of biological material on a short time scale, i.e. the immediate collapse of living cells eventually resulting in the death of the individual within hours or days but also effects appearing on a much longer time scale. Instead of a complete damage of cells, radiation can result in genotoxic or mutagenic effects, i. e. strand breaks in DNA or change of its sequence (genetic expression).

Countless numbers of articles have been published in relation to radiation damage since decades. Here we only refer to an older[1] and a recent monograph[2]. It appears that phenomenological aspects of radiation damage are well documented. There is, however, a very fundamental lack in the description of the initial steps following irradiation. Recently, there is much interest in the interaction of low energy electrons with biomolecules since it was shown that sub-ionization electrons can effectively induce single and double strand breaks in plasmid DNA[3].

2. PRIMARY AND SECONDARY PROCESSES IN LIVING CELLS

The most important component of the cell nuclei is DNA in which genetic information is stored. To understand the effect of high energy radiation to a living cell (DNA and its environment) one may follow this interaction from the primary through the secondary reactions in a chronological and spacial order. As an example, consider a bunch of photons at energies in the MeV range interacting with DNA and its environment. The *primary* photon interaction (absorption, scattering) removes electrons from essentially any occupied state, from valence orbitals to core levels. Depending on the energy of these ionized electrons they induce further ionization events, etc. thereby being slowed down. The estimated

Institut fuer Chemie -Physikalische und Theoretische Chemie, Freie Universitaet Berlin
Takustrasse 3, D-14195 Berlin

Gaseous Dielectrics X.
Edited by Christophorou *et al.*, Springer, New York, 2004.

quantity is 10^4 *secondary* electrons per 1 MeV primary quantum[3]. These electrons are usually assigned as *secondary* although they are the result from primary, secondary, tertiary etc. interactions, including electrons from Auger processes generated during relaxation of the core holes. Taking a snapshot at some femtoseconds after the primary interaction we have then multiple charged sites within the complex molecular network (eventually undergoing Coulomb explosion), single ionized and electronically excited sites and, last but not least, an exceeding number of low energy secondary electrons with an energy distribution extending to a few tens of eV[3]. Although the double and single ionized sites as well as electronic excitation can result in the rupture of chemical bonds, the major effects are induced by the large number of secondary electrons. In the course of successive inelastic collisions within the medium they are thermalized within picoseconds before they reach some stage of solvation, then as chemically rather inactive species.

Damage of the genome in a living cell by ionizing radiation is about one third *direct* and two third *indirect*. Direct damage concerns reactions by energy deposition directly in the DNA and its closely bound water molecules. Indirect damage results from energy deposition in water and other biomolecules in the surrounding of the DNA. It is believed that almost all the indirect damage is due to the attack of the highly reactive hydroxyl radical OH.

3. ELECTRON INITIATED REACTIONS IN DNA BASES

DNA is a biopolymer consisting of two chains (strands) consisting of the 4 heterocyclic bases thymine (T), adenine (A), cytosine (C) and guanine (G), each of them bound to the DNA backbone which itself is composed of phosphate and sugar units. Both strands are connected through reciprocal hydrogen bonding between pairs of bases in opposite positions in the two strands. The composition is such that adenine pairs with thymine (A-T) and guanine with cytosine (G-C) resulting in the well recognized double helix form. The DNA itself directly interacts with other adjacent biomolecules (proteins) and water.

In order to reveal the mechanism of DNA damage, the first step would be to investigate the interaction of low energy electrons with single DNA bases representing the building blocks of the large polymer. In an effort to describe these effects on a molecular level, different laboratories started programs to probe building blocks of biomolecules in the gas phase[4-7]. The question then is to which degree these *intrinsic* properties (as revealed by gas phase studies) can be transferred to its analogue in solution. This problem has been a longstanding issue in many areas of *Physical Chemistry.* One has to keep in mind that the solvent represents a dissipative environment, and in reactions where charged particles are involved, solvation can considerably modify the energy profile of the reaction coordinate.

For the study of electron attachment to gas phase molecules two methods have been applied so far, crossed electron/molecular beam experiments with mass spectrometric detection of the anions formed and electron transmission (ET) through a gaseous sample of biomolecules. While ET mirrors the initial Franck-Condon transition and hence the energy of the precursor ion, the mass spectrometric detection reflects the two step process of DEA, the initial transition forming the TNI and its decomposition into the particular ionic fragment.

In our experiment the molecular beam is generated by moderately heating the powder sample containing the DNA bases to 150 - 200°C and effusing the molecules through the collision region. The ions resulting from the electron – molecule collisions are extracted from the interaction region and focused onto the entrance of a mass filter where they are analyzed and detected.

As an example Fig. 1 shows electron attachment to the DNA base thymine (T) leading to the most dominant fragment $(T\text{-}H)^-$ formed by abstraction of a neutral hydrogen atom with the negative charge remaining on the ring, *viz.*,

$$e^-(\approx 2\,eV) + T \rightarrow T^{\#-} \rightarrow (T\text{-}H)^- + H \qquad (1)$$

with a resonance maximum near 1 eV and a threshold close to 0.5 eV. The absolute cross section for this dissociative electron attachment (DEA) cross section can be estimated as $\sigma_{DEA} \approx 4$ Å^2 which is in the same order of magnitude as the geometrical cross section of the molecule.

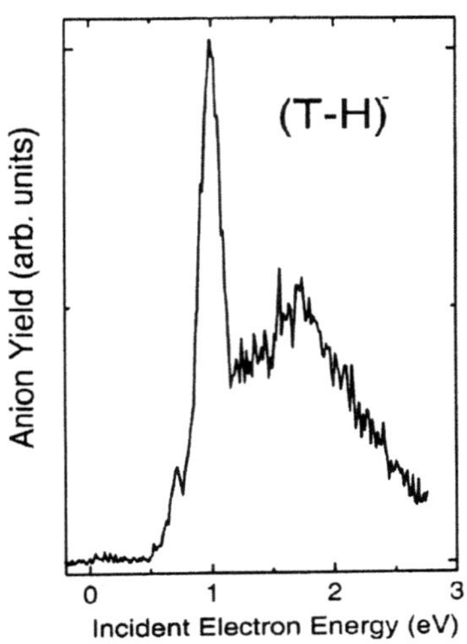

Effective dehydrogenation is a general feature in low energy electron induced reactions to DNA bases and also other biologically relevant molecules like amino acids. The reaction is driven by the large electron affinity of the corresponding radical (M-H) which in the case of the DNA bases is in the range of 3.0-3.5 eV. Electron impact to DNA bases at higher energy results in DEA fragmentations involving opening of the cyclic unit.

Figure 1. Dissociative electron attachment to thymine (T) leading to the dominant channel of dehydrogenation by ejection of a neutral hydrogen atom with the excess charge remaining on the molecule creating the closed shell anion $(M\text{-}H)^-$ (adapted from ref. [4]).

While from these experiments the route to strand breaks is not directly obvious and may involve a sequence of reactions, the present experiments clearly show the propensity of electrons to induce fragmentation and produce reactive radicals at sub-excitation energies (< 3 eV) and also to initiate ring opening reactions at sub-ionization energies (< 10 eV).

4. RADIATION DAMAGE AND RADIOSENSITIZERS FOR TUMOR THERAPY

While high energy radiation can damage biological material, radiation is, on the other hand, successfully used in tumor therapy. The problem here is to only expose the affected material but keep the other areas non-irradiated. One way of potential treatment is the use of *radiosensitizers* with the effect that the sensitized cells will collapse already at dosages leaving the healthy material virtually unaffected. The necessary prerequisite for effective therapy strategies, however, is the understanding of the molecular mechanisms of the underlying processes. We shall illustrate this issue for the case of radiation damage which is in turn directly associated to the question of the molecular mechanism by which a radiosensitizer operates.

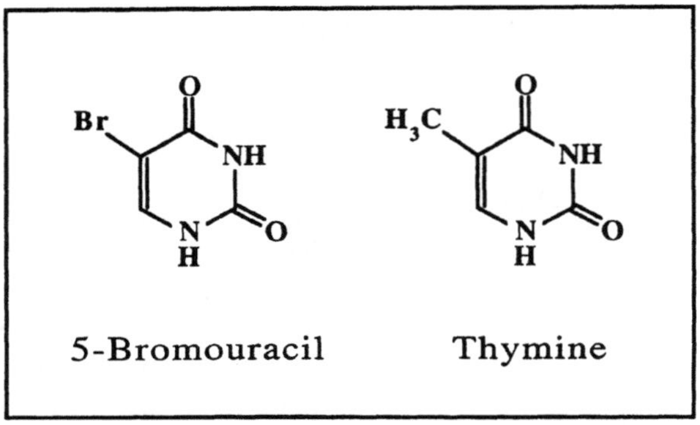

Figure 2. Molecular structure of 5-Bromouracil (BrU) and Thymine (T). In BrU the CH_3 group is replaced by Br.

We shall consider here a prototype gas phase result to illuminate the effect when the DNA base thymine (T) is replaced by bromouracil (BrU), see Fig. 2. It has been known for many years that substitution of T by BrU in the genetic sequence of cellular DNA leads to a greater sensitivity to ionizing radiation without changing the gene expression in non-irradiated cells. Hence bromouracil possesses potential application as a tumor specific sensitizer in cancer therapy. On proceeding from higher energies to low energies the following features in electron impact to T and BrU become apparent:

In thymine (T) the most abundant channel was identified as dehydrogenation (reaction (1)). In contrast to that, the radiosensitizer bromouracil exhibits a very intense and narrow low energy resonances (Fig. 3) located close to 0 eV[8] and associated with the processes

$$e^{-}(0\ \text{eV}) + \text{BrU} \rightarrow \text{BrU}^{\#-} \rightarrow (\text{U-yl}) + \text{Br}^{-} \qquad (2a)$$

$$\rightarrow (\text{U-yl})^{-} + \text{Br} \qquad (2b)$$

which are complementary with respect to the negative charge. (U-yl) denotes the fragment formed by the loss of bromine. (2a) is the most abundant channel with an estimated cross section of 600 Å^2. Reaction (2b) is also operative at zero eV (though at only 6% of the intensity of (2a)). Due to the appreciable electron affinities of Br and (U-yl) exceeding 3 eV, these reactions have low energy thresholds. Figure 3 also shows that the undissociated anion weakly appears within the time scale of the experiment (ca. 50 μs).

It is interesting to note that both T and BrU are damaged at electron energies below 3 eV. The absolute cross sections, however, differ by more than two orders of magnitude. The conclusion then is that the initial mechanism for *direct* DNA damage is bond cleavage by low energy secondary electrons which is much more effective in the radiosensitizers.

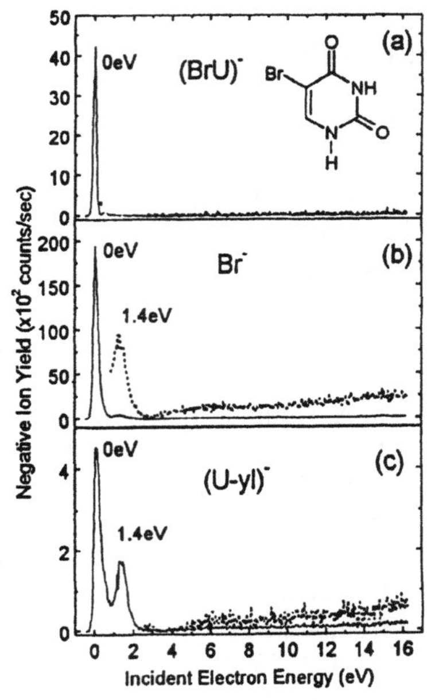

Figure 3. Electron attachment to the radiosensitizer BrU. In addition to the complementary DEA channels Br- and (U-yl)- the metastable parent anion is also observed (adapted from ref. [8]).

5. GAS PHASE RESULTS AND BIOLOGICAL REALITY

The general question remains to which degree such gas phase results are relevant for a real (*in vivo*) biological system. DNA as a polymer is embedded in a medium while the present reactions are observed from isolated gas phase components. In the following we consider a few points.

(a) Coupling of the nucleobases to the backbone and the opposite chain will modify the energy of the involved molecular orbitals to some degree but one can assume that the essential features of the isolate bases will remain in the polymer.
(b) In a condensed environment the reactivity (bond cleavage) is affected and can lead to both a decrease or an enhancement in reactivity.
(c) The reaction route from dissociative electron capture to single and double strand breaks is not directly obvious and has to be explored.
(d) Electrons in the hydrated stage can be viewed as chemically rather inactive since their binding energy is above 3 eV and hence any reaction is associated with a large activation barrier and a correspondingly low reaction rate

On a superficial view it seems paradoxical that the damage of high energy radiation in the MeV range is actually the result of the interaction of secondary electrons at very low energies. Capture of electrons into antibonding molecular orbitals, however, is a very effective means to transfer energy of the light electron into motion of the heavy nuclei.

Work supported by the Deutsche Forschungsgemeinschaft (DFG), the European Union via the EPIC Network, The Humboldt-Foundation and the FU Berlin.

REFERENCES

1. M. Burton and J. I. Magee (eds.) *Advances in Radiation Chemistry*, Wiley, New York, 1969.
2. A. Mozumder and Y. Hatano (eds.) *Charged Particle and Photon Interactions with Matter. Chemical, Physicochemical and Biological Consequences with Applications*, Marcel Dekker, New York, 2004.
3. B. Boudaiffa, P. Cloutier, D. Hunting, M. A. Huels and L. Sanche, Resonant Formation of DNA Strand Breaks by Low Energy (3 to 20 eV) Electrons, *Science* **287**, 1658-1660 (2000).
4. H. A.-Carime, S. Gohlke and E. Illenberger, Site Selective Dissociation of DNA Bases with Subexcitation Electrons, *Phys. Rev. Lett.* (in print).
5. A. M. Scheer, K. Aflatooni, G. A. Gallup and P. D. Burrow, Bond Breaking and Temporary Anion States in Uracil and Halouracils: Implications for DNA Bases, *Phys. Rev. Lett.* **92**, 068102-1 - 068102-4 (2004).
6. R. Abouaf, J. Pommier and H. Dunet, Negative Ions in Thymine and 5-Bromouracil Produced by Low Energy Electrons, *Int. J. Mass Spectrom.* **226,** 397-403 (2003).
7. G. Hanel, B. Gstir, S. Denifl, P. Scheier, M. Probst, B. Farizon, M. Farizon, E. Illenberger and T. D. Märk, Electron Attachment to Uracil: Effective Destruction at Subexcitation Energies, *Phys. Rev. Lett.* **90** 188104-1 - 188104-4 (2003).
8. H. Abdoul-Carime, M. A. Huels, L. Sanche and E. Illenberger, Sensitizing DNA to Secondary Electron Damage: Formation of Oxidative Radicals from 5-Halouracils, *J. Am. Chem. Soc.* **123**, 5354-5355 (2001).

LOW ENERGY (< 3eV) ELECTRONS AS A SOFT TOOL FOR SURFACE MODIFICATION

Richard Balog and Eugen Illenberger

1. INTRODUCTION

One of the ultimate goals in chemistry has always been to prepare a molecular system in a particular way in order to induce a specific bond cleavage which is one of the necessary prerequisites to control a chemical reaction. We mention mode selective multi-photon excitation by IR lasers or state selective excitation in the UV region. The selectivity for bond breaking, however, is often deteriorated by effective energy redistribution within the molecule or, in the case of molecules at surfaces and in the condensed phase, by energy dissipation. A very timely subject of research along that direction is the control of a reaction by the application of ultra-short laser pulses. The idea here is to find the optimum pulse shape (and hence the optimum phase correlation between the frequencies) by means of evolutionary algorithms to push a reaction into the desired direction[1,2]. Here we demonstrate that the direction and extend of a condensed phase reaction can effectively be controlled by low energy electrons.

In the framework of electron driven reactions in condensed material we mention here ozone synthesis in an oxygen film[3], CO formation in condensed acetone[4], activation of adsorbed cyclopropane and transformation into propene[5], bond selective dissociation of alkanethiole based self assembled monolayers[6], strand breaks in supercoiled DNA[7], desorption of neutrals from chemisorbed biological molecular systems[8] or the modification of H-Si surfaces relevant for industrial applications[9]. In all these studies the reaction was observed at electron energies essentially *above* the threshold for electronic excitation and with the synthesized product often in minor amounts with respect to the initial molecule. Of particular interest for the present results is a recent study on Cl_2 synthesis from condensed $C_2F_4Cl_2$[10]. It was shown that Cl_2 synthesis is also operative at *subexcitation* energies. Here we demonstrate that at sufficiently high electron dosage the chemical transformation can be made complete.

2. EXPERIMENTAL

The electron stimulated desorption experiments were carried out in a UHV apparatus consisting of a *trochoidal electron monochromator* (TEM), a cryogenic cooled monocrystalline Au substrate mounted on a manipulator and a commercial quadrupole mass spectrometer with an extraction system[10] (Fig. 1).

Institut fuer Chemie - Physikalische und Theoretische Chemie, Freie Universitaet Berlin, Takustrasse 3, D-14195 Berlin, Germany

Gaseous Dielectrics X.
Edited by Christophorou *et al.*, Springer, New York, 2004.

The substrate can be cooled down to approximately 45 K (measured by a thermocouple directly mounted at the crystal) by means of a closed cycle He

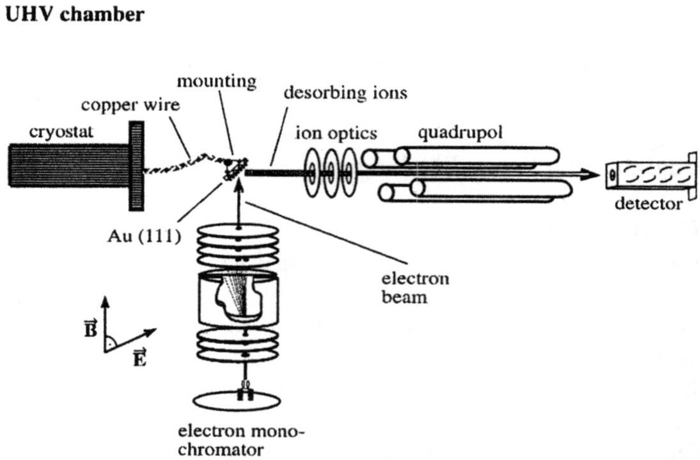

Figure 1: Experimental arrangement for the study of electron induced desorption of ions from adsorbed and condensed molecules.

refrigerator and resistively heated up to several hundred K. The molecules are condensed on the metallic substrate by exposing it to a volumetrically calibrated gas quantity effusing from a capillary located 0.7 cm from the crystal. The TEM operates with a weak homogeneous magnetic field (~ 30 Gauss) which is appropriate to study electron induced processes in the low energy domain. The base pressure in the chamber is in the 10^{-10} mbar range. The electron energy is calibrated by the onset of electron transmission to the substrate (the electron injection curve) representing the vacuum level (0 eV). From the steepness of the onset curve the energy width of the electron beam can be estimated as 0.2 eV at a current of ≈ 50 nA. Exposure of the film to the electron beam generally leads to accumulation of negative charges due to stabilized anions and otherwise trapped electrons. This results in a shift of the electron injection curve with respect to the monochromator potential. All desorption spectra shown are calibrated with respect to the onset of the injection curve.

3. RESULTS AND DISCUSSION

In the gas phase, Cl^- is predominantly formed from a low energy feature peaking at 0.3 eV[11]. At that energy Cl^- formation is *selective* (since other DEA channels are energetically inaccessible) and also *effective* with a cross section of about 2 x 10^{-19} m^2. In the condensed phase, desorption is strongly suppressed at low energies in favor of an intense signal in the energy range between 7 and 13 eV. Suppression of the *desorption* intensity at low energies does not necessarily imply suppression of DEA in the condensed phase. In the contrary, DEA may even be enhanced, but the majority of the ions do not possess sufficient kinetic energy to leave the surface. At higher energies, core excited resonances are involved and it appears a general phenomenon that DEA (leading to desorption) via these states is strongly enhanced in a condensed phase environment[12,13]. We use the intensity of this desorption feature peaking around 8 eV as a measure for the concentration of $C_2F_4Cl_2$.

We have studied the behavior of a 5 ML $C_2F_4Cl_2$ film in the course of extended irradiation with electrons at different energies up to dosages of several mC[14] (corresponding to irradiation times up to about 20 hours). The general trend is that at energies above the threshold for electronic excitation some equilibrium composition between $C_2F_4Cl_2$ and Cl_2 is achieved (essentially completed at dosages of several μC) accompanied with an overall degradation of the film. At an energy around 4 eV the chemical transformation is very weak (equilibrium appreciably on the side of the initial compound) with a moderate overall degradation while at energies below 3 eV a complete transformation can be achieved leaving the new film nearly unaffected upon further irradiation.

Figure 2 shows a sequence of desorption spectra performed at 9.5 eV electron energy and Fig. 3 a comparable set taken at 1 eV. As has been demonstrated[10] irradiation of the $C_2F_4Cl_2$ film creates a new feature around 5 eV which is identified as Cl^- desorption from molecular chlorine. From our previous data an explicit relative cross section for the transformation of $C_2F_4Cl_2$ into Cl_2 versus electron energy could be extracted after irradiating the sample with a constant dosage of 10 mC. This cross section showed two maxima, one below 2 eV and one near 10 eV with the value virtually dropping to zero at an energy near 4 eV.

During irradiation the energy of the beam is kept constant and only by performing desorption scans (scan time 25s corresponding to a dosage of about 1μC) the sample is exposed to the varying electron energy in the range 0 – 13 eV. Figure 2 shows that irradiation results in a considerable decrease of the $C_2F_4Cl_2$ feature in favor of that due to Cl_2 peaking near 5 eV. Further irradiation at that energy leaves the relative composition virtually unaffected but leads to a degradation of film, i. e. evaporation of material from the irradiated area.

Upon irradiation with 1 eV electrons (Fig. 3) the situation changes considerably in that at sufficiently large dosage (4.45 mC corresponding to an irradiation time of 19 hours) there is no further indication of the presence of $C_2F_4Cl_2$ and no evidence of a significant degradation of the molecular film. The 4.45 mC desorption peak is a nearly perfect image of the desorption resonance obtained from a pure Cl_2 layer.

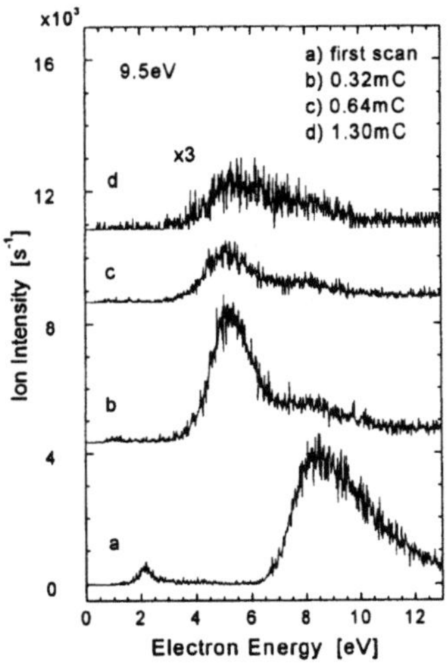

Figure 2. Evolution of Cl^- desorption from a 5 monolayer $C_2F_4Cl_2$ film during irradiation with electrons at 9.5 eV. The new structure near 5 eV ist due to formation of Cl_2 (adapted from Ref. [14]).

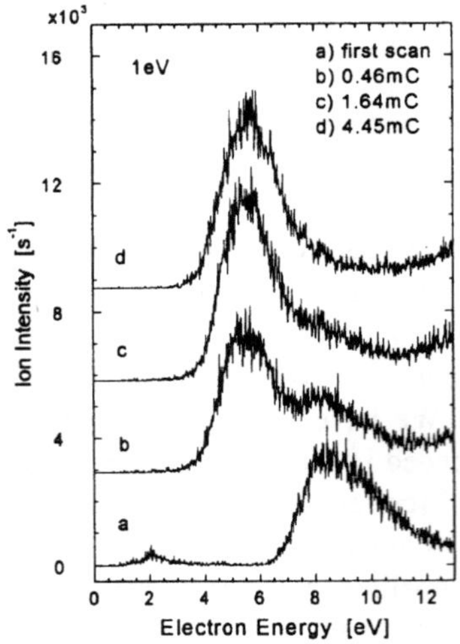

Figure 3. Evolution of Cl^- desorption from a 5 monolayer $C_2F_4Cl_2$ film during irradiation with electrons at 1 eV. The remaining spectrum is a perfect image of Cl^- desorption from a pure Cl_2 film [14].

The number density of the initial molecules at the surface is in the range of 10^{15} cm^{-2}. The *irradiated* area of the sample can be estimated as 0.2 – 0.5 cm^2 (dependent on the electron energy) at a total sample area of 1 cm^2. In the case of irradiation with 9.5 eV the equilibrium is obtained after about 300 s corresponding to a number of electrons which roughly equals that of $C_2F_4Cl_2$ molecules in the interaction area. This indicates that the initial cross section to transform $C_2F_4Cl_2$ into Cl_2 must be at least in the range of the geometrical cross section of $C_2F_4Cl_2$. Further irradiation does not change the relative composition, but degrades the film. On the other hand, in the case of 1 eV electron interaction the reaction can be made complete at sufficiently high dosage. During the largest dosage the number of electrons interacting with the surface is about 3 x 10^{16}.

From the molecular point of view, electrons at 9 eV create a variety of electronically excited states including electronically excited resonances which may decompose into fragments. Cl_2 can then be formed via secondary reaction sequences which are not easy to reveal. Apart from the synthesis of Cl_2, electrons of 9.5 eV energy impinging the surface will degrade both, the initial molecule, the product and also the yet not identified byproducts.

Electrons of 1 eV energy, on the other hand, can only induce a bond cleavage via the DA process

$$e^- + C_2F_4Cl_2 \rightarrow C_2F_4Cl + Cl^- \qquad (1)$$

which is slightly exothermic by 0.2 eV. Possible secondary pathways for Cl_2 synthesis may include polymerization of the radicals formed in the first step and also reactive scattering of Cl^- as discussed[10]. From the thermodynamic point of view we have to consider the reaction

$$C_2F_4Cl_2 \rightarrow C_2F_4 + Cl_2 \quad (2)$$

which is the most direct way for Cl_2 synthesis. For C_2F_4 representing perfluoroethene, the reaction is endothermic by 3.0 eV[15] and thermodynamics hence predict that the chemical equilibrium is completely on the left hand side ($C_2F_4Cl_2$). As demonstrated above, the presence of low energy electrons shifts the chemical equilibrium in the condensed phase virtually to the right hand side, more precisely to, Cl_2 plus byproducts carrying the negative charge. Since any form of C_2F_4 may not posses a sufficiently high electron affinity to drive such a reaction we can assume that larger byproducts are formed. Possible scenarios are (formulated at a dimer unit)

$$e^- + 2C_2F_4Cl_2 \rightarrow (c\text{-}C_4F_8)^- + 2Cl_2 \quad (3a)$$

and

$$2e^- + 2C_2F_4Cl_2 \rightarrow 2Cl^- + c\text{-}C_4F_8 + Cl_2 \quad (3b)$$

Reaction (3a) is endothermic by 3.9 eV minus the (unknown) electron affinity of $(c\text{-}C_4F_8)^-$ while (3b) is exothermic by 0.7 eV .Thermodynamically the large electron affinity of Cl (3.6 eV) represents the driving force of reaction (3b) which we hence consider as the basic reaction relevant at subexcitation energies.

One essential point necessary to achieve complete transformation remains noteworthy. It is based on the fact that the initial molecule ($C_2F_4Cl_2$) is strongly affected by low energy electrons while this is not the case for the product (Cl_2). This is based on the fact that gas phase Cl_2 shows *p-wave* behavior with a comparably low cross section in the threshold region[16].

This work has been supported by the Deutsche Forschungsgemeinschaft (DFG), the European Union via the EPIC Network, and the Freie Universitaet Berlin.

REFERENCES

1. C. Daniel, J. Full, L. Gonzales, C. Lupulescu, J. Manz, A. Merli, S. Vajda, L. Wöste, Deciphering the Reaction Dynamics Underlying Optimal Control Laser Fields, *Science* **299**, 536-539 (2003).
2. T. Brixner, N. H. Damrauer, P. Niklas and G. Gerber, Photoselective Adaptive Femtosecond Quantum Control in the Liquid Phase, *Nature* **414**, 57-60 (2001).

3. S. Lacombe, F. Cemic, K. Jacobi, M. N. Hedhili, Y. Le Coat, R. Azria and M. Tronc, Electron-Induced Synthesis of Ozone in a Dioxygen Matrix, *Phys. Rev. Letters* **79**, 1146-1149 (1997).
4. M. Lepage, M. Michaud, and L. Sanche, Low-Energy Electron Scattering Cross Section for the Production of CO within Condensed Acetone, *J. Chem. Phys.* **113**, 3602-3608 (2000).
5. H. Winterling, H. Haberkern and P. Swiderek, Electron-Induced Reactions in Thin Solid Films of Cyclopropane, *Phys. Chem. Chem. Phys.* **3**, 4592-4599 (2001).
6. C. Olsen and P. A. Rowntree, Bond-Selective Dissociation of Alkanethiol Based Self-Assembled Monolayers Adsorbed on Gold Substrates, Using Low-Energy Electron Beams, *J. Chem. Phys.* **108**, 3750-3764 (1998).
7. B. Boudaïffa, P. Cloutier, D. Hunting, M. A. Huels and L. Sanche, Resonant Formation of DNA Strand Breaks by Low-Energy (3 to 20 eV) Electrons, *Science*, **287**, 1658-1660 (2000).
8. H. Abdoul-Carime, P. C. Dugal and L. Sanche, Desorption Induced by Electronic Transitions (DIET) of Neutral Fragments from Chemisorbed Biological Molecular Systems, *Surf. Sci.* **451**, 102-107 (2000).
9. W. Di, P. Rowntree and L. Sanche, Energy-Selective Reaction of the Hydrogen-Passivated Si Surface with Carbon Tetrafluoride via Dissociative Electron Attachment, *Phys. Rev.B,* **52**, 16618-16622 (1995).
10. R. Balog, N. M. Hedhili, F. Bournel, M. Penno, M. Tronc, R. Azria, E. Illenberger, Synthesis of Cl_2 Induced by Low-Energy (0-18 eV) Electron Impact to Condensed 1,2-$C_2F_4Cl_2$ Molecules, *Phys. Chem. Chem. Phys*, **4**, 3350-3355 (2002).
11. J. Langer, M. Stano, S. Gohlke, A. Rosa, W. Barszczewska, S. Matejcik, and E. Illenberger, Low Energy (0-15 eV) Electron Stimulated Reactions in Single 1,2-$C_2F_4Cl_2$ Molecules and Clusters, *Int. J. Mass Spectrom Ion Proc.* **223-224**, 193-204 (2003).
12. I. Le Coat, N. M. Hedhili, R. Azria, M. Tronc O. Ingólfsson, and E Illenberger, Medium-Enhanced, Electron Stimulated Desorption of CF_3^- from CF_3I, *Chem. Phys. Letters* **296**, 208-214 (1998).
13. E. Illenberger, Formation and Evolution of Negative Ion Resonances at Surfaces, Surf. Sci. **528**, 67-77 (2003).
14. R. Balog and E. Illenberger, Complete Chemical Transformation of a Molecular Film by Subexcitation Electrons, *Phys. Rev. Letters*, **91**, 213201-1-213201-4.
15. http://webbook.nist.gov/
16. S. Barsotti, M.-W. Ruf and H. Hotop, Clear Experimental Evidence for p-Wave Attachment-Threshold Behavior in Electron Attachment to Chlorine Molecules, *Phys. Rev. Letters*, **89**, 083201-1-083201-4 (2002).

NEGATIVE ION MOTION IN PURE SF_6 AND ITS MIXTURES WITH ATMOSPHERIC, HALOCARBON, AND RARE GASES

Jaime de Urquijo*

1. INTRODUCTION

The study of ion transport in gases has become increasingly important in view of the new and growing areas of applied research such as ion mobility mass spectrometry, in which the identification of the ionic species in the gas (pure or mixture) is carried out through the measurement of the ion mobility. In particular, SF_6 is studied extensively because of its widespread use in the power and semiconductor plasma processing industries, despite its high global warming potential that has raised concerns on its release into the atmosphere. Thus, research is currently aimed at finding replacements or mixtures with other gases that would retain many of its outstanding characteristics while reducing its atmospheric impact[1].

Negative ions play an important role in the development of discharges, via attachment, ion clustering, and detachment-induced breakddown. Research on ion transport in SF_6 has been concentrated on the measurement of the mobility and diffusion of several positive ion species in pure SF_6 and a few binary mixtures, and on the drift of SF_6^-, SF_5^- and F^- in its parent gas[2,3], though little is yet known about the transport properties of any of these ions in the gas mixtures of SF_6 with atomic and molecular gases, even though some of these, like SF_6-N_2, are widely used. Mixtures of SF_6 with N_2, CF_4 or Ar are either currently used for practical applications, or have been suggested as potential replacements to pure SF_6 in electrical power insulation and control.

* Centro de Ciencias Físicas, UNAM, P.O. Box 48-3, 62251, Cuernavaca, Mor. México. E-mail address: jdu@fis.unam.mx

Gaseous Dielectrics X.
Edited by Christophorou *et al.*, Springer, New York, 2004.

Ion transport is usually studied on drift tubes with mass spectrometry, since these provide an almost unambiguous identification of the species under investigation. However, the upper pressure limit of operation of these devices, in the 0.1-1 torr range, limits their capabilities for studying ion transport and reactions at low E/N. Alternately, the Pulsed Townsend Method (PTM) may be useful for studiyng ion transport under some special circumstances –such as the present one- where a strongly attaching gas like SF_6, low E/N, and high pressures (>50 torr) are operative. In such case, even though the PTM lacks mass spectrometry means, the identity of the ion can be readily inferred to be SF_6^-, as it will be shown below.

The very large electron attachment cross section, σ_a, of SF_6 for slow electrons leads to the formation of SF_6^- with $\sigma_a=1.5x10^{-13}$ cm^2 for 1 meV electrons, and $\sigma_a=5x10^{-15}$ cm^2 for 0.7 eV electrons, in contrast with a fairly constant, much lower σ_a value for SF_5^- formation of $2x10^{-16}$ cm^2 over the same energy range[4]. In the present study, since the E/N range was kept fairly low, well below the ionisation threshold of SF_6 or of any of its mixture partners, the negative ion signals arising from the PT experiment can be ascribed to the presence of SF_6^-, at least as the overwhelming majority species. Moreover, for O_2, CH_4 and CF_4, the other electronegative gases mixed SF_6 reported in this contribution, negative ion formation is almost negligible in comparison to that of SF_6 at very low energies, since it proceeds by dissociative attachment at energies peaking at several eV.

The research reported here deals with the measurement of the mobility of SF_6^- in the binary mixtures of SF_6 with He, Ne, Ar, Xe, O_2, N_2, CF_4, and with the ternary mixtures with CH_4-Ar and dry air. Also, the measured low-field mobilities are compared successfully with those calculated from Blanc's law. Further evidence that the drifting ion in this experiment is predominantly SF_6^- will be given below.

2. EXPERIMENTAL DETAILS

2.1 The negative ion current waveforms from a PT experiment

The essentials and of the Pulsed Townsend method and the details of the apparatus used in this research are given elsewhere[5-7]. Briefly, n_o photoelectrons are released from the cathode of a parallel-plate discharge gap by the incidence of an ultrafast pulse of UV light upon its surface. The uniform electric field between the electrodes will cause the photoelectrons move towards the anode. Under the present conditions of a strongly electronegative gas mixture, low E/N and high gas pressures, these photoelectrons will soon become attached to the SF_6 molecules within a very small fraction of the gap of width *d* (ionisation processes are absent). Since the electrons are more mobile than the ions (typically by a factor or 100 or more), then the latter will remain virtually in their positions of formation during the electron transit and, on a much slower time scale, these will move toward the anode. In the absence of ionisation, it has been shown that the displacement current due to the negative ions is[8]

$$i_n(t) \approx \frac{n_o q_o}{T_n}, \tag{1}$$

where q_o is the elementary charge, v_n and $T_n=d/v_n$ are the negative ion drift velocity and transit time, respectively. Figure 1(a) shows a sample transient of the type predicted by Eq. (3). As indicated in the same figure, the measurement of T_n is straightforward, and from it the calculation of the drift velocity

$$v_n = d/T_n. \tag{2}$$

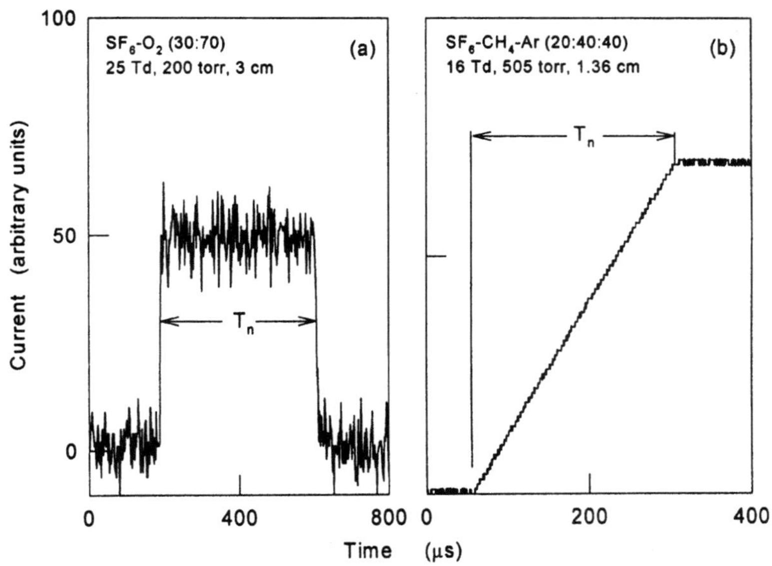

Figure 1. Sample negative ion transients in SF_6 mixtures. (a) Differentiated pulse condition; (b) Integated pulse condition. T_n is the transit time of the negative ion.

When the displacement current is integrated at the input of a very high impedance voltage amplifier, the stray capacitance connected effectively at its input terminals is charged, and the voltage generated at the input of the amplifier is[5,6,9]

$$V_n(t) = \frac{1}{C}\int_0^t i_n \, dt \approx \frac{n_o q_o}{C T_n} t, \tag{3}$$

which describes a linear current growth from zero to a maximum of $V_n(T_n)=n_o q_o/C$. Figure 1(b) illustrates this behaviour very clearly for a negative ion transient.

2.2. The apparatus

The two apparatus used for these measurements were of similar design, although slightly different in size and in the measurement technique[7,9]. For the first apparatus, the differentialed pulse condition [Fig. 1(a)] discussed in the previous section was used [Eq. (1)]. The ionic avalanche was detected by a transimpedance amplifier. For the second experiment, the integrated pulse condition was used [sample transient of Fig. 1(b)]. The ionic avalanche was detected by a $1G\Omega$ input resistance, voltage amplifier. The transient pulses were registered by means of digital oscilloscopes and stored on personal computers for further analysis.

SF_6 and the other gases were mixed inside the discharge vessel, over the pressure range 50-1000 torr. Gas purities are quoted elsewhere[8,9]. The measurements were made at room temperatures over the range 296-303 K.

3. MOBILITY OF SF_6^- IN THE SF_6 MIXTURES

The uncertainty in the drift velocities discussed below lies within the range 2-4%. These drift velocities were converted into reduced mobilites by the formula

$$K_0 = (v_n / N_0)/(E / N), \tag{4}$$

where N_o=2.69x10^{19} cm^{-3} is the gas density at STP.

3.1. Binary mixtures of SF_6 with He, Ne, Ar and Xe

The reduced mobility of SF_6^- in the SF_6-He mixture is shown plotted in Fig. 2 as a function of E/N[10]. The negative ion mobility data for pure SF_6, obtained with the present apparatus, compare very well over 50<E/N<150 Td with the drift tube-mass spectrometer data of SF_6^- in SF_6 (de Urquijo et al, 1991), thereby leading us to reassure that the most abundant negative ion species with pure SF_6 was SF_6^-. The only mobilities of SF_6^- in He available in the literature are the calculations quoted by Viehland and Mason[2]. Our measurements for 0.01% SF_6 in He are about 10% smaller than the calculated ones. We thus regard the measured mobilities as those pertaining to the experimental values for SF_6^- in pure He, since little change is observed when one attempts even lower concentrations.

The above mobilities, extrapolated to zero field, have been tested by means of Blanc's law[12,13]

$$\frac{1}{K_{00}} = \sum_{j=1}^{n} \frac{x_j}{K_{i,j}}, \tag{5}$$

where $K_{i,j}$ is the mobility of ionic species i (SF_6^- in the present case) in the pure gas j, and x_j is the molar fraction of gaseous species j in the mixture. In order to use the above expression meaningfully, a knowledge of the mobility of the

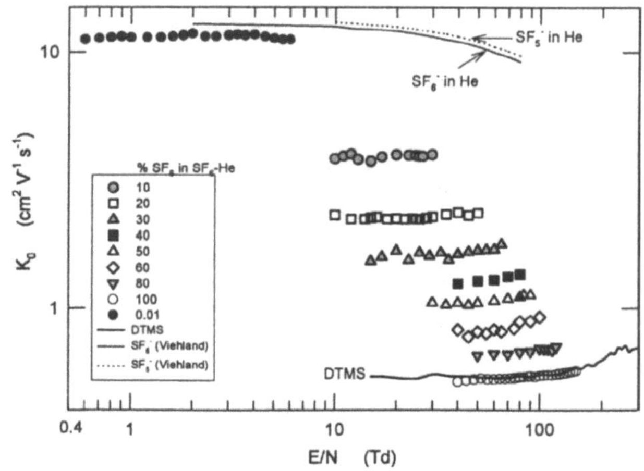

Figure 2. The reduced mobility of SF_6^- in SF_6-He (de Urquijo et al, 2001b). DTMS: Drift tube data[11]; the upper solid and dotted lines are the calculated values[2] for SF_5^- and SF_6^-.

drifting ion both in both pure SF_6 and in the noble gas is needed. In order to accomplish this goal, we measured the mobility of SF_6^- in the buffer gas containing only small fractions of SF_6 (0.005%-0.1%), as it was done for He. These values are summarised in Table 1.

Table 1. Zero-field mobilities of SF_6^- in the pure gases[a]

SF_6^- in	K_{00} ($cm^2V^{-1}s^{-1}$)	SF_6^- in	K_{00} ($cm^2V^{-1}s^{-1}$)
He	11.5	N_2	1.89
Ne	4.32	O_2	1.95
Ar	1.68	Air (dry)	1.89
Xe	0.658	CH_4	1.87
CF_4	0.825	SF_6	0.542[b]

[a] Measured by mixing only ppm SF_6 in the buffer gas
[b] Drift tube-mass spectrometer data[11]

The zero-field reduced mobility, K_{00}, of SF_6^- in He, Ne, Ar, and Xe is given in Fig. 3 as a function of the SF_6 content in the mixture. The solid curves are the calculated values of K_{00} according to Blanc's law, showing fair to good agreement with the measured values.

3.2. Binary mixtures of SF_6 with N_2, O_2 and CF_4

The reduced mobility of SF_6^- in the SF_6-N_2 mixture[8] is plotted in Fig. 4, for concentrations ranging from 0.02 to 50%. The mobility curve for SF_6-O_2 differs

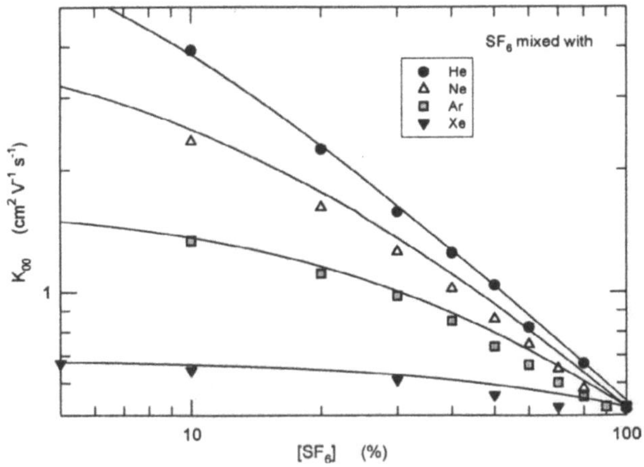

Figure 3. The zero-field mobility of SF_6^- in the mixtures of SF_6 with He, Ne, Ar and Xe. The solid lines are the calulated values according to Blanc's law (Eq. 4)

slightly form that for SF_6-N_2. The reduced mobility of SF_6^- in the SF_6-N_2 mixture[8] is plotted in Fig. 4, for SF_6 concentrations ranging from 0.02% to 50%.

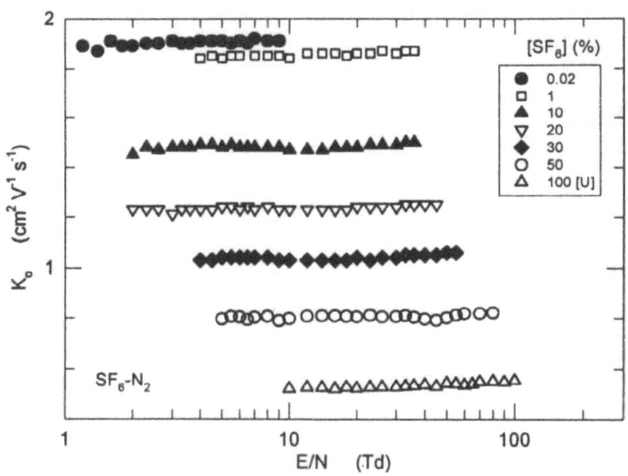

Figure 4. The reduced mobility of SF_6^- in SF_6-N_2 [Ref. 8]; 100% SF_6 values [U] from Ref. [11].

A smilar mobility curve for the SF_6-O_2 mixture differs slightly from that of SF_6^-. The fairly constant behaviour of these curves with E/N is characteristic of the low-field regime, dominated by the long-range, attractive part of the interaction potential, as the ion transport theory predicts[12]. Figures 5 and 6 show the variation of the low-field mobility of SF_6^- in the SF_6-N_2 and SF_6-O_2 mixtures

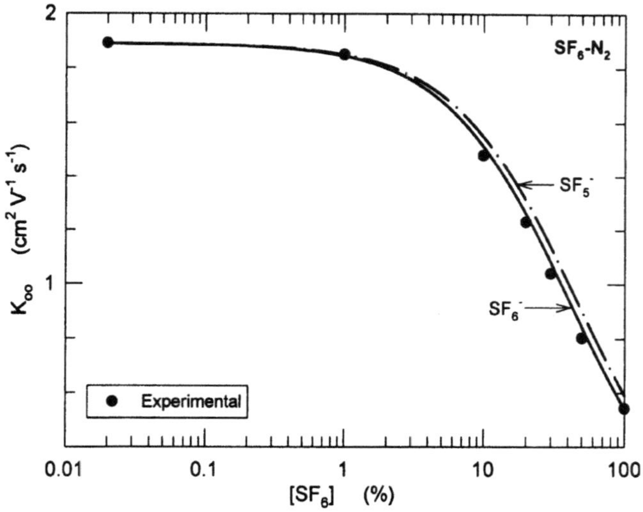

Figure 5. The zero-field mobility of SF_6^- in SF_6-N_2 as a function of the SF_6 concentration. The solid curve is the calculated value from Blanc's law for SF_6^- in SF_6, while the chain line corresponds to SF_5^- in SF_6.

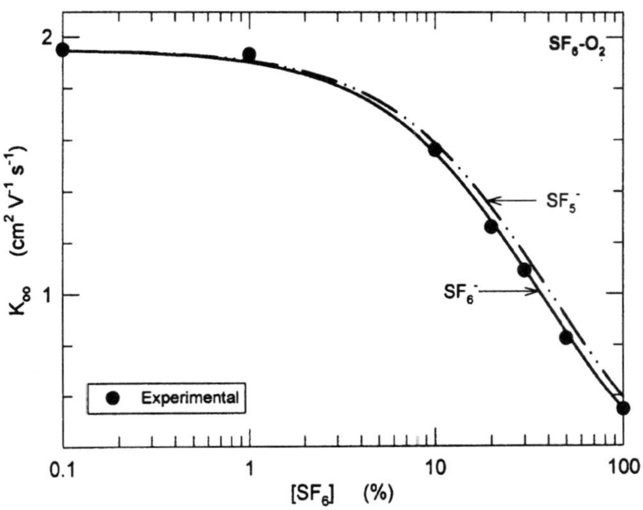

Figure 6. The zero-field mobility of SF_6^- in SF_6-O_2 as a function of the SF_6 concentration. The solid curve is the calculated value from Blanc's law for SF_6^- in SF_6, while the chain line corresponds to SF_5^- in SF_6.

with the SF_6 content, respectively. The agreement between the calculated values from Blanc's law and the measured ones is excellent when the mobility of SF_6^- in pure SF_6 is used in the calculations, in contrast to that of assuming SF_5^-, which shows a large departure for large SF_6 concentrations.

Resorting to Langevin's polarisation limit formula[12] for small E/N

$$K_{00} = 13.56 \left(\mu_r \alpha_d\right)^{-1/2} \quad (5)$$

where μ_r is the reduced mass of the ion-neutral system and α_d is the dipole polarisability of the molecule (buffer gas), then we calculate the ratio between the zero-field mobility of N_2 and that of O_2 to obtain

$$\frac{K_{N_2}}{K_{O_2}} = \sqrt{\frac{\mu_r(O_2 - SF_6)\alpha_{O_2}}{\mu_r(N_2 - SF_6)\alpha_{N_2}}} \quad (6)$$

Using $\alpha_d(O_2)$=1.57 Å^3 and $\alpha_d(N_2)$=1.74 Å^3 [Ref.12], a calculated value for this ratio of 1.00 compares very well with the experimental value of 0.969. The reduced mobilities of SF_6^- in SF_6-CF_4 will be shown in the next section.

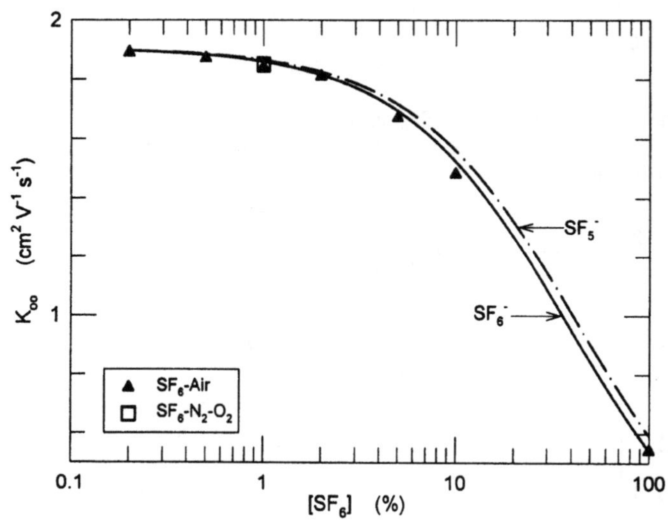

Figure 7. The zero-field mobility of SF_6^- in SF_6-O_2-N_2 (square, 1% SF_6-78.21% N_2-20.79%O_2) and of SF_6^- in dry air, as a function of the SF_6 concentration[8]. The solid curve is the calculated value from Blanc's law for SF_6^- in SF_6, while the chain line corresponds to SF_5^- in SF_6.

3.3. Ternary mixtures of SF_6 with CH_4-Ar and air

One further test to our technique was the measurement of the low-field mobility of SF_6^- in the ternary mixtures of SF_6 with dry air and with SF_6-CH_4-Ar (20:40:40). The reduced mobility of SF_6^- in air for low SF_6 concentrations, up to E/N=10 Td[8], shows only slight differences with respect to those for SF_6-N_2 and SF_6-O_2. Since atmospheric air contains small amounts of Ar and CO_2 (<1%), we performed a measurement of the ternary mixture SF_6-N_2-O_2 for 1% SF_6-78.21% N_2-20.79%O_2, (see the the square in Fig. 7), which was in excellent agreement with the value obtained for the SF_6-air mixture, thereby indicating that the influence Ar and CO_2 in atmospheric air affects the mobility only slightly. Also shown in Fig. 7 are the zero-field mobilities of SF_6^- in dry air. The agreement between measured and calculated values from Blanc's law is excellent and, once more, the calculated curve for SF_5^- (assumed) shows a clear deviation from the experimental ones.

The reduced mobility of SF_6^- in the mixture of SF_6 with CH_4-Ar[9] (20:40:40) is given in Fig. 8 for SF_6 concentrations between 5 and 50%, together with the curve for the zero-field mobility of SF_6^- in the SF_6-CF_4 mixture. In the former case, the agreement between measurements and calculations is excellent.

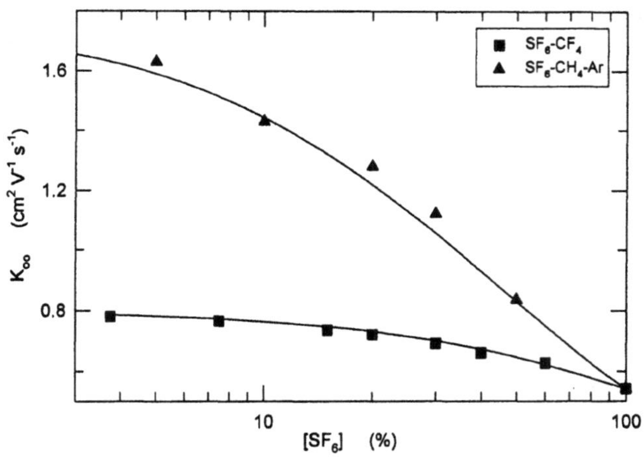

Figure 8. The zero-field mobility of SF_6^- in SF_6-CH_4-Ar (20:40:40) (triangles) and of SF_6^- in CF_4, as a function of the SF_6 concentration. The solid curve is the calculated value from Blanc's law for SF_6^- in SF_6 [Ref. 9].

5. CONCLUSIONS

Thus far, we have presented a set of mobility data for SF_6^- in mixtures of SF_6 with rare, atmospheric and halocarbon gases. One of our main concerns is to give as ample evidence as possible that, even though our experimental technique lacks mass spectrometry, the identity of the ion can be stated clearly. To summarise, the various reasons that support the above statement are as follows. (a) The photon energy used to produce the initial photoelectrons is very close to the work function of the cathode material (Al). Then, these electrons leave the cathode with nearly zero kinetic energy, and therefore attach readily to the SF_6 neutrals, therby forming SF_6^-. (b) We have kept the E/N range low enough so that the average energy of the electron swarm peaks at values where SF_6^- is produced predominatly. (c) The fairly constant mobility of the mobility curves with E/N supports the fact that there is, predominatly, one ionic species drifting across the gap. (d) The extrapolated, zero-field measured mobilities of SF_6^- in all the mixtures can be explained well by means of Blanc's law, provided that the mobilities of this ion in the pure gases were also measured.

These measurements are further validated by considering that, when SF_6 is present only in small amounts in a buffer gas, then the collisions of SF_6^- will occur mostly with the noble gas atoms, thereby approximating the conditions of the experiment in a drift tube mass spectrometer very closely, with the additional advantage that the PT experiment can be operated at buffer gas pressures sufficiently high so as to ensure full equilibrium of the drifting ion with the neutrals and with the electric field. For instance, if one assumes a typical momentum transfer cross section of 10^{-16} cm^2, and for a typical gas mixture pressure of 100 torr, the probability for a collision of SF_6^- with a SF_6 molecule is, order of magnitude, one in 10^6. Then, for the gas pressure ranges used, this probability would lie between $5x10^{-5}$ and 10^{-7}, a figure that reinforces our assumption that SF_6^- practically collides with the buffer neutrals during its passage to the anode of the gap.

In view of the importance of modelling gas discharges in SF_6 with as much accuracy as possible, it is hoped that the transport data presented in this contribution be useful for such purpose. Besides the practicall applications, we believe that data of this kind would be useful for the development of sensors of SF_6 present in atmospheric air through the measurement of the ion mobility. In connection with this, reliable values of the mobility of SF_6^- in air have been given. It is hoped that the present measurements could be extended to other electronegative gases in which negative ion formation proceeds via electron attachment at energies in the meV range or below.

ACKNOWLEDGEMENTS

This work was supported by DGAPA-UNAM, Project IN 104501.

REFERENCES

1. L.G. Chirstophorou and J.K. Olthoff, NIST Technical Report 1325, 1997
2. L.A. Viehland and E.A. Mason, *Atom. Data Nucl. Data Tables* **60**, 37 (1995)
3. J. de Urquijo, *Gaseous Dielectrics IX*, edited by L.G. Christophorou et al, (Kluwer, New York, 2001) pp. 25-36
4. L.G. Chirstophorou and J.K. Olthoff, *J. Phys. Chem. Ref. Data* **29**, 267 (2000)
5. H. Raether, *Electron Avalanches and Breakdown in Gases* (Butterworths, London, 1964)
6. S.R. Hunter, J.G. Carter and L.G. Christophorou, *J. Appl. Phys* **60**, 24 (1986)
7. J. de Urquijo, C.A. Arriaga, I. Alvarez and C. Cisneros, J. Phys. D **32**, 41 (1999)
8. G. Hinojosa and J. de Urquijo, *J. Phys. D* **36**, 2510 (2003)
9. J. de Urquijo and F.B. Yousif, *Phys. Rev. E* **68**, 046406 (2003)
10. J. de Urquijo, E. Basurto and J.L. Hernández-Ávila, *J. Phys. D* **34**, 2151 (2001)
11. J. de Urquijo, C. Cisneros and I. Alvarez, *J. Phys. D,* **24**, 664 (1991)
12. E.W. McDaniel and E.A. Mason, E.A, *The Mobility and Diffusion of Ions in Gases* (Wiley, New York, 1973)
13. E.A. Mason and E.W .McDaniel, *Transport Properties of Ions in Gases* (Wiley, New York, 1988)

ELECTRON TRANSPORT AND IONIZATION PROPERTIES OF C_2F_6 AND CHF_3 MIXED WITH Ar AND N_2

J.L. Hernández-Ávila, E. Basurto and J. de Urquijo *

1. INTRODUCTION

Among other applications, hexafluoroethane (C_2F_6) and fluoroform (CHF_3) are used for the plasma processing of semiconductors, usually diluted with Ar or other inert gases. Even though there exist several measurements of the electron drift velocity v_e, the density-normalized ionization α/N, attachment η/N and effective ionization $(\alpha-\eta)/N$ coefficients for C_2F_6 [1], and only a few for $(\alpha-\eta)/N$ in CHF_3 [2], there are only a few data on v_e and $(\alpha-\eta)/N$ for the mixtures of these gases with Ar and N_2. As regards the density-normalized longitudinal diffusion coefficients ND_L, there is only one measurement for C_2F_6-Ar [1] and none for CHF_3-Ar. Besides the interest in these coefficients for the above practical applications, where they are essential for plasma simulation, these can be used for testing the cross sections sets for the pure halocarbon gases.

This paper deals with the measurement of v_e, $(\alpha-\eta)/N$, ND_L, and the ratio of the longitudinal diffusion coefficient to the electron mobility D_L/K for the C_2F_6-Ar, C_2F_6-N_2, CHF_3-Ar and CHF_3-N_2 mixtures, over a wide range of the density-reduced electric field strength E/N from 50-500 Td (1 Townsend=10^{-17} V cm^2). The mixture data presented in this contribution range between 1-50% Ar and N_2.

2. DETAILS OF THE EXPERIMENT

* J.L. Hernández-Ávila, Universidad Autónoma Metropolitana, CBI- Energía, 02220, México, D.F., E. Basurto, Universidad Autónoma Metropolitana, CBI-Ciencias Básicas, 02220 México, D.F., J. de Urquijo, Centro de Ciencias Físicas, UNAM, 62151 Cuernavaca, Mor. México.

The above transport coefficients were obtained from a Pulsed Townsend experiment, the details of which can be found elsewhere[3-5]. The purity of the gases used was 99.9%, 99.9%, 99.999% and 99.99% for CHF_3, C_2F_6, N_2 and Ar, respectively, and were mixed inside the vacuum vessel. The measurements were performed at room temperatures in the range 293-303 K.

3. SWARM AND TRANSPORT COEFFICIENTS

The data presented here are the result of averaging the measured coefficients at several pressures. The typical uncertainties in the data are 1-2% for v_e, 6-9% for $(\alpha-\eta)/N$ and 15-20% for ND_L. The latter data are presented in this paper as the ratio D_L/K, where K is the electron mobility.

3.1 Electron drift velocities

The variation of the electron drift velocity in the mixtures of C_2F_6 with N_2 and Ar with E/N is shown in Fig. 1. For the sake of clarity, the electron drift velocities for the C_2F_6-Ar mixture have been shifted by a factor of 10, so that its values must be read off the left axis scale.

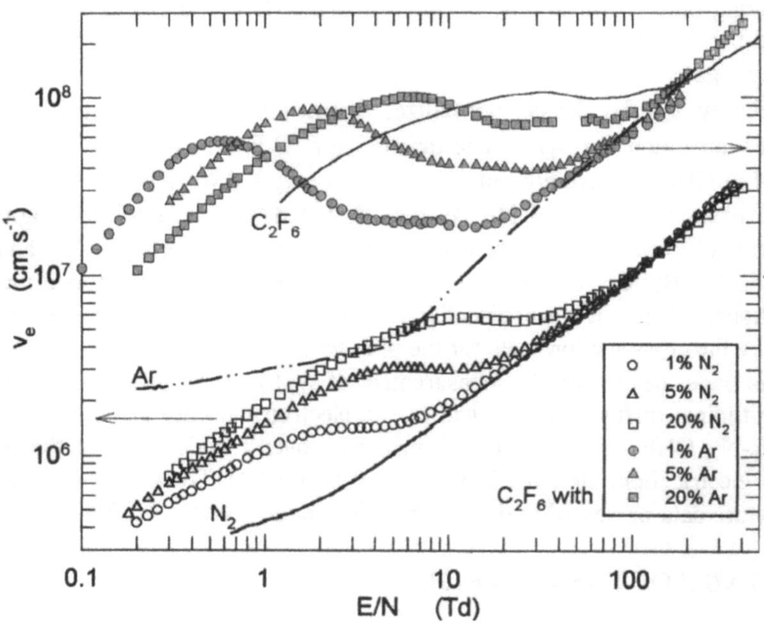

Figure 1. The electron drift velocity in the mixtures of C_2F_6-N_2 (lower, open symbols) and C_2F_6-Ar (upper, closed symbols). The C_2F_6-N_2 data are read from the left-hand axis and the C_2F_6-Ar data from the right-hand axis. The lines correspond to our measurements for pure C_2F_6, Ar and N_2, and are labelled correspondingly.

In this figure we have also included the measured data for pure C_2F_6 (solid line, upper curves) and the Ar data (chain curve), which illustrate very clearly the large effect that the presence of Ar causes at low E/N, where v_e can be over one order of magnitude higher. In this respect, we note that the C_2F_6-Ar curves bear a more pronounced region of negative differential conductivity (NDC) than those for C_2F_6-N_2. This is due to the fact that N_2 does not possess a minimum in its momentum transfer cross section as Ar does. It is also seen that at relatively high E/N the trend of the curves follows that of Ar and N_2.

The corresponding values for the electron drift velocity in the mixtures of CHF_3 with Ar and N_2 are shown plotted in Fig. 2. As for the above case, the NDC regions are well defined for CHF_3-Ar, and less pronounced for CHF_3-N_2. Moreover, it seems that for CHF_3 concentrations exceeding 20%, the NDC effect is practically absent in the latter, since the interaction is more dominated by the presence of CHF_3, the values of which were taken from Ref. [4].

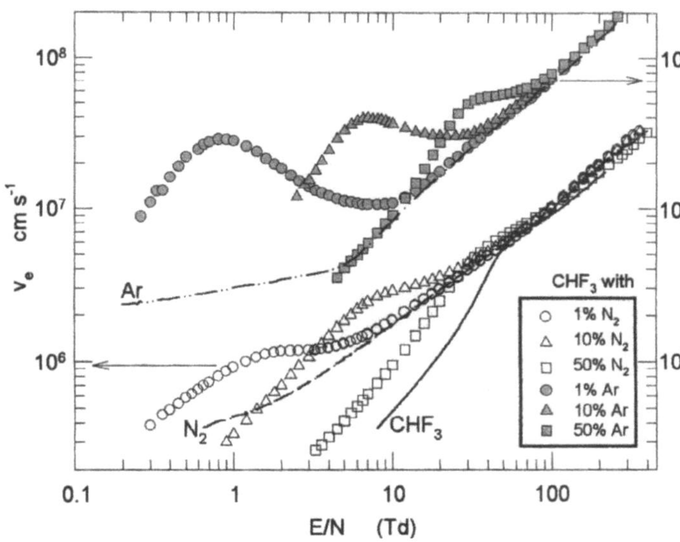

Figure 2. The electron drift velocity in the mixtures of CHF_3-N_2 (lower, open symbols) and CHF_3-Ar (upper, closed symbols). The CHF_3-N_2 data are read from the left-hand axis and the CHF_3-Ar data from the right-hand axis. The lines correspond to previous measurements for pure CHF_3 (Ref. [4]), and the present ones for Ar and N_2, and are labelled correspondingly.

3.2 The ratio D_L/K

The originally measured longitudinal diffusion coefficients of electrons are presented here as the ratio D_L/K. Figure 3 shows the D_L/K values for pure C_2F_6 and the mixtures of C_2F_6 with Ar and N_2 in the same manner as in the previous section. Again, it is noted the strong effect of the presence of Ar in the mixture at low E/N due to the Ramsauer-Townsend effect. Note that the C_2F_6 curves fan out for different Ar concentrations at mostly the whole E/N range covered, while those for the C_2F_6-N_2 mixture practically merge into a single curve for E/N>100Td.

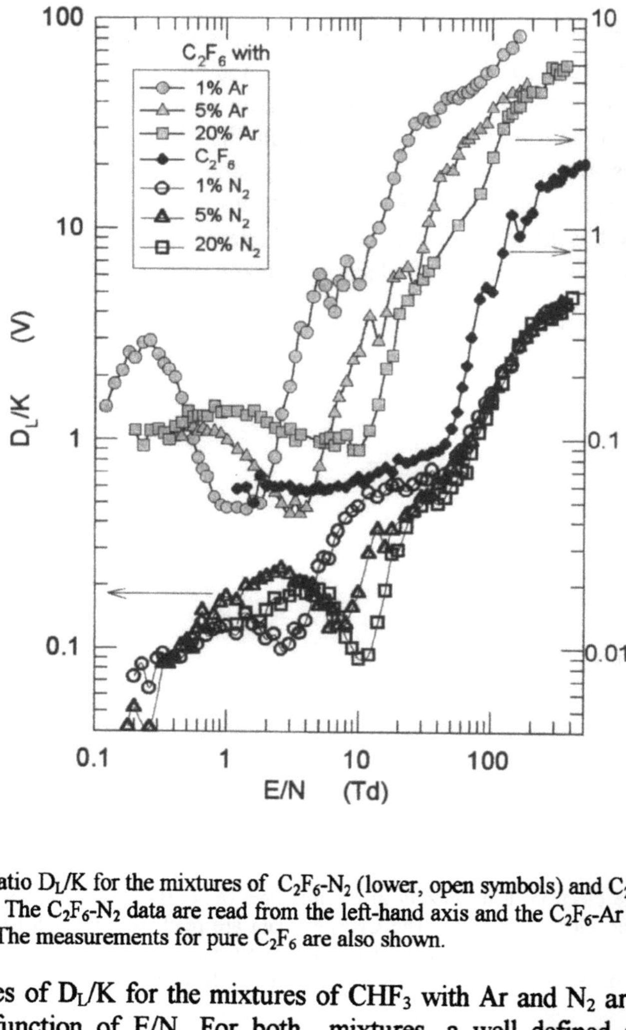

Figure 3. The ratio D_L/K for the mixtures of C_2F_6-N_2 (lower, open symbols) and C_2F_6-Ar (upper, closed symbols). The C_2F_6-N_2 data are read from the left-hand axis and the C_2F_6-Ar data from the right-hand axis. The measurements for pure C_2F_6 are also shown.

The values of D_L/K for the mixtures of CHF_3 with Ar and N_2 are plotted in Fig. 4 as a function of E/N. For both mixtures, a well defined structure is suggested for low CHF_3 concentrations. For the CHF_3-Ar mixture, the pure Ar curve seems to be the upper limit, while that for CHF_3, shown together with the CHF_3-N_2 curves, indicates that for E/N>100 Td, the interaction is fully dominated by CHF_3. Apart from the ND_L values of Nakamura[6] for C_2F_6-Ar at two different concentrations in the range 0.5-5%, we have not found any longitudinal diffusion coefficients for these two gas mixtures in the literature.

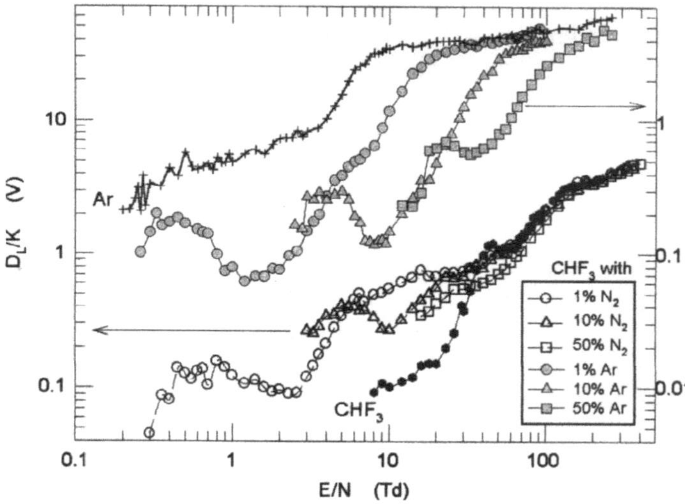

Figure 4. The ratio D_L/K for the mixtures of CHF_3-N_2 (lower, open symbols) and CHF_3-Ar (upper, closed symbols). The CHF_3-N_2 data are read from the left-hand axis and the CHF_3-Ar data from the right-hand axis. The measurements for pure C_2F_6 and for Ar are also shown.

3.3 Effective ionization coefficients

The effective ionization coefficients $(\alpha-\eta)/N$ for C_2F_6 and its mixtures with Ar and N_2 are shown in Figs. 5a and 5b, respectively.

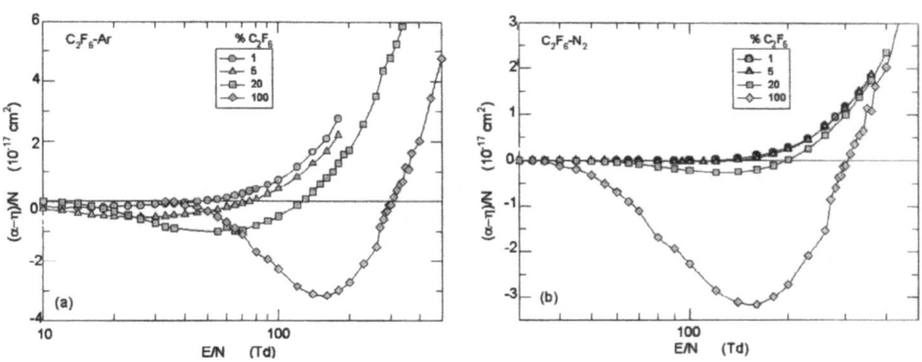

Figure 5. The effective ionization coefficients of C_2F_6 and its mixtures with Ar (a) and N_2 (b).

The ionization coefficients for pure C_2F_6 have been measured previously by other authors[1] down to E/N=180 Td. We were able to measure down to E/N=30 Td. Even though the trend of our values is similar to most of these measurements, ours are slightly lower. It is seen that the mixtures of C_2F_6 with Ar and N_2 retain only slightly electronegative with respect to pure C_2F_6; moreover, the $(\alpha-\eta)/N$ values for the mixtures of CHF_3 with Ar and N_2, displayed in Fig. 6, bear essentially an electropositive character.

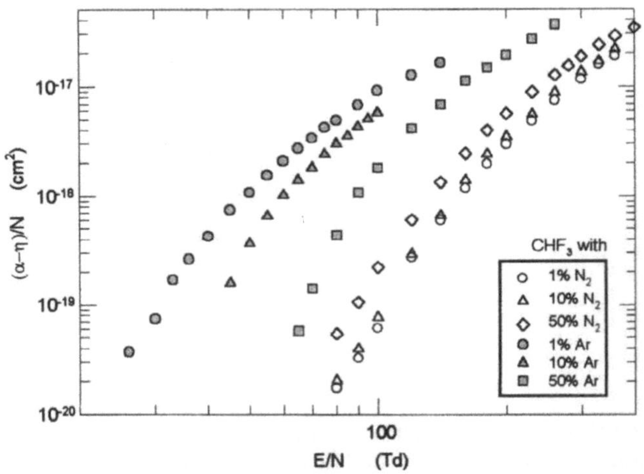

Figure 6. The effective ionization coefficients of CHF_3 and its mixtures with Ar (closed symbols) and N_2 (open symbols).

4. CONCLUSIONS

Good agreement has been found between our v_e values and those summarized in the literature[1,2]. However, to our knowledge, most of $(\alpha-\eta)/N$ and D_L/K data are the first to be reported. Thus, we hope that the present coefficients be useful for the modeling of practical plasma applications and cross section tests.

ACKNOWLEDGEMENTS

This work was supported by DGAPA, IN104501 and by Conacyt. Thanks are due to A. Bustos and A. Castrejón for their assistance.

REFERENCES

1. L.G. Christophorou and J.K. Olthoff, *J. Phys. Chem. Ref. Data* **27**, 1 (1998)
2. L. G. Christophorou, J.K. Olthoff, and M.V.V.S. Rao, *J. Phys. Chem. Ref. Data* **26**, 1 (1997); L.G. Christophorou and J.K. Olthoff, *J. Phys. Chem. Ref. Data* **28**, 967 (1999)
3. J. de Urquijo, C.A. Arriaga, I. Alvarez and C. Cisneros, *J. Phys. D* **32**, 41 (1999)
4. J. de Urquijo, I. Alvarez and C. Cisneros, *Phys. Rev. E* 60, 4990 (1999)
5. J. de Urquijo, E. Basurto and J.L. Hernández-Ávila, *J. Phys. D* **34**, 2151 (2001)
6 H. Okumo and Y. Nakamura, in *Proc. Int. Conf. Atomic and Moelcular Data and their Applications*, NIST Special Publ. 926, editedd by W.L. Wiese and .J. Mohr, 1998, p. 265

RECENT MEASUREMENTS OF ELECTRON SWARM COEFFICIENTS IN THE MIXTURES OF SF_6 WITH HALOCARBON AND RARE GASES

E. Basurto, J.L. Hernández-Ávila and J. de Urquijo*

1. INTRODUCTION

Besides the well known SF_6-N_2 mixture, new ones are being studied with rare and molecular gases in view of the recent concerns raised by the emission of large amounts of SF_6 into the atmosphere, since its global warming potential one of the highest[1]. This paper deals with the measurement electron drift velocity v_e, the density-normalized longitudinal diffusion coefficient ND_L, and the density-normalized effective ionization coefficient $(\alpha-\eta)/N$ as a function of the density-normalized electric field strength, E/N, for the mixtures of SF_6 with He, Ar, Xe , CHF_3 and CF_4 with overall SF_6 concentrations in the range 1-90%. Additionally, the limiting field strength of the mixtures E/N_{lim} could be derived.

2. EXPERIMENTAL DETAILS

The pulsed Townsend experiment that was used to measure the above swarm parameters has been described in detail previously, as well as the techniques for the derivation of the above coefficients [2-5]. Detailed accounts are given elsewhere [5-7].

3. TRANSPORT AND IONIZATION COEFFICIENTS

* E. Basurto, CBI-Ciencias Básicas, UAM, Av. San Pablo 180, 02200, México, DF., México; J.L. Hernández-Ávila, CBI-Energía, UAM, Av. San Pablo 180, 02200, México DF. México; J. de Urquijo, Centro de Ciencias Físicas, UNAM, P.O. Box 48-3, 62251 Cuernavaca, Mor., México

Gaseous Dielectrics X.
Edited by Christophorou *et al.*, Springer, New York, 2004.

The ionisation and electron transport parameters presented in this paper are the average values of at least three different measurements at the same E/N and different gas pressures. The lines joining the experimental points on all the figures are intended only to guide the eye through the points.

3.1. Electron drift velocities

The electron drift velocities for the mixtures of SF_6 with He, Ar, Xe, CHF_3 and CF_4 as a function of E/N, are shown in Figures 1-3 (see Ref. [4-7]). The overall accuracy for this parameter was estimated to be between 1- 2 %.

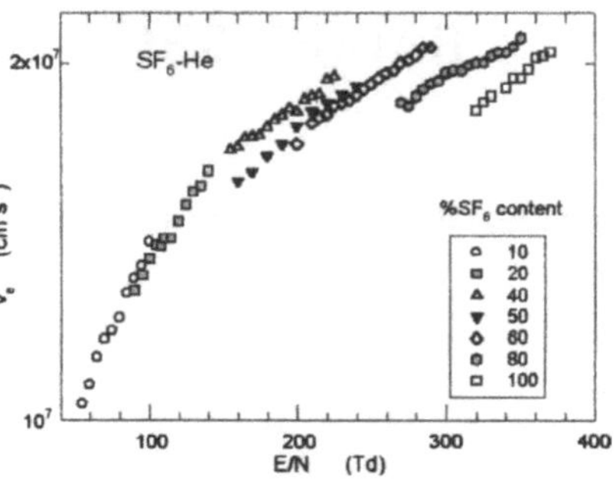

Figure 1. Electron drift velocity for SF_6-He [4] mixtures and pure SF_6 [8].

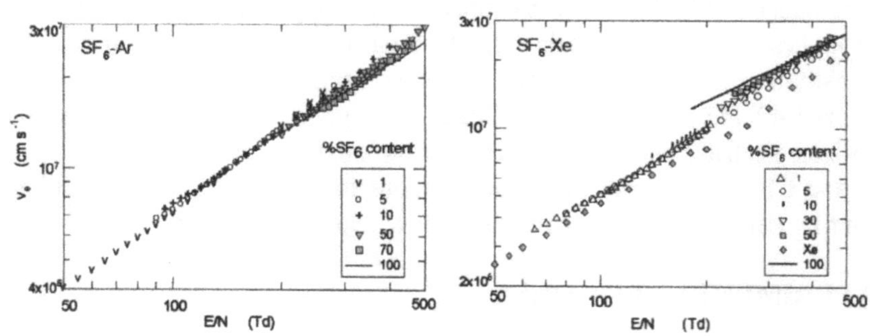

Figure 2. Electron drift velocity for SF_6-Ar [6] and SF_6-Xe [6] mixtures. Pure SF_6 [8].

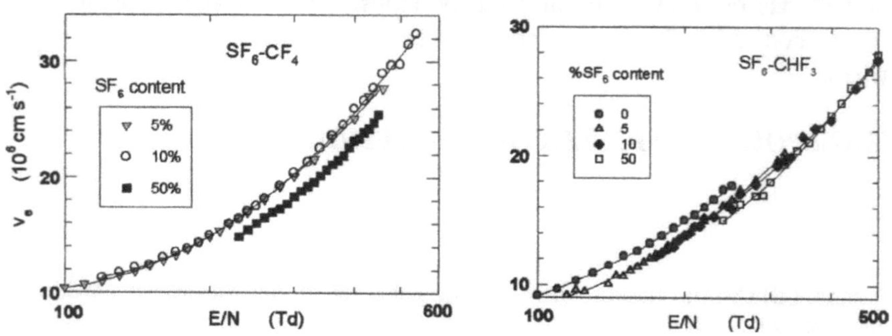

Figure 3. Electron drift velocity for and for SF_6-CF_4 [7] and SF_6-CHF_3 [7] mixtures.

The electron drift velocities in SF_6-He show a monotonic decrease with the SF_6 content for fixed E/N, while those for SF_6-Ar seem to converge to those for pure SF_6 for large SF_6 concentrations in the mixture; for SF_6-Xe mixture, given E/N, v_e becomes increasingly smaller for decreasing SF_6 concentrations. The latter behaviour is consistent with the relative size of the momentum transfer cross sections for Xe and Ar, since that of Xe is between five to ten times larger than that of Ar over an energy range between 0.001 and 30 eV. The electron drift velocities in SF_6-CHF_3 show a clear trend towards lower v_e values for increasing SF_6 concentrations in the mixture. A common feature of these curves is the fairly small changes in the drift velocity for low SF_6 concentrations (5% and 10%). This condition prevails for the case of SF_6-CHF_3, where the variation in v_e is small even for 50% SF_6 mixture, in contrast with the other two mixtures, where the difference is more apparent. The overall accuracy for ND_L is 12-25%.

3.2 The longitudinal diffusion coefficient

Figure 4 display the dependence of ND_L as a function of E/N for the mixtures of SF_6 with Ar and Xe, respectively (see Ref. [6]). For fixed E/N, the ND_L values for SF_6-Ar decrease with the increase in the SF_6 content in the mixture. Indeed, the limiting curves for pure SF_6 indicate that the above behaviour is consistent with the growing influence of the electron-SF_6 interaction as the SF_6 content increases. On the other hand, the ND_L values for the SF_6-Xe vary little with the change in the SF_6 content, normally within the uncertainty of the data (8-15%).

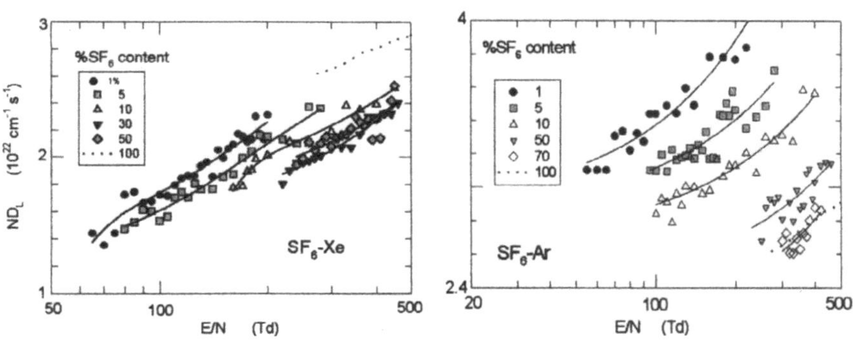

Figure 4. The longitudinal diffusion coefficient for SF_6-Ar [6] and SF_6-Xe [6] mixtures. Pure SF_6 [8].

Figure 5 shows the dependence of ND_L with E/N for the mixtures of SF_6 with CHF_3 and CF_4 respectively (see Ref. [7]). In spite of the fairly large scatter of the ND_L data for SF_6-CHF_3 and SF_6-CF_4, the trend lines suggest that there is a weakly defined maximum, shifting toward higher E/N values with the increase of the SF_6 content. The same shift is observed for the SF_6-CF_4 mixture, although the diffusion coefficients rise smoothly toward a well defined plateau.

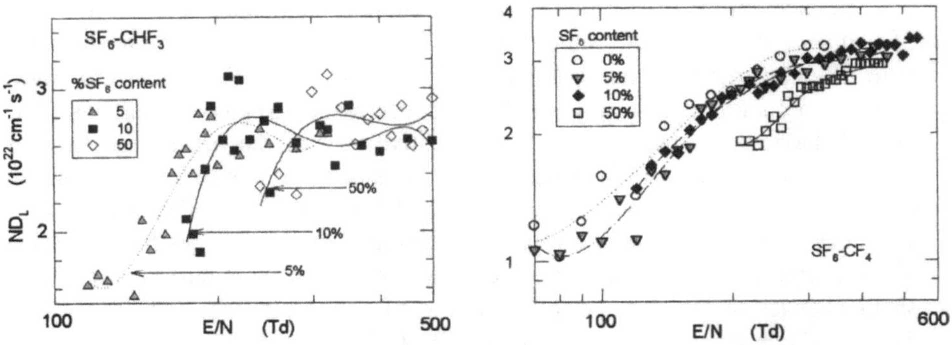

Figure 5. The longitudinal diffusion coefficient of electrons for SF_6-CHF_3 [7] and SF_6-CF_4 [7] mixture.

3.3. Effective ionization coefficients

The density-normalised effective ionisation coefficients as a function of E/N, for the SF_6 mixtures with He, Ar and Xe are shown plotted in Figs. 6-7, respectively (see Ref. [4,6]). The overall uncertainties for this parameter were estimated to be between 2-3 %. In figure 6-7 the values of $(\alpha-\eta)/N$ shows a fairly uniform decrease for a fixed E/N, as the SF_6 content in mixture increases. This is a consequence of the increasingly electronegative character of the mixture. Regardless of the composition for the SF_6-He mixture, $(\alpha-\eta)/N$ varies almost linearly with E/N over the E/N range studied.

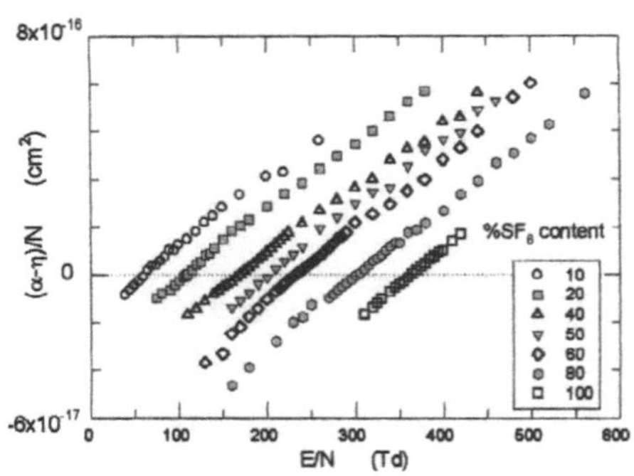

Figure 6. The effective ionization coefficients for SF_6-He [4] mixture. Pure SF_6 [8].

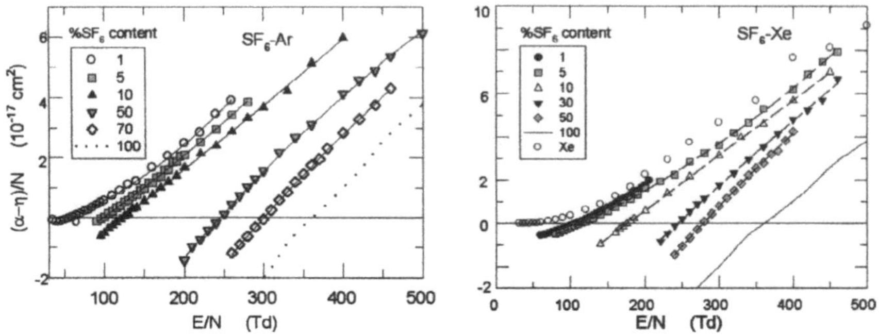

Figure 7. Effective ionization coefficients for SF_6-Ar [6] and SF_6-Xe [6] mixtures. Pure SF_6 [8].

In Figure 8, the trend of the values of $(\alpha-\eta)/N$ for the SF_6-CHF_2 and SF_6-CF_4 mixture (see Ref. [7]) shows the influence of CHF_3 or CF_4 for low SF_6 concentrations; however, for the case of 50% SF_6 content in the mixture, the dependence of $(\alpha-\eta)/N$ with E/N resembles that of pure SF_6, although the absolute values are smaller at the same E/N value. As the SF_6 content in the mixture is increased, the effective ionisation coefficients decrease in value for the same E/N. Likewise, the attachment coefficient increases.

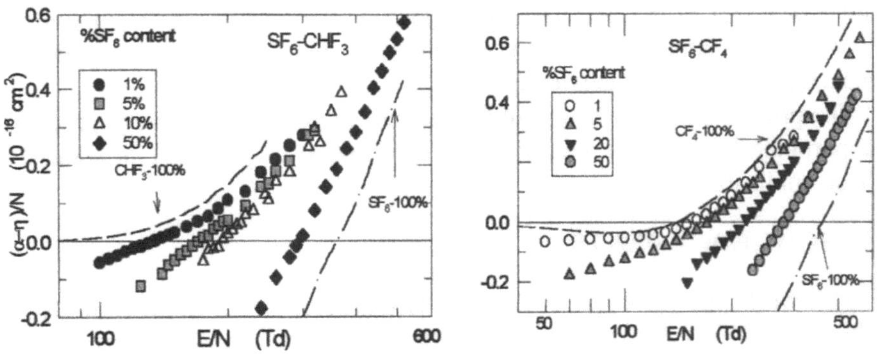

Figure 8. Effective ionization coefficients for SF_6-CHF_3 [7] and SF_6-CF_4 [7] mixtures. Pure SF_6 [8].

3.4 The critical field strength of the mixtures

A parameter of importance in gas discharge, is the critical field strength, E/N_{crit}, defined as the E/N value for which $\alpha-\eta=0$, the values of E/N_{crit} obtained from Figures 6-8 are shown plotted in figure 9 as a function of the SF_6 content in the mixture[4-7]. The SF_6-CHF_3 mixture has a slightly higher value of E/N_{crit} than the other gas mixtures. It is noticed that E/N_{lim} value for SF_6-Xe for 50% SF_6 is only 10% lower than that for SF_6-N_2. Thus, in some applications demanding a non-reactive gas additive to SF_6, Xe may be a possible candidate.

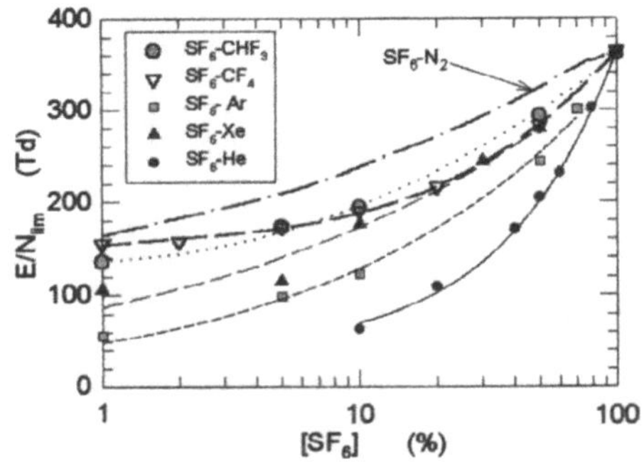

Figure 9. Variation of the critical field strength, E/N_{crit} as a function of the SF_6 content in the SF_6-CHF_3, SF_6-CF_4, SF_6-Ar, SF_6-Xe, SF_6-He mixtures[4-7]. The E/N_{crit} values for the mixture of SF_6-N_2 [8] is also shown.

4. CONCLUSIONS

We have presented a self consistent set of ionisation and transport coefficients derived from Pulsed Townsend experiment, under pre-breakdown conditions. Of the three rare gas mixtures studied, it appears that the SF_6-Xe combination is the one with a higher critical field strength, although it is still smaller than that of the SF_6-N_2 combination. On the other hand, the mixtures of SF_6 with CHF_3 and CF_4 are only advantageous for low SF_6 concentrations in comparison to those with the rare gases.

The present data may be useful for the modelling of gas discharge processes, and particularly for those involving the substitution of pure SF_6 in some high voltage applications.

ACKNOWLEDGEMENTS

This work was supported by DGAPA-UNAM IN104501, UAM 2250216 and CONACYT 36645-A. Thanks are due to F. Ramírez and A. Bustos for their assistance.

REFERENCES

1. J Castonguay and J-M Gauthier, *"Study of the decomposition of traces of SF_6 in the lower atmosphere: the kinetic results"*, in Gaseous Dielectrics IX, L. G. Christophorou and J.K. Olthoff, Eds., Plenum-Kluwer, 413-418 (2001); L G Christophorou, J K Olthoff and D S Green, *NIST Technical Report* 1425, (1997).
2. J de Urquijo, C Arriaga, I Alvarez and C Cisneros, *J. Phys. D* **32** 41-45 (1999).
3. J de Urquijo, I Alvarez, C Cisneros, *Phys. Rev. E* **60** 4990 (1999).
4. J. de Urquijo, E. Basurto and J L Hernández-Ávila, *J. Phys D* **34** 2151–59 (2001).
5. J L Hernández-Ávila, E. Basurto and J. de Urquijo, *J. Phys D* **35** 2264-69 (2002).
6. J de Urquijo, J.L. Hernández-Ávila, E. Basurto and F. Ramírez, *J. Phys D* **36** 1489-14 (2003).
7. J de Urquijo, E. Basurto and J.L. Hernández-Ávila, *J. Phys D* **36** 3132-37 (2003).
8. Th. Aschwanden, Ph. D. Dissertation, ETHZ Zürich, (1985).

MOBILITY ESTIMATES OF SF_6^- ION IN PARENT SF_6 GAS

Celal KORASLI[1] and Vedat KARSLI[2]

1. INTRODUCTION

In high-voltage transmission systems and plasma technology it is of primary interest to understand various aspects of corona discharge phenomenon leading to corona stability regime in non-uniform gaps filled with high-pressure SF_6 gas. This invokes determination of swarm parameters for modeling kinetic behavior of ions and ionic clusters. To estimate mobilities of SF_6^- ion in the parent SF_6 gas a new semi-quantitative approach is made utilizing space-charge current flow and field measurements in concentric cylindrical-gap geometry.

1. THEORY

The classical approach followed in the solution of field problems involving space charge the usual method is to divide the inter-electrode space into two regions as shown in Fig.1. (i) Ionization region is a narrow band close to the coronating electrode, and is treated as an extension of the electrode because of its highly energetic molecule and ion content. (ii) The space-charge drift region starts from the boundary of the ionization region and extends up to the outer electrode boundary, which is characterized by the transport of neutral molecules and composite ion species forming a space charge maintaining continuity of electric field fluxes throughout the boundaries[1].

[1]University of Bahrain, Department of Electrical Engineering, P.O. Box 32038, Isa Town, State of Bahrain. korasli@eng.uob.bh
[2]University of Gaziantep, Department of Electrical Engineering,Town, 27300 Gaziantep, Turkiye. vkarsli@gantep.edu.tr

Gaseous Dielectrics X.
Edited by Christophorou *et al.*, Springer, New York, 2004.

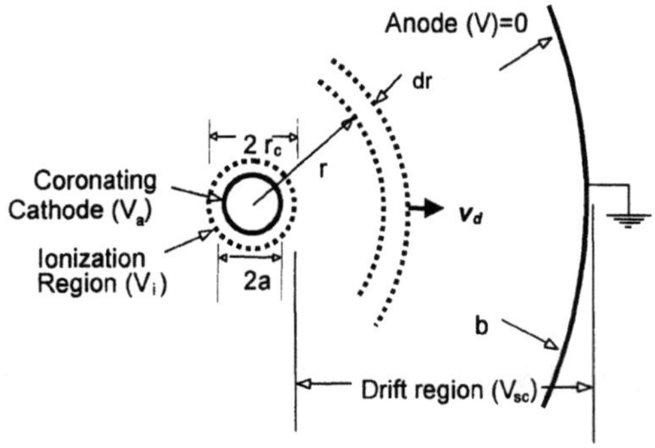

Figure 1. Model for negative corona in cylindrical discharge gap.

In the drift region in the existence of electric fields of arbitrary strength including all ion-neutral interaction potentials and mass ratios, Vieland and Mason[2] proved that reduced drift velocities of gaseous ions are independent of gas temperature, but strongly dependent on ionization cross section $\sigma(T_e)$ and $(E/n)^{1/2}$, and related by

$$v_d = \left(\frac{3\pi}{2\mu M}\right)^{1/4}\left(\frac{3q}{8\sigma(T_e)}\right)^{1/2}\left(\frac{E}{n}\right)^{1/2} \qquad T_e > T \qquad (1)$$

In an earlier work of Skullerud[3] this relation was proved to produce accurate results with exhaustive Monte Carlo trials over a large range of the parameter (E/n).

Using Eq.(1) ion mobility may be expressed as

$$K = \frac{v_d}{E} = \left(\frac{3\pi}{2\mu M}\right)^{1/4}\left(\frac{3q}{8\sigma(T_e)}\right)^{1/2}\frac{1}{n}\left(\frac{E}{n}\right)^{-1/2} \qquad (2)$$

where μ is the reduced mass $(=mM/(m+M))$ for the ion and neutral molecule masses m and M respectively, q is the ion charge, $\sigma(T_e)$ is the effective ionization cross section, T_e is the effective ion temperature and T is the gas temperature.

For the purpose of the present analysis Eq.(2) may be rewritten as

$$v_d = \alpha(T_e)\left(\frac{E}{n}\right)^{1/2} \qquad (3)$$

where, $$\alpha(T_e) = \left(\frac{3\pi}{2\mu M}\right)^{1/4}\left(\frac{3q}{8\,\sigma(T_e)}\right)^{1/2} \qquad (4)$$

In general, in the presence of ionized unipolar space charge, the composite space-charge density ρ is related to the background field strength $\boldsymbol{E}$ by Poisson's equation

$$\nabla.\boldsymbol{E} = \frac{\rho(r,\boldsymbol{E})}{\varepsilon} \qquad (5)$$

where ε is permitivity of the space. Here, ρ is a complex function of both physical and gaseous parameters. In the drift region of a coaxial cylindrical gap (Fig.1), charge density can be expressed in an incremental cylindrical shell of thickness dr at a radius r and carrying a differential charge dq as

$$\rho = \frac{dq}{2\pi r\, dr\, L} \tag{6}$$

where L is the axial length of the cylinder. Since ions will drift across dr in a time duration of dt with a drift speed of $v_d(r, E)$, i.e. $dr = v_d\, dt$, and letting $i_L(t)=i(t)/L$, Eq.(6) now reads

$$\rho = \frac{i_L(t)}{2\pi r v_d} \tag{7}$$

Substituting Eq.(3) in Eq.(7) we obtain

$$\rho = \frac{i_L(t)}{2\pi r \alpha(T_e)}\left(\frac{E}{n}\right)^{-1/2} \tag{8}$$

Now, from Eq.(5) divergence of the field in the drift region may be expressed as

$$\nabla.\boldsymbol{E} = \frac{i_L}{2\pi \varepsilon r \alpha}\left(\frac{\boldsymbol{E}}{n}\right)^{-1/2} \tag{9}$$

or along the radial direction in differential form as

$$\frac{d\boldsymbol{E}}{dr} + \frac{1}{r}\boldsymbol{E} = \frac{i_L}{2\pi \varepsilon r \alpha}\left(\frac{\boldsymbol{E}}{n}\right)^{-1/2} \tag{10}$$

Hence, the solution of radial space-charge electric field at any position $\boldsymbol{E(r)}$ may be found by direct integration of Eq.(10) using boundary conditions; for $r \geq r_c$ in the region outside of the ionization region, $E=E(r)$ and for $r=b$ on the outer boundary of the drift region, i.e., at the outer electrode surface, $E=E_b$.

After two-step parametric substitution, the solution of Eq.(10) becomes

$$E(r) = \left[\left(\frac{i_L n^{1/2}}{2\pi \varepsilon \alpha}\right) - \left(\frac{i_L n^{1/2}}{2\pi \varepsilon\, \alpha} - E_b^{\,3/2}\right)\left(\frac{b}{r}\right)^{3/2} \right]^{2/3} \tag{11}$$

Substituting Eq.(11) in the space-charge voltage drop equation

$$V_{sc} = -\int_b^{r_c} \boldsymbol{E(r)} \cdot d\boldsymbol{r} \tag{12}$$

and performing the integration across the space-charge drift region, i.e., from the boundary of ionization region $r=r_c$ for V_c to the outer electrode inner surface $r=b$ for $V_b=0$, we obtain the space-charge potential function as

$$V_{sc} = -\frac{1}{D}\left[F(x_b) - F(x_c)\right] \tag{13}$$

where F may be identified after the parametric substitution $y=1-Ax^{-3}$ as

$$F = \frac{1}{2y^{2/3}} + \left\{ \frac{1}{6}\ln\left[\frac{(-1+y^{1/3})^2}{y^{2/3}+y^{1/3}+1}\right] + \frac{1}{\sqrt{3}}\arctan\left(-\frac{2y^{1/3}+1}{y^{1/3}}\right) \right\} \tag{14}$$

and where $x_b=(A+C\,b^{-3/2})^{1/3}$ and $x_c=(A+C\,r_c^{-3/2})^{1/3}$ in which the coefficients are

$A=\dfrac{i_L n^{1/2}}{2\pi \varepsilon \alpha}$, $B=A-E_b^{3/2}$, $C=-Bb^{3/2}$ and $D=-\dfrac{1}{2C^{2/3}}$

If the radial field strength, E_b, at the outer electrode surface is known, the iterative solution for determining the gaseous function $\alpha(T_e)$ follows minimization of the difference function

$$\Delta V = |V_a - (V_{sc} + V_i)| \qquad (15)$$

where V_a and V_i are the applied and corona inception voltages, respectively.

The complex problem evaluating the drift velocity function $v_d(\sigma(Te), E/n)$ is reduced to solving difference equation, (Eq.(15)) in each iteration cycle for ionization cross section function $\sigma(Te)$. In the present work, the radius of the ionization region was photographically recorded (Table I) and the field at the outer electrode surface is measured during current-voltage measurements, starting from corona inception to breakdown voltage.

For the iterative solution, an initial value for the lowest ionization cross-section $\sigma = 10^{-17}$ m^2 is assigned to $\sigma(Te)$, which corresponds to the lowest ionization cross-section lowest energy10^{-3} eV level[4].

Table 1. Variation in radius of ionization region.

p (MPa)	0.02	0.05	0.08	0.12	0.14
r_c (mm)	0.82±0.053	0.82±0.053	0.71±0.077	0.62±0.086	0.52±0.091

4. EXPERIMENTAL

The details of the test vessel have been described in the earlier work of the Author[5]. Cylindrical electrodes made of stainless steel having radii of 0.5/25-mm and polished to initial CLA roughness 0.7 μm were used throughout the test.

A biased current probe[6] of 4-mm in radius in flush with the inner surface of the outer electrode was used for measuring space charge induced electric field strengths on the outer electrode for several phases of corona discharge. The probe was designed for a radius-to-gap clearance ratio of 40. The current was recorded with Keithley 614 digital electrometer.

5. RESULTS AND DISCUSSION

Fig.2 shows the variations of corona discharge current with voltage measured at pressures 0.02, 0.05, and 0.08 MPa. These plots are computer processed digitized traces of storage CRO (HP-7044B) and expressed as per unit length of the electrode gap.

The results of field measurements on the outer electrode, shown in Fig.3 were measured at the same voltage and pressure values of I-V characteristics. Each test point corresponds to the average of six measurements. The maximum uncertainty is in the order of ±5% which was observed at pressures 0.08 MPa.

Mobility estimates of SF_6^- ion were based on the iterative solutions of the Poisson and continuity equations given in Eqs. (1), (5) and (6). The results of ion

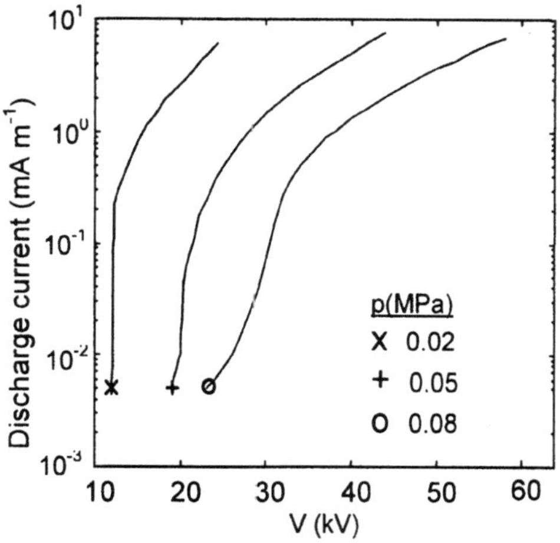

Figure 2. Corona discharge current variations.

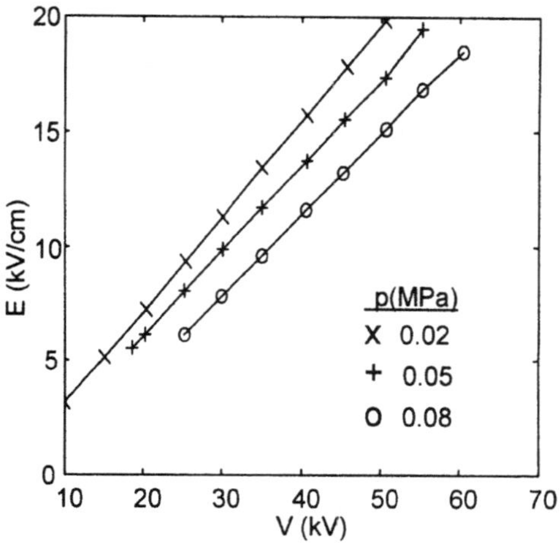

Figure 3. Induced electrostatic field variations on the outer electrode surface.

mobilities are reported as reduced ion mobilities $K_o=(v_d/n_o)(E/n)$, where $n_o=2.69\text{x}10^{25}$ m^{-3} is the gas density at standard temperature and pressure conditions. Results of mobility estimates of SF_6^- obtained at pressures 0.02, 0.05 and 0.08 MPa are illustrated in Fig.4 as a function of the parameter *(E/n)*.

In the same figure low-pressure mobility data of other researchers are also included in the form of curves after fitting a sixth order polynomial to their data points. Curve 1 displays the drift-time results measured in drift-tube experiments compiled by Brand and Jungblut[7] in the pressure range 0.0002-0.4 MPa and. Curve 2 is also the drift-time mobility data of uniform field gap in the pressure range 0.0066-0.093 MPa[8].

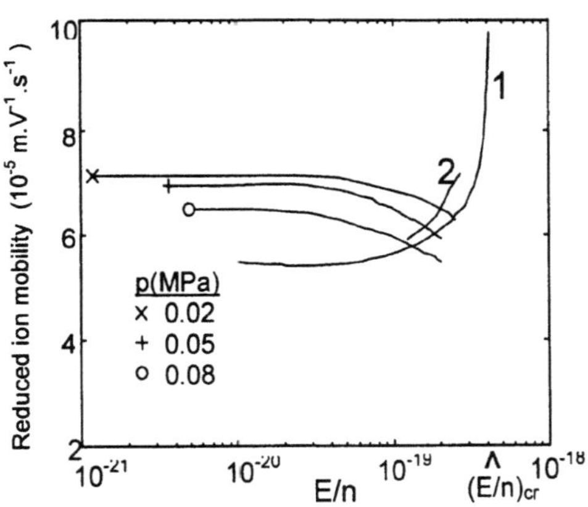

Figure 4. Reduced SF_6^- ion mobilities. 1: Brand and Jungblut[6]. 2: Blair et al[8].

The apparent disagreements of the present estimates compared to the low-pressure drift-time measurement data (Curve 1) is due to the fact that the types of ions existing in the drifting space-charge cloud may not be the same as those found under the present pressure conditions. This is because of charge transfer and dissociation of ions leading to different formations of 'labile clusters' [9,10,11].

Further, ionic species in the time-drift measurements are generated in a space charge drift region is free of any type of ions or clouds, whereas in the present measurements the drift region is full of continuously replenishing ionic species. Combinational measurement uncertainties and heating hysteresis may also be considered other effects causing disagreements.

7. CONCLUSION

The iterative solution of the Poisson and space-charge current continuity equations based on field dependent mobility expression of Vieland and Mason[2] appears to work well to estimate mobility coefficients of SF_6^- ion in the discharging parent gas. The technique employs ion-current flow produced by corona discharge and space-charge field measured on the outer electrode surface of a coaxial-cylindrical gap in the pressure range (0.02-0.08) MPa. In general, the present mobility estimates follow a similar trend to the existing low pressure data compiled by Brand and Jungblut[6].

Eliminating the heating effect of ion-current flow on the drifting space charge during field measurements, and making corrections for unaccountable electronic processes and other effects, the method provides mobility estimates of SF_6^- ion over a pressure range in the existence of corona discharge.

8. REFERENCES

1. 0. M. P. Sarma and W. Janischewskyj, Analysis of Corona Losses on Transmission Lines:I-Unipolar Lines, *IEEE Trans. on Power Apparatus* **PAS- 88** 718-731(1969).
2. L. A. Viehland and E. A. Mason, Gaseous Ion Mobility and Diffusion in Electric Fields of Arbitrary Strength, *Annals. Phys.*, **110,** 287-328 (1978).
3. H. R. Skullerud, J. Phys., **B 6,** 728-731(1973).
4. A. V. Phelps and R. J. van Brunt, Electron Transport, Ionization, Attachment, and Disociation Coefficcients in SF_6 and its Mixtures, *J. Apply. Phys.* **64**(9), 4269-4277(1988).
5. C. Korasli, Statistical Inference for Breakdown Voltage in SF_6 GIS from First Breakdown Data, *IEEE Trans. on Dielectrics and Electrical Insulation,* **DIE-5**(4), 596-602(1998).
6. K. P. Brand and H. Jungblut, The interaction potentials of SF_6 ions in SF_6 Parent Gas Determined from Mobility Data, J. *Chem. Phys.* **78,** 1999-2007 (1983).
7. E. O. Selim and R. T. Waters, Static Probe for Electrostatic Field Measurement in the Presence of Space Charge, *IEEE Tran. on Industry Applications,* **IA-16**(3) 458-462(1980).
8. D. T. Blair, B. H. Crichton, F. J., Al-Kindi and T. L. Sharma, Drift Velocities of Positive Ions and Negative Ions in Cylinder SF_6, *J. Phys. D: Appl.*Phys. **22** (1989) 755.
9. W. F. Schmidt, H. Jungblut, D. Hansen and H. Tagashira, Ion Mobility in Compressed Electro-Negative Gases, *Proc. 2nd Int. Symp.on Gaseous Dielectrics, Knoxville,1980* , pp. 12-15.
10. W. F. Schmidt, H. Jungblut, D. Hansen and H. Tagashira Ion Mobility in Compressed Electro-Negative Gases, *Proc. 6th Int. Conf. on Gas Disch., London, 1982,* pp. 12-15.
11. K. P. Brand, *Proc of 15th Int Conf. on Phenomena in Ionized Gases, Part I, Minsk*, 389-390(1981).

SECTION 2: GAS-DISCHARGE BEHAVIOR: BASIC MECHANISMS

SPATIO-TEMPORALLY RESOLVED SPECTROSCOPIC DIAGNOSTICS OF THE BARRIER DISCHARGE IN N_2-O_2 MIXTURES AT ATMOSPHERIC PRESSURE

H.-E. Wagner (1), K.V. Kozlov (2), R. Brandenburg (1), A.M. Morozov (2)

1. INTRODUCTION

Dielectric-barrier discharges (DBDs) belong to the group of cold non-equilibrium plasmas, characterized by a kinetic electron temperature of 1...10 eV and a near room gas temperature. They have found numerous applications in many fields. Usually DBDs operate in the so-called filamentary mode, namley in N_2-O_2 mixtures. If the local electric field strength in the gas gap arrives the ignition level, the breakdown starts at many points of the electrode followed by the development of sets of microdischarges (MDs). Experimental and theoretical studies on the breakdown in DBDs have a long history [1-8]. However, despite of the considerable progress in the understanding, the knowledge of this subject nowadays appears to be insufficient to provide an adequate quantitative description. Main reasons for this situation are the difficulties to investigate experimentally the dynamics of the filamentary plasmas, requiring an sub-ns temporally and a sub-mm spatially resolution. Therefore it is not surprising that a great deal of effort has been devoted to computer simulation [5-8].

The cross correlation emission spectroscopy (CCS) has been applied to the diagnostics of DBDs [9-11,14,15]. It is a powerful tool to analyze the spatio-temporal structure of these plasmas. The array of experimental results are sufficient for the quantitative estimation of the local electric field strength and the relative electron density in air. The knowledge of these basic plasma parameters allows a deep understanding of the process of electrical breakdown, including discharge physics and chemistry.

(1) Ernst-Moritz-Arndt University of Greifswald, 17489 Greifswald, Domstr. 10a, Germany
(2) Moscow State University, 119899 Moscow, Russia

2. CROSS-CORRELATION SPECTROSCOPY

Basic idea

The main idea of the CCS is to replace a direct measurement of the single pulse luminosity of a repetitive light pulse emitter by a statistically averaged determination of the correlation function between two optical signals, both originating from the same source. The first one of these signals (so-called "synchronizing signal") is used to define a relative time scale, the second one ("main signal") has to be detected with a probability at least one order of magnitude lower, than that one of the synchronizing signal. The measured quantity is actually a time delay between these two signals, and the recorded quantity is a probability density function for the light pulse intensity evolution. If the repetitive light pulses reproduce each other sufficiently exactly, and if the synchronizing signal detection is adjusted in such a way as to occur always at the same moment of a single light pulse evolution, then the recorded probability density function is proportional to the light intensity I(t) of the source under consideration. A detailed description of the theoretical foundations of CCS technique is given in the review paper [12]. Extremely high sensitivity of the CCS combined with the temporal resolution in a sub-nanosecond range is the main advantage of this experimental technique, as compared to other methods of emission spectroscopy.

Experimental set-up

A general scheme of the experimental set-up and the measurement procedure is presented in Figure 1. The successive MDs arising between the electrode tips were used as an emitter of repetitive light pulses. Spatial resolution and scanning over the MD axis was provided by the optical system shown in the left upper part of the figure. Monochromatic light of the main signal was detected by the highly sensitive photomultiplier from HAMAMATSU (type H5773-04), which operated in a single photon counting mode. In order to increase the sensitivity of this detector by reducing the signal-to-noise ratio, the detector is cooled down to 12 °C. The second detector for the synchronizing signal, (HAMAMATSU, type H5773-06, shown in the figure at the right) was adjusted so as to reach the maximum possible temporal resolution of the steep front of a light pulse, since the derivative of this signal is actually used to define a zero-point of the relative time scale. The measurements of the correlation functions and accumulation of the results were carried out by the Time Correlated Single Photon Counting Module SPC-530 from BECKER & HICKL GmbH. Two Constant Fraction Discriminators are used to select electric pulses which belong to a pre-defined range of the amplitudes. A selected pulse of the main signal initiates a linear rise of the voltage of the Time-to-Amplitude Converter and a first following selected pulse of the synchronizing signal stops this rise.

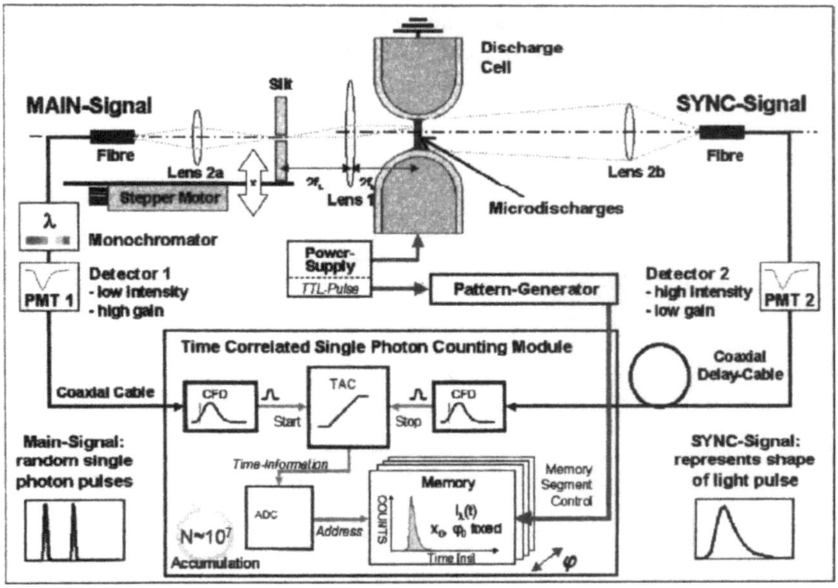

Figure 1. A general diagram of the apparatus and measurement procedure. Abbreviations: SYNC – Synchronizing; PMT – Photo-multiplier; TTL – Transistor-Transistor Logic; CFD – Constant Fraction Discriminator; TAC – Time-to-Amplitude Converter; ADC – Analogue-to-Digital Converter; φ is the phase of the feeding sinusoidal voltage (see Figure 2).

The measured quantity is the final voltage amplitude, whose value is converted to the number of the corresponding time channel of the device (actually, to a memory address). The entire scale of a memory segment consists of 1024 channels. If no synchronizing signal comes and the voltage of the Time-to-Amplitude Converter reaches its upper limit, then no count occurs. A correspondence between the time value and the number of the time channel is determined by the duration of the voltage rise, which is variable from 50 ns to 2 μs. A shift of zero-point to the left over time axis is possible due to the insertion of the coaxial Delay-Cable between the detector for synchronizing signal and the Module SPC-530. An additional coarse time scale was provided by the Pattern-Generator which was synchronized with the phase of the feeding voltage φ by a standard scheme of Transistor-Transistor Logic (see figures 1 and 2). A voltage period (T=154 μs) was divided into 16 equal time intervals. The CCS measurements have been recorded by a separate memory segment for each of these intervals. A special electrode arrangement of the type "glass-glass" (Figure 2) was used, in order to localize the DBD to a definite position. A further advantage of such a geometry of the discharge cell is the possibility to investigate the volume part of the DBD as well as the surface discharges. The DBD was driven by a sinusoidal alternative voltage (frequency: 6.5 kHz, peak-to-peak amplitude: 12 kV).

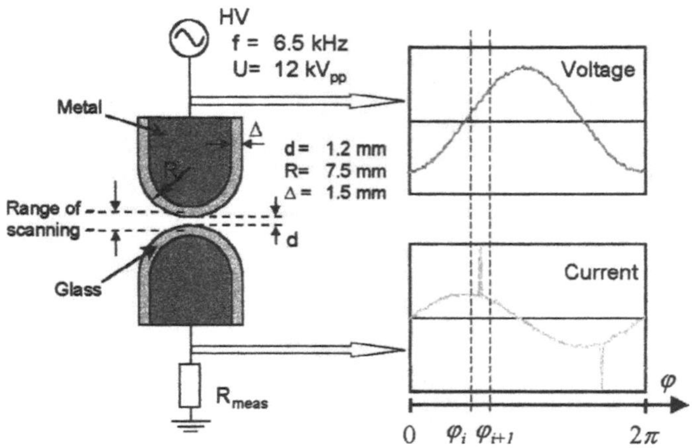

Figure 2. Electrode arrangement of barrier discharge with the indicated range of axial scanning, and a typical example of the oscillograms of voltage and current for the "one-to-one MD" operation mode.

The main characteristics of our CCS measurements are summarized in the Table 1. It should be noted that the values of the fine resolution over time given there refer to the apparatus parameters. Actually, these values depend on the quality of synchronization. Under the conditions of our experiments, the real time resolution for 50 ns scale has been found to be not worse than 0.15 ns.

Table 1. Resolution parameters of the CCS measurements [10].

Quantity	Resolution
Time (50 ns scale)	< 0.15 ns
Space (axial direction)	0.1 mm
Wavelength	0.3 nm
Phase of feeding voltage	$\varphi = 2\Pi/16$

3. RESULTS AND DISCUSSION

Spatio-temporal structure of the luminosity

The experiments in DBDs were performed in open gas flow systems, in synthetic air and N_2/O_2 mixtures at ambient pressure. The gas flow rate was maintained so high as to avoid any noticeable influence of ozone (which is produced by the discharge itself) upon the spatio-temporal structure of MDs.
In Figure 3 examples of the spatio-temporal distributions of the MD luminosity in synthetic air are shown for the $N_2\ (C^3\Pi_u)_{v'=0} \rightarrow N_2\ (B^3\Pi_g)_{v''=0}$ transition of the second positive system and the $N_2^+(B^2\Sigma_u^+)_{v'=0} \rightarrow N_2^+ (X^2\Sigma_u^+)_{v''=0}$ transition of the first negative system of nitrogen, also covering the pre-breakdown phase as well as the afterglow. These selected transitions correspond to extremely different excitation energies of the nitrogen molecule (ΔE = 11 eV) and N_2^+ ion (ΔE = 18.7 eV), respectively. Each of these plots contains two peaks situated near the electrodes and related to nearly the same time coordinate. The peaks near the cathode are preceded by the "waves" of light intensity, that start directly from the anode surface and propagate towards the cathode with increasing velocity and growing amplitude.
To interpret these results, it is necessary to consider the origin of the light sources within a MD channel. For the spectral bands C(0-0) and B(0-0) the emission of light of the corresponding wavelengths is caused by the elementary processes (1-4):

$$e + N_2\left(X^1\Sigma_g^+\right)_{v=0} \rightarrow N_2\left(C^3\Pi_u\right)_{v'=0} + e \qquad \left(\Delta E = 11.0eV\right) \tag{1}$$

$$N_2\left(C^3\Pi_u\right)_{v'=0} \rightarrow N_2\left(B^3\Pi_g\right)_{v''=0} + h\nu \ \left(\lambda = 337.1nm; \tau_0^C = 40ns\right) \tag{2}$$

$$e + N_2\left(X^1\Sigma_g^+\right)_{v=0} \rightarrow N_2^+\left(B^2\Sigma_u^+\right)_{v'=0} + 2e \qquad \left(\Delta E = 18.7eV\right) \tag{3}$$

$$N_2^+\left(B^2\Sigma_u^+\right)_{v'=0} \rightarrow N_2^+\left(X^2\Sigma_g^+\right)_{v''=0} + h\nu \ \left(\lambda = 391.4nm; \tau_0^B = 60ns\right) \tag{4}$$

In air at atmospheric pressure, a stepwise excitation of molecular and ionized nitrogen is negligible because of the very effective collisional quenching of excited states by molecular oxygen. That is why only direct electron impact has to be taken into account in the kinetic scheme (1-4). For the same reason, the characteristic life-times of the considered excited states in air are less than 1ns, while their radiation life-times (τ_o) are within the range 40-60 ns. Therefore, the spatio-temporal distributions of the intensity of light emission (2,4) follow the reaction rates (1,3), respectively, with a time lag less than 1ns. These reaction rates are directly proportional to electron density, while the rate constants are strongly dependent on electric field. A qualitative analysis of the results presented in Figure 3 leads to the following conclusions: The wave of light intensity, which is observed most clearly for the spectral band B(0-0) with a threshold energy of 19 eV, reveals a wave of electric field (ionizing wave, or streamer). Two peaks of the luminosity distributions are of different physical nature, namely, there is a higher electric field and a lower electron density near the cathode, than near the anode.

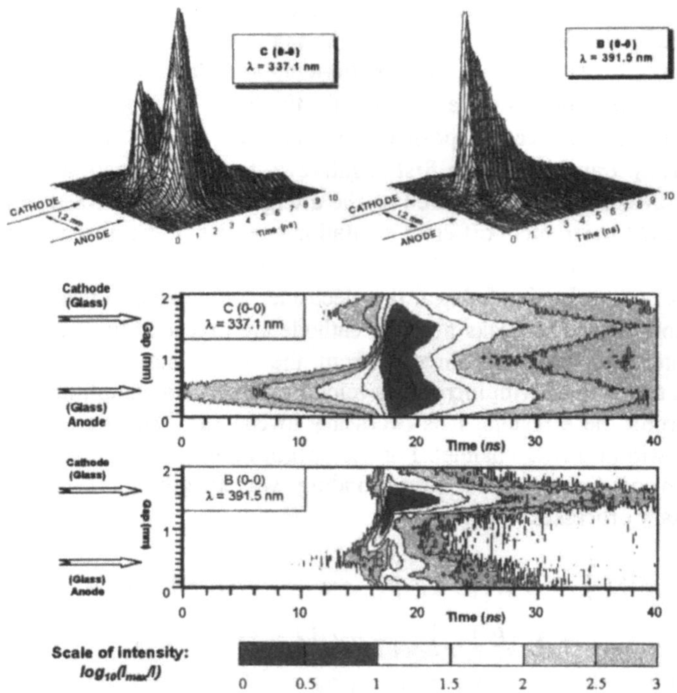

Figure 3. Spatially and temporally resolved spectral intensities of the $N_2\ (C^3\Pi_u)_{v'=0} \rightarrow N_2\ (B^3\Pi_g)_{v''=0}$ transition of the second positive system and the $N_2^+\ (B^2\Sigma_u^+)_{v'=0} \rightarrow N_2^+\ (X^2\Sigma_u^+)_{v''=0}$ transition of the first negative system of nitrogen (conditions: synthetic air, p=1bar, flow=700sccm , V_{pp}= 13kV, f=6.5kHz, d=1.2mm, glass dielectrics, comp. fig. 2).

Derivation of the electric field strength and electron density

On the base of the measurements presented in Figure 3 the derivation of the electric field strength in air at atmospheric pressure was possible. The used kinetic model and the details of the calculations have been discussed in very detail in former articles [10,11]. The spatially and temporally resolved values of the reduced electric field strength as well as the relative density of electrons for the conditions of Figure 3 are summarized in Figure 4.

These results confirm the conclusions derived from the qualitative analysis of the luminosity distributions. A cathode-directed wave of the electric field is clearly seen in Figure 4. A high electric field strength is observed mostly near the cathode, and high electron density, on the contrary, near the anode. Thus, our qualitative consideration and quantitative analysis of the results of CCS measurements of the BDs in air at atmospheric pressure lead to the following conclusions.

1. There are three distinct phases of the MD development. The pre-breakdown phase lasts for a period of more than 0.5 μs. It is characterized by a continuous glow on the glass surfaces of both electrodes. During the last 100 ns of the pre-breakdown phase, the maximal light intensity is observed on the anode.
2. The streamer starts directly from the surface of the anode and crosses the gap with an increasing velocity that reach 10^6 m/s. Electric field strength of the streamer grows respectively from 120 Td at the anode to 240 Td at the cathode.
3. The third phase of the MD is a phase of decay of the light and current pulses.

There are two distinct regions with essentially different properties of the plasma within the MD channel. The electric field near the cathode is higher than near the anode, and the electron density, on the contrary, is considerably lower. The different physical properties of these regions result in distinct differences in the chemical kinetics of plasma chemical processes (JPhD).

Admixture of oxygen to nitrogen

In pure nitrogen, a diffuse ("glow", Townsend-like) DBD is generated [13]. But, already an admixture of even some hundred ppm of oxygen leads to the transition to the filamentary mode. In the case of more than 1vol. % O_2, a profound filamentary discharge is investigated [14]. Qualitatively, no significant difference in the microdischarge behaviour can be found here. Under all the conditions studied, the discharge evolution is characterised by a cathode directed luminosity wave followed by a glow at the anode. The velocity of the cathode directed luminosity wave (ionization wave) seems to be independent of the gas composition, as shown in Figure 5. However the higher the oxygen content, the faster the decay of the SPS-signal. Assuming the relaxation of the $N_2(C)$ density to due spontaneous radiation and collisional quenching, a reasonable agreement between measured and calculated data could be obtained.

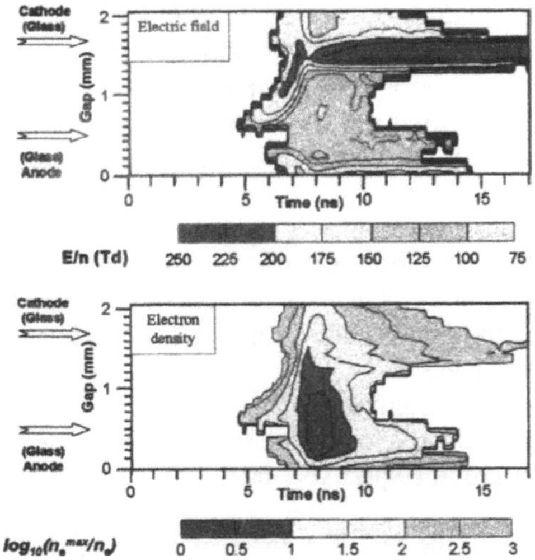

Figure 4. Spatially and temporally resolved local reduced field strength and relative electron density of single microdischarge in synthetic air (conditions of figure 3).

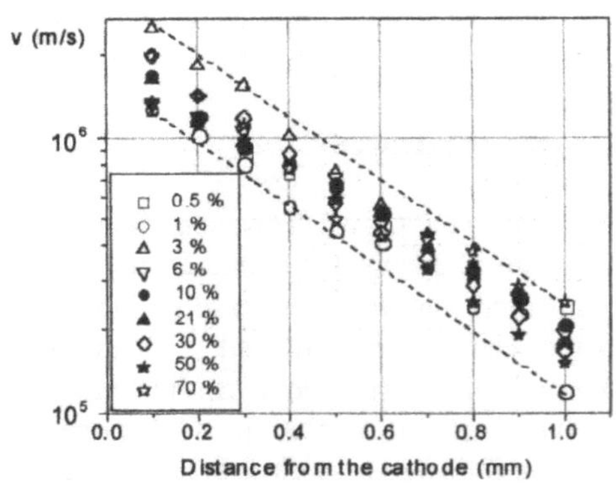

Figure 5. Velocity of the cathode directed ionization wave as a function of the axial coordinate and gas composition (parameter: O_2-admixture to N_2 in vol.%).

Two-dimensionally resolved CCS

Recently, the 2-dimensionally resolved CCS of the single filament DBD has been started [15]. An example is shown in Figure 6 for discharge operation conditions, comparable to those of Figs. 3 and 4. The results visualize the phases of the discharge development in impressive manner, imaging the volume situation as well as the discharge propagation on the dielectric surfaces. The discharge starts with an Townsend-phase followed by a cathode directed ionisation wave and a glow near the anode. The microdischarge channel has a thickness of about 0.2 mm. It expands towards both electrodes up to a value comparable with the discharge gap. On both dielectric surfaces discharges with outward propagating intensity maxima are observed, indicating the accumulation of charge carriers on the surfaces. From those investigations detailed experimental data on the local plasma parameters (reduced electric field strength, electron density) of DBDs in the volume as well as on the electrode surfaces can be expected.

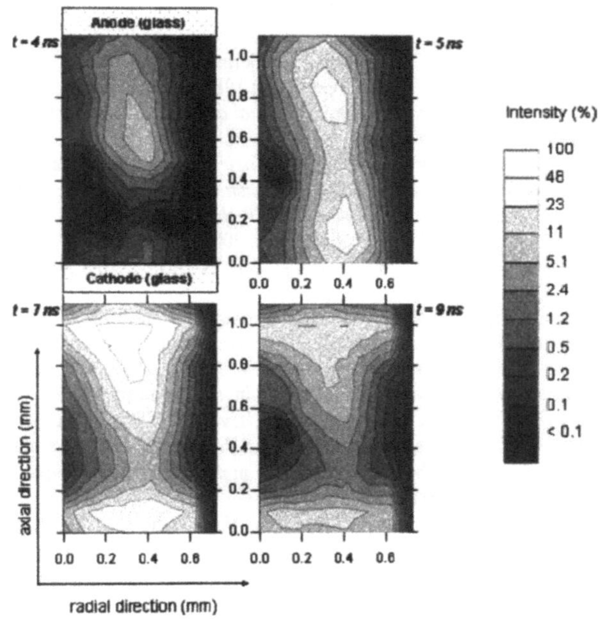

Figure 6. Experimental results of the 2-dimensional spatially, spectrally and temporally resolved CCS (N_2 ($C^3\Pi_u$)$_{\Theta}$ → N_2 ($B^3\Pi_g$)$_{\Theta}$, transition of the second positive system of N_2 (λ=337 nm); gas mixture N_2/O_2=88/2).

4. CONCLUSIONS

The cross correlation emission spectroscopy (CCS) is a powerful tool to analyze the spatio-temporal structure of DBDs in the filamentary mode with a sub-ns and sub-mm resolution. The array of experimental results are sufficient for the quantitative estimating of the local electric field strength and the relative electron density. Quantitatively, no profound influence of the

N_2 / O_2 gas composition on the MD behavior has been found. The differences my be explained by the changing role of the collisional quenching of nitrogen excited species by oxygen. First investigations of the temporally resolved 2D structures of the nitrogen molecular bands allow a detailed insight into the discharge development in the volume as well as on the dielectric surfaces.

6. ACKNOWLEDGEMENTS

The work was supported by the DFG –Sonderforschungsbereich 198 "Kinetics of Partially Ionized Plasmas".

7. REFERENCES

1. K. Buss. Arch. Electrotechnik 1932, 26, 261 (in German).
2. C. Heuser, PhD-thesis RWTH Aachen (Germany) 1985 (in German).
3. V. Samoilovich, V. Gibalov, K.V. Kozlov, *Physical chemistry of the barrier discharge,* Düsseldorf 1997: DVS-Verlag GmbH.
4. U. Kogelschatz, *Dielectric-barrier Discharges: Their History, Discharge Physics, and Industrial Applications*, Plasma Chemistry and Plasma Processing 2003; 23(1):1-453.
5. K. Yoshida, H. Tagashira, Memoirs Kitami Inst. Technol. 1986; 18: 11.
6. V.I. Gibalov, G.J. Pietsch, *J. Phys. D: Appl. Phys.* 2000; 33: 2618.
7. Yu. B. Golubovskii, V. A. Maiorov, J. Behnke, J.F. Behnke, *J. Phys.D: Appl. Phys.* 2002; 35: 751.
8. C. Khamphan, P. Segur, F. Massines, M.C. Bordage, N. Gherardi, Y. Cesses, in: *Proc. XVIthInt. Symp. on Plasma Chemistry ISPC (Italy, Taormina)* 2003.
9. O.S. Shepelyuk, K.V. Kozlov, A.P. Monyakin, V.V Dobryakov, V.G. Samoilovich, Fast kinetic spectroscopy of single microdischarge in an ozonizer, *4th Int. Symp. on High Pressure Low Temp. Plasma Chem.*, p.153-157, University of Bratislava, Slovakia, 1993.
10. K.V. Kozlov, H.-E. Wagner, R. Brandenburg, P. Michel, Spatio-temporally resolved spectroscopic disgnostics of the barrier discharge in air at atmospheric pressure, *J. Phys. D: Appl. Phys.*, 34, 3164-76 (2001).
11. K.V. Kozlov, V.V. Dobryakov, A.P. Monyakin, V.G. Samoilovich, O.S. Shepeliuk, and H.-E. Wagner, R. Brandenburg, P. Michel, In: Selected Research *Papers on Spectroscopy of Nonequilibrium Plasma at elevated Pressures,* ed. by V.N. Ochkin, Proceedings of SPIE, Washington (USA) 2002, vol. 4460, p.165.
12. W.R. Ware, Technique of pulse fluorometry, In: *Time-Resolved Fluorescence Spectroscopy in Biochemistry and Biology, NATO ASI Series A: Life Sciences,* ed. by Cundall R.B. and Dale R.E., 69, Plenum Press, New York, London, 1983, p.23-57.
13. N. Gherardi, G. Gouda, E. Gat, A. Ricard, F. Massines, *Plasma Sources Sci. Technol.*, (2000) 340.
14. R. Brandenburg, K.V Kozlov, N. Gherardi, P. Michel, C. Khampan, H.-E. Wagner, F. Massines, *Proc. of 8th Int. Sym. on High Press. Low Temp. Plasma Chemistry,* Pühajärve/ESTONIA, 2002, vol. 1, p. 28.
15. R. Brandenburg, K.V. Kozlov, H.-E. Wagner, P. Michel, Yu.V. Yurgelenas, In: *Proc. XXVIth ICPIG (Germany, Greifswald)* 2003: IV-39.

IMPULSE PARTIAL DISCHARGE PROPAGATION AND BREAKDOWN CHARACTERISTICS IN N_2/SF_6 GAS MIXTURES

N.Hayakawa[1] Y.Yoshitake[1] N.Koshino[1] and H.Okubo[2]

1. INTRODUCTION

Since leader discharge is a kind of partial discharge (PD) via streamer discharge as a precursor to breakdown (BD) in electronegative gases[1], its inception and propagation mechanisms are important for fundamental electrical insulation design and insulation diagnosis, including breakdown prediction of gas-insulated power apparatus for next generation.

We have been investigating the discharge characteristics and mechanisms in N_2/SF_6 gas mixtures through the simultaneous measurement system of current and light intensity waveforms, streak and still light emission images under lightning impulse voltage application[2-4].

In this paper, we discussed the impulse streamer / leader discharge and BD characteristics and their transition mechanisms of a needle-plane gap in N_2/SF_6 gas mixtures. The 50% probability PD inception voltage ($PDIV_{50}$), BD voltage (BDV_{50}) and leader onset voltage (LOV) were measured, and impulse discharge type was classified into streamer / leader discharge regime for different SF_6 contents and electrode diameters in N_2/SF_6 gas mixtures.

2. EXPERIMENTAL SETUP AND PROCEDURE

Figure 1 shows the experimental setup. The needle electrode with the tip diameter of ϕ=1, 2, 3mm was fixed on the sphere electrode and the gap length of the needle-plane electrodes was 40mm. The experimental chamber was filled with N_2/SF_6 (SF_6 content: 0~100%) gas mixtures at the gas pressure of P=0.1~0.4MPa. The $PDIV_{50}$ and BDV_{50} were measured by the up-and-down method under positive standard lightning impulse voltage application.

The applied impulse voltage and PD current pulse waveforms were fed into

[1] Department of Electrical Engineering, Nagoya University, Nagoya, 464-8603, Japan

[2] Center for Integrated Research in Science and Engineering, Nagoya University, Nagoya, 464-8603, Japan

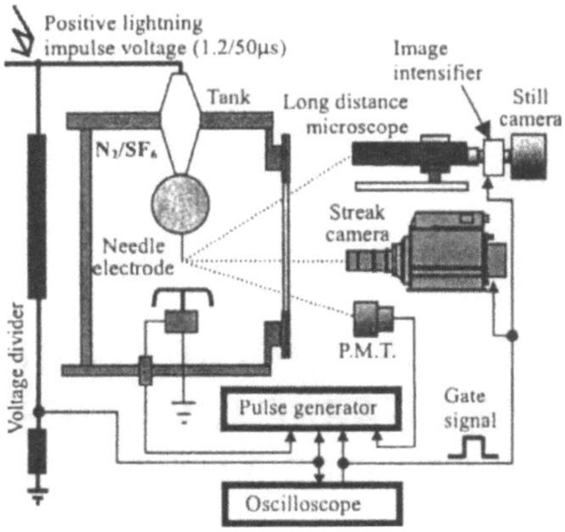

Fig. 1 Experimental setup.

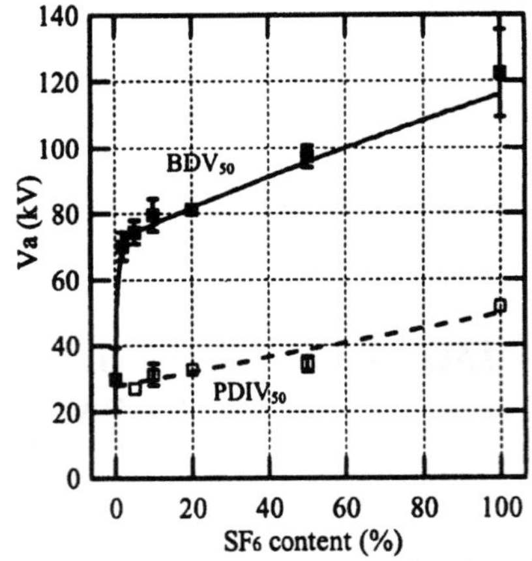

Fig.2 $PDIV_{50}$ and BDV_{50} as a function of SF_6 content in N_2/SF_6 gas mixtures at P=0.1MPa.

a digital oscilloscope (2GS/s, 1GHz) through a matching circuit. The still image and the intensity of PD light emission were observed using a digital camera with an image intensifier and a photomultiplier tube (P.M.T.), respectively. The streak image of PD light emission was recorded by an ultra high speed streak camera (time resolution < 250ps) to observe the propagation process of the impulse PD. The gate signal from a pulse generator enabled us to measure these electrical and optical
5 signals simultaneously and synchronously.

3. IMPULSE PD AND BD CHARACTERISTICS

Figure 2 shows positive impulse $PDIV_{50}$ and BDV_{50} as a function of SF_6 content in N_2/SF_6 gas mixtures at P=0.1MPa. $PDIV_{50}$ increased linearly with the SF_6 content. On the other hand, BDV_{50} increased drastically with the SF_6 content; BDV_{50} reached 240% of that in pure N_2 gas by only 2% addition of SF_6 gas, which exhibited the strong synergistic effect.

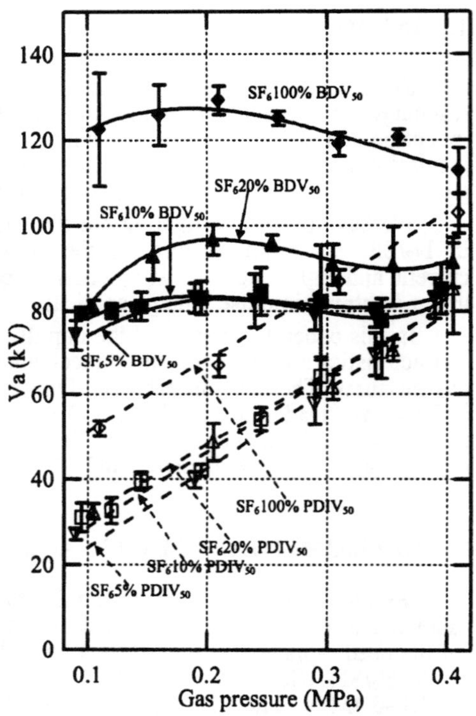

Fig.3 $PDIV_{50}$ and BDV_{50} as a function of gas pressure for different SF_6 content in N_2/SF_6 gas mixtures.

Figure 3 shows $PDIV_{50}$ and BDV_{50} as a function of gas pressure for different SF_6 contents. $PDIV_{50}$ increased linearly with the gas pressure, whereas BDV_{50} exhibited non-linear characteristics with the gas pressure; maximum at P=0.20~0.25MPa and minimum at P=0.3~0.35MPa. Thus, there is a significant effect of corona stabilization even under impulse voltage application.

4. IMPULSE PD INCEPTION AND PROPAGATION CHARACTERISTICS

Figures 4 and 5 show the applied voltage, PD current and light intensity pulse waveforms, streak and still light emission images in $N_2$95%/$SF_6$5% gas mixtures for the applied voltage of V_a=68.1kV and 69.4kV at P=0.3MPa, respectively. PD

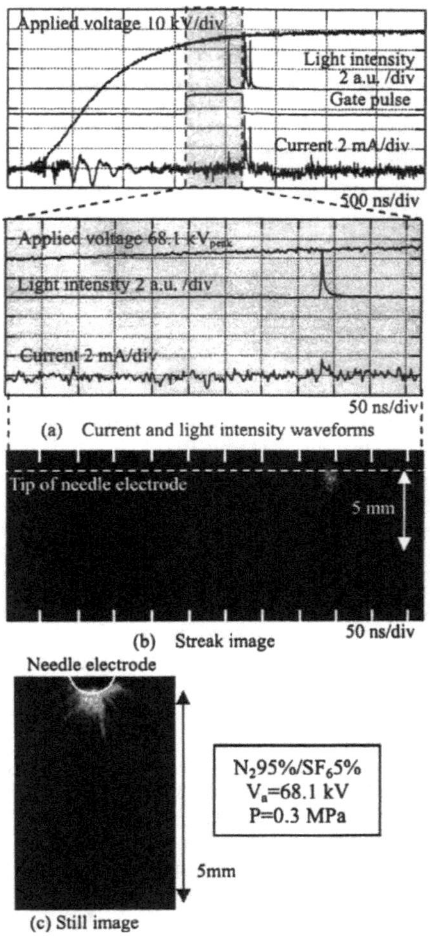

Fig.4 PD current, light intensity, streak and still images of streamer discharge.

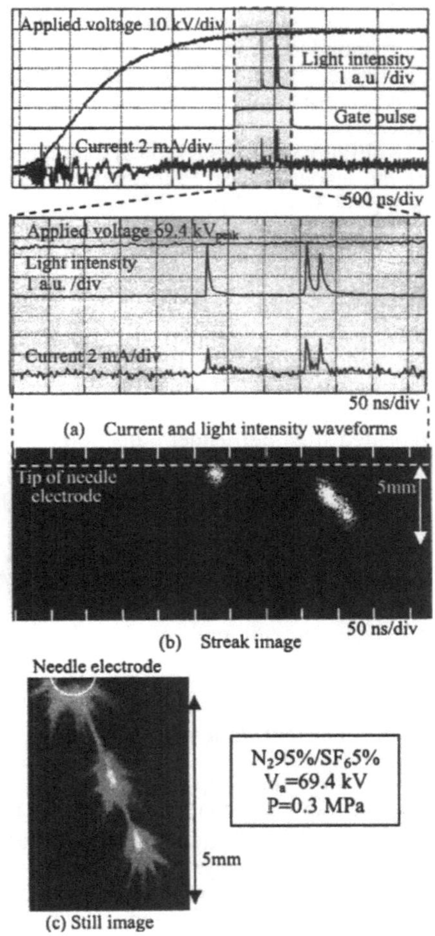

Fig.5 PD current, light intensity, streak and still images of leader discharge.

current and light intensity waveforms in Fig.4 (a) and streak image in Fig.4 (b) revealed that only 1st PD was measured within the gate pulse width. The still image in Fig.4 (c) verified that the 1st PD with brush-like image corresponded to the streamer discharge extending from the needle electrode. On the other hand, 3-peak pulses were measured within the gate pulse width in Fig.5 (a). Together with the streak and still light emission images in Figs.5 (b) and (c), the impulse PDs were verified to propagate stepwise. The 1st PD corresponded to the streamer discharge, whereas the subsequent PDs corresponded to the leader discharge.

Figure 6 shows a typical example of the leader discharge leading to BD at V_a=68.3kV and P=0.3MPa. In Fig.6 (a), the PD current and light intensity waveforms were composed of the superimposed pulses of the leader discharge, and the intensity was getting increased. Figure 6 (b) shows the still image of the PD propagation just before BD. The impulse PDs led to BD via stepwise propagation due to repeated streamer/leader transition.

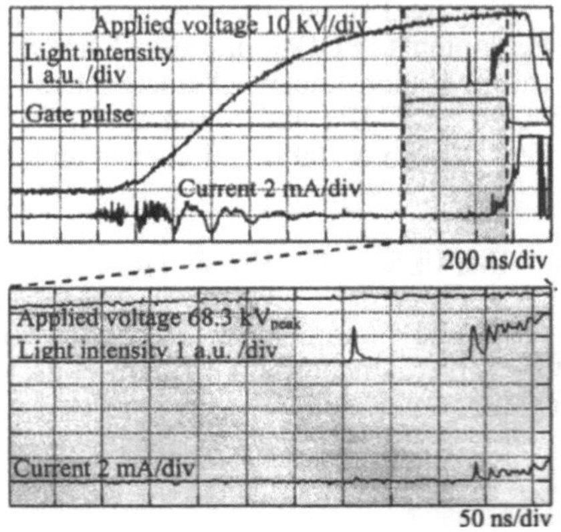

(a) Current and light intensity waveforms

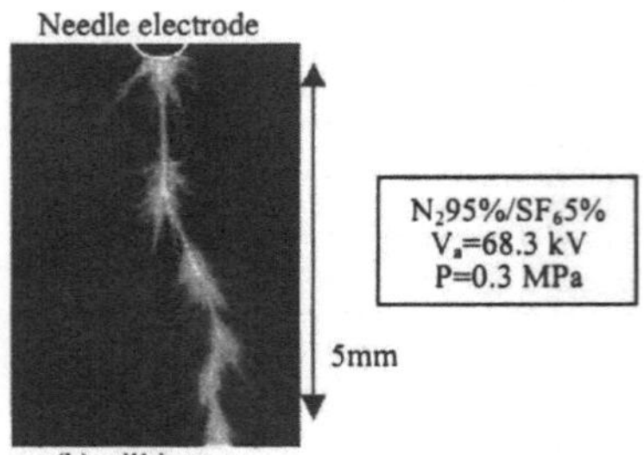

(b) still image

Fig.6 PD current, light intensity and still image immediately before BD.

5. CLASSIFICATION OF IMPULSE PD TYPE

Figure 7 (a)~(d), 8 (a) and (b) show the leader discharge onset voltage (LOV) together with $PDIV_{50}$ and BDV_{50} in N_2/SF_6 gas mixtures, respectively. LOV was defined as the minimum value of the applied voltage where the stepwise PD propagation with multi-peak pulses were observed, as shown in Fig.5. These results enabled us to classify the impulse PD type in N_2/SF_6 gas mixtures into streamer discharge or leader discharge.

5.1 SF_6 content dependence

In Fig.7, LOV decreased with the gas pressure, but increased at the higher gas pressure because $PDIV_{50}$ reached LOV. Comparing Figs.7 (a)~(d), the leader discharge regime became larger with the gas pressure and the SF_6 content, whereas the streamer discharge regime became smaller. Moreover, leader discharge regime also increased with the SF_6 content at the lower gas pressure. These are attributed to the enhancement of electronegativity in N_2/SF_6 gas mixtures with the higher gas pressure and SF_6 content, which could induce the shrinkage and heating of leader channel and the activation of the streamer / leader transition. Such classification of impulse PD type has been reported in pure SF_6 gas[5], and those in Fig.7 is obtained in N_2/SF_6 gas mixtures as a function of SF_6 content.

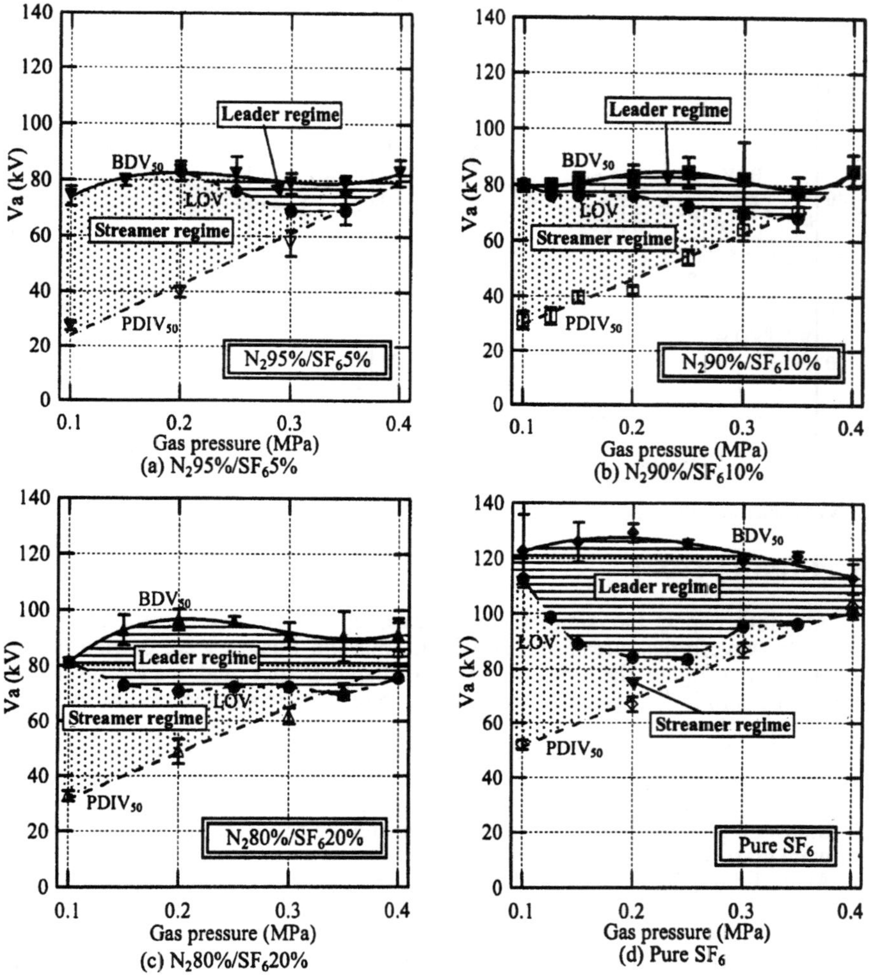

Fig.7 Classification of impulse PD type in N_2/SF_6 gas mixture for different SF_6 contents (ϕ=1mm).

5.2 Electrode diameter dependence

Figures 7 (b), 8 (a) and (b) show the $PDIV_{50}$, LOV and BDV_{50} as a function of gas pressure in $N_2$90%/$SF_6$10% gas mixtures for different electrode diameters of ϕ=1,2,3 mm, respectively. $PDIV_{50}$ increased with the electrode diameter, whereas LOV were almost constant at 70~75kV. This may be because the electric field strength at leader onset depended on the tip radius of the previous PD channels, which could be irrespective of the needle-tip radius. Therefore, streamer discharge regime decreased, i.e. corona stabilization effect became weaker with the electrode diameter.

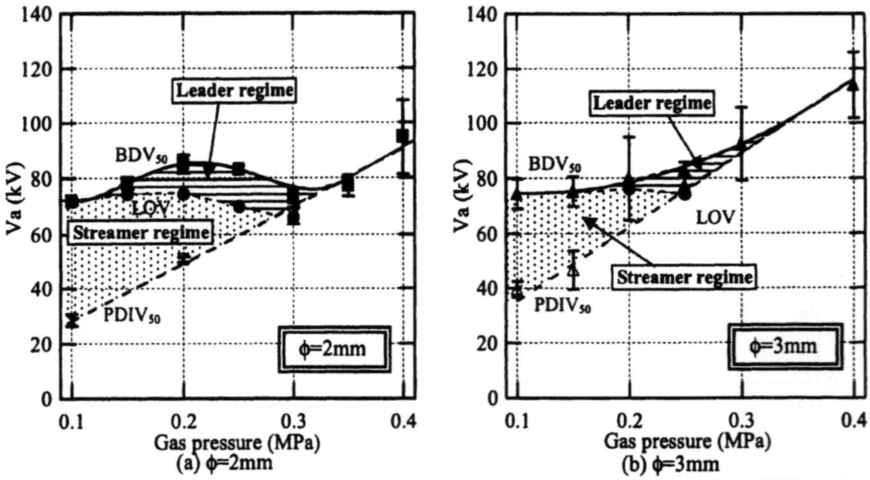

Fig.8 Classification of impulse PD type in $N_2$90%/$SF_6$10% gas mixtures for different electrode diameters.

6. CONCLUSIONS

In this paper, the simultaneous measurements of PD current and light intensity waveforms, streak and still light emission images were carried out in order to investigate the impulse PD/BD characteristics and leader discharge mechanisms in N_2/SF_6 gas mixtures. The main results are summarized as follows:

1. Impulse PD inception voltage ($PDIV_{50}$), leader onset voltage (LOV) and breakdown voltage (BDV_{50}) were measured as a function of gas pressure for different SF_6 contents and electrode diameters.
2. Stepwise propagation of leader discharge and the resultant breakdown process were clarified in N_2/SF_6 gas mixtures.
3. According to $PDIV_{50}$, LOV and BDV_{50}, impulse PD type was classified into streamer or leader discharge regime in N_2/SF_6 gas mixtures.
4. The larger leader discharge regime in N_2/SF_6 gas mixtures with the higher gas pressure and SF_6 content was attributed to the enhancement of electronegativity, which could induce the shrinkage and heating of leader channel and the activation of the streamer/leader transition.
5. The smaller streamer discharge regime for the larger electrode diameter was due to the increased $PDIV_{50}$ and the constant LOV, where LOV depended on the tip radius of the previous PD channel, and not on the needle-tip radius.

REFERENCES

1. I.Gallimberti and N.Wiegart: "Breakdown Mechanisms in Electronegative Gases", *Gaseous Dielectrics V*, pp.61-79, (1987)
2. T.Yamada, T.Ishida, N.Hayakawa and H.Okubo:• Partial Discharge and Breakdown Mechanisms in Ultra-dilute SF_6/N_2 Gases Mixtures•••*IEEE Trans. on Elect. Insul.*, Vol.8, No.1, pp.137-142, (2001).
3. H.Saitoh, K.Morita, T.Kikkawa, K.Kato, N.Hayakawa and H.Okubo: "Impulse Partial Discharge and Breakdown Characteristics of Rod-Plane Gaps in N_2/SF_6 Gas Mixtures", *CEIDP*, pp.397-400 (2001).
4. Y.Yoshitake, N.Hayakawa, T.Ueda and H.Okubo: "Impulse Partial Discharge Propagation Mechanism under Non-uniform Electric Field in N_2/SF_6 Gas Mixtures", *13th ISH*, p.471 (2003).
5. F.Pinnekamp and L.Niemeyer: Qualitative Model of Breakdown in SF_6 in Inhomogeneous Gaps• ,*J. Phys. D: Applied Physics*, Vol.16, pp.1293-1302 (1983).

Microwave Breakdown Studies of He – N_2 Mixtures in a Pillbox Cavity for Repetitively Pulsed High Power Systems

David Hemmert, Andreas Neuber, and James Dickens*

Abstract

Microwave breakdown of various He-N_2 mixture combinations is investigated for the prevention of electrical and thermal breakdown in repetitively pulsed high power systems. The objective is to determine the best combination of the two gases to maximize the mixture's ability to cool a dielectric window surface while maintaining high electrical breakdown thresholds. Helium is known to have a high thermal conductivity for heat transport off the dielectric window surface, but it also has a low microwave breakdown threshold. Nitrogen does not transport heat as well, but has a much higher electrical breakdown threshold. Initial studies focused on generating a series of microwave power versus gas mixture pressure curves for electrical breakdown to help identify optimum He – N_2 mixture ratios. The gas mixtures flow through ports in a 333 cm^3 S-band double window pillbox cavity. The double window cavity is placed in a traveling wave resonant ring (TWRR) coupled to a 2.85 GHz, 4 MW, magnetron. This combination of double window pillbox and TWRR allows for testing power levels up to 40 MW. High speed diagnostics are used to measure the incident/reflected power and discharge luminosity. An ICCD camera provides time integrated images of the breakdown event. Coupled mass flow controllers maintain the gas mixture ratio and continuous gas flow through the cavity. Investigations can be conducted in either single or multiple pulsed configurations. Results included several gas mixtures from pure helium to pure nitrogen at 760, 1520, and 3040 torr, and showed a significant increase in breakdown levels by adding a small percentage of nitrogen to helium.

1. INTRODUCTION

Dielectric windows are commonly used to transport high power microwaves from one environment or device to another. Failure of these windows can be catastrophic to the system. Typical failures of such dielectric windows are by either electrical breakdown as a discharge or mechanical breakdown from thermal stress. Numerous studies have been made to understand and overcome these problems[1,2]. Typically, SF_6 is used to prevent electrical breakdown due to its high flashover properties. However, SF_6 released by a window puncture or

* P3E Center, Dept of Elec and Comp Engr, Texas Tech University, Lubbock, TX 79410.

Gaseous Dielectrics X.
Edited by Christophorou *et al.*, Springer, New York, 2004.

crack may react adversely with a high power microwave device, damaging the device permanently or creating toxic byproducts. Therefore it is desired to find a relatively inert gas mixture which exhibits both a high microwave breakdown threshold and high heat transport capability.

Electrical Breakdown

Electrical breakdown at a gas/dielectric interface has previously been found to be dominated by different mechanisms, depending upon the pressure of the gas. At low pressures, breakdown is described by high energy free electron processes at the interface leading to a saturated secondary electron emission avalanche, desorption of adsorbed gases from the dielectric surface, and a discharge in the mixture of desorbed gases and volume gas. Interface breakdown at high pressures, however, was found to be dominated by volume breakdown associated with the interface gas and initiated at the dielectric surface[3].

From the diffusion theory of rf breakdown in gas, the critical criteria for breakdown is;

$$\sigma_i - \sigma_a - \frac{D}{\Lambda^2} \equiv 0 \tag{1}$$

where σ_i and σ_a are the ionization and attachment frequencies for the gas, D is the diffusion coefficient, and Λ is the diffusion length[4]. To prevent breakdown, it is desired to find a gas mixture and pressure such that the criteria of Eq. (1) is not met. Such a gas mixture is characterized by a large collisional cross section to minimize the electron mean free path, resulting in lower electron energies and a lower collisional ionization rate.

The usual derivation of the diffusion length is to consider all surfaces of a cavity as a loss mechanism. For a right cylindrical cavity the characteristic diffusion length, Λ, is defined as;

$$\Lambda \equiv \left[\left(\frac{2.405}{R}\right)^2 + \left(\frac{\pi}{X}\right)^2\right]^{-1/2}, \tag{2}$$

where R is the radius, and X is the height of the cavity[4]. However, for a dielectric pillbox cavity, the dielectric windows do not necessarily act as a loss mechanism. If it is assumed that a saturated secondary electron emission avalanche develops on the dielectric surface from single surface and two surface multipactors, then the yield at the interface is unity and the right cylinder geometry appears to have infinite height, modifying Eq. (2);

$$\Lambda \equiv \frac{R}{2.405} . \tag{3}$$

For a cavity with a small height relative to the dielectric window radius, this greatly increases Λ and decreases the influence of the cavity geometry in the breakdown criteria. Breakdown therefore becomes more dependent upon the ionization and attachment frequencies of the gas in a dielectric pillbox than for a similar metallic cavity.

Thermal Breakdown

Thermal breakdown at a dielectric window usually appears as window pinholes, punctures, cracking, or surface evaporation. These are caused by either rf dissipation within the bulk material of the dielectric or phonon production and excitation of trapped electrons in F- centers by multipactor.

Texas Tech University estimated that 0.1 MW of power was deposited on a planar alumina surface in an S-band waveguide for a 10 MW TE_{10}-mode microwave at 2.85 GHz. This translates into a temperature rise of about 1.5°C/kW for the dielectric surface region. Thus multipactor plays a significant role in localized heating of the dielectric near the surface compared to rf heating for high purity alumina[1]. Similar results were found in a 500 MHz accelerating cavity at Daresbury Laboratory in the United Kingdom for uncoated alumina windows, which showed the greatest rise in temperature (2°C/kW at 20 kW rf power), and alumina windows with lossy coatings which performed almost as poorly (1.5°C/kW at 20 kW rf power)[5]. Additional thermal stress modeling at Daresbury Laboratory predicted that the temperature rise due to dielectric losses for coated alumina to be 0.32°C/kW and corroborated the results with well performing windows with temperature rises of 0.2°C/kW to 0.6°C/kW .[5]

To reduce the effects of thermal stress on the dielectric windows, the windows must be cooled. This is possible by flowing a gas across the window surface to transport the rf heat away by forced convection. The peak surface temperature of the window, T, is then defined by:

$$T \equiv \theta_i + \frac{Q}{V * \rho_o * C_p} + \frac{q_{\max}}{\kappa/\delta} + \frac{q_{\max}}{h} \tag{4}$$

where θ_i is the gas inlet temp, Q, the power removed by the gas, V, the volumetric flow rate, ρ_o, the outlet gas density, C_p, the gas specific heat, q_{max}, the peak heat flux, κ, the window thermal conductivity, δ, the window thickness, and h, the heat transfer coefficient[6].

2. EXPERIMENTAL SETUP

An S-band double window pillbox cavity was placed in an S-band traveling wave resonant ring (TWRR) coupled to a 4 MW magnetron, shown in Figure 1. Peak power within the TWRR can be up to 100 MW. A vacuum system coupled to the TWRR provides sufficiently low pressure, down to 2×10^{-8} torr. High speed and resolution diagnostics recorded the TWRR forward and reverse power, luminosity, and spatial imaging. The TWRR was checked for tuning utilizing an HP8719C network analyzer.

The double window pillbox cavity is 1.5 cm in height by 4.2 cm in radius for a 333 cm^3 volume. The windows are high purity alumina discs with an 11.7 cm diameter. The vacuum side of the windows is coated with TiN to prevent multipactor. The test structure has a VSWR of 1.15 for a TE_{10}-mode, 3.85 GHz microwave, as measured by an HP8719C network analyzer. A greater than 100% safety margin of the estimated fracture pressure was assumed and 3040 torr set as the maximum test pressure. The calculated diffusion length for the cavity using Eq. (4) is 1.8 cm.

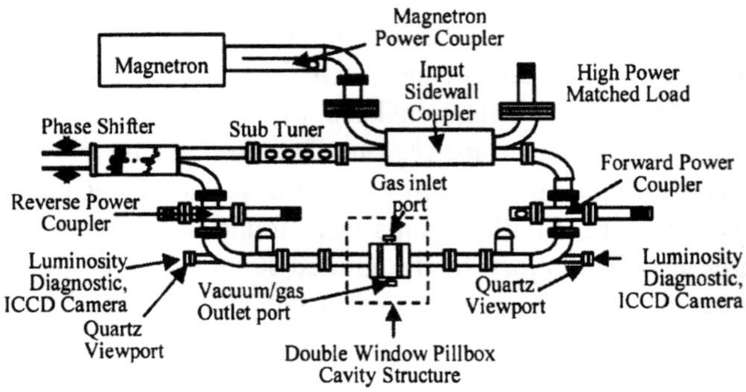

Figure 1. *Experimental Setup*

The cavity between the windows in the test structure can be pressurized with a regulated gas flow rate or placed under high vacuum. The gas flow rate was controlled with MKS 500 sccm and 2000 sccm mass flow controllers. Helium was chosen for the tests due to its high thermal conductivity of 1.499 mW cm^{-1} $^{o}K^{-1}$ and ionization potential of 24.6 eV. Nitrogen was selected because of its relatively high ionization potential, 15.6 eV, and molecular size, and N_2 is a relatively non-reactive molecule[7].

3. RESULTS

Electrical Breakdown

Figure 2 shows the microwave power at breakdown for various gas mixtures of He and N_2. External radiation leakage and breakdown of the pillbox cavity made the collected data for TWRR power levels greater than about 22 MW unreliable. A line in Figure 3 denotes the 22 MW power level above which the data becomes unreliable. The Paschen-like dependence on pressure for breakdown can clearly be seen in the data for helium. The breakdown threshold rises significantly from 5 MW at 760 torr to 19 MW at 3040 torr. More importantly, note the significant increase in breakdown when the gas mixture is 80% He and 20% N_2. The power breakdown threshold increases more than three-fold at 760 torr. At 60% He and 40% N_2, the breakdown level is comparable to that of N_2. At 1520 torr, a significant increase in breakdown threshold is again seen for a mixture of 80% He to 20% N_2. Exterior breakdown of the double window cavity structure from radiation leakage at the window-wall junction occurred above 22 MW resulting in an overlap and wide scatter of data at 1520 torr and 3040 torr for other mixtures, limiting further analysis.

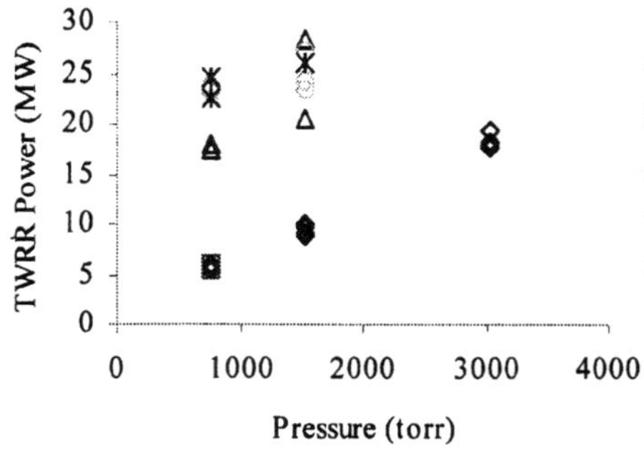

Figure 2. *Breakdown power versus pressure for different helium/nitrogen mixtures in the double window cavity.*

Figure 3 shows time integrated spatial images for the different gas mixtures taken from the downstream viewport. The translucence of the alumina blurs the appearance of the actual discharge. For pure He, the breakdown is very diffuse. However, for partial mixtures of He and N_2, the images appear to be a more localized discharge. Finally, the pure N_2 breakdown indicates a much more distinct discharge. Both partial mixture images show a much closer resemblance to the localized discharge of N_2 than to the diffuse discharge of He.

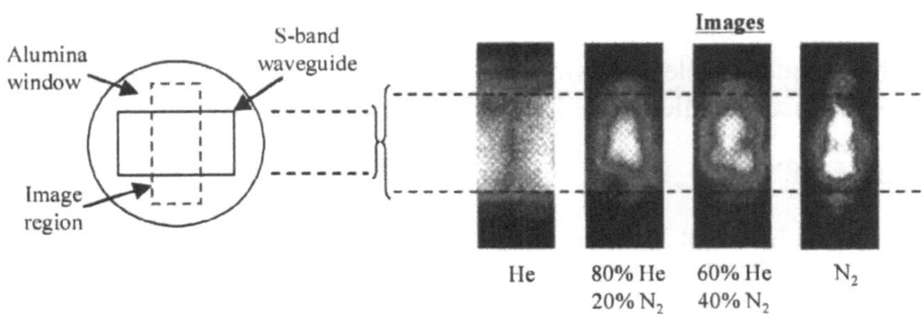

Figure 3. *Time integrated images of breakdown as seen through the downstream alumina window for various gas mixtures at 760 torr.*

Thermal Breakdown

A qualitative analysis of the change in the heat transport capability of the cooling gas for different gas mixtures can be made from Eq. 4 showing that the power removed by the cooling gas at a constant flow rate is proportional to the specific heat the gas, $Q \propto C_p$. The specific heat of a gas mixture is:

$$C_{p\_mix} \equiv \sum_i Y_i * C_{p\_i} \tag{5}$$

where Y_i is the molar fraction of each gas. This results in a linear relation for He-N_2 mixtures. For an 80% He, 20% N_2 mixture with specific heats of 5.19 kJ/kg °K and 1.04 kJ/kg °K respectively, C_{p_mix} of the mixture is 4.36 kJ/kg °K This is only 16% less than the maximum specific heat for He.

4. CONCLUSIONS

The experimental setup of the double window and TWRR combination proved suitable for high power testing of gas/interface breakdown up to 22 MW in the TWRR. Current redesign of the structure to prevent exterior breakdown may allow testing up to the full capabilities of the TWRR of 100 MW.

Using a small percentage of nitrogen in helium increases the breakdown significantly at 760 torr and 1520 torr. The presence of the large nitrogen molecule significantly increases the collision frequency and decreases the overall ionization frequency of the system. The breakdown images of the He-N_2 mixtures showed the discharge to be more like N_2 indicating that the actual breakdown properties of nitrogen become significantly dominant at low mixture percentages relative to He. Also, as determined qualitatively for an 80-20 mixture of He-N_2, the addition of only a small amount of N_2 to He does not significantly affect the heat transport capabilities of the gas mixture while the breakdown power level changes significantly. Further experimental study is required in this area to determine the optimal mixture for both electrical breakdown holdoff and window cooling.

5. ACKNOWLEDGEMENTS

The S-band double window pillbox cavity was designed and provided by Patrick Ferguson of Microwave Device Systems, Inc.

6. REFERENCES

1. A. A. Neuber, L. Laurent, Y. Y. Lau, and H. Krompholz, "Windows and RF Breakdown," in High-Power Microwave Sources and Technologies, R. J. Barker and E. Schamiloglu, Eds., New York: IEEE Press, 2001, pp.325-375.
2. S. Michizono and Y. Saito, "RF Windows Used at S-Band Pulse Klystrons in KEK Linac," Vacuum, vol 47, pp. 625-628, 1996.
3. D. Hemmert, A. A. Neuber, J. C. Dickens, H. Krompholz, L.L. Hatfield, and M. Kristiansen, "High-power Microwave Window Breakdown under Vacuum and Atmospheric Conditions," in Proc. Aerosense 2000, SPIE vol. 4031, Jul 2000, pp. 90-98.
4. A. D. MacDonald, Microwave Breakdown in Gases. New York: John Wiley & Sons, Inc., 1966.
5. R. A. Rimmer, "Development of a Strategy for Modelling Operation and Failure of High Power R.F. Windows," in Proc. EPAC, 1988, pp.1349-1351.
6. C.B. Baxi, K.M. Redler, and J.P. Smith, "Design, Fabrication, and Testing of Helium-Cooled High Heat Flux Module," Proc. High Heat Flux Engineering III, SPIE vol. 2855, Aug 1996, pp. 291-298.
7. The CRC Handbook of Chemistry and Physics, 61st Edition, R.C. Weast, Ed., The CRC Press, Inc., 1980.

STUDY OF TRANSIENT PROCESSES BY MEASUREMENTS OF DYNAMIC BREAKDOWN VOLTAGES

V. Lj. Marković, S. R. Gocić, S. N. Stamenković and Z. Lj. Petrović[1]

1. INTRODUCTION

Statistically based measurements of dynamic breakdown voltages U_b and breakdown delay times t_d and their variations in transient regimes are a convenient method for to study stochastic processes of electrical breakdown in gases. Accompanied by approximate analytical and exact numerical models, they can be used to study the gas-phase and surface processes during the transient regimes of establishment and relaxation of discharges. In a series of papers it has been shown how measurements of the breakdown delay times were applied to investigate collision and transport processes (surface recombination of nitrogen atoms, decay of charged particles, effective secondary electron yield, etc.). In this paper we present analysis of these processes based on measurements of the breakdown voltages U_b for linearly rising pulses. Some preliminary results of applying this method were presented[1]. The experimental details are shortly summarized in section 2, while the theoretical expressions for $\overline{U_b}(k)$ and statistical time delay $\overline{t_s}(k)$ are derived in section 3. In section 4, the experimental breakdown voltage distributions are fitted by theoretical predictions based on approximate analytical and numerical models and dependence of the effective secondary electron yield γ on the overvoltages is obtained.

[1] V. Lj. Marković, S. R. Gocić and S. N. Stamenković, Department of Physics, University of Niš, P. O. BOX 224, 18001 Niš, Serbia and Montenegro; Z. Lj. Petrović, Institute of Physics, P. O. BOX 57, Belgrade, Serbia and Montenegro.

2. EXPERIMENTAL DETAILS

The breakdown voltage measurements have been performed for a gas tube made of molybdenum glass with plane-parallel electrode system made of copper. The diameter of electrodes was $D = 22\ mm$, area $S = 3.8\ cm^2$, gap $d = 3\,mm$ and pressure $2.66\ mbar$. The static breakdown voltage was $U_s = 247\ V$ DC. The dynamic breakdown voltages were determined as the maximum values before the collapse of the applied voltage, at glow current of $I_g = 100\,\mu A$, at different afterglow or relaxation times and the rates of voltage rise. The measurements consisted of 100 single shots in series by using storage oscilloscope and its sawtooth pulses that were further amplified[1].

3. THE DEPENDENCE OF BREAKDOWN VOLTAGE ON THE RATE OF VOLTAGE RISE

The time that elapses from the moment of application of minimum breakdown voltage to the breakdown is the breakdown time delay t_d. It consists of the statistical time delay t_s (from the application of voltage to the appearance of a free electron that may produce the breakdown) and the formative time t_f (from this moment to the collapse of the applied voltage and occurrence of a self-sustained current)[2]. The formative time has negligible statistical spread and, moreover can be neglected in most cases[3]. The mean value of statistical time delay is $\overline{t_s} = 1 / Y\,P$, where Y represents a number of electrons generated in the interelectrode space per second (electron yield), and P is the probability that one electron will cause a breakdown[2].

In the case of ramp voltages, the statistical time delay can be approximately expressed as:

$$\overline{t_s} \approx \frac{1}{\overline{Y\,P}} \approx \frac{1}{\dfrac{d\left(\overline{Y\,P}\right)}{dU}\dfrac{d\,U}{d\,t}\overline{t_s}}, \tag{1}$$

and, therefore, it is proportional to:

$$\overline{t_s} \propto 1/\sqrt{k}\,. \tag{2}$$

On the other hand, for linear voltages:

$$\overline{U_b} = U_{b0} + k\,\overline{t_d}\,, \tag{3}$$

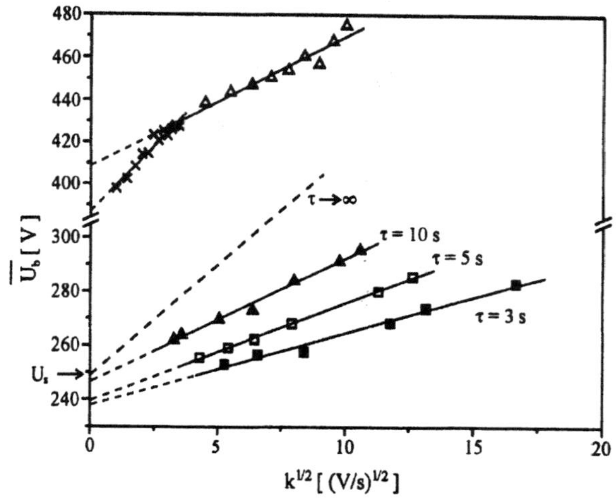

Figure 1. The mean values of dynamic breakdown voltages versus the square roots of the rates of voltage rise for different relaxation times of our measurements and measurements[4] (upper scales).

where U_{b0} is the extrapolated value of dynamic breakdown voltage at $k = 0$ and determined by the level of preionization or residual active states that remain from the preceding glow. In the range of rates of voltage covered in this experiment, Eq. (3) can be approximated by:

$$\overline{U_b} = U_{b0} + k\,t_f + k\,\overline{t_s} \approx U_{b0} + k\,t_s \,. \tag{4}$$

By taking Eq. (2) into account, it follows that

$$\overline{\Delta U_b} = \overline{U_b} - U_{b0} \propto \sqrt{k} \,. \tag{5}$$

In Fig. 1, the mean values of dynamic breakdown voltages versus the square roots of the rates of voltage rise $\overline{U_b}\left(\sqrt{k}\right)$ for the different relaxation times are presented. The curves are linear in accordance with Eq. (5). It can be seen that $\overline{U_b}$ and U_{b0} increase with increasing rates of voltage rise as well as the relaxation times and always $U_{b0} < U_s$ (due to preionization or residual states). In order to regard U_{b0} as U_s (or $U_{b0} \to U_s$), it is necessary and sufficient that $k \to 0$ and $\tau \to \infty$ (preionization or residual states should vanish down to the cosmic rays and natural radioactivity level). In paper[4] the $U_b(k)$ dependence was fitted by third order polynomial, by obtaining values of U_s with the spread of 24 volts (Fig. 1), without citing τ which is determining parameter.

4. FITTING OF EXPERIMENTAL DATA BY THEORETICAL DISTRIBUTIONS

The distribution function of the dynamic breakdown voltages U_b can be expressed by relation[5]:

$$F(U_b)=1-exp\left[-\int_{U_{b,min}}^{U_b}\frac{Y(U)P(U)}{k}dU\right], \tag{6}$$

where $U_{b,min}$ is the minimum dynamic breakdown voltage and Y, P and k are already defined. The upper limit of the integral is $U_b - k\,t_f \approx U_b$, since formative time delay t_f can be neglected compared to the statistical time delay t_s under our conditions[3]. In Townsend's model[2], the breakdown probability is given by relation[2]:

$$P=\begin{cases}0, & for \quad q<1\\ 1-1/q, & for \quad q>1\end{cases} \tag{7}$$

where $q=\gamma\left[exp(\alpha d)-1\right]$, α - electron ionization coefficient and γ - effective secondary emission coefficient. Value of γ is determined from the breakdown criterion[2] $\gamma_s=\left[exp(\alpha d)-1\right]^{-1}$, for the static breakdown voltage U_s, and in our case for the minimum dynamic breakdown voltage $U_{b,min}$, it is designated by $\gamma_{b,min}$.

4.1. Approximate analytical model

If Y and P are constant, or the effective electron yield is averaged over the range ΔU of the voltage pulse $\overline{YP}$, the fraction of voltages exceeding U_b is:

$$n_i/N=exp\left(-\frac{\overline{YP}}{k}\Delta U\right), \tag{8}$$

where $\Delta U = U_b - U_{b,min}$ is the overvoltage. Equation (8) is linear in the semi-log scale, and from the slope we may derive:

$$\overline{YP}=\frac{-k\,ln(n_i/N)}{\Delta U}. \tag{9}$$

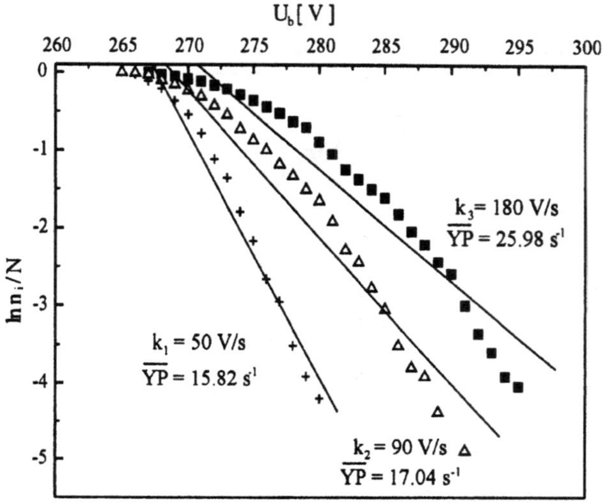

Figure 2. The linear least square fits based on the approximate analytical model.

The experimental data (represented by symbols in Fig. 2) are curved, showing that Y and/or P cannot be regarded as constants when fitting the experimental data by theoretical distribution Eq. (6), but as increasing functions of voltage. Moreover, the approximate values of the mean effective electron yield can be obtained from the simple linear least square fits in Fig. 2 using Eq. (9), which are approximately equal to the corresponding values from the numerical model.

4.2. Numerical model

In this model, it is allowed for Y, γ and P to be voltage-dependent. The secondary electron yield γ is linearly increasing function of the overvoltage $\gamma = \gamma_{b,min}\left(1 + C_1 \Delta U\right)$, causing the electron yield to rise according to relation $Y = C_2 \gamma$, where C_1, C_2 are constants. The fit of experimental data (Fig. 3) is obtained for the initial electron yield of $Y = 95 s^{-1}$, for all series, initial $\gamma_{b,min} \approx 0.02829, 0.02797, 0.028132$, and the relative γ and Y increase of $13\%, 15\%$ and 20%, respectively, according to increasing range of overvoltages. It should also be noted that no fit is possible with a constant γ and Y, contrary to published results[4,6]. Moreover, the initial electron yield Y in all series must be the same when the measurements were carried out at the same relaxation time, contrary to results[6] where Y actually decreases with increasing k. Thus, the variation of γ may be best approximated by linear dependence on

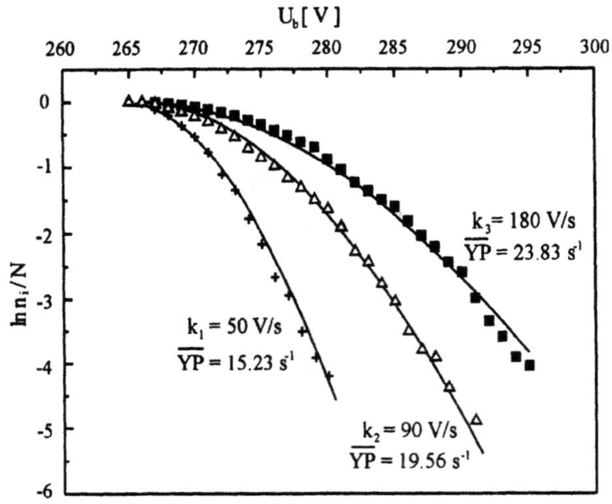

Figure 3. The fits of experimental data with numerical model with linearly rising γ and Y.

the overvoltage (i.e. with the reduced electric field E/N) within the series with the same k. For different series with different voltage ramps it is proportional to $\propto \sqrt{k}$. There is no unique set of fitting parameters Y and $\Delta\gamma$, i.e. when initial Y increases, $\Delta\gamma$ decreases and vice versa, but always preserving $\overline{YP}$ as invariant. Thus, our simple models give the basis for explanation of the experiments and a possibility for the study of gas-phase and surface processes by measurements of dynamic breakdown voltages.

Acknowledgments. This work was partially supported by MNTRS project 1478.

1. V. Lj. Marković, S. R. Gocić, S. N. Stamenković and Z. Lj. Petrović, Study of gas-phase and surfaces processes by measurements of dynamic breakdown voltages, *21th Symp. on the Physics of Ionized Gases (Sokobanja, Yugoslavia)*, 426-429 (2002)
2. C. G. Morgan, Irradiation and Time Lag, in *Electrical Breakdown of Gases* edited by J. M. Meek and J. D. Craggs (John Wiley & Sons, Chichester, 1978)
3. V. Lj. Marković, Z. Lj. Petrović and M. M. Pejović, Modeling of charged particle decay in nitrogen afterglow, *Plasma Sources Sci. Technol.* **6**, 240-246 (1997)
4. M. M. Pejović, G. S. Ristić and J. P. Karamarković, Electric breakdown in low pressure gases, *J. Phys. D: Appl. Phys.* **35**, R91 – R103 (2002)
5. R. V. Hodges , R. N. Varney and J. F. Riley, Probability of electrical breakdown: Evidence for a transition between the Townsend and streamer breakdown mechanisms, *Phys. Rev. A* **31**, 2610 –2620 (1985)
6. M. K. Radović and Č. A. Maluckov, Statistical analysis of the dynamic voltage electrical breakdown in nitrogen, *IEEE Trans. Plasma Science*, **29** (5), 832-836 (2001)

INFLUENCE OF METAL VAPOR AND COMPOSITE MATERIAL ON THE DECAYING CONDUCTANCE OF A LOW -VOLTAGE CIRCUIT BREAKER

A. Gleizes, A. Mercado Cabrera, Ph. Teulet and J.J. Gonzalez[1]

1. LOW VOLTAGE CIRCUIT-BREAKER ARC

In a low-voltage circuit-breaker (LVCB), the contact opening leads to the formation of an arc in air, that is moved to an extinction chamber. During its displacement, largely due to external electromagnetic forces, the arc interacts with rails and with the walls made of composite material. The material ablation creates an increase of the local pressure that favors the arc movement. But the presence of metal and carbon vapors that have rather low ionization potentials, may increase the electron number density and, in certain cases, it has been observed an arc restriking near the contacts. In order to study this phenomenon we have been developing a numerical study for calculating the arc movement, the interaction with the walls and all the problems related to the extinction.

In this communication we present one step of this large study, that deals with a chemical kinetics model in the mixture air + metal vapor + organic material vapor (here *PA6,6*, containing the species C, H, O and N) coupled with a 1D hydrodynamic decaying arc model. The aim of this study was to put in evidence the influences of possible departures from equilibrium and of the presence of organic and metallic vapors, on the interruption capability of the system. The general model needs first the calculation of the equilibrium plasma composition and of the material properties.

[1] CPAT, UMR CNRS5002, Université Paul Sabatier, 118 route de Narbonne, F31062 Toulouse cedex 4, France alain.gleizes@cpat.ups-tlse.fr

2. MATERIAL PROPERTIES

Some papers dealing with the determination of equilibrium compositions and thermodynamic properties of plasmas formed of ablated material of circuit breaker walls were already published [1-2]. In these papers, the equilibrium composition was obtained using the well-known Gibbs free enthalpy minimisation technique. In our work, the calculations of equilibrium compositions of Air – *PA6,6* – Cu mixtures in the temperature range 300 K – 30000 K have been performed with a method based on the mass action law and the chemical base concept described by Godin and Trépanier [3]. The basic monomer of *PA6,6* is $C_{12}H_{22}O_2N_2$. We took into account 103 chemical species.

With the equilibrium composition, it is possible to calculate the thermodynamic properties. The mass density ρ ($Kg.m^{-3}$) and the enthalpy H ($J.Kg^{-1}$) are directly obtained from the equilibrium composition and the specific heat at constant pressure Cp ($J.Kg^{-1}.K^{-1}$) is calculated as a numerical derivative of the enthalpy. The specific heat at constant pressure of Air and pure *PA6,6* plasmas are compared in fig. 1a. In the case of air, the four peaks located at 3500 K, 7000 K, 15000 K and 30000 K are due respectively to the dissociation of O_2, the dissociation of N_2 and the first and second ionization of N and O atoms. For the 100% *PA6,6* plasma, the first peak situated at 1000 K comes from the dissociation of the C_2H_4 and CH_4 polyatomic molecules. The two following at 3700 K and 5000 K are due respectively to the dissociation of H_2 and the dissociation of C_2, C_3, CN and C_2H. The peak located around 7000 K comes from the dissociation of N_2 and CO. The two last at 15000 K and 26500 K are the consequence of the first ionization of carbon and hydrogen atoms and of the second ionization of carbon respectively.

The transport coefficients (viscosity, electrical and thermal conductivities) have been calculated by the Chapman-Enskog method [4] and we do not give any details here. Figure 1b shows the comparison of the thermal conductivity obtained for pure Air and pure *PA6,6* plasmas. The existence of peaks coming from the reaction thermal conductivity K_{reac} can be explain as for specific heat C_P. In the case of the *PA6,6* plasma, the increase of the peak located at 15000 K is due to elevated values of the collision integrals involving C and H species (e-C, e-H, e-C^+, e-H^+, C-C^+, H-H^+, ...). In the low temperature range, the main peak due to the dissociation of N_2 is situated near 7000 K in the case of air plasma while for *PA6,6*, the most important peaks coming from the dissociation of H_2 and C_2, C_3, CN and C_2H are located at lower temperatures ($T \approx 3700$ K and $T \approx 5000$ K respectively). This difference in the location of the low temperature peaks is very important. As we will see later, it will conduce to an increase of the cooling speed during the extinction of the arc in the case of a *PA6,6* plasma.

For radiation treatment, we have calculated the net emission coefficient by the method proposed by Lowke [5] assuming an isothermal plasma. The results are useful for calculating the temperature in the hottest regions of the plasma.

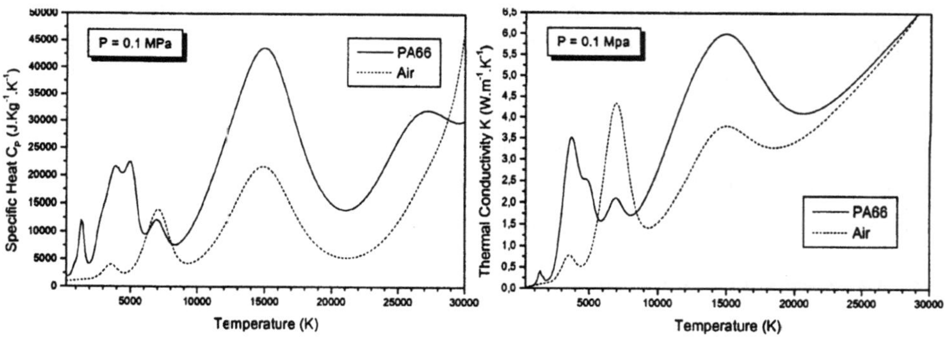

Figure 1a-b: Specific heat and thermal conductivity for air and PA66, at atmospheric pressure

3. BASES OF THE KINETICS MODEL

For the calculation of the chemical kinetics, several species were neglected because these species are minority, i.e. they are present in small quantities for temperature ranges going from 1000 K to 3000 K. The aim of the kinetics model is to calculate the composition of the plasma, i.e. the number densities of all the species taking into account the chemical reactions between the species. Let us consider the chemical process:

$$A + B \underset{k_r}{\overset{k_f}{\Leftrightarrow}} C + D \qquad (1)$$

k_f and k_r are the direct and reverse reaction rates. The main assumptions of our kinetics model are: 1) Maxwellian energy distribution functions; 2) thermal equilibrium. With these conditions the reverse reaction rates can be deduced from the direct ones by microreversibility. We took into account 52 species and 114 chemical reactions, the details and reaction rates of them being given in [6]. The main difficulty of the study is the determination of the reaction rates, that was made by bibliographic analysis and by particular calculations.

Special emphasis was carried out to the calculation of ionisation and recombination rates for atomic atoms and ions. Indeed, in thermal or quasi-thermal plasmas, these processes must be considered globally as it is schematically shown in Fig. 2. The total mechanism of recombination describes that after a direct recombination between an atomic ion and an electron, an atom is created on an excited level. Owing to the high electron number density, the excitation or de-excitation frequencies are very high so that the bounded electron is moved through several transient excited states, until it reaches a stable state corresponding either to the ground state or to the ionisation. Thus, the global or total coefficients of ionisation and recombination are related to the transition rate of the bounded electron through the excited states of the atom. This method needs a collisional radiative model. In the case where departures from equilibrium are not very important, we can develop an approximate method called "bottleneck" model [7] based on the fact that there exists a minimum on the total rate of de-excitation of atoms for a particular transition between two adjacent levels. Figure 3 shows results for C atoms.

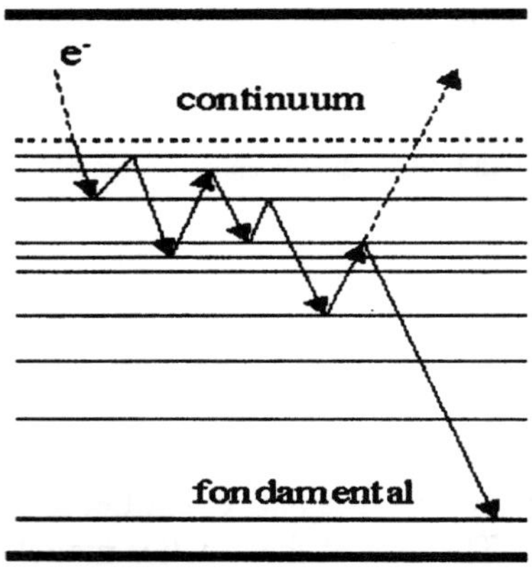

Figure 2: principle of global ionisation and recombination

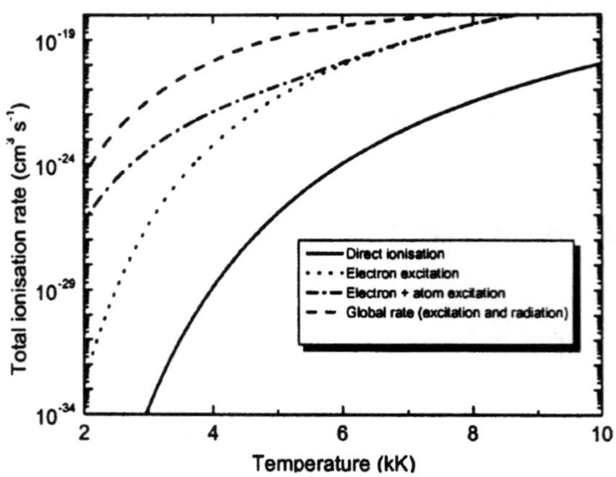

Figure 3: Ionisation rate for carbon atoms

4. TOTAL 1D MODEL OF A DECAYING ARC

4.1 Hypotheses

The aim of the model was not to describe exactly the behaviour of the arc in a real circuit breaker, but to study the possible role of departures from equilibrium and of the composition in a situation rather analogous with that of a LVCB, i.e. regarding the extinction of an arc. We have thus considered a simplified 1D model where the hydrodynamic part corresponds to the extinction of a wall stabilised arc [8] but coupled with the kinetics model. The main hypotheses are:

- Plasma medium made by $C_{12}H_{22}O_2N_2$ or air with or without metallic vapours.
- The plasma is considered to be in thermal equilibrium, so only one temperature is calculated, but departures from chemical equilibrium can exist.
- A cylindrical symmetry exists and only radial convection occurs.

- The fluid is laminar and the flow is generated during the arc cooling to compensate the pressure variation.
- Diffusion and demixing phenomena are neglected, what supposes that the velocities of the chemical processes are preponderant.
- In the transient state model, the electric field is assume to be equal to zero.

4.2 Equations

The species conservation:

$$\frac{\partial n_i}{\partial t} + \frac{1}{r}\frac{\partial}{\partial r}(n_i v_r) = S_i \qquad (2)$$

n_i represents the number density of species i, v_r the radial velocity and S_i the net production rate of the species by inelastic collisions (chemical kinetics). This equation multiplied by m_i and summed to all the other species conservation equations allows to represent the mass conservation equation.

The radial momentum equation

$$\rho\frac{\partial v_r}{\partial t} + \rho v_r \frac{\partial v_r}{\partial r} = \frac{2}{r}\frac{\partial}{\partial r}(\mu r\frac{\partial v_r}{\partial r}) - \frac{\partial P}{\partial r} - 2\frac{\mu v_r}{r^2} \qquad (3)$$

where ρ represents the total mass density and μ the viscosity.

The energy conservation equation

$$\rho C_p \frac{\partial T}{\partial t} + \rho C_p v_r \frac{\partial T}{\partial r} = \frac{1}{r}\frac{\partial}{\partial r}(r\kappa\frac{\partial T}{\partial r}) - S_r \qquad (4)$$

S_r represents the source term of the energy equation. It only includes the radiation losses through the use of the net emission coefficient. The interruption capability is represented by the evolution of the conductance:

$$G = 2\pi \int \sigma r dr \qquad (5)$$

5. RESULTS AND CONCLUSION

Among the lot of results, the main interesting ones are the following:

Temperature evolutions.

Figure 4 show the transient temperature evolution in air and in *PA66* vapour (the role of the presence of 1% Cu vapour is negligible in this evolution). We can observe a stronger cooling in the case of *PA66*, due to the thermal conductivity peaks at low temperature (see Fig 1b).

Departures from equilibrium. These departures remain rather weak, with a local overpopulation of electron always inferior to a factor of 2.

Interruption capability. This property is represented in Fig. 5 by the transient evolution of the conductance:

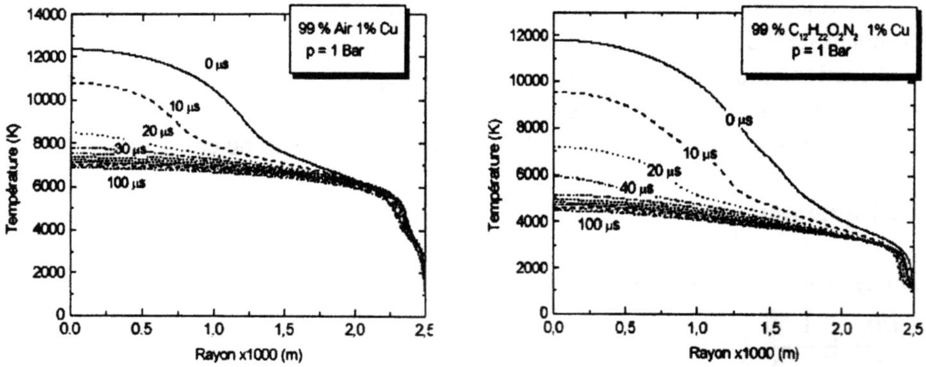

Figure 4 : Transient evolution of temperature in air (left) and *Pa6,6* (right)

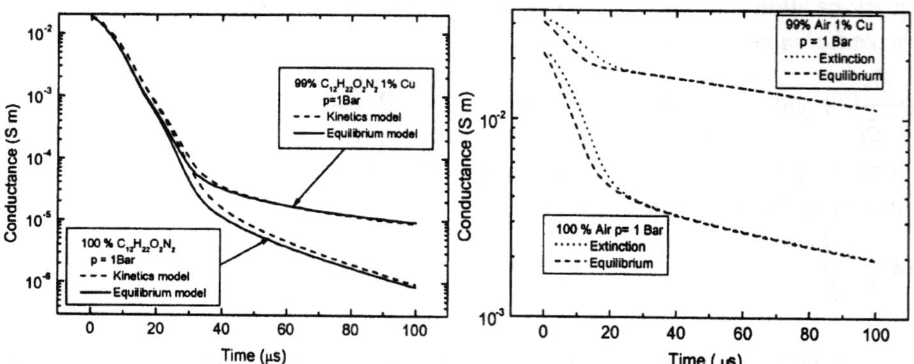

Figure 5: Transient evolutions of conductance for various conditions

The main conclusions of these results are: the departures from equilibrium are almost negligible; the influence of the plasma composition (presence of organic or metal vapours) on the interruption capability is very important.

Acknowledgements: Partial financial support of Schneider Electric is gratefully acknowledged.

References

1. P. André *J. Phys. D*, **30**, 475 (1997)
2. P. André *J. Phys. D*, **29**, 1963 (1996)
3. D. Godin and J.Y. Trépanier *Proceedings ISPC14*, vol. 1 p. 239, Praha, Czech Republic, August 2-6 (1999)
4. J.O. Hirschfelder, C.F. Curtis and R. Byron Bird, *Molecular Theory of Gases and Liquids*, John Wiley and Sons, New York (1964)
5. J.J. Lowke *JQSRT*, **14**, 111 (1974)
6. A. Mercado, Thèse de l'Univerité Paul Sabatier (2003)
7. E. Hinnov, J. G. Hirschberg *Phy. Rev.* **125** 795-801 (1962)
8. B. Chervy, JJ. Gonzalez, A. Gleizes IEEE Trans. Plasma Sc., **24**, 210 (1996)

SECTION 3: PARTIAL DISCHARGES: MECHANISMS AND APPLICATIONS

PARTIAL DISCHARGE AND ITS APPLICATION TO GAS INSULATED SWITCHGEAR

Tokio YAMAGIWA and Fumihiro ENDO *

1. INTRODUCTION

Higher reliability is demanded with gas insulated apparatuses such as gas insulated switchgear (GIS) and gas insulated circuit breaker (GCB). Since gas insulated apparatuses are sealed structures, it is difficult to find internal defects from outside. Therefore, development of preventive maintenance techniques is necessary for GIS to prevent failures and to detect defects early. Detection of partial discharge (PD) is one effective means for this purpose[1]. PD signals in SF_6, especially, have radio frequency components of high and wide bandwidth in the UHF region, which is more than a few hundred MHz. We have developed a UHF based GIS diagnostic system, which includes a neural network for improved and accurate fault assessment[2].

UHF partial discharge monitoring system (UHF PDM) has become popular for PD diagnosis of GIS in recent years, and has been regulated as one of the PD measuring methods in IEC-60270. Many experiences are reported to clarify its performance[3]. For accurate diagnosis, extremely high sensitivity of sensors and proper algorithms for PD assessment are necessary. Moreover, PD is measured in dBm units by UHF PDM, which is quite different from the conventional pC unit (apparent charge). The conversion between dBm and pC is demanded to help understanding of PD severity.

The present paper describes partial discharge characteristics in GIS, diagnostic techniques using the UHF method, sensitivity and calibration of UHF internal and external sensors. Furthermore, UHF PDM is experimentally compared with a conventional PD measuring system described in IEC 60270.

* Tokio YAMAGIWA, Japan AE Power Systems Corporation, Ibaraki, Japan 316-8501.
Fumihiro ENDO, Hitachi, Ltd., Ibaraki, Japan 319-1221.

Gaseous Dielectrics X.
Edited by Christophorou *et al.*, Springer, New York, 2004.

2. DIAGNOSTIC TECHNIQUES USING UHF METHOD FOR GIS

Dielectric breakdowns that are almost fatal to GIS are often preceded by PDs. Thus, techniques of detecting and diagnosing PDs are employed as forecasting maintenance techniques. Recently, attention has been directed to the development of techniques that not only detect but also locate PDs.

The main causes of insulation abnormalities are protrusions of electrodes, depositions on insulators, voids in insulators and poor contact, as well as free metallic particles. Insulation abnormalities due to such defects generate PDs of particular levels, phase and frequency.

One of the methods for detecting PDs with the highest sensitivity is the UHF method, is the detection of electromagnetic waves in the GIS.

A sensitive detection of PDs is possible in the GIS with a high S/N ratio by analyzing the electromagnetic waves having frequencies above several 100MHz that are largely free of external noise. Figure 1 shows the basic measuring system and typical measurement results[1]. A sensor for detecting electromagnetic waves is installed in the GIS and a spectrum analyzer is used to check the detected signal by frequency analysis. While electromagnetic waves due to external PDs (400pC) from a conductors in the air are within a frequency range up to a little above 300MHz ((1)), PDs in the GIS are in the frequency range above 600MHz ((2)). If internal and external PDs are concurrent, the spectrum pattern obtained is composed of (1) plus (2), as shown in (3), and detection can be performed by observing the high-frequency zone of the spectrum in the UHF band.

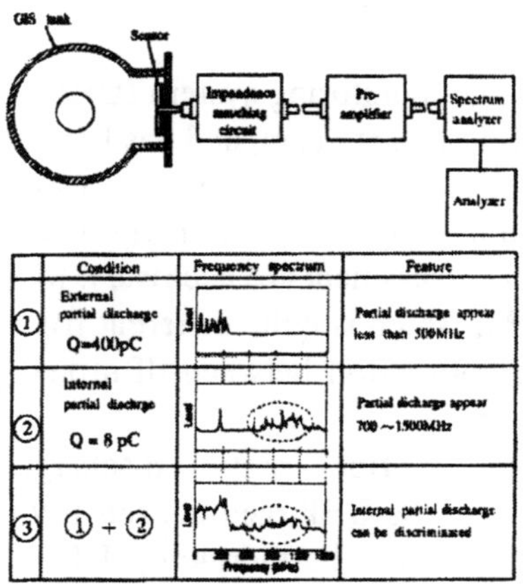

Figure 1. Detecting method of small partial discharge in GIS by frequency spectrum analyzer (UHF method)

3. ESTIMATION OF ABNORMALITY TYPE THROUGH PATTERN

Once PD occurs in a GIS the inherent danger they pose to the GIS depends on their location and cause of the PD. For prevetative maintenance it is important to know the reason and location of a PD, and the PD measuring circuit should be able to discriminate from internal generated signals due to PDs and external generated signals.

3.1. Pulse-phase and Spectrum of Particle-initiated Partial Discharge

It is also proposed to attempt a more accurate diagnosis by acquiring a database on PDs from phase analyses and spectral analyses of PD pulse waveforms simultaneously[4]. With the causes of PDs confined to a metallic needle like particle and its location varied, The PD phase patterns and PD pulse spectrum patterns were measured. The results are shown in Figure 2. The phase patterns agree with the conventional results. The spectrum patterns have main component at around 300, 800 and 1300MHz.

If the particle is fixed on the spacer surface, the spectrum pattern component at around 1300MHz is stronger than at around 800MHz. If the particle is fixed on the high-voltage electrode or grounding electrode, the component at around 800MHz is stronger than at around 1300MHz. A free metallic particle caused intermittent PDs nearly 100 times stronger than those caused by a fixed metallic particle. It was also found that peaks of spectrum patterns were proportional to the pulse intensity under the conventional PD measuring method discribed in IEC 60270.

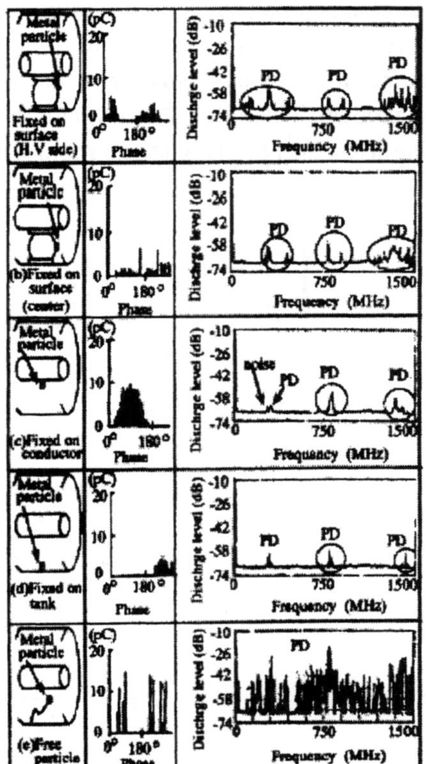

Figure 2. Pulse-phase and spectrum of particle-initiated partial discharge.

3.2. Time Dependence of Particle-initiated Partial Discharge

The model gas insulated bus bar (coaxial cylinder electrode) of $\phi 15 / \phi 60$

mm diameter was used. Long duration V-t characteristics are shown in Figure 3. The phase patterns of the typical partial discharge before breakdown are also shown in Figure 3. V-t characteristics are expressed by the equation (1).

$$V = V_0 \cdot t^{-1/n} \qquad (1)$$

Values of n were obtained and listed in Table 1. For the protrusion, these values ranged between 79-160 and for the spacer surface with particle n=25-51. V-t characteristics curve of spacer surface indicates steeper inclination compared with that of conductor.

In the case of a protrusion, the phase patterns of PD do not depend for at voltage applied time, but those are an almost fixed pattern. On the other hand, in the case of a spacer surface with particle, although generating phases of PD do not change, pattern and magnitude of PD change as voltage applied time. Moreover, there is a tendency for polarity difference to become larger just before breakdown although polarity difference is very small when it is not near the breakdown. Therefore, it is expected that these phenomena are applicable to diagnosis and the risk assessment for GIS.

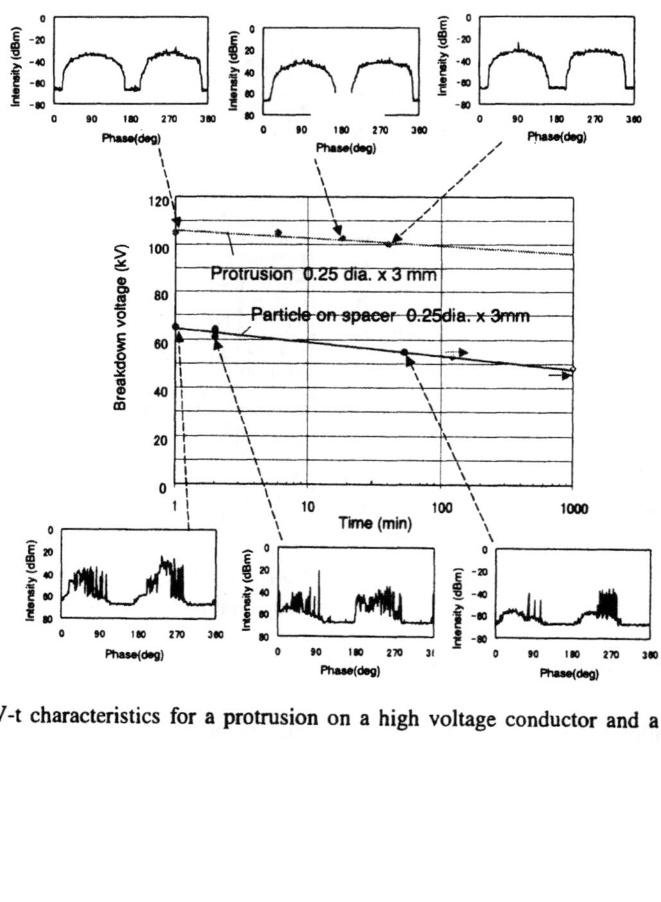

Figure 3. V-t characteristics for a protrusion on a high voltage conductor and a particle on a spacer.

Table 1. n value of V-t characteristics for various defects.

Defect	Protrusion on conductor		Spacer surface with particle	
Shape	φ0. 25 x3mm	φ0. 25 x5mm	φ0. 25 x3mm	φ0. 25 x5mm
n value	79	158	25	51

4. HIGHLY SENSITIVE UHF SENSOR

A high sensitivity UHF sensor is required to detect PDs in detail by the UHF method. The UHF sensor used for this measurement is explained below.

4.1. Various Antennas

A high sensitive UHF sensor was developed. Various antennas (disc, dipole, log-period, spiral, and so on) were investigated from view of the sensitivity, frequency characteristics, and antenna structure.

As shown in the result of Table 2, the semicircular dipole antenna proofed to have be most suitable property in almost all comparison items including sensitivity.

Table 2. Various antennas for UHF method.

shape					
antenna	disk	semicircular dipole	monopole	logarithmic periodical	spiral
sensitivity	○	◎	◎	◎	◎
freq. band	◎	◎	×	◎	◎
directivity	◎	○	○	○	○
size(small)	○	○	○	×	×
assembly	◎	◎	×	×	×
Evaluation	○	◎	×	×	×

◎: Excellent, ○: Good, × :NG

4.2. Internal UHF Sensor

The new internal UHF sensor consists of two half-disc plates (semicircular dipole antenna in Table 2). Their size and arrangement were optimized based on dipole antenna technology. Sensitivity of the new UHF sensors is shown in Figure 4. This sensitivity confirmed to be 0.3 pC (apparent charge) at S/N = 3 and

0.1 pC at S/N = 1. The dBm value was measured by spectrum analyzer. Signal Intensity dBm define the maximum amplitude of frequency region from 500 MHz to 1500 MHz. Noise level define the magnitude of back ground noise which consist of white noise of measuring apparatuses and/or external noise. The pC value was measured at the same time with a conventional method according to IEC-60270.

4.3. External UHF Sensor

The external UHF sensor was designed to be fit around the outer surface of a spacer. UHF waves emitted from PDs leak through the outer edge of a spacer and its intensity is usually small, the structure of the sensor was designed based on the antenna theory to obtain high sensitivity. The developed sensor has the sensitivity of 2 pC at S/N = 3 as showing in Figure 4.

4.4. Relationship Between pC and dBm

PDs are usually measured using a conventional instrument based on IEC-60270, and estimated by the amplitude of PD charge (pC). Since measurements are made for a low frequency band of several 100 kHz to several MHz in the case of the conventional method, we must prevent the penetration of unwanted UHF noise signals. Although this is possible in a fully shielded laboratory, it is difficult in actual substations. However, the UHF PDM method, which has excellent sensitivity and S/N ratio, now serves as the global standard.

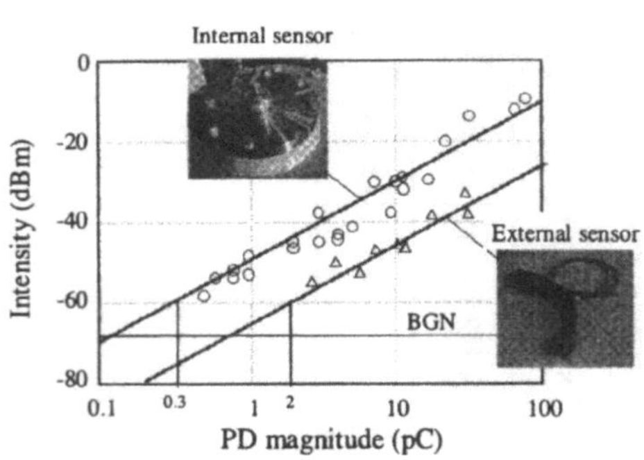

Figure 4. Sensitivity of developed UHF sensor.

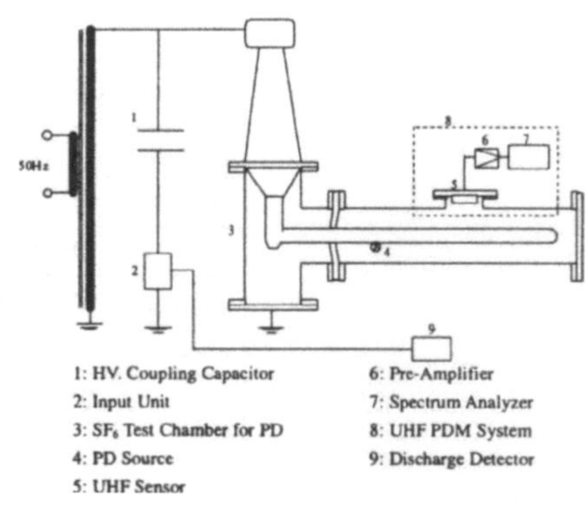

Figure 5. Measuring circuit.

(1) Experimental method

Figure 5 shows the measuring circuit for pC – dBm conversion of PD magnitude. Various PD sources were placed in the coaxial cylinder electrodes of 60/250 mm diameter.

Both the discharge detector by conventional instrument based on IEC-60270 and the UHF PDM system were connected to the calibration circuit. PD sources included protrusion, free metallic particles, particle on spacer, void in spacer, etc. PD signals were measured simultaneously by both measuring systems.

(2) Experimental Results

PDs were generated using typical PD sources in SF_6 insulation GIB. PDs were measured simultaneously by both detectors. Figure 6 shows examples of measured results for free metallic particles. Figure 6 (a) and (b) show frequency spectrum and phase spectrum measured by UHF PDM system, and (c) and (d) show phase(ϕ) characteristics, and ϕ –q characteristics measured by the conventional system. Comparing (b) and (d), it can be observed that the phase characteristics are almost identical. In this case, the magnitude, q, of the PD pulse was 17.7 pC by the conventional method, and –27.6 dBm by the UHF PDM system.

Figure 7 plots the magnitude q(pC) of PD charges by the conventional method and the maximum intensity (dBm) of 400 - 1500 MHz by the UHF method. In Figure 7, a remarkable difference for various defects was not observed. Even for different kinds of defect, measured data converges well to a single curve. This curve can be expressed by equation (2).

$$q = 10^{(x+50)/20} \qquad (2)$$

That is, pC is easily converted from the measured dBm value from equation (2).

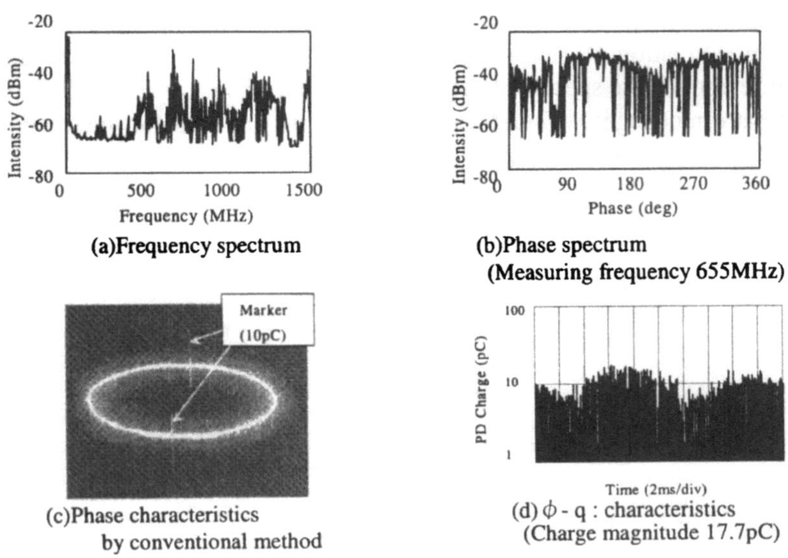

(a)Frequency spectrum

(b)Phase spectrum (Measuring frequency 655MHz)

(c)Phase characteristics by conventional method

(d) ϕ - q : characteristics (Charge magnitude 17.7pC)

Figure 6. Typical PD characteristics in free particle case.

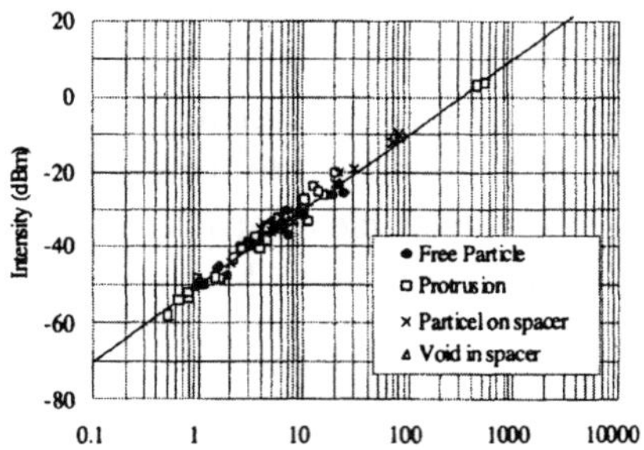

Figure 7. pC-dBm characteristics for various defects.

5. DISCRIMINATION OF PD SOURCES BY NEURAL NETWORKS

5.1. Evaluation of Insulation Diagnosis Algorithm

Because we need to make an insulation diagnosis in GIS, it is necessary to evaluation various algorithm techniques. As for the general pattern recognition technique, examples include neural network, finger print method, fuzzy system, wavelet and so on. There are merits and demerits to each technique. From comparison of various algorithms used, it was proved that the neural network was the best in point of examination precision, ability of knowledge acquirement, computation time and so on. Below, we used it to examine a neural network. The structure concept is shown in Figure 8. As for the PD signal, which was gotten with the UHF sensor set in GIS, the signals are analyzed in the frequency domain and the voltage phase domain by the spectrum analyzer. Insulation diagnosis in GIS such as the existence or non-existence of defects, identification of kind of defects, and estimation of lifetime, are possible by the neural network using the above. For example, the spectra of various defects are shown in Figure 9. Since a spectrum pattern changes with defects, these patterns are used for judgment of a defective using the neural network.

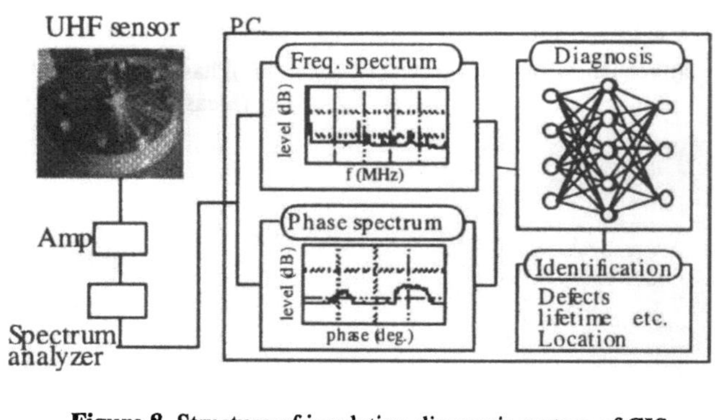

Figure 8. Structure of insulation diagnosis system of GIS

5.2. Identification of Kind of Defects by Neural Net work

A sufficiently correct answer percentage was not achieved when calculating from the existence or non-existence of defects to judge the causes of detects by the standard neural network. As a result of investigation of various examination conditions, the method of analyzing with a branching neural network was able to judge the defect with a high correct answer probability. An outline is shown in Figure 10.

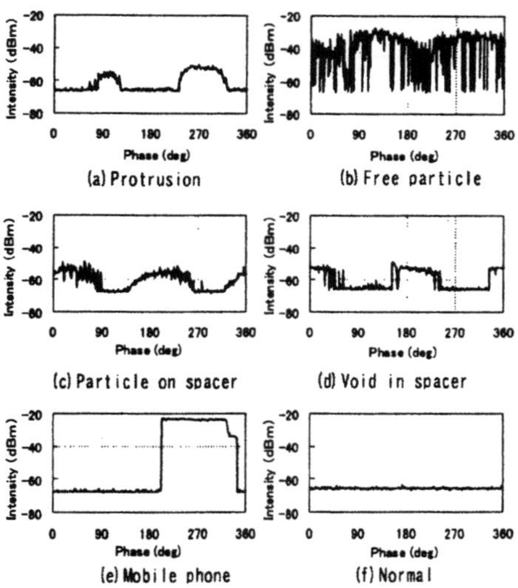

Figure 9. Phase spectra for various defects.

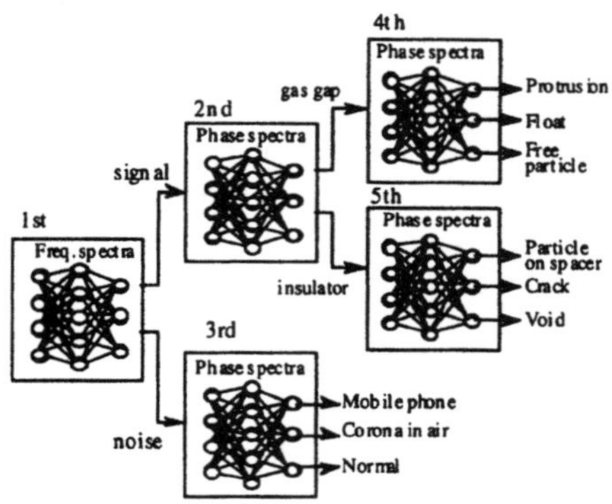

Figure 10. Diagnosis method of branch structure of neural network.

Judgment is made by division to five neural networks. A rough judgment, normal or abnormal, using frequency spectrum is done in the first step. The reason for using the frequency spectrum utilizes the fact that separation of the normality, the noise, and the inner defects differs in the frequency band where each occurs. The second step judge kind of defects when judged to be abnormal in the first step. In the second step, it judges three kinds of discharges in gas, free particle, or defects of spacer by using the phase spectrum. Moreover, the third step, the fourth step, and the fifth step judge 1 in detail. A diagnosis result by the branch structure of neural network mentioned from Figure 10 is shown in Table 3. The correct answer percentage of the first step was 48 - 91 %. Mistaken answer was limited when the PD signal was near the PD inception voltage. It occurred for the case of a pattern that was difficult to judge even for an expert. The finally correct answer percentage was equal to or more than 90 %. Diagnosis algorithm had developed functions sufficiently, as it judged the noise of a mobile phone and so on was not PD.

Table 3. Rate of diagnosis accuracy of various defects.

	diagnoses by freq. and phase		cor. rate by freq.
	kinds of defects	cor. rate	
existence of PD	protrusion on cond.	90%	91
	particle on tank	100%	
	floating particle	93%	
	free particle	100%	75
	particle on spacer	99%	48
	crack of spacer	92%	
	void in spacer	90%	
	floating electrode	100%	85
normal or noise	corona in air	100%	
	mobile phone	100%	ND*
	normal	100%	90

* not diagnosed

6. CONCLUSIONS

(1) UHF method proves to be very effective means to detect partial discharges in GIS.

(2) We developed two new sensors, in order to carry out exact diagnosis. One was an internal sensor for new GIS and its sensitivity was 0.3 pC. The other was an outer sensor for existing GIS, whose sensitivity was 2 pC.

(3) We clarified the calibration method of the UHF sensor and relation between pC and dBm. Moreover, we introduced the developed diagnosis system using the neural network. The Rate of diagnosis accuracy equal to or more than 90 %.

7. REFERENCES

1. T. Yamagiwa, H. Yamada, F. Endo, Y. Oshita, S. Izumi and I. Yamada, Development of Preventive Maintenance System for Highly Reliable Gas Insulated Switchgear; *IEEE Trans. On Power Delivery, Vol.6, No.2*, June (1991).
2. T. Kato and F. Endo, Development of UHF Insulation Diagnosis System of GIS; *T. IEE Japan, Vol.119-B, No.4*, April (1999).
3. CIGRE TF 15/33.03.05, Partial Discharge Detection System for GIS: Sensitivity Verification for the UHF Method and the Acoustic Method, *ELECTRA, No.183*, pp.75-87, (1999).
4. T. Utsumi, F. Endo and T. Yamagiwa, Partial Discharge Diagnosis by Spectrum Analysis and Pulse-Phase Analysis, *Proceedings of the 4th Annual Conference Power & Energy Society (1993), IEE of Japan Paper No.598* (in Japanese).

INFLUENCE OF THE POWER FREQUENCY ON PARTIAL DISCHARGE PHENOMENA

Sander Meijer, Johan J. Smit, Alain Girodet[1]

1. INTRODUCTION

For commissioning testing of GIS, a voltage level around 0.8 times the power frequency withstand level (PFWL) is applied according to IEC 60517. This voltage level is applied at least for 1 minute and longer in case of pre-conditioning when partial discharge tests are combined with power frequency tests. Otherwise, partial discharge measurements are performed at different voltage steps until the maximum test voltage is reached. In any case leads the test procedure to a voltage level much higher than the maximum operating voltage of voltage measurement transformers (VT's). As a result, the core will saturate during these test over-voltages, giving rise to high currents flowing in the windings of the VT. If the time such a current flows is too long a thermal failure of the voltage transformer can occur. Therefore, voltage measurement transformers are disconnected during commissioning testing.

The magnetic flux density is proportional to the applied test voltage, but moreover, also inversely proportional to the power frequency. So, application of a higher power frequency results in a lower magnetic flux density and no saturation takes place at higher test voltages. A short theoretical approach is given in section 2.

However, performing tests at higher power frequencies should have no influence on the results of the PD measurements, which are often performed simultaneously. In this paper, three typical insulation defects are investigated. The measuring circuit is described in section 3. Results obtained from a void, a free particle and a fixed protrusion are described in section 4.

---

[1] Sander Meijer and Johan J. Smit, Delft University of Technology, High Voltage Technology and Management, Delft, The Netherlands.
Alain Girodet, Alstom T&D, Villeurbanne Cedex, France.

Gaseous Dielectrics X.
Edited by Christophorou *et al.*, Springer, New York, 2004.

2. THEORY

From the second law of Maxwell (law of Faraday), the relation of the magnetic flux density and the applied test voltage can be deduced:

$$\begin{cases} \oint_C \vec{E} \cdot d\vec{l} = -\dfrac{d\phi_m}{dt} \\ \phi_m = \int_S \vec{B} \cdot \hat{n} dA \end{cases} \tag{1}$$

From (1) follows

$$\begin{cases} V = -\dfrac{d\phi_m}{dt} \\ \phi_m = NBA \end{cases} \tag{2}$$

If we assume the following voltage V:

$$V = \hat{V} \sin \omega t \tag{3}$$

then we can calculate the magnetic flux density B by integrating the voltage:

$$\begin{cases} \phi_m = NBA = \int \hat{V} \sin \omega t dt = -\dfrac{\hat{V}}{\omega} \cos \omega t \Leftrightarrow \\ B = \dfrac{\phi_m}{NA} = -\dfrac{\hat{V}}{NA\omega} \cos \omega t = K \dfrac{\hat{V}}{\omega} \cos \omega t \end{cases} \tag{4}$$

From (4) can be concluded that the magnetic flux density is proportional with the amplitude of the applied test voltage and inversely proportional with the power frequency. The current through the windings of the transformer can be calculated from the first law of Maxwell (law of Ampere):

$$\oint_C \vec{H} \cdot d\vec{l} = I \tag{5}$$

So, the current only depends on **H**. As known, the magnetic field and magnetic flux density are related by the magnetization curve. Normally, the VT is operating in the linear part of this curve. However, due to the increase of **B**, the core can get saturated resulting in a large increase in **H** and thus in the current, which can harm the transformer. To keep the maximum magnetic field constant, the maximum magnetic flux density also has to be constant. As shown before, this can be achieved by increasing the power frequency and the voltage pro rata.

3. MEASURING SET-UP

For proper judging the PD measuring results obtained at higher power frequencies, the influence of the power frequency on the following partial discharge phenomena has to be investigated:

1. Inception voltage level (U_{inc});
2. Extinction voltage level (U_{ext});
3. PD magnitude;
4. Phase-resolved PD pattern.

For this purpose the following measuring set-up has been used, see figure 1. The measuring set-up comprises the following parts: a wave generator, power amplifier, variable resistor and inductance, transformer and coupling capacitor.

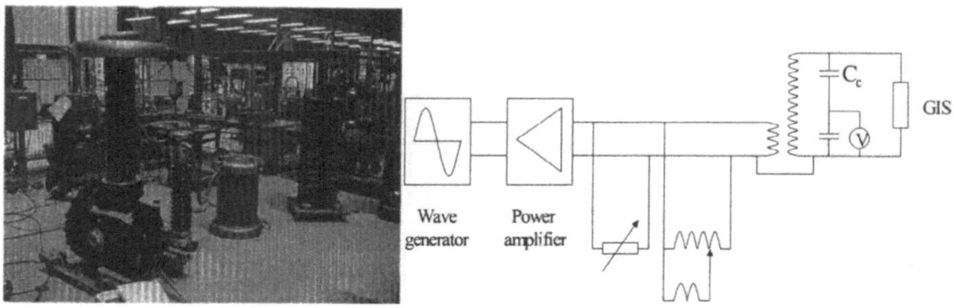

Figure 1: Photograph and line scheme of the measuring set-up.

To measure the inception and extinction voltage level, the PD magnitude and the phase-resolved PD patterns, a conventional measuring circuit was used, according to the IEC 60270 recommendations.

4. MEASURING RESULTS

In this section the following types of defects have been investigated, see figure 2: a void in polyester resin, a free particle and a fixed protrusion. The defects have been tested at different power frequencies (50 Hz, 120 Hz and 200 Hz).

4.1 Void in polyester resin

To simulate a void in GIS, a void in polyester resin (with ϵ_r=3.45) has been inserted inside the test vessel. The test vessel was filled with 4.3 bar pure SF_6. Phase-resolved PD patterns have been obtained at above-mentioned frequencies.

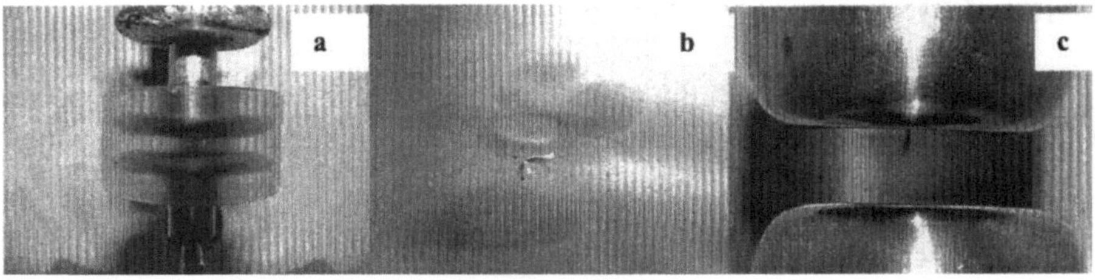

Figure 2: Details of the investigated defects:

a	void in polyester resin
b	free particle
c	fixed protrusion

Table 1 shows the influence of the power frequency on the main PD quantities of interest.

Table 1: Influence of the power frequency on the basic PD quantities: inception and extinction voltage and PD magnitude.

	U_{inc} [kV]	U_{ext} [kV]	PD magn. [nC]
50 Hz	14.9	10.4	26
120 Hz	15.0	10.6	32
200 Hz	15.0	10.3	27

From the phase-resolved PD patterns shown in figure 3 no clear differences can be concluded. So the test frequency has no influence on the main PD quantities and the shape of the PD pattern[2].

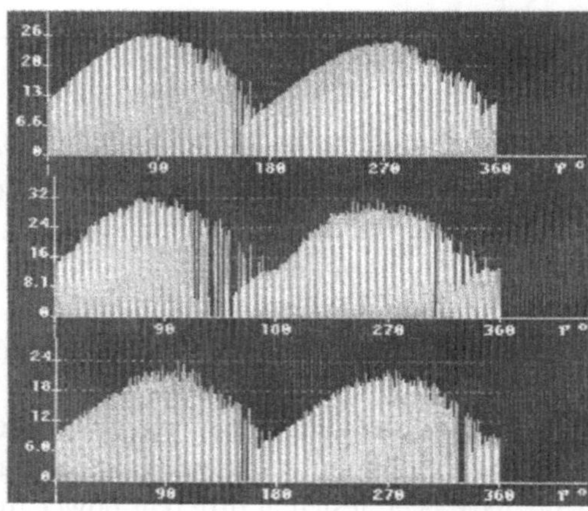

Figure 3: The influence of power frequency on the PD pattern measured at 1.2 times the inception voltage: a) at 50 Hz, b) at 120 Hz and c) and 200 Hz.

4.2 Free particle

A free particle has been inserted in the test vessel, see figure 2b. Pure SF_6 pressurized to 4.5 bar has been used as insulating gas.

Due to the fact that the inception voltage (start of particle movement) strongly depends on the particle geometry, the location of the particle and the electric field, it is difficult to determine this voltage. Therefore the inception voltage is determined by averaging the results of five experiments, see table 2. The PD magnitude at 20 kV is also shown. It can be concluded that the power frequency has no significant influence on the main PD quantities.

Table 2: Influence of the power frequency on the basic PD quantities: inception and extinction voltage and PD magnitude.

	U_{inc} [kV]	U_{ext} [kV]	PD magn. [pC]
50 Hz	17.8	15.8	18
120 Hz	16.8	14.1	18
200 Hz	18.9	14.0	20

Phase-resolved PD patterns have been obtained at above-mentioned frequencies. Figure 4 shows the patterns obtained at different frequencies, measured at 18 kV. In order to reduce the probability for breakdown, the patterns have only been measured slightly above the inception voltage.

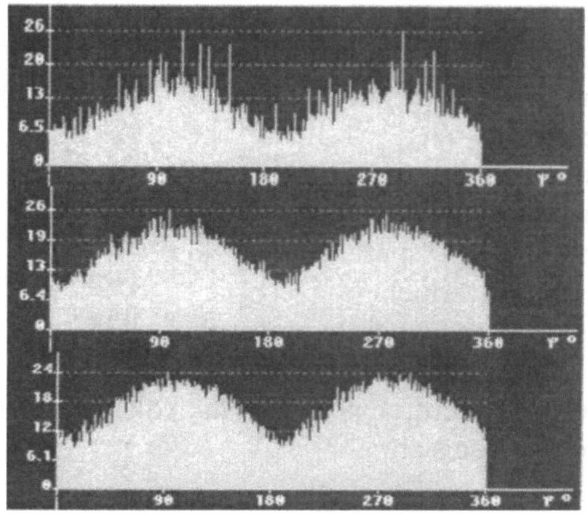

Figure 4: The influence of power frequency on the PD pattern measured at 1.2 times the inception voltage: a) at 50 Hz, b) at 120 Hz and c) and 200 Hz.

No significant difference in the PD patterns can be concluded from figure 4. It shows that the frequency has no influence on the shape of the pattern. However, because a fixed measuring time (2 minutes) has been applied, the number of measured periods increases with increasing frequency. As a result, the number of detected pulses is different and increases also with increasing frequency. Therefore more smooth patterns are obtained at higher power frequencies.

4.3 Fixed protrusion

A protrusion fixed to the high-voltage conductor has been inserted in the test vessel, see figure 2c. The protrusion was 5.7 mm long and the distance between both electrodes was 14 mm. Pure SF_6 at 4.5 bar has been used as insulating gas.

Table 3: Influence of the power frequency on the basic PD quantities: inception and extinction voltage and PD magnitude.

	U_{inc} [kV]	U_{ext} [kV]	PD magn. [pC]
50 Hz	17.8	17.3	10
120 Hz	17.9	17.2	13
200 Hz	17.9	17.3	12

Table 3 shows the PD inception and extinction voltage levels and the maximum PD magnitude. No significant differences in these PD quantities can be concluded. Figure 5 shows the patterns obtained at different frequencies. In this case, some clear differences in the patterns can be seen. In particular, the early inception of discharges around the positive peak of the sine wave if higher power frequencies are applied. This is confirmed by the results described in[2]. It can be explained by the influence of space charges. It takes some time for the space charges to disappear, which is a constant factor in the whole process. So at higher frequencies, not all electrons are disappeared at the voltage reversal and this means that discharges are easier to start.

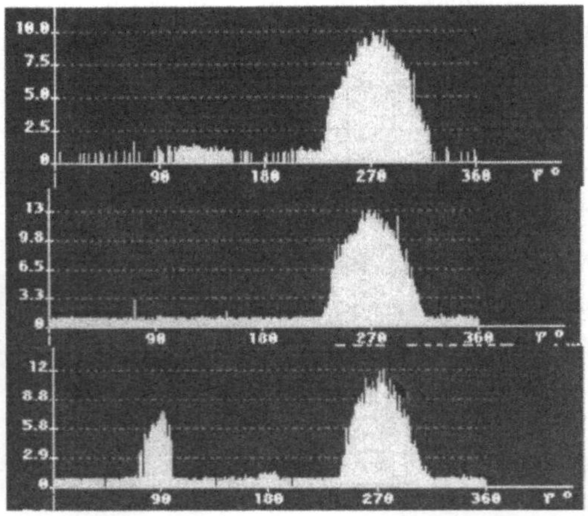

Figure 5: The influence of power frequency on the PD pattern measured at 1.2 times the inception voltage: a) at 50 Hz, b) at 120 Hz and c) and 200 Hz.

5. CONCLUSIONS

In this paper the influence of the power frequency on the basic PD phenomena are investigated. The following can be concluded:

1) The power frequency has no significant influence on the PD inception and extinction voltage and PD magnitude.
2) Except for the fixed protrusion, there is no significant influence of the power frequency on the PRPD patterns.
3) Due to space charge phenomena in case of fixed protrusions, inception of partial discharges at the positive peak occurs sooner with increasing frequency (less time for the space charge to spread out).

6. REFERENCES

1. R. Bodega, A. Cavalini, P.H.F. Morshuis, F.J.Wester, *The Effect of Voltage Frequency on Partial Discharge Activity*, 2002 Annual Report CEIDP, pp. 685-689.
2. T. Takahashi, T. Yamada, N. Hayakawa, H. Okubo, *Power Frequency Dependence of Space Charge Behavior and Partial Discharge Characteristics in SF_6 Gas*, IEEE Tr. on DEI, Vol. 7, No. 1, pp. 152-160.

TIME-DEVELOPING PARTIAL DISCHARGE PROPERTIES LEADING TO SPARKOVER IN SF_6, N_2, CO_2 AND THEIR GAS MIXTURES UNDER LIGHTNING IMPULSE VOLTAGE APPLICATION

Shinya Ohtsuka, Kazuharu Kiyama, Mengu Cho,
Satoshi Matsumoto and Masayuki Hikita

1. INTRODUCTION

A lot of studies have been made on insulation properties of gas mixtures including or not including SF_6 gas, high-pressurized natural gas like N_2 or CO_2, gas-solid hybrid insulation system, all solid insulation system, and so on to seek SF_6 substitute [1-4]. Among them, CO_2 gas is focused on due to its relatively good insulation performance as well as arc quenching ability [5,6]. The authors have investigated insulation properties of CO_2 gas with and without a metallic particle in GIS [7]. Moreover, we have experimentally shown that adding CO_2 with a suitable ratio into SF_6/N_2 gas mixture, *e.g.* 1% for 50%SF_6/N_2 mixture, increases sparkover voltage under both uniform and non-uniform fields [8]. We measured partial discharge (PD) light emission properties up to sparkover with applying lightning impulse voltage. The experimental results revealed that sparkover is less likely to occur in the gas mixture including CO_2 gas, *e.g.* 50%SF_6/49%N_2/1%CO_2 mixture and the time elapsed from PD generation to sparkover is much longer as compared with those in the gas mixture not including CO_2 gas, *e.g.* 50%SF_6/50%N_2 mixture and SF_6 gas [9]. Additionally, we have investigated the mechanism of the increase in the sparkover voltage by adding CO_2 gas and tried to identify the gas species giving rise to light emission and discharge type of PD pulses in the gas mixture by optical observation [10].

From this point of view, in this paper we investigate PD properties leading to sparkover in SF_6, CO_2, N_2 gases and their gas mixtures, especially 50%SF_6/49%N_2/1%CO_2 gas mixture, by applying positive lightning impulse voltage to the needle to plane electrode system and changing the gas pressure up to 0.6 MPa. In particular, the sparkover properties of CO_2 gas are investigated in detail based on PD phenomena with a simultaneous measurement of PD light emission and current. From these results, we discuss the light emission gas species responsible for successive PD pulses in $SF_6/N_2/CO_2$ gas mixture and the effect of adding CO_2 gas on the sparkover properties.

2. EXPERIMENTAL SETUP

Fig.1 shows experimental setup to simultaneously measure PD phenomena, such as PD light emission intensity, still and streak images and time-resolved spectrum as well as PD current waveform. Electrode system used in the experiment is the same as before [8-10]: a needle to plane electrode with the gap length of 10 mm. The tip radius of the needle electrode is 0.5 mm, *i.e.* the field enhancement factor is 14.5. A positive standard lightning impulse (LI) (1.2/50 μs) was applied to the needle electrode to generate PDs and sparkover. Waveforms of PD current and PD light emission intensity are measured with a digital oscilloscope through an impedance matching circuit and a photomultiplier tube (PMT) [9,10], respectively. Still and streak images of PD light emission are measured with an image intensified CCD (ICCD) camera and a streak camera (Hamamatsu, C7700). Time-resolved PD light emission spectrum is measured with a spectroscope (Hamamatsu, C5094) and the streak camera.

Kyushu Institute of Technology, Sensui-cho, Tobata-ku, Kitakyus, 804-8550 Japan.

Gaseous Dielectrics X.
Edited by Christophorou *et al.*, Springer, New York, 2004.

These simultaneous measurements are carried out by controlling the gate time of ICCD and the streak cameras as well as the trigger timing *Tr* with a pulse generator (P.G.).

CO_2 and 50%SF_6/49%N_2/1%CO_2 gas mixture [8] were mainly used as tested gas. SF_6 and N_2 gas were also used in experiments as comparison. Sparkover voltage V_S was measured by applying a positive LI voltage according to the so-called step-up method with 2 kV step. From the results, 50% probability sparkover voltage V_{50} was calculated based on Weibull distribution function. Sparkover voltages under UV irradiation condition were also measured with a Hg-Xe lamp which has wavelength of 200-1000 nm and output power of 200 W.

3. RESULTS

3.1 Simultaneous Measurements of PD Phenomena

Fig.2 shows PD current and light emission intensity till sparkover in 50%SF_6/49%N_2/1%CO_2 gas mixture at 0.2 MPa. The results of the current or light emission intensity waveforms show that several PD pulses are generated until sparkover occurs [9,10]. The 2nd PD pulse is observed by PMT (light intensity, *L*) while the corresponding current is not seen in the figure. Similar results were also obtained in pure CO_2 gas. Fig.3 shows the results of two successive PD pulses observed by streak and still images and intensity waveform of PD light emission, and PD current in 50%SF_6/49%N_2/1%CO_2 gas mixture at 0.2 MPa. This figure indicates that the 2nd PD pulse observed by PMT occurs at the side of the needle electrode. The disappearance of PD current pulse corresponding to the 2nd PD light emission pulse can be interpreted in terms of the current flowing not through the electrode gap but dominantly through the path between the needle electrode side and the wall of the pressure chamber. As can be mentioned above, the simultaneous measurement allows us to identify the time-variant shape, position and intensity of PD light emission and the corresponding current for each PD event and the evolution of PD up to sparkover.

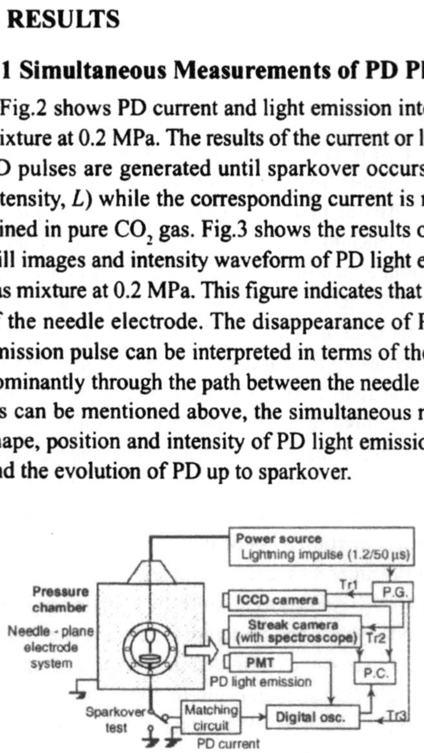

Figure 1. Experimental setup to measure PD phenomenon simultaneously

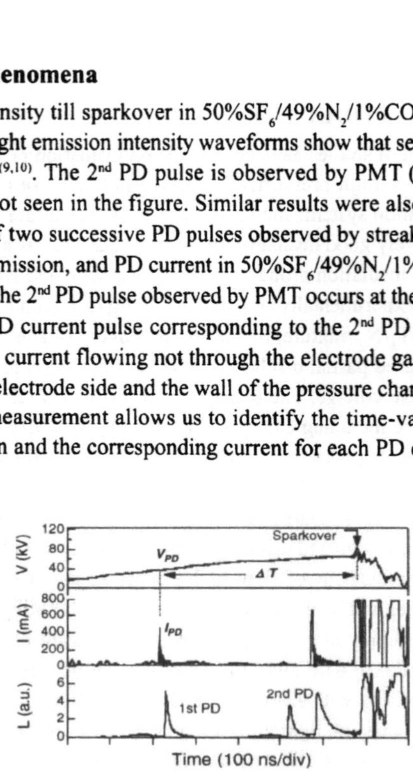

Figure 2. PD current and PD light emission intensity until sparkover in 50%SF_6/49%N_2/1%CO_2 mixture at 0.2 MPa.

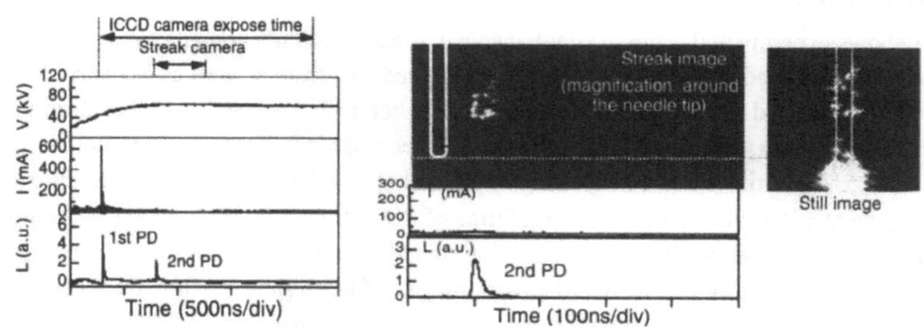

Figure 3. PD occurring from the side of the needle electrode: this can be confirmed by simultaneous measurement of streak and still images, and waveforms of PD light emission intensity and PD current in 50%SF_6/49%N_2/1%CO_2 gas mixture at 0.2 MPa.

3.2 Insulation Properties of CO_2 Gas under Non-Uniform Field

3.2.1 Conditioning Effect of Sparkover Voltage

Fig.4 shows test number dependence of sparkover voltage V_S in CO_2 gas at 0.6 MPa. Below 31 times of the test number, V_S slightly scatters around 40 kV (which is defined as V_L hereafter (State L)) while abruptly increases up to 60 kV. As the test number exceeds 31 times, V_S remains stable around 60 kV (state H). We call these characteristics of the change of V_S with the test number as conditioning effect . Since the sparkover voltage in CO_2 shows the conditioning effect, we define average values of V_S in state L and H as V_L and V_H, respectively. Fig.5 shows gas pressure dependence of an increment in V_S, *i.e.* $\Delta V_S = V_H - V_L$, to quantity the conditioning effect. From Fig.5, ΔV_S due to the conditioning effect increases with the gas pressure. Additionally, the sparkover voltage measured while depressurizing from 0.6 MPa (white square symbol) is larger than V_L (black circle) and V_H (white circle) measured while keeping the gas pressure constant. In other words, the sparkover voltage has hysteresis depending on the measurement procedure. Therefore, a special attention should be paid for V_S measurement in CO_2 gas with the gas pressure change. We hereafter refer to V_S measured while depressurizing from 0.6 MPa as V_H. Fig.6 shows V_S vs the test number properties in CO_2 gas at 0.6 MPa where some sparkover tests were performed with UV irradiation (white circle). We refer to V_S measured with UV irradiation as V_{UV}. From Fig.6, UV irradiation diminishes V_H to the initial value V_L before the conditioning effect emerges, *i.e.* $V_{UV} < V_H$.

3.2.2 Gas Pressure Dependence of V_{50}

Fig.7 shows a typical example of Weibull distributions of V_S given in Fig.4 and V_{UV} in CO_2 gas at 0.6 MPa. The sparkover voltages can be classified into two groups: V_{UV} and V_H , both of which obey Weibull distribution. Namely, V_L agrees well with V_{UV}, and V_S appearing in the state L with magnitude larger than V_L also agrees with V_H. Note that the shape parameters m for V_H and V_{UV} measured at each gas pressure were identical about 20. Fig.8 shows the gas pressure dependence of V_{50} for V_H and V_{UV} in

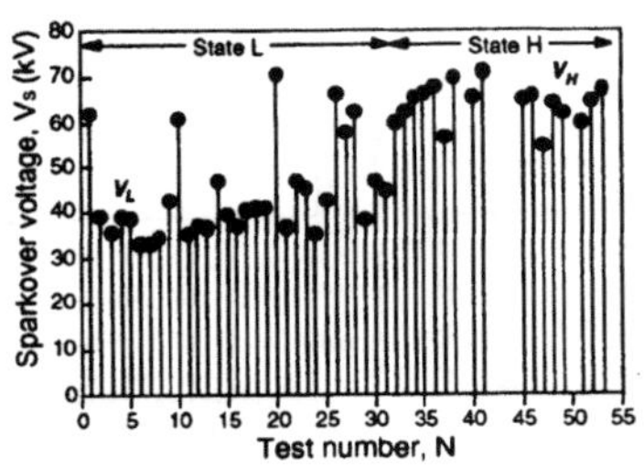

Figure 4. Conditioning effect of V_S in CO_2 gas at 0.6 MPa.

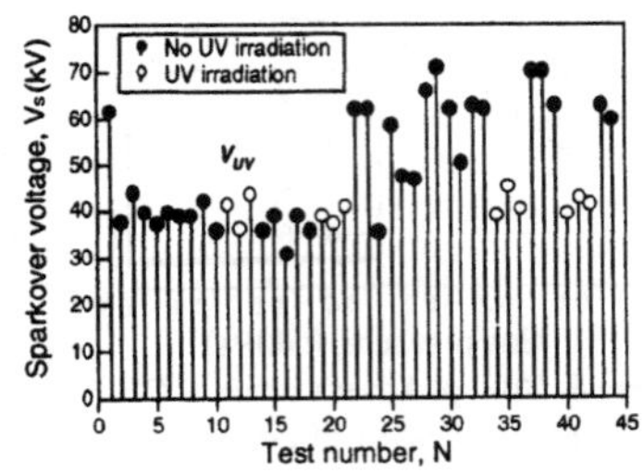

Figure 6. Effect of UV irradiation on sparkover voltage in CO_2 gas at 0.6 MPa before and after conditioning.

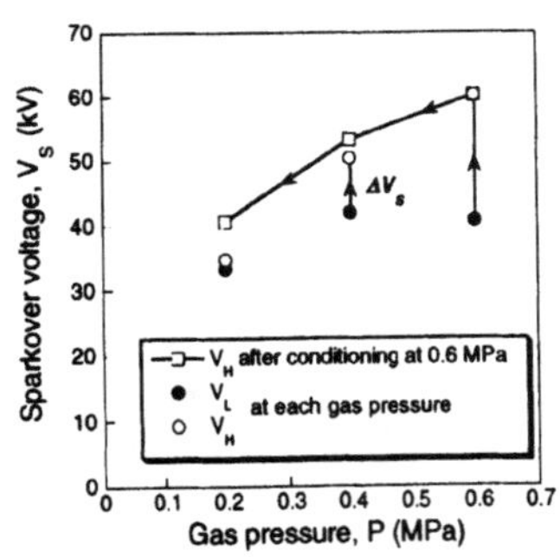

Figure 5. Increment in sparkover voltage due to the conditioning effect in CO_2 gas as a function of gas pressure.

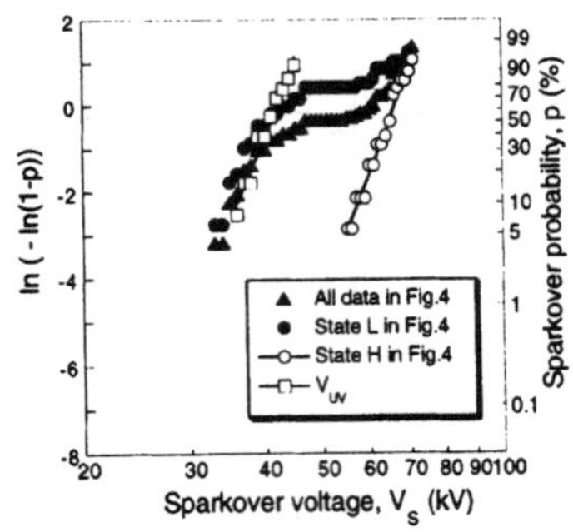

Figure 7. Weibull distribution of sparkover voltage in CO_2 gas at 0.6 MPa.

CO_2 gas together with V_{50} in SF_6 gas. In CO_2 gas, V_{50}s of both V_H and V_{UV} increase with the gas pressure. V_{50} for V_H is 1.4 to 1.6 times as much as that for V_{UV}. As compared with SF_6 gas, V_{50}s for V_H and V_{UV} in CO_2 gas are reduced by 53.1% and 33.6% at 0.1 MPa and by 92.3% and 62.7% at 0.6 MPa, respectively. It can be concluded from these results that the sparkover voltage in CO_2 gas significantly differs depending on the experimental condition and procedure due to the conditioning effect.

4. Discussion

4.1. Sparkover Voltage of CO_2 Gas Based on PD Phenomena

4.1.1 Relationship between PD Inception and Sparkover Voltages

Fig.9 shows typical waveforms of PD current and applied LI voltage in CO_2 gas under various conditions. Under UV irradiation condition sparkover follows successive PD pulses and takes place at the wavefront of the applied LI voltage as shown in Figs.9 (a) and (b). Above 0.4 MPa, sparkover also occurs by following successive PD pulses but occurs at the wavetail of the applied LI voltage as shown in Fig.9 (c). Note that the time ΔT [9] to sparkover from the 1st PD generation as defined in Fig.2 exceeds 100 ns. On the other hand, without UV irradiation (state H) sparkover takes place at the wavetail of the applied voltage without such PD pulses, *i.e.* sparkover occurs instantly by following a PD with a relatively larger current peak as shown in Fig.9 (d). Here, Fig.10 shows gas pressure depen-

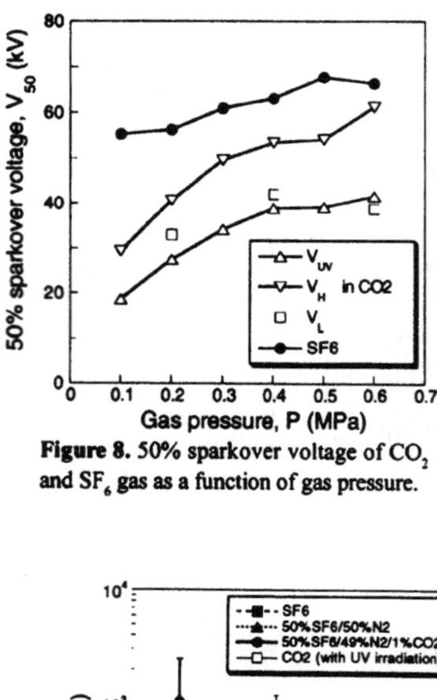

Figure 8. 50% sparkover voltage of CO_2 and SF_6 gas as a function of gas pressure.

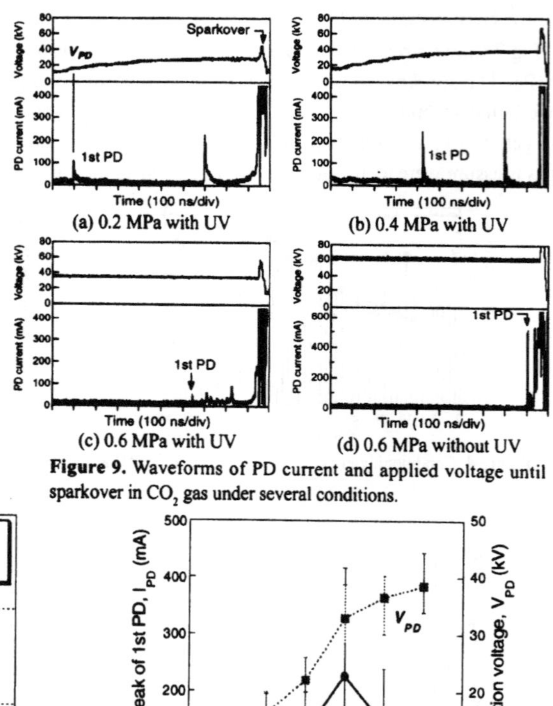

Figure 9. Waveforms of PD current and applied voltage until sparkover in CO_2 gas under several conditions.

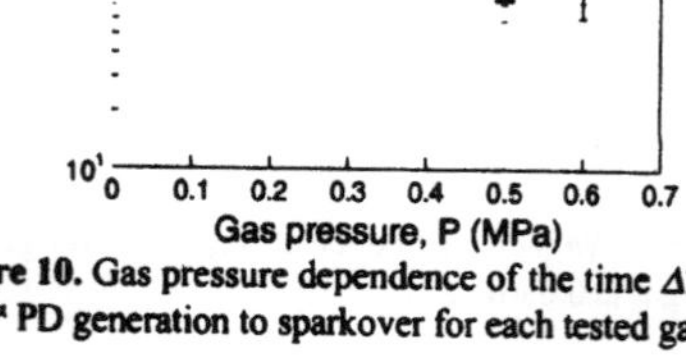

Figure 10. Gas pressure dependence of the time ΔT from the 1st PD generation to sparkover for each tested gas [9,10].

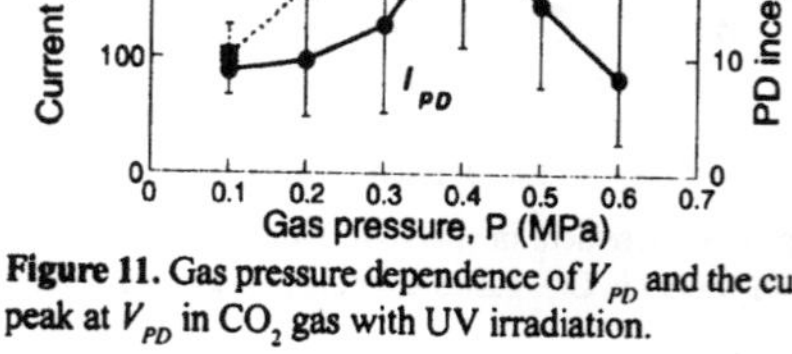

Figure 11. Gas pressure dependence of V_{PD} and the current peak at V_{PD} in CO_2 gas with UV irradiation.

dence of ΔT for each tested gas [9]. ΔT for CO_2 gas is the longest among those of tested gas. Thus, the experimental result that ΔT for $SF_6/N_2/CO_2$ gas mixture was longer than that for SF_6/N_2 mixture can be attributed to the characteristics of CO_2 gas. Fig.11 shows gas pressure dependence of PD inception voltage V_{PD} and the current peak I_{PD} of the 1st PD pulse at V_{PD} in CO_2 gas under UV irradiation condition. It is seen from Fig.11 that V_{PD} increases with the gas pressure while I_{PD} has a peak at 0.4 MPa. This implies that PD phenomena change around 0.4 MPa. In the next section, the change of PD phenomena will be discussed together with PD light emission properties.

Figs.12 (a) and (b) depict relationship between V_{PD} and V_S with or without UV irradiation in CO_2 and SF_6 gas, respectively. For both CO_2 and SF_6 gas, a linear relation between V_S without UV irradiation (V_H) and V_{PD} is obtained irrespective of the gas pressure. On the other hand, a plot of V_S with UV irradiation (V_{UV}) and V_{PD} deviates the linear dotted line below 0.4 MPa, while V_{UV} also increases with V_{PD}. The differences in the sparkover voltage with or without UV irradiation arise from PD generation properties as mentioned in Fig.9. Namely, the time when PD generation and the resultant sparkover occur was changed from the wavetail of the applied LI voltage to the wavefront by UV irradiation. It should be noticed in Fig.12 that the difference between V_{UV} and V_{PD} (which is defined as ΔV) in SF_6 gas is much larger than that in CO_2 gas. This ΔV relating to the PD evolution during which the sparkover occurs from the PD generation would be used as an index representing the degree of the corona stabilization effect.

4.1.2. Gas Pressure Dependence of PD Properties Measured by Optical Technique

As mentioned above, PD phenomena in CO_2 gas depend on the gas pressure and are classified into two types at the critical pressure 0.4 MPa. Typical results of PD light emission and current waveform of the two types in CO_2 gas are shown in Figs.13 (a) and (b), respectively. These results are obtained by the simultaneous measurement of still and streak images of PD light emission and waveforms of PD current and applied voltage at 0.2 and 0.6 MPa. As the gas pressure increases, the light emission image of the 1st PD pulse is changed from the expanded fan-shape emerging at the tip of the needle electrode as shown in Fig.13 (a) to the small pointed-shape as shown in Fig.13 (b). Above 0.4 MPa, successive PD pulses are observed without a sparkover. As can be seen in Fig.13 (b), when the stepwise developing light emission images of successive four PD pulses were observed with the streak camera, these corresponding four current pulses were also detected. On the other hand, the length of the light emission still image observed with the ICCD camera was longer than that observed with the streak camera (mainly arising from the 4th PD pulse) even though the corresponding current pulse after the 4th PD generation was not found within the exposure time of the ICCD camera. This result implies that PD light emission in CO_2 gas would develop even without such stepwise evolutions, resulting in no

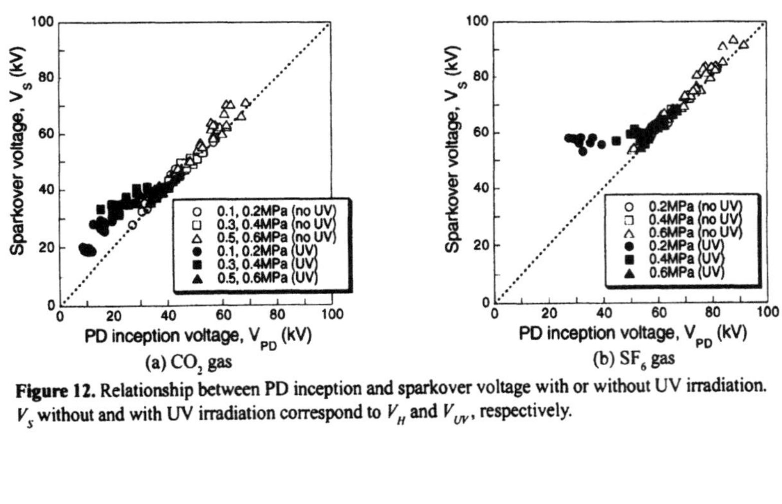

Figure 12. Relationship between PD inception and sparkover voltage with or without UV irradiation. V_S without and with UV irradiation correspond to V_H and V_{UV}, respectively.

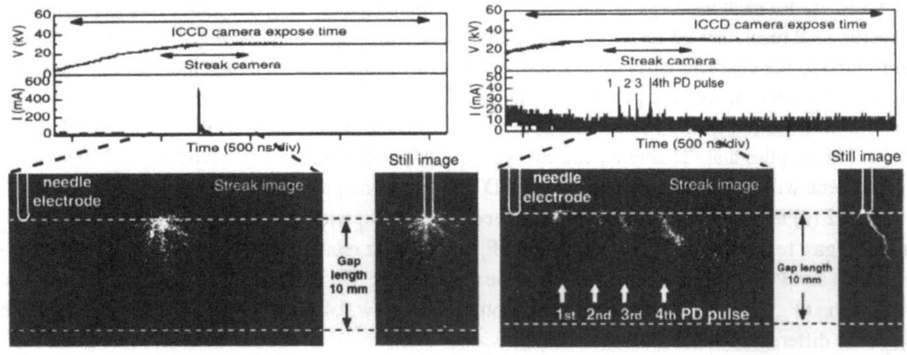

(a) 0.2 MPa (b) 0.6 MPa

Figure 13. Streak and still images of PD light emission and the corresponding current in CO_2 gas.

detection of the pulsed current.

PD inception voltage in CO_2 gas was also measured under ac voltage with the same electrode system. As a result, it was found that instantaneous value of PD inception voltage in the positive half cycle of applied ac voltage agrees with V_{PD} for positive LI voltage with UV irradiation. It should be noticed that the sparkover voltage differed between ac and LI voltages while both V_{PD}s were almost the same. These experimental results can be interpreted in terms of the difference in PD evolution properties relating to the space charge behavior. The gas pressure giving the maximum V_S under ac voltage (around 0.35 MPa), *i.e.* the peak appearing in the *N*-shape characteristics of the V_S vs *P* plot, is almost the same as the gas pressure at which PD phenomenon changed under LI voltage. These results obtained under ac voltage application indicate that PD generation and the PD properties at the PD inception in CO_2 gas are almost the same as those under LI voltage with UV irradiation.

4.2. Light Emission Measurement in $SF_6/N_2/CO_2$ Mixture

Fig.14 shows time-resolved spectrogram and still image of the light emission arising from successive PD pulses together with the light intensity and the current waveforms in 50%SF_6/49%N_2/1%CO_2 mixture at 0.4 MPa. As can be seen in the figure, the still image of PD light emission consists of four successive PD pulses. Two spectra of the 3rd and 4th pulses passing through the slit of the spectrometer and streak camera were detected in the spectrogram. As can be recognized from this result, the simultaneous measurement is needed for the time-resolved spectrum analysis because it gives us significant information on the evolution of discharge activity as well as the location and shape of the discharge (this information is needed to calibrate the wavelength in spectrum measurement).

Fig.15 shows time-resolved PD light emission spectra given in Fig.14 together with the results measured in (a) open-air, (b) N_2 and (c) 50%SF_6/49%N_2/1%CO_2 mixture under the same experimental conditions as those in Fig.14 (0.1 MPa) except for the gas pressure. From Fig.15, remarkable spectral peaks in the wavelength between 300 and 400 nm for the gas mixture agree well with those observed in N_2 and open-air irrespective of the gas pressure. This result suggests that these spectral peaks in the gas mixture are mainly attributed to the second positive band of N_2. Thus, one has to pay a special attention when PD light emission in the gas mixture including N_2 is observed with a spectrometer. It is because there is a possibility that spectra caused by other gases except for N_2 are masked with those by N_2 due to its strong light emission intensity. Although we have not yet identified the gas species except for N_2 gas in the successive PD pulses in $SF_6/N_2/CO_2$ mixture, we have observed the spectrum with different peaks as shown in Fig.15 (e) which was given from the 4th PD pulse shown in Fig.14. Thus, the gas species in the successive PD pulses would be identified by investigating the spectrum which does not arise from N_2. This investigation is under way.

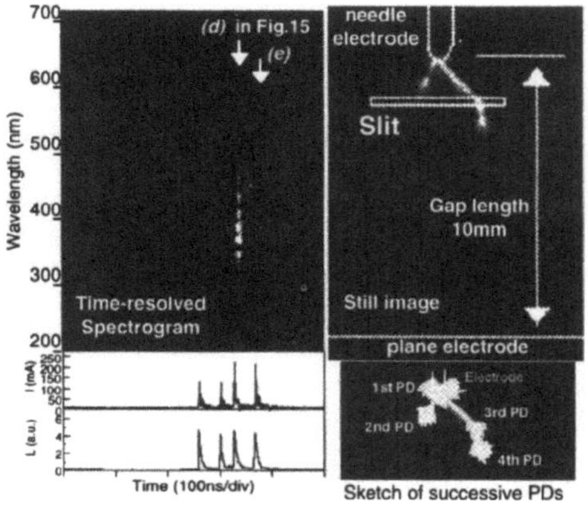

Figure 14. Time-resolved spectrum and the still image of light emission by successive four PD pulses in 50%SF_6/49%N_2/1%CO_2 gas mixture at 0.4 MPa.

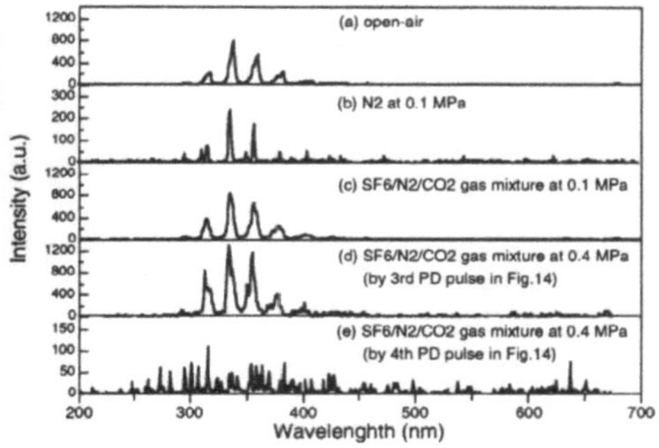

Figure 15. Time-resolved light emission spectra of one PD pulse for each tested gas. (d) and (e) are of the results for the 3 rd and 4 th PD pulses given in Fig.14 and calibrated according to the PD position in the slit.

5. CONCLUSIONS

We investigated PD phenomena leading to sparkover in SF_6, N_2, CO_2 gas and their gas mixture by simultaneous measurements of streak and still images of light emission as well as waveforms of the light emission intensity and current of PD pulses occurring under a positive LI voltage with the needle to plane electrode system. From the obtained results, we discussed the sparkover properties, particularly the conditioning effect in CO_2 gas, based on PD phenomena. Time-resolved spectrum of PD light emission in $SF_6/N_2/CO_2$ gas mixture was also discussed in terms of identification of gas species giving rise to the light emission and the evolution of successive PDs to sparkover.

6. REFERENCES

1. M.Finkel, W.Boeck and Ch.Lederle, Statistical Behavior of GIS and GIL Insulation, *Proc. of the 7th ICPADM*, pp.37-40 (2003)
2. Y. Yoshitake, N. Hayakawa, T. Ueda, H. Okubo, Impulse Partial Discharge Propagation Mechanism under Non-Uniform Electric Field in N_2/SF_6 Gas Mixtures, *Proc. of the 13th ISH* (2003)
3. H.Goshima, H.Shinkai and M.Yashima, Lightning Impulse Breakdown Characteristics of High-Pressure N_2 as an Alternative Insulation Gas to SF_6, *Gaseous Dielectrics IX*, pp.359-364 (2001)
4. T.Okamoto, M.Shibuya, T.Takahashi, Y.Mizutani and T.Takahashi, Proposal of All Solid Insulated Substation for Future Substation, *Proc. of the 7th ICPADM*, pp.1190-1193 (2003)
5. H.Goshima, H.Shinkai and H.Fujinami, Lightning Impulse Breakdown Characteristics of High-Pressure CO_2 Compared with Those of N_2 as Alternative Insulation Gases to SF_6, *Proc. of the 13th ISH* (2003)
6. T.Uchii, Y.Hoshina, T.Mori, H.Kawano, T.Nakamoto and H.Mizoguchi, Development of 72kV Class Novel CO_2 Gas Circuit Breaker Model, *IEEJ the Papers on Joint Technical Meetings on ED,SP, HV*, pp.141-146 (2003)
7. S.Ohtsuka, M.Onomoto, T.Nishimura, S.Matsumoto, M.Shibuya and M.Hikita, Insulation Properties of High-Pressure CO_2 Gas under Existence of a Particle, *Proc. of the 3rd IASTED Int. Conf. PES*, pp.721- 726 (2003)
8. S.Ohtsuka, M.Koumura, M.Cho, Y.Hashimoto, M.Nakamura and M.Hikita, Insulation Properties of CO_2/N_2 Gas Mixture with a Small Amount of SF_6, *Gaseous Dielectrics IX*, pp.295-300 (2001)
9. S.Ohtsuka, M.Koumura, M.Cho, M.Nakamura and M.Hikita, PD Time Sequential and Light Emission Properties as Pre-Breakdown Phenomena of $SF_6/N_2/CO_2$ Gas Mixture, *2002 Annual report CEIDP*, pp.801-804 (2002)
10. S.Ohtsuka, K.Eguchi, M.Cho, S.Yuasa, S.Okabe and M.Hikita, Light Emission Properties of SF_6 Gas Mixture in Impulse Breakdown Process, *Proc. of the 13th ISH* (2003)

7. ACKNOWLEDGMENTS

The authors thank to Mr. N.Yokoyama for his support to perform the experiments. This study was partially supported by a Grand-in-Aid for Scientific Research from the Ministry of Education, Science, Sports and Culture, Japan.

PARTIAL DISCHARGES WAVESHAPE CHARACTERISATION

M. A. Elborki, Z. D. Wang and P. A. Crossley*

1. INTRODUCTION

Power transformers have a significantly more complex insulation structure than other types of power devices. The economic penalties incurred by transformer failure, and the resulting outage cost, are considerable.

Over the years, the level of investigative effort in PD field has varied considerably both as regards to the type of electrical apparatus under consideration as well as the type of discharge behavioural aspect being examined.

Partial discharges are in general a consequence of local electrical stress in the insulation or on the surface of the insulation. The inception of partial discharge gives information on the limit of the electrical strength of the insulating material before a complete discharge takes place.

Normally the partial discharge tests are simple if the background noise is low enough to permit a sufficient and accurate measurement, which is the case in well shield high voltage laboratories.

The monitoring of all pulses allows the determination of the largest partial discharge impulse within a given time, the determination of the apparent charge per pulse and the relation between the pulses and the phase of the applied power frequency voltage is determined at a specified test voltage [1,2].

The intent of this paper is to examine the behavioural characteristics and to delineate and compare the PD detection and measurement.

2. CIRCUIT AND ARRANGEMENTS

Tests were performed to assess the ability of various parameters to identify the types of PD source by using electrode configurations that reproduce the PD phenomena.

Magdi Elborki and Zongdong Wang, UMIST, UK. Peter Crossley, Queen's University of Belfast, UK.

Gaseous Dielectrics X.
Edited by Christophorou *et al.*, Springer, New York, 2004.

A test circuit for creating discharge is shown in figure1. Prior to measuring discharge activity various tests were performed on the test system, without the specimen Xs to determine the maximum voltage at which the test system was PD free.

The single–phase test voltage was derived from the AC 50Hz mains via a regulator transformer. The test voltage is monitored on the voltmeter on Robinson discharge detector via the resistive voltage divider. The secondary of the test supply transformer generated the desired test voltage.

This feeds the resistive voltage divider, capacitive voltage divider, the test specimen and a discharge-free coupling capacitor.

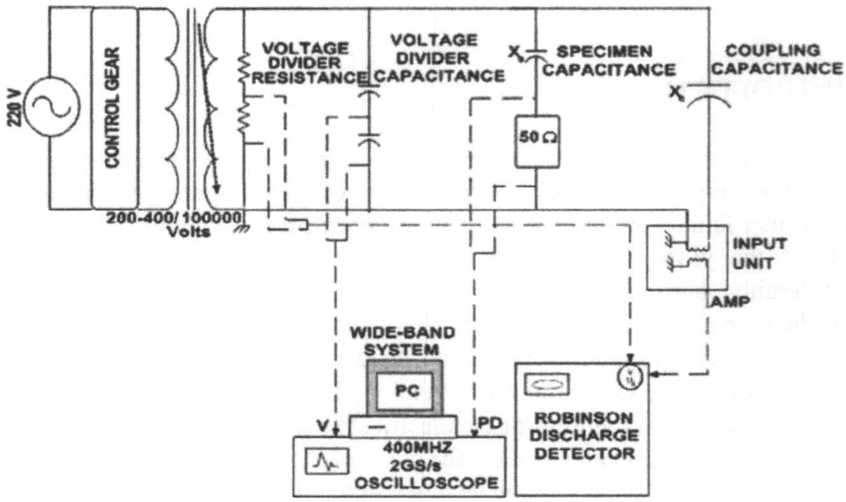

Figure 1. Schematic circuit diagram for the partial discharge measurement

The PD detector system has the following basic functions; Display and Record the wave-shape data of each individual PD pulse. During the tests, PD signals were taken from the input unit to Robinson discharge detector for various discharge pattern displays.

3. ELECTRODES CONFIGURATIONS

These mechanisms were produced experimentally in the laboratory by using electrode configurations that reproduce the basic PD phenomena; tests were also performed to assess the ability of various parameters to identify the type of PD source.

3.1 Corona in air

Corona can be caused, by a small protrusion at the surface of an HV conductor. This can be represented by a simple rod-to-plate electrode arrangement in air, as shown in Figure 3. The electrodes are made from copper / steel material. A plate electrode will be set at different distance (5 to 20mm) below the needle tip of a rod electrode to make an inhomogeneous electric field configuration.

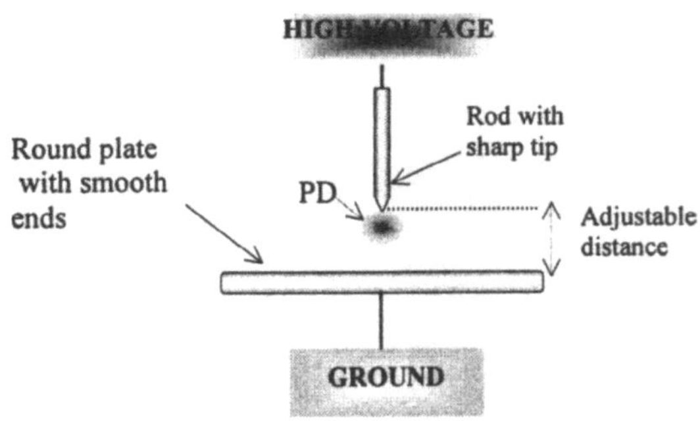

Figure 3. Corona in air as electrical interference

3.2 Surface discharge along pressboard in oil

A point-to-plane electrode arrangement immersed in oil was used to produce surface discharges Figure 4. The plate electrode is covered by pressboard. The rod tip touches the surface of the pressboard, which is laid on the plate. This arrangement creates an oil/surface discharge.

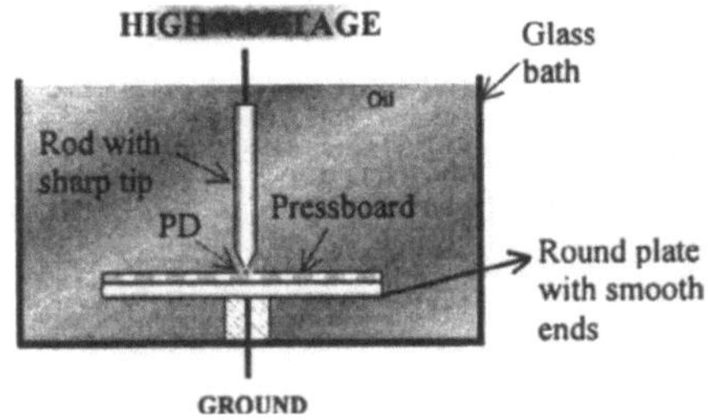

FIGURE 4. SURFACE DISCHARGE ALONG PRESSBOARD IN OIL

3.3 Surface discharge along bushing

A bushing electrode arrangement was used to produce surface discharge, see Figure 5. The discharges were along the outer surface. Salty water was spread over the bushing to reduce the inception PD voltage.

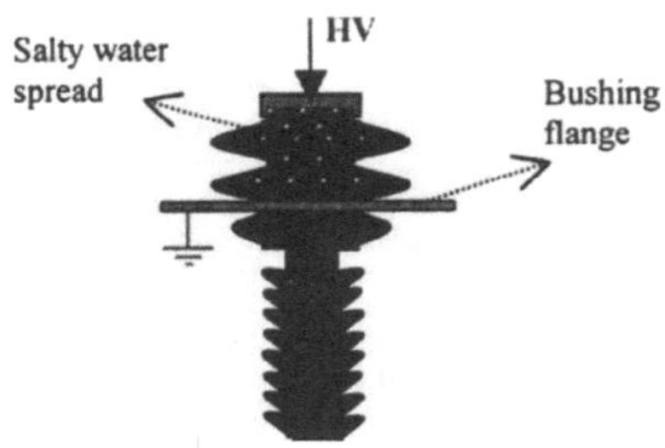

Figure 5. 33kv transformer bushing

4. EXPERIMENTAL RESULTS AND ANALYSIS

One of the ways to identify defects inside an insulation system is to analyse the shape of the PD signals. The rise-time and spectral characteristics of the signal were measured for different types of PD.

4.1 Corona

The behaviour of the insulation defect model shown in Figure 3 was investigated using the test circuit in Figure 2. The applied voltage was increased above the discharge inception voltage. This resulted in corona in the negative half power cycle. The signal measured by a resistive shunt has negative polarity and is referred to as "negative corona". When the voltage is further increased, corona can be produced in the positive half cycle. The measured signal has a positive polarity and is referred to as called "positive corona".

4.2 Surface discharge along pressboard in oil

Observation of surface discharge along pressboard in was carried out using the insulation defect model shown in Figure 4. The discharge was generated on 5mm thick pressboard surfaces immersed in oil using a rod-to-plate electrode. A series of tests were used to identify the difference in the waveforms generated by corona and surface discharge along pressboard in oil.

4.3 Surface discharge along bushing

The tests were carried out by applying different voltages to the porcelain surface of a 33kV bushing that was coated in salty water. A surface discharge on the clean porcelain surface could not be achieved. Salty water was used to increase the conductivity in the discharge current bath and reduce the inception PD voltage.

4.4 Comparisons

A comparison between the signals obtained from corona, a surface discharge along the pressboard in Oil and a surface discharge along the bushing are shown in Figure 6. The normalised time domain signals for: Corona at 3.7kV, a surface discharge along pressboard in oil at 22.5kV and a surface discharge on bushing at 24kV are shown in Figure 6(a).

Figure 6(b) shows the frequency spectra of these signals. The main frequency components for corona and a surface discharge on a bushing are below 150MHz.The main frequency components for a surface discharge in oil is below 50MHz.

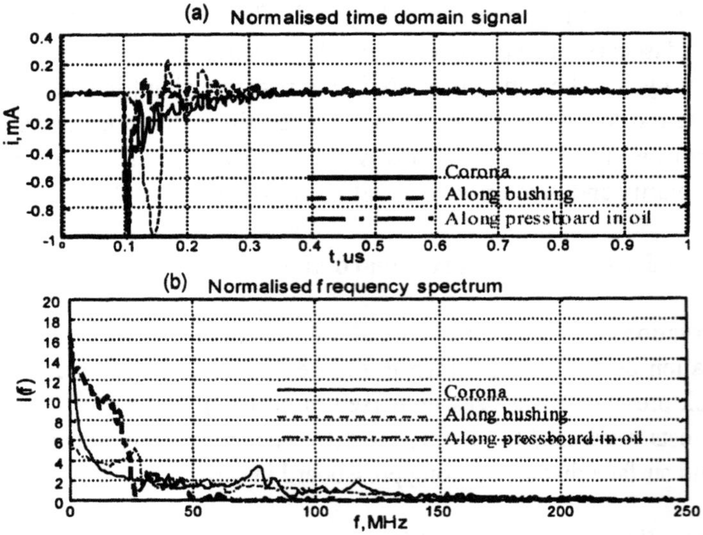

solid line: Corona, dashed line: Bushing discharge, dot-dashed line: Discharge along pressboard in oil

Figure 6. Different types of Discharges

5. CONCLUSION

Corona and surface discharge on a bushing have a fast rising slope, whilst the surface discharge on pressboard on oil has a slow rising slope. The surface discharge on bushing has a sharper falling slope. The falling slope of the PD signal on pressboard in oil also has a sharp slope. This can be compared with corona, which has a longer falling slope. A comparison between the wave shapes of corona, a surface discharge along pressboard in oil and surface discharges along a bushing were used to identify the type of defect.

6. REFERENCES

1. E. Carminati, A. Gandelli and M. Lazzaroni, "A past hybrid system for PD measurement," *IEEE Trans. Dielectrics and Electrical Insulation*, vol.7, pp. 440-445, June 2000.

2. Z.D.Wang, P.A.Crossley, K.J.Cornick and D.H.Zhu, "An algorithm for partial discharge location in distribution power ransformers," in *Proc. 2000 IEEE Power Engineering Society Conf.*, pp.2217-2222.

ION MOBILITY IN A NEGATIVE CORONA DISCHARGE CONTAINING MIXTURES OF NITROUS OXIDE AND OXYGEN

Jan D. Skalny, Stefan Matejcik, Nigel J. Mason

1. INTRODUCTION

Recently an analysis of the physico-chemical properties of a negative corona discharge in pure CO_2 and its mixtures with oxygen has been reported [1]. Since both CO_2 and nitrous oxide, N_2O, are both linear molecules containing oxygen, analogous effects might be expected when N_2O gas is mixed with oxygen in a negative corona discharge. However, there are several distinct differences in the electron interactions with CO_2 compared to N_2O molecules which may affect the electron induced chemistry in the drift region of negative corona discharges containing these two molecules. Dissociative electron attachment (DEA) to carbon dioxide has a maximum cross section at relatively high electron energies (around 4 eV) [2], hence in $CO_2 + O_2$ discharges the DEA process is confined to the glow region. In contrast DEA in N_2O exhibits two maxima in the O^- yield at low energies (0.55 and 2.4 eV) [3] making DEA processes prevalent throughout the discharge and not just within the glow region. Moreover, the permanent dipole moment of nitrous oxide molecule can lead to the formation of ion clusters in the drift region substantially changing the mobility of charge carries in this region. Therefore considerable differences in the electrical parameters of CO_2 compared to N_2O negative corona discharges might be expected.

In this paper we present results of an experimental study of electrical parameters of the negative corona discharge in pure N_2O and its mixtures with oxygen.

Jan D., Skalny, Department of Plasma Physics, Comenius University, Mlynska dolina F-2, Bratislava, Slovakia 842 48. Stefan Matejcik, Department of Plasma Physics, Comenius University, Mlynska dolina F-2, Bratislava, Slovakia 842 48. Nigel J. Mason Department of Physics and Astronomy, Centre of Molecular and Optical Sciences, The Open University, Walton Hall, Milton Keynes, United Kingdom MK7 6AA.

Gaseous Dielectrics X.
Edited by Christophorou *et al.*, Springer, New York, 2004.

2. EXPERIMENTAL APPARATUS AND DIAGNOSTICS

The experiments were carried out at atmospheric pressure and at ambient temperatures using the experimental apparatus described in detail in earlier publications[1], hence only a brief description is presented here. The discharge system is composed of an inner wire electrode with a radius r_0 = 62.5 μm and an outer cylinder with radius R = 7.5 mm. The active length of the outer electrode was 100 mm. The experiments were performed at a constant flow rate 100 cm^3 per minute. Product ozone concentrations in the gas mixture were monitored at the outlet of the discharge reactor using a FTIR spectrometer.

3. EXPERIMENTAL RESULTS AND DISCUSSION.

The onset voltage U_0 for the corona discharge was found to be only slightly influenced by the composition of the gas mixture being 4.35 ± 0.05 kV in pure oxygen and slightly increasing to a value of 4.85 ±0.05 kV in pure nitrous oxide. However in contrast the discharge current changes dramatically with gas mixture. The experimentally measured current-voltage characteristics (CV) are shown in Figure 1 and follow the well-known quadratic dependence of the current per unit length of the outer electrode I on the voltage U applied on electrodes [4]. The Townsend formula for the current-voltage (CV) characteristic was then used to calculate the average mobility of negatively charged particles in the drift region of the discharge [5]

$$I = \frac{8\pi\varepsilon_0\mu}{R^2 \ln R/r_0} U(U - U_0) \tag{1}$$

where ε_0 is the vacuum permittivity and μ is the mobility of the electric charge carriers in the drift region. The mobility μ was calculated at constant gas composition from the experimental data in Figure 1. The results are shown in Fig. 2. The mobility is found to fall with increasing N_2O concentration with a mean value of typically 1 $cm^2.V^{-1}.s^{-1}$.

In pure nitrous oxide the primary O^- ions are formed via dissociative electron attachment

$$e + N_2O \longrightarrow O^- + N_2 \qquad k_2 = f(E/N) \tag{2}$$

Process (2) is effective across the whole discharge gap because the maximum in the attachment cross-section reaches a maxima at 0.55 eV and 2.4 eV [3]. In pure nitrous oxide the ions O^- can undergo very fast ion-molecule reactions via

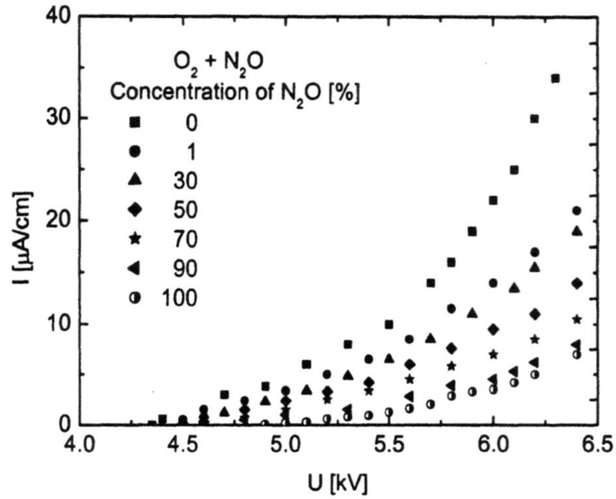

Figure 1. CV characteristics in flowing mixtures of oxygen with different admixtures of N_2O.

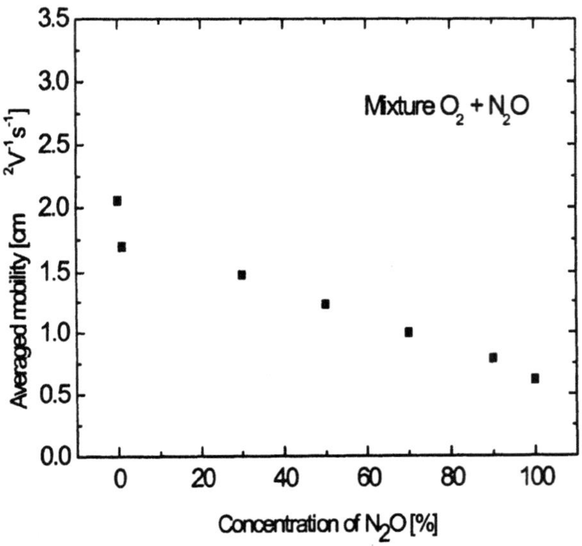

Figure 2. The dependence of averaged mobility on the content of N_2O in mixtures with oxygen.

$$O^- + N_2O \longrightarrow NO^- + NO \qquad k_3 = 2.2 \times 10^{-10}\ cm^3/sec\ ^6 \qquad (3)$$

$$O^- + N_2O \longrightarrow NO_2^- + N \qquad k_4 = 2.0 \times 10^{-10}\ cm^3/sec\ ^6 \qquad (4)$$

The lifetime of O^- ions is therefore less than 1 nsec at atmospheric pressure. At low pressures, close to 10 mbar, the reaction time of processes (3) and (4) becomes comparable with the time required for the transport of O^- ions to the outer electrode. Hence O^- ions should be prevalent at low pressures. This is in agreement with the reported results of Gardiner and Craggs [7] who found a strong yield of O^- ions in pure nitrous oxide at a pressure of 13 mbar.

NO^- ions may undergo a fairly fast detachment process

$$NO^- + N_2O \longrightarrow N_2O + NO + e \qquad k_5 = 5.1 \times 10^{-12}\ cm^3/sec\ ^8 \qquad (5)$$

or may be converted to NO_2^- ions via the ion-molecule reaction

$$NO^- + N_2O \longrightarrow NO_2^- + N_2 \qquad k_6 = 2.0 \times 10^{-10}\ cm^3/sec\ ^8 \qquad (6)$$

The lifetime of NO^- ions at atmospheric pressure in N_2O is therefore of the order of 10 nsec, thus the NO_2^- ions should be the dominant ions in the drift region of the discharge at high pressures.

The mobility of NO_2^- ions at nearly zero reduced electric field E/N and ambient gas temperature 300 K and atmospheric pressure has been found by Perkins et al.[9] to be 2.55 $cm^2/V.s$. However the present experimental data shows that the mobility of charge carriers in the drift region of the negative corona discharge fed by nitrous oxide is much less than this value (Figure 2). One explanation for these lower mobility values be the formation of $O^-.(N_2O)$ and $NO^-.(N_2O)$ clusters [7, 10, 11]. Relatively stable clusters can be formed not only in the drift region of the discharge but also in the glow region. Unfortunately there is no data on the production rates of such clusters via processes

$$O^- + N_2O + N_2O \longrightarrow O^-.(N_2O) + N_2O \qquad (7)$$

$$NO^- + N_2O + N_2O \longrightarrow NO^-.(N_2O) + N_2O \qquad (8)$$

The only available rate constant is for the three body process forming $O^-.(N_2O)$ [8] at presence of CO_2

$$O^- + N_2O + CO_2 \longrightarrow O^-.(N_2O) + CO_2 \quad k_9 = 1.0 \times 10^{-27}\ cm^6/sec\ ^8 \qquad (9)$$

However it is reasonable to assume that the rate constant for process (7) is comparable with that of (9) Therefore the rate for process (7) at atmospheric pressure is two orders of magnitude greater than that for either process (3) or (4). A high efficiency for formation of $O^-.(N_2O)$ clusters is also likely due to the permanent dipole moment of N_2O (0.167 D). There are no data on the mobility of

$O^-.(N_2O)$ clusters but one can surmise that these should be smaller than the mobility of the parent O^- ions and NO^- and NO_2^- ions formed by ion-molecule reactions in the drift region of the discharge. Hence we tentatively conclude that low mobility of charge carriers calculated from our CV data is due to the presence of $O^-.(N_2O)$ cluster ions in the drift region of negative corona discharge fed by pure N_2O. The role of $NO^-.(N_2O)$ clusters is likely to be marginal because the formation of such ions is a two-step process. In the first process NO^- ions are formed via process (3) and $NO^-.(N_2O)$ subsequently formed in the second step (8). Therefore the total rate of generation of $NO^-.(N_2O)$ clusters is smaller in comparison with the direct process (7). The low abundance of $NO^-.(N_2O)$ clusters observed in the experiments performed by Gardiner and Craggs [7] supports this statement.

The addition of nitrous oxide into a stream of pure oxygen and air leads to a decrease in the discharge current at the constant voltage. The extremely high rate of process (7) is one possible explanation of this effect. In pure oxygen gas O^- ions are predominately converted to O_3^- ions [6] by the associative process

$$O^- + O_2 + O_2 \longrightarrow O_3^- + O_2 \qquad k_{10} = 1.5 \times 10^{-30}\ cm^6/sec \text{ [12]} \qquad (10)$$

If the concentration of nitrous oxide in the mixture with oxygen is higher than 1 % the rate of processes (7) and (10) are comparable and the formation of $O^-.(N_2O)_n$ clusters prevails. Hence the mobility of charge carriers is dramatically reduced. Further decreases at higher concentrations of N_2O can be ascribed to the reduction in the number of free electrons present in the drift region due to electron attachment to N_2O molecules. At low concentrations of nitrous oxide in oxygen the primary source of O^- is the dissociative electron attachment to oxygen molecules

$$e + O_2 \longrightarrow O^- + O \qquad k_{11} = f(E/N) \qquad (11)$$

but with increasing content of nitrous oxide in the mixture process (2) becomes dominant. Since the rate constant for process (11) is evidently smaller than that of process (2), especially at low electron energies [12, 13], the number of free electrons in the discharge gap is reduced due to more effective electron attachment via (2). This is a second reason why the values of calculated mobility are found to decrease with increasing content of N_2O in mixture (Figure 2).

4. CONCLUSIONS

The current of the negative corona discharge in oxygen is considerably affected by the addition of nitrous oxide. The decrease in discharge current may be explained by formation of $O^-.(N_2O)$ and $NO^-.(N_2O)$ clusters and the effective dissociative electron attachment to N_2O molecules, especially in the drift region of the discharge. Further experiments to test this hypothesis should be performed in which anions are extracted form the discharge, these are currently under preparation.

5. ACKNOWLEDGEMENTS

The authors are pleased to acknowledge the financial assistance of Slovak Grant Agency VEGA under project 1/1267/04, the UK EPSRC research council (Grant GM/98944) and the CEEPUS project A 103.

6. REFERENCES

1. T. Mikoviny, M. Kocan, S. Matejcik, N. J. Mason, and J. D. Skalny, Experimental study of negative corona discharge in pure carbon dioxide and its mixtures with oxygen, *J. Phys. D.: Appl. Phys.* **37**(1), 64-73 (2004).
2. P. Cicman, G. Senn, G. Denifl, D. Muigg, J. D. Skalny, P. Lukac, A. Stamatovic, and T. D. Märk, Dissociative electron attachment to CO_2, *Czech. J. Phys.* **48**(10), 1135-1145 (1998).
3. F. Brüning, S. Matejcik, E. Illenberger, Y. Chu, G. Senn, D. Muigg, G. Denifl, and T. D. Märk, Effects of temperature on the dissociative electron attachment to N_2O, *Chem. Phys. Lett.* **292**(1-2), 177-182 (1998).
4. M. Goldman and A. Goldman, in: *Gaseous Electronics,* edited by M. N. Hirsh and H. J. Oskam (AcademicPress, New York, San Francisco, London, 1978), pp. 219-290.
5. J. Q.Feng , An analysis of corona currents between two concentric cylindrical electrodes, *J of Electrost.***46**(1) 37-48 (1999).
6. D. L. Albriton Ion-neutral reaction-rate constants measured in flow reactors through 1977, 1978 *Atom. Data and Nuclear Data Tab.* **22**(1), 1-1001 (1978).
7. P. S. Gardiner, J. D. Craggs, Mass spectrometry of negative ions in Trichel corona, in: *Proc. 5th International Conference on Gas Discharges* (Liverpool, 1978) pp. 81-84.
8. L. I. Virin, R. V. Dyagaspanyan, G. V. Karatchevsev, V. K. Potapov, and V. L. Talrose, *Ion-molekularnyie reakcii v gazach* (Isdatelstvo NAUKA, Moscow, 1978)
9. M. D. Perkins, F. L. Eisele, E. W. McDaniel, Temperature dependent mobilities: NO_2^-, NO_3^-, CO_3^-, CO_4^-, and O_2^+ in O_2, *J. Chem. Phys.* **74**(7), 4206-4207 (1981).
10. J. L. Moruzzi and J. T. Dakin, Negative-Ion-Molecule Reactions in N_20, *J. Chem. Phys.* **49**(11), 5000-5006 (1968).
11. J. Dutton, F. M. Harris, D. B. Hudges, Ionization, electron-attachment and negative-ion reactions in N_2O, *J. Phys B.: Atom. Molec. Phys.* **8**(2), 313-324 (1975).
12. B. Eliasson, *Electrical discharge in oxygen, Part 1: Basic data, rate coefficients and cross sections, Report KLR 83/40 C,* (Brown Boveri Forschungszentrum, Baden-Dättwil,, 1985).
13. L. G. Lee and F. Li, Shortering of electron conduction pulses by electron attachers O_2, N_2), and CF_4, *J. Appl. Phys.* **56**(11), 3169-3174 (1984).

NEGATIVE CORONA DISCHARGE IN POINT-PLANE GEOMETRY: TWO SPECIES MODEL IN OXYGEN

Jiacheng Zhang, Kazimierz Adamiak, and G. S. Peter Castle*

1. INTRODUCTION

The electric corona discharge is generated in many industrial and research applications.[1] Often it is necessary to numerically simulate the process and many papers have been published on this topic.[2, 3, 4] Most algorithms for the corona simulation published so far assume that only one species of ions is injected from the ionization layer and moves towards the ground plane. While ions of only one polarity can leave the ionization layer, there is usually more than one species; in particular some electrons are always present in the negative discharge. In negative corona discharge near sparkover, free electrons may lead to the abnormally high particle charge.[5] Also, while the ionization of neutral molecules and recombination of ions is, by definition, limited to the ionization layer, the electron attachment is not and may occur in the transport zone.

This paper describes a numerical algorithm for simulation of the corona discharge in oxygen which includes two different ionic species, electrons and O^-_2.

---

* Authors are with the Dept. of Electr. and Comp. Eng., Univ. of Western Ontario, London, Ontario, Canada N6A 5B9, e-mail: jzhang52@uwo.ca, Tel: (519)661-2111 ext.88358, Fax: (519)661-3488.

Gaseous Dielectrics X.
Edited by Christophorou *et al.*, Springer, New York, 2004.

The Boundary Element Method (BEM) and Finite Element Method (FEM) are utilized to calculate the electric field distribution. The charge conservation equations for the electrons and negative ions are converted into a set of two ordinary differential equations by the Method of Characteristics (MOC). The Finite Difference Method (FDM) is used to solve these equations.

2. MATHEMATICAL MODEL

In this paper the corona discharge in the point-plane geometry is analyzed; consisting of a hyperboloidal needle with a tip radius curvature R and length L, perpendicular to an infinitely large plane at a distance D. The needle is supplied with a negative DC potential V_0 and the ambient gas is pure oxygen at room temperature and atmospheric pressure. It is assumed that only electrons are injected to the transport zone, but they can attach to the oxygen molecules. The amount of the electrons is determined so that Kaptzov's hypothesis is satisfied.[4]

The charge conservation equations for electrons and negative oxygen ions O^-_2 can be written as[6]

$$\frac{\partial N_e}{\partial t} + \nabla \cdot \boldsymbol{J}_e = -N_e \eta |\boldsymbol{W}_e| \tag{1}$$

$$\frac{\partial N_n}{\partial t} + \nabla \cdot \boldsymbol{J}_n = N_e \eta |\boldsymbol{W}_e| \tag{2}$$

where N_e and N_n are the charge densities of electrons and negative ions, $\boldsymbol{W}_e$ is the drift velocity of electrons and η is the attachment coefficient. $\boldsymbol{J}_e$ and $\boldsymbol{J}_n$ are the current densities due to electrons and negative oxygen ions:

$$\boldsymbol{J}_e = -k_e N_e \boldsymbol{E} \qquad \boldsymbol{J}_n = -k_n N_n \boldsymbol{E} \tag{3}$$

where k_e and k_n are the mobility of electrons and negative ions O^-_2.

The Poisson equation governs the electric field

$$\varepsilon \nabla^2 V = N_e + N_n \tag{4}$$

where $\boldsymbol{E} = -\nabla V$, V is the scalar electric potential and ε is the permittivity of the ambient gas.

3. NUMERICAL ALGORITHM

An iterative algorithm has been used to solve Eqs.(1)-(4). For given space charge density, the Poisson equation is solved to calculate the electric field. A hybrid FEM-BEM technique is used for this part, as it can provide a good accuracy for the strongly divergent electric field near the electrode tip in reasonable computing time.[4]

With known electric field, the space charge densities of electrons and negative oxygen ions O^-_2 are solved by the MOC.[4] Assuming two new variables

$$\alpha_e = k_e N_e \quad \text{and} \quad \alpha_n = k_n N_n \tag{5}$$

the following equations can be derived from Eqs.(1)-(3)

$$\frac{d\alpha_e}{dt} + [k_e \eta |\boldsymbol{E}| + \frac{k_e}{\varepsilon}(\frac{\alpha_e}{K_e} + \frac{\alpha_n}{K_n})]\alpha_e = 0 \tag{6}$$

$$\frac{d\alpha_n}{dt} + \frac{k_n}{\varepsilon}(\frac{\alpha_e}{K_e} + \frac{\alpha_n}{K_n})\alpha_n = \eta k_n \alpha_e |\boldsymbol{E}| \tag{7}$$

along the characteristic lines given by

$$\frac{d\boldsymbol{r}}{dt} = k_e \boldsymbol{E} \tag{8}$$

Some number of characteristic lines is traced from the surface of the corona electrode to the ground plane. Equations (6)-(7) form a set of two ordinary differential equations along these characteristic lines and the FDM is used to solve them. Then the space charge densities of electrons and negative ions can be calculated from the Eq.(5).

There is a double iterative loop for this algorithm.[4] First, some space charge density is assumed and Eqs.(4), (6), and (7) are solved iteratively (inner iterative loop). The charge density of electrons on the surface of the corona electrode cannot be derived from the equations governing the problem. As a result, an outer iteration is necessary, in which the charge density of electrons is updated until the electric field on the electrode surface is equal to the Peek's experimental value.

4. RESULTS

The space charge densities of electrons and negative ions along the axis of symmetry were calculated for R=100 μm, L=1 cm, and V=6, 8, and 10 kV. The results are shown in Figs. 1 and 2. The density of electrons decreases from the corona electrode to the ground plate because of the attachment to oxygen molecules, that results in generation of the negative O^-_2 ions. The distance of the penetration of electrons is only a few millimeters from the corona electrode. The concentration of negative oxygen ions reaches a maximum value at around 0.83mm, 0.76mm, and 0.69mm, respectively, and then decreases due to spreading of the ion trajectories caused by their mutual repulsion.

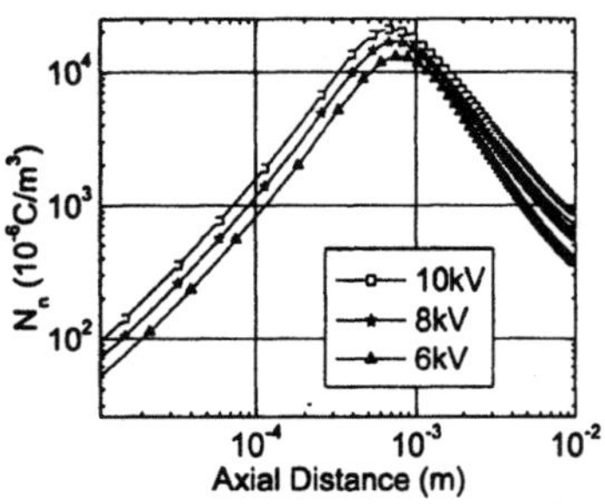

Figure 1. Density of negative ions along the axis of symmetry

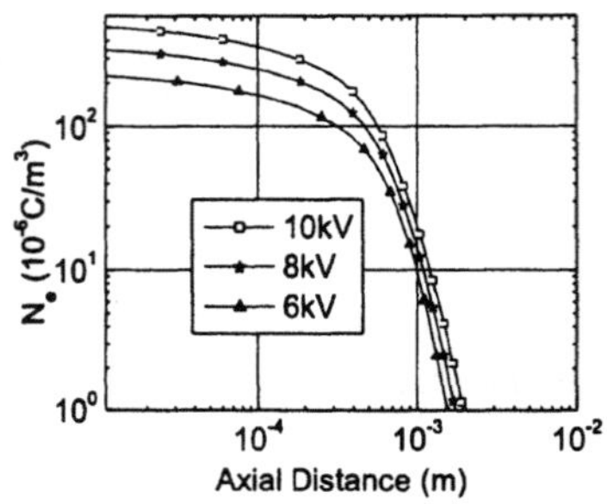

Figure 2. Density of electrons along the axis of symmetry

Figure 3 presents the lines of equal charge density for the negative ions at V=8kV. Along each characteristic line, the concentration of O^-_2 increases rapidly in the area close to the corona electrode and decreases slowly near the plate.

The presence of electrons and negative ions in the air gap significantly affects the electric field distribution; the ratio of the actual electric field and the Laplacian field along the line of symmetry for V=6, 8, and 10 kV are shown in Fig. 4. The space charge decreases the magnitude of the electric field in the vicinity of the discharge electrode, while increasing it near the plate.

The current densities on the ground plate for V=6, 8, and 10 kV are shown in Fig. 5. Each of them is the sum of the current densities for electrons and negative

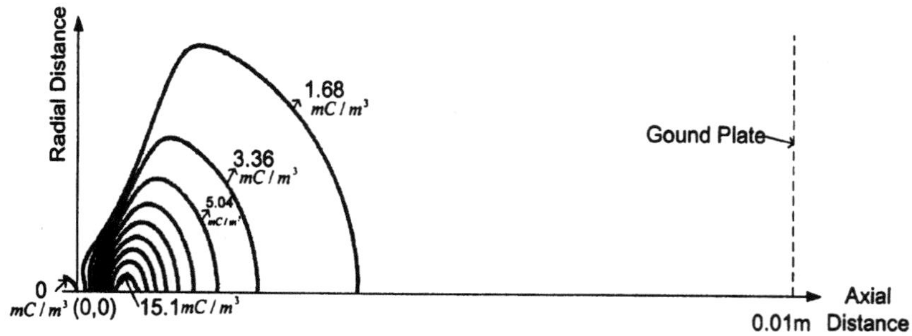

Figure 3. Lines of constant charge density of negative ions (cathode voltage = 8 kV)

ions. These results are compared with that by the Warburg formula[7]

$$J(\theta) = J(0)\cos^{m}\theta \quad , \quad \theta \leq 60^{o} \tag{9}$$

in which θ is the angular coordinate of a point on the ground plate and m=4.65. According to the Warburg formula, the current densities decrease abruptly to zero at the point, where θ is about 60° degrees. The numerical curves follow rather closely the Warburg curve, although the current abruptly decreases to zero at the smaller radial distance.

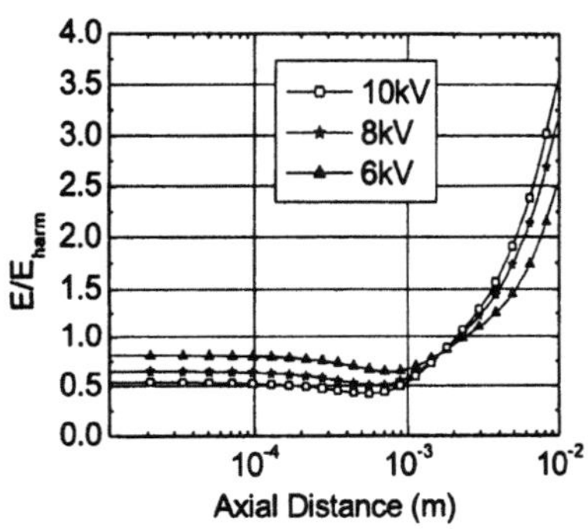

Figure 4. The ratio of the actual and the Laplacian field along the axis of symmetry

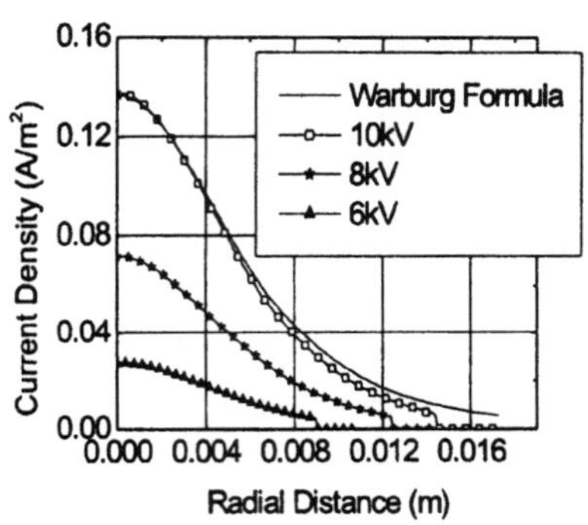

Figure 5. Current density distribution on the ground plate

5. CONCLUSIONS

A double-species model of the negative point-to-plane corona discharge is presented in this paper. The results of calculations show that at normal pressure and temperature the extent of the area penetrated by the electrons is only a few millimeters. The ion charge density rises quickly around the corona electrode, then reaches a maximum at some point in the gap depending upon the applied voltage and later decreases slowly close to the plate. Due to the presence of electrons and negative ions, the resulting electric field is smaller than the Laplacian field around the corona electrode and larger than it in the area close to the ground plate. The results of simulation agrees well with the experimental Warburg law for higher values of the current density on the ground plate, however, the numerically predicted value of current density reaches zero value at a much smaller radial distance from the center line.

ACKNOWLEDGMENTS

This work is partially supported by the Natural Sciences and Engineering Research Council (NSERC) of Canada.

REFERENCES

1. J. Cross, *Electrostatics: Principles, Problems and Applications* (Adam Hilger, Bristol, 1987).
2. R. Morrow, The theory of positive glow corona, *J. Phys. D: Appl. Phys.* **30**, 3099-3114 (1997).
3. M. N. Horenstein, Computation of corona space charge, electric field, and V-I characteristic using equipotential charge shells, *IEEE Trans. on Ind. Appl.* **IA-20**(6), 1607-1612 (1984).
4. K. Adamiak, J. Zhang and L. Zhao, Finite element versus hybrid BEM-FEM techniques for the electric corona simulation, *Electrotechnical Review* **10**, 753-756 (2003).
5. R. R. Crynack and G. W. Penney, Charging of fine particles in negative corona near spark over, *IEEE Trans. on Ind. Appl.* **IA-10**, 524-531 (1974).
6. B. Eliasson and U. Kogelschatz, Basic Data for Modeling of Electrical Discharges in Gases: Oxygen, Bown Boveri Technical Report, 1986 (unpublished).
7. R. S. Sigmond, Simple approximate treatment of unipolar space-charge-dominated coronas: the Warburg law and the saturation current, *J. Appl. Phys.* **53**, 891-898 (1982).

SECTION 4: GASEOUS DIELECTRICS AS PLASMA PROCESSING GASES

ELECTRONEGATIVE PLASMA REACTOR ENGINEERING

Demetre J. Economou[1]

1. INTRODUCTION

Low pressure (0.1mtorr to 10 torr), cold (gas temperature ~500 K), weakly ionized (degree of ionization 10^{-6}-10^{-1}) glow discharge plasmas are used extensively in the processing of electronic materials, especially for etching and deposition of thin films[1]. Such plasmas also find application in surface modification (e.g., hardening, corrosion resistance), lighting, and even environmental remediation. Figure 1 is a schematic of a plasma etch process carried out in a capacitively-coupled reactor. The case of polysilicon etching in a chlorine plasma is shown as an example. The plasma is generated by applying radio frequency power between a pair of parallel plates in a low pressure chamber (Fig. 1a). The Cl_2 feedstock gas is attacked by plasma electrons to produce Cl radicals and Cl_2^+ ions. Radicals diffuse or are convected by gas flow towards the wafer where they adsorb on the surface. Ions accelerate in the sheath naturally occurring over the wafer, and bombard the wafer vertical to its surface (Fig. 1b). The combination of radical and ion bombardment produces $SiCl_4$ product which desorbs and is removed by the gas flow. It is this directional ion bombardment which promotes anisotropic etching of microscopic features (Fig. 1c), whereby the film etches much faster in the vertical as compared to the horizontal direction. At the atomic level, ion bombardment produces a modified surface layer in which the reactant (Cl) is mixed within the silicon lattice (Fig. 1d) to a depth depending on the ion energy (~ 10s of Å). The energy deposited by ions promotes the formation of products that are either sputtered away or desorb spontaneously in the gas phase. Figure 1 also demonstrates the disparity in length scales encountered in plasma processing. Wafers are now 300 mm in diameter, so the reactor scale is of the order of 10s of cm. The sheath thickness ranges from 0.1-10 mm depending on the Debye length and the voltage applied to the electrode. The feature size is rapidly moving below 100 nm. Finally the lattice has to be described on the Å length scale.

The goals of any plasma etch process are to achieve high etch rate, uniformity, selectivity, controlled shape of the microscopic features etched into the film (anisotropy), and no radiation

[1] Plasma Processing Laboratory, Department of Chemical Engineering, University of Houston, Houston, TX 77204-4004.

damage. Choosing the reactor design and operating parameters are crucial for meeting these goals. Given a reactor type, its dimensions and materials of construction, one can manipulate operating parameters (pressure, power, frequency, etc.) to influence the process output (rate, uniformity, anisotropy, etc.).

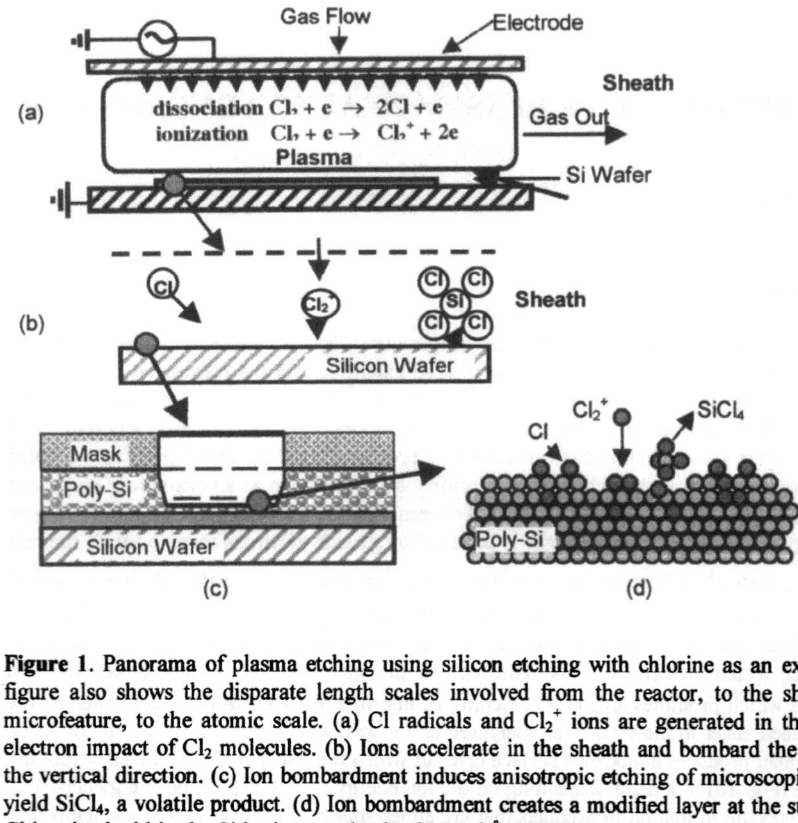

Figure 1. Panorama of plasma etching using silicon etching with chlorine as an example. This figure also shows the disparate length scales involved from the reactor, to the sheath, to the microfeature, to the atomic scale. (a) Cl radicals and Cl_2^+ ions are generated in the plasma by electron impact of Cl_2 molecules. (b) Ions accelerate in the sheath and bombard the wafer along the vertical direction. (c) Ion bombardment induces anisotropic etching of microscopic features to yield $SiCl_4$, a volatile product. (d) Ion bombardment creates a modified layer at the surface where Cl is mixed within the Si lattice to a depth of 10s of Å depending on the ion energy.

2. PLASMA REACTORS

Low Plasma Density Reactors - An example of a low plasma density system is the capacitively-coupled diode shown in Fig. 1a. Typical operating parameters are pressure of 10-1000 mtorr, power of 50-2000 W, frequency of 0.05-60 MHz, plasma density of 10^9-10^{10} cm^{-3}, electron temperature of 1-5 eV, ion bombardment energy of 100-1000 eV, fractional ionization of 10^{-6}-10^{-3}.

High Density Plasma (HDP) Reactors - A disadvantage of the diode capacitively-coupled reactor is that the plasma (electron and ion) density can't be controlled independently of the ion bombardment energy. As the applied RF voltage amplitude is increased to increase the plasma density, so does the sheath potential and consequently the ion bombardment energy. Excessive ion energy is not beneficial as it can lead to unwanted sputtering and heating of the wafer. Furthermore, etching of modern 100 nm device structures requires extreme directionality of the impinging ions. This can only be achieved by avoiding ion collisions in the sheath. Finally, etching of wafers with ever increasing diameter demands a uniform plasma over large areas. In an effort to satisfy these requirements, high (charge) density, low (gas) pressure plasmas have been developed[2]. Typical operating parameters are pressure of 0.5-50 mtorr, power of 100-5000 W, frequency of 1-2450 MHz, plasma density of 10^{10}-10^{13} cm^{-3}, electron temperature of 2-12 eV, ion bombardment energy of 20-300 eV, and fractional ionization of 10^{-4}-10^{-1}.

Inductively coupled plasma (ICP) reactors (Fig. 2) are particularly attractive variants of HDPs because their design is relatively simpler and they are scaleable to large diameter substrates[3]. In ICPs, the plasma is excited in a cylindrical chamber (r, z, θ) by a helical (solenoidal) or planar (stovetop-type) coil powered at radio frequencies, for example 13.56 MHz. The coil current induces a time-varying magnetic field which in turn induces an azimuthal (in the θ direction) electric field that couples power to the plasma, i.e., heats the plasma electrons. For common excitation frequencies (less than the plasma frequency), the electromagnetic fields are absorbed by the plasma within the *skin depth*. For typical conditions, fields penetrate a few cm into the plasma. The power is deposited non-uniformly in the shape of a toroid.

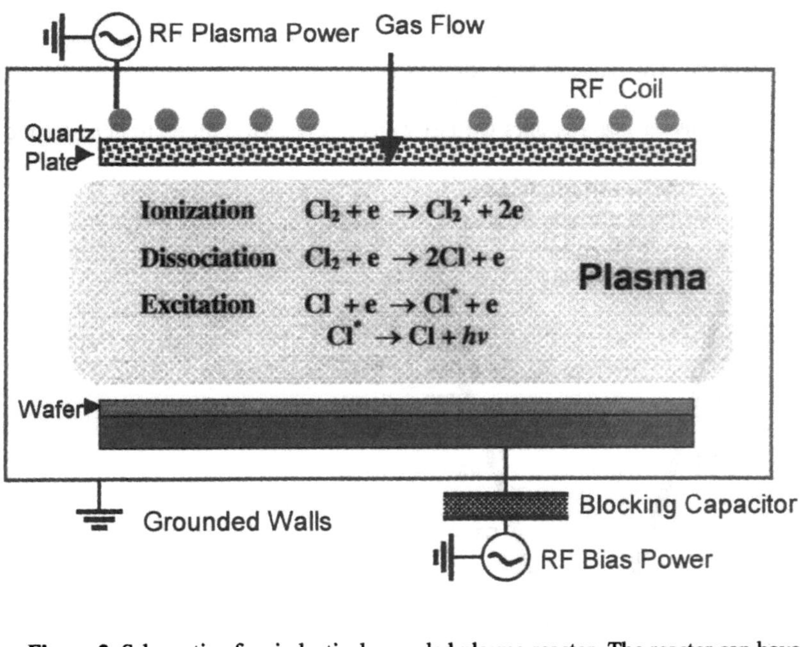

Figure 2. Schematic of an inductively coupled plasma reactor. The reactor can have a helical (solenoidal) coil or a planar (stove-top) coil shown here. The substrate electrode can be biased independently to adjust the energy of ions bombarding the wafer.

Because of the low pressure, however, species diffusion is facile and the plasma fills the whole reactor. In the absence of any capacitive coupling from the coil, the plasma potential is relatively low (~ 20 V) thus minimizing unwanted sputtering of the reactor walls. Capacitive coupling from the coil can result in larger plasma potentials. The wafer platen can be biased independently by a separate RF power supply in order to control the energy of ions bombarding the wafer. As in the case of capacitive coupling from the coil, larger (but now controllable) time-dependent plasma potentials are then established.

3. PLASMA REACTOR SIMULATION

Modeling and simulation of plasma systems has emerged as a tool for enhancing one's intuition about the physicochemical processes occurring in the plasma, for understanding the complex spatiotemporal plasma dynamics, and for assisting in the design of new reactors or the optimization of existing ones. Collections of papers on plasma modeling and simulation can be found in special issues of IEEE Transactions on Plasma Science[4].

Disparate Length and Time Scales - One of the central issues in modeling the behavior of plasma processes is the disparity in length and time scales (Fig 3; see also Fig. 1). Length scales range from atomistic, to microscopic (microfeature width), to mesoscopic (sheath, cluster of features) to macroscopic (reactor, wafer). Even if one focuses on reactor scale simulations, the presence of the extremely thin sheath in high density plasmas introduces a wide range of length scales from 100s of microns to 10s of cm. The range of time scales is also extremely wide, from ps for the ion bombardment collision cascade, to the ns response time of electrons, to μs for positive ions in the plasma, to 10-100s of ms for heavy species chemistry and gas residence times, to mins for the duration of etching processes.

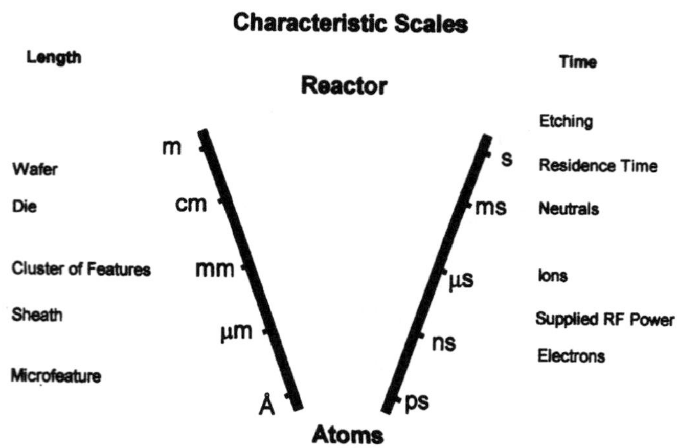

Figure 3. Disparity in length and time scales in plasma processing. From atoms to reactor.

Reactor Scale Simulations - The reactor scale simulation task is split into "modules" (see Fig. 4) to separate the disparate time scales of electron, ion, and neutral transport. This is essentially an equation splitting approach. Calculation of the electron velocity distribution function (EVDF) by solving the Boltzmann equation is part of the electron transport and reaction module. The EVDF determines the space- and time-dependent electron energy and transport properties (mobility, diffusivity) , as well as the electron impact reaction rate coefficients for given electric and magnetic fields and plasma gas composition profiles. The electromagnetics (EM) module solves for the electric and magnetic field distributions in the reactor using the Maxwell equations. The neutral and ionic species concentrations are obtained from the corresponding modules. Information is cycled among the modules until a converged solution has been obtained. There is no guarantee that the modular approach will converge, although it usually does so in practice.

Plasma reactor simulations range from zero-dimensional (well-mixed) to three-dimensional. Well mixed and one-dimensional models (including plug flow models) are best for sorting out the complicated gas and surface chemistry to arrive at a reduced reaction set for use in multidimensional simulations. Two-dimensional simulations can address the important aspect of reaction uniformity across the wafer radius. Three-dimensional simulations are useful for studying azimuthal asymmetries in the reactor due to non-axisymmetric power deposition, or non-axisymmetric gas inlets and pumping ports[5,6].

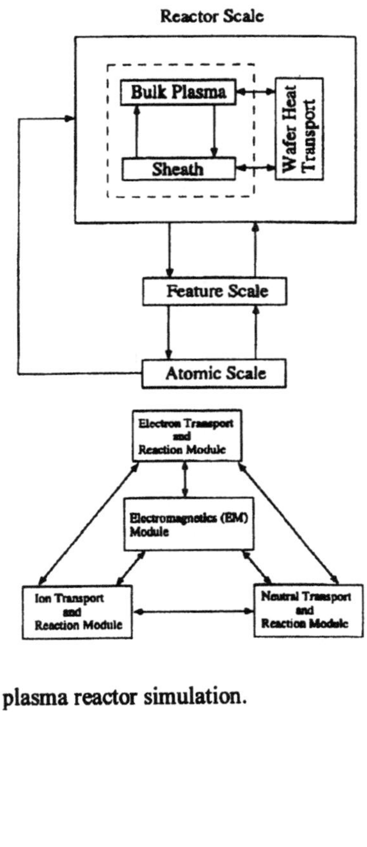

Figure 4. Modules used for plasma reactor simulation.

There are three kinds of glow discharge simulations: fluid, kinetic and hybrid. *Fluid* simulations use moments of the Boltzmann equation describing species density, momentum and energy conservation[7]. They require some assumptions regarding the species distribution function to achieve closure of the equations. An electromagnetics module is also required to compute the power depositions profiles. The set of equations along with the corresponding boundary conditions describing this type of simulation has been given elsewhere[8,9]. For most conditions of interest, the flow is laminar (Reynolds number <1) and diffusive (Peclet number <1) except for the gas inlet ports where convective velocities can be high. Judicious approximations can be made[10] to reduce the computational burden and arrive at TCAD (Technology Computer-Aided Design) simulation tools. *Kinetic* simulations yield the particle distribution functions as an output of the simulation. They are considered more accurate than fluid simulations at low pressures when the species mean free path is comparable to or longer than a characteristic length scale (Kn > 0.2) or for highly non-equilibrium situations. Kinetic simulations are computationally intensive as compared to fluid simulations. They include Particle-In-Cell with Monte Carlo Collisions (PIC-MCC) and Direct Simulation Monte Carlo (DSMC) approaches. *Hybrid* simulations have been developed[8], in an attempt to preserve the accuracy of kinetic simulations and at the same time reduce the computational burden. A common hybrid procedure is to obtain the electron transport (mobility, diffusivity) and reaction rate coefficients (e.g., ionization rate) by solving the Boltzmann equation via a Monte Carlo method. These are then used in the species mass and momentum continuity equations to obtain species densities and fluxes.

4. PULSED ELECTRONEGATIVE PLASMAS

Recently, it has been recognized that pulsed plasma operation (in which the plasma power is modulated with a given period and duty cycle) can ameliorate anomalous etch profiles (e.g., notching) and other undesirable effects that occur in conventional continuous wave (cw) discharges. Due to the large difference in electron and ion temperatures in a cw discharge, positive ions bombarding the wafer have a strongly anisotropic velocity distribution, while that of electrons is almost isotropic. As a result, positive ions can penetrate deeply into a trench charging the trench bottom positively, while electrons deposit their charge on the sidewalls of the trench. Such differential charging, and the associated etch profile distortion, are thought to be a major problem for the fabrication of the next generation of microdevices.

The situation becomes more favorable in pulsed-power discharges[11]. Studies indicate that extraction of charged species from the afterglow plasma and their acceleration towards a radio frequency (RF) biased electrode could dramatically reduce charging damage and improve etching characteristics. For instance, greatly improved etch performance has been reported in pulsed-power chlorine discharges by applying a low frequency bias to the substrate in the afterglow of an inductively coupled plasma source. It was suggested that low frequency bias results in alternate acceleration of positive and negative ions extracted out of the plasma. Since positive and negative ions have almost equal masses and similar velocity anisotropy under the influence of RF bias, the net charge deposited on the wafer surface should be greatly reduced, ameliorating charge-induced damage.

Figure 5 shows the results of a 1-dimensional simulation of a high density pulsed plasma in chlorine[12]. The pulse period was 100 μs and the duty cycle was 50 %. The power density was 1.0 W/cm^3 and the pressure was 20 mtorr. The negative ion Cl^- density is comparable to the electron density during the plasma-on phase, and exceeds greatly the electron density during plasma-off (Fig. 5, left). When power to the plasma is turned off (at 50 μs), electrons cool down (T_e decreases) and are lost by dissociative attachment to molecular chlorine to form Cl^-, and by diffusion to the walls. At the same time the sheath potential collapses, and late enough in the afterglow the electron density has decreased to very low values, leaving behind what is essentially a positive ion - negative ion (ion-ion) plasma (phase IV shown in the figure). When the sheath potential becomes comparable to the negative ion thermal ene gy, negative ions can leave the plasma. This is reflected in the time evolution of the particle fluxes (Fig. 5, right). When the plasma is on, the sheath potential is large enough (several kT_e) o be able to contain all negative ions within the plasma.

5. ION-ION PLASMAS

Ion-Ion plasmas are plasmas in which negative ions replace electrons as the only (or the dominant) negative charge carriers. The electrostatic fields in an ion-ion plasma are determined by negative ions instead of electrons. In conventional electron-ion plasmas, the origin of electrostatic fields is to balance the wall losses of the lighter and more energetic electrons from the plasma with that of the heavier and colder positive ions. By replacing electrons with negative ions in the plasma, the mass and energy of the negatively-charged and positively-charged species become comparable. Ion-ion plasmas, therefore, are characterized by significantly weaker electrostatic fields with a plasma potential of the order of the ion temperature. After the transition to an ion-ion plasma (Fig. 5), electrons are no longer confined in the plasma and are lost at a ever faster rate which eventually approaches the limit of free diffusion of electrons to the walls in the absence of electrostatic fields. The positive-ion and negative-ion densities decay together much more slowly due to ion-ion recombination and diffusion losses to the walls.

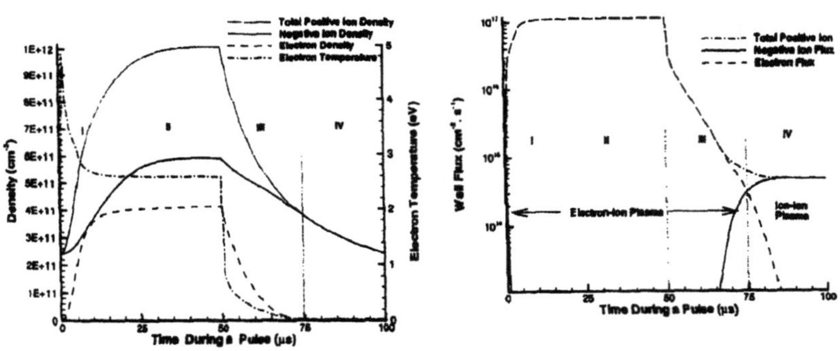

Figure 5. Time variation of species densities (left) and fluxes (right) predicted by a 1-D model of a high density pulsed plasma through chlorine. Power 1.0 W/cm^3, pressure 20 mtorr, pulse period 100 μs, and duty cycle 50%.

Without applied bias in an ion-ion plasma with equal ion mass (e.g., Cl^+ and Cl^-) and temperature, the profile of positive-ion density simply overlaps with the negative-ion density, and there is no sheath formed near the electrodes. In the presence of a bias potential, however, the positive and negative ions separate to form ion-ion sheaths near the electrode. Processing with ion-ion plasmas in which a bias is applied in the late afterglow may offer unique possibilities. For example, charging and notching effects associated with conventional cw electron-ion plasma processing may be reduced due to alternate bombardment of the substrate by positive and negative ions[11,13].

Results shown below are for a one-dimensional ion-ion (Cl^+-Cl^-) plasma simulation (no electrons) at 20 mtorr and 130 V peak-to-peak applied bias[14]. The bias (sinusoidal waveform) was applied to the left electrode, such that the electrode had a negative potential during the first half of the cycle ($0<\tau<0.5$) and a positive potential during the second half of the cycle ($0.5<\tau<1$). The right electrode was kept at zero potential at all times. The peak ion density was 10^{11} cm^{-3}.

Figure 6 (top left) shows the evolution of ion fluxes at the left electrode during a RF cycle for a bias frequency of 100 KHz. For this low operating frequency, the left electrode is bombarded alternately by positive ions (dashed line) and negative ions (solid line) during each half of the RF cycle. The peak in the negative-ion flux is slightly less than the peak in positive-ion flux due to continuous depletion of the plasma. At this relatively low operating frequency, a significant fraction of the plasma is depleted during the course of each half cycle. At any instant of the cycle, the flux of positive ions exiting the plasma at one electrode is equal to the flux of negative ions exiting the other electrode (not shown). Figure 6 shows that the peak ion flux during a half cycle is comparable to twice the diffusion flux of the ions in the absence of a bias.

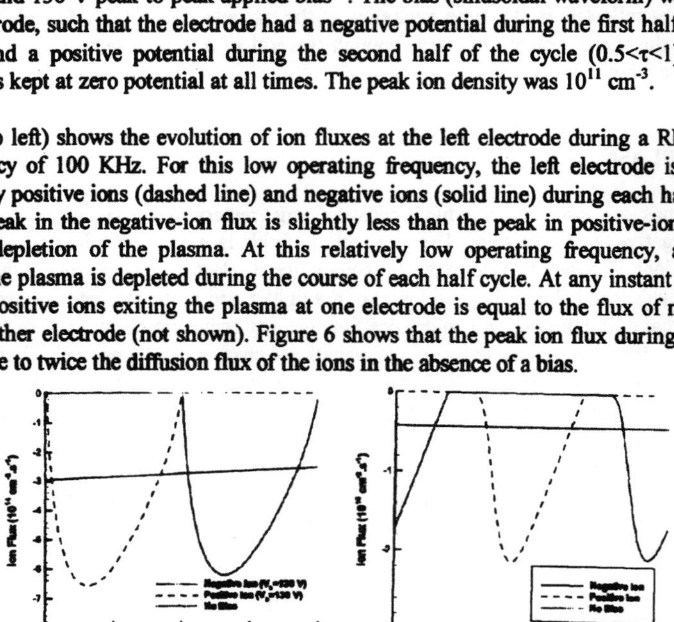

Figure 6. Biasing an ion-ion plasma: positive and negative ion flux on the biased electrode vs. time in RF cycle for a frequency of 100 kHz (top left), 10 MHz (top right) and 60 MHz (bottom).

The total ion flux integrated over a full cycle is equivalent to the diffusion flux of ions without a bias. Therefore, for low bias frequencies, the flux of ions bombarding the electrode is independent of magnitude of the applied bias and is limited by the ion diffusion flux. The only effect of the applied bias is to temporally redistribute this diffusion flux between the two electrodes. Instead of exiting both electrodes simultaneously, the positive ions exit only the negatively-biased electrode while the negative ions exit only the positively-biased electrode during each half cycle. The RF bias, however, increases the ion energy.

Figure 6 (top right) shows the evolution of the ion fluxes at the left electrode during the course of a RF cycle for a bias frequency of 10 MHz, which is close to the ion plasma frequency evaluated at the sheath edge. The effects of the transit time of ions through the sheath are clearly evident in the evolution of fluxes. The positive-ion flux continues till $\tau=0.7$ even when the electrode is positively biased with respect to the plasma. Negative-ion flux does not appear till $\tau=0.8$ representing the transit time of the negative-ion fronts through the sheath. As opposed to the low frequency case, the ion flux is significantly greater than the corresponding diffusion flux (dotted line) without an applied bias. At this bias frequency, there is insufficient time for the applied bias potential to be fully shielded in the sheath regions. Due to the presence of relatively high electric fields in the bulk plasma, the ion flux to the walls exceeds the diffusion flux. Consequently, a relatively high bias frequency, approaching the (ion) plasma frequency is required to extract ions at a faster rate from the plasma.

Figure 6 (bottom) shows the evolution of the ion flux at the left electrode during a RF cycle for a bias frequency of 60 MHz. Since the ions are no longer able to respond to the applied frequency, the perturbation in the ion velocities due to the applied bias is small. The ion fluxes, therefore, are only weakly modulated by the applied bias. The electrode is now <u>continuously</u> bombarded by both positive and negative ions throughout the cycle with an average flux equal to the diffusion flux of the ions.

Based on the ion flux evolution at the electrode, operation in the intermediate frequency regime appears advantageous for rapidly extracting ions from the plasma at the base conditions (20 mtorr pressure and bias potential of 130 V). At low frequencies (e.g., 100 kHz), alternate bombardment by positive ions and negative ions is possible but the ion flux is limited by the diffusional flux in the bulk plasma. In addition, if the bias period is comparable to the residence time of the charge on device features, devices will still have adequate time to charge during each half of the bias cycle. Therefore, charge damage and etch profile distortions may still occur during each half cycle of the pulse. Alternate bombardment of the electrode by positive and negative ions is also possible with intermediate values of the frequency (e.g., 10 MHz), albeit with a finite phase lag with respect to the applied bias. There is less time for collisional drag to limit the ion flux at intermediate frequencies and ions may be extracted at a faster rate from the plasma compared to low frequencies. In addition, less time is available during each half cycle for charging and etch profile distortions to occur. Increasing the bias frequency (e.g., 60 MHz) beyond the ion plasma frequency, results in trapping the ions in the plasma and the ion flux at the electrode is essentially reduced to the diffusion flux of the ions.

6. CONCLUDING REMARKS

Multidimensional simulations of plasma reactors can serve as powerful tools for improving our understanding of reactive plasmas, and for helping in the design of new and improved plasma processes. Further understanding of plasma physics and chemistry as well as developments in numerical methods and in parallel computing will have a profound impact on multidimensional simulations of glow discharges. Efficient treatment of the disparate length and time scales will be the key to the further development of predictive computational models for plasma engineering. In general, a combination of well-defined experiments, plasma diagnostics, and mathematical modeling/simulation will continue to be the best combination to unravel the intricacies of plasmas and the plasma-surface interactions. Electronegative discharges are of particular interest since they find widespread application in microelectronics manufacturing. Ion-ion plasmas can be formed in the late afterglow of pulsed electronegative discharges. Application of a RF bias to an ion-ion plasma can result in alternate bombardment of the substrate by positive ions and negative ions, which can reduce charging damage.

Acknowledgments
The author would like to thank the National Science Foundation (CTS-0072854 and CTS-0227232), and the State of Texas (Texas Advanced Technology Program) for financial support of this work.

7. REFERENCES

1. S. M. Rossnagel, J. J. Cuomo, and W. D. Westwood (eds.), *Handbook of Plasma Processing Technology*, Noyes Publications, Park Ridge, NJ (1990).
2. M.A. Leiberman and A.J. Lichtenberg, *Principles of Plasma Discharges and Materials Processing*, John Wiley and Sons, New York, NY(1994).
3. J. Hopwood, *Plasma Sources Sci. Technol.* **1**, 109 (1992).
4. See papers in special issue of *IEEE Trans. Plasma Science*, **31**(4), 2003; **27**(5), 1999; **23**(4) 1995; **19**(2) 1991; **14** 1986.
5. M. J. Kushner, *J. Appl. Phys.* 82, 5312 (1997).
6. T. Panagopoulos, V. Midha, D. Kim and D. J. Economou, *J. Appl. Phys.*, **91**, 2687 (2002).
7. M. Meyyappan, ed., *Computational Modeling in Semiconductor Processing*, Artech House, Boston, MA (1994).
8. P. L. G. Ventzek, R. J. Hoekstra and M. J. Kushner, *J. Vac. Sci. Technol. B* **12**, 461 (1994).
9. D. J. Economou, "Plasma Reactor Engineering," in *Advances in Electrochemistry and Electrochemical Eng.*, Vol. **6**, p. 237, ed. R. C. Alkire and D. M. Kolb, Wiley-VCH (1999).
10. D. J. Economou, J. Feldsien, and R. S. Wise in *Electron Kinetics and Applications*, U. Kortshagen and L. D. Tsendin (eds.), p. 367, Plenum, New York, NY (1998).
11. S. Samukawa and T. Mieno, *Plasma Sources Sci. Technol.* **5**, 132 (1996); T. H. Ahn, N. Nakamura, and H. Sugai, *Plasma Sources Sci. Technol.* **5**, 139 (1996).
12. V. Midha and D. J. Economou, *Plasma Sources Sci. Technol.*, **9**, 256 (2000).
13. S. K. Kanakasabapathy, L. J. Overzet, V. Midha, and D. J. Economou, *Appl. Phys. Lett.*, **78**, 173 (2001).
14. V. Midha and D. J. Economou, *J. Appl. Phys.*, **90**, 1102 (2001).

PLASMA POLYMERIZATION IN AN ATMOSPHERIC PRESSURE DIELECTRIC BARRIER DISCHARGE IN A FLOWING GAS

Rüdiger Foest, Martin Schmidt, and Jürgen F. Behnke*

1. INTRODUCTION

The plasma of atmospheric pressure dielectric barrier discharges (DBD) has been used for technical applications starting with ozone generation by Siemens 1857. Non-thermal atmospheric pressure discharges are of considerable interest because the industrial application requires no vacuum devices. In recent years, the use of the DBD for surface treatment has been reported, e.g. for the treatment of foils to improve printability, for surface cleaning and thin film deposition[1,2]. The DBD is known as a strongly inhomogeneous, filamentary discharge. A non filamentary, homogeneous DBD was observed in He, other gases, and mixtures. This type is called atmospheric pressure glow (APG) discharge and is of importance for surface treatments. In this contribution, hexamethyldisiloxane ($(CH_3)_3SiOSi(CH_3)_3$, HMDSO), diluted in He as carrier gas, is used as precursor for the deposition of thin films by plasma polymerization in an APG. The static deposition profiles are measured and compared with results of a model which includes monomer activation and thin film growth.

2. EXPERIMENTAL

The experiments were carried out in flowing He (2.5 slm) with small admixtures of HMDSO ($3 \cdot 10^{-6}$) under atmospheric pressure conditions with a DBD housed in a vacuum tight cylindrical glass vessel. The upper plane electrode con-

* Rüdiger Foest, Martin Schmidt, Institut für Niedertemperatur-Plamaphysik Greifswald, Germany,
Jürgen F. Behnke, Institut für Physik, Ernst Moritz Arndt Universität Greifswald, Germany

Gaseous Dielectrics X.
Edited by Christophorou *et al.*, Springer, New York, 2004.

sists of two parallel metallic stripes (80 mm x 15 mm) covered with glass as the dielectric barrier. The lower electrode is a plain aluminum substrate. The gas flows through a slit of 2 mm between the upper electrodes into the discharge region. The discharge gap width was kept at 0.8 mm. The discharge was operated with a sinusoidal Voltage at 100 kHz. The time dependent voltage and current signals, measured with oszilloscope, served to calculate the power consumption and gap voltage[3,4]. The sustaining voltage varied from 1.0 – 1.6 kV_{pp} and the power from 1 - 5 W. The light emission is measured by a photo multiplier. The gas phase is analyzed by a quadrupole mass spectrometer (Balzers Omnistar, mass range 300 u), linked by a heated capillary to the chamber, which is positioned at the outer side of the electrode, directly at the discharge gap where the gas leaves the discharge region. The duration of the deposition experiments is about 60 min. The deposition profile on the Al surface is measured by a DEKTAK 3ST profilometer.

3. RESULTS

Deposition profiles for different powers are presented in Figure 1. They are characterized by a sharp maximum near the electrode edge at the gas flow input and decrease in flow direction.

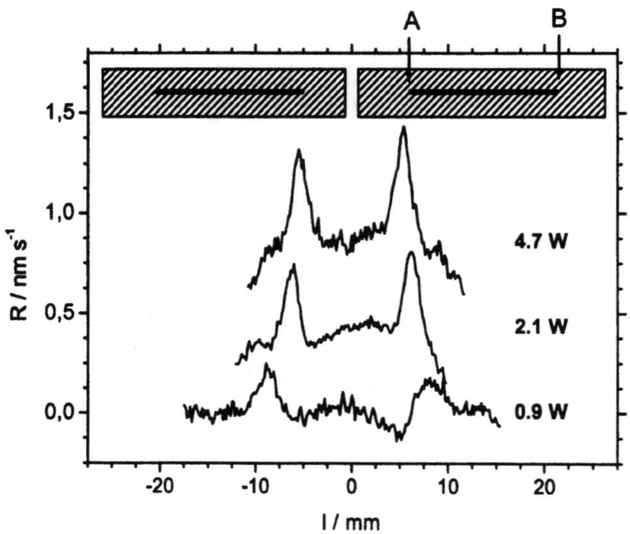

Figure 1. Experimental deposition profiles for different powers. The curves are shifted vertically for better visibility. The position of the upper electrode construction with dielectrics and embedded metal (from A to B) is indicated schematically. Gas flows down the central slit and in both directions (left and right).

The maximum deposition rate increases with power. The film deposition is associated with a consumption of monomer, verified by mass spectrometry. The monomer concentration decreases by nearly one order of magnitude.

4. DISCUSSION

A known model[5,6] is applied to calculate the film deposition at the wall of a flow system based on monomer activation inside the plasma and the film formation by the activated species. In our case the deposition occurs at the electrode surfaces. The activated particles, here designed as radicals, are neutral radicals or ions. An example of an electron impact activation process of HMDSO is the dissociative ionization[7]

$$(CH_3)_3SiOSi(CH_3)_3 + e^- \rightarrow (CH_3)_3SiOSi(CH_3)_2^+ + CH_3 + 2\,e^-$$

The balance equations for the monomer *M*, the radicals *R*, and equation for the deposition rate *ds*/*dt* are:

$$v\frac{dM}{dz} = D_1\frac{d^2M}{dz^2} - k\ n\ M \tag{1}$$

$$v\frac{dR}{dz} = D_2\frac{d^2R}{dz^2} + k\ n\ M - \frac{1}{\tau}R \tag{2}$$

$$\frac{ds}{dt} = \frac{m\ a}{2\ \rho\ \tau}R \tag{3}$$

The solution is:

$$M = M_0 \exp(\Gamma_1 z) \tag{4}$$

$$R = kn M_0 \frac{\exp(\Gamma_2 z) - \exp(\Gamma_1 z)}{D_2\Gamma_1^{\ 2} - v\Gamma_1 - 1/\tau} \tag{5}$$

$$\frac{ds}{dt} = \frac{ma\,kn\,M_0}{2\rho\tau}\frac{\exp(\Gamma_2 z) - \exp(\Gamma_1 z)}{D_2\Gamma_1^{\ 2} - v\Gamma_1 - 1/\tau} \tag{6}$$

Where Γ_1 and Γ_2 are defined as:

$$\Gamma_1 = \frac{v}{2D_1} - \sqrt{\left[\frac{v}{2D_1}\right]^2 + \frac{kn}{D_1}} \qquad \Gamma_2 = \frac{v}{2D_2} - \sqrt{\left[\frac{v}{2D_2}\right]^2 + \frac{1}{D_2\tau}} \qquad (7,8)$$

Here v is the gas flow velocity in the gap a, z the flow direction, M the monomer concentration, and M_0 its concentration at the gas input. D_1 and D_2 are the diffusion coefficients in the carrier gas of the monomer and the radicals, respectively. k is the rate coefficient for radical formation, n the concentration of the collision partner (electrons, excited particles). τ is the life time of the radicals, m the mass which is transported into the film per radical, and ρ the density of the film of the thickness s.

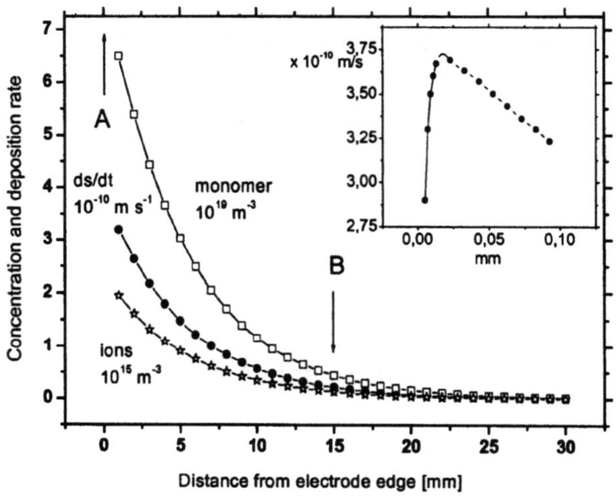

Figure 2. Calculated deposition rate along with monomer and ion concentrations in dependence on the distance of the electrode edge for kn = 50 1/s. A, B indicate the electrode position (see Fig. 1)

The assumption of thin film deposition by ions is justified by the essentially higher ionization probability of the HMDSO molecule in relation to the He atoms[7]. The small admixture of HMDSO to the He carrier gas leads to a dominant HMDSO ion population in the plasma. The results of equations (4-6) are presented in Figure 2. The monomer concentration M, the deposition rate ds/dt and the ion concentration are shown in dependence on the distance of the electrode edge. The region at the electrode edge is zoomed in the upper left diagram where the progression of the deposition rate at its maximum is plotted. A and B indicate the plasma region below the electrode (see also Figure 1). D_1 ($2.3\cdot10^{-6}$ m^2/s) is determined by $c\lambda/3$, with the mean thermal velocity c and the mean free path λ of the monomer molecules. For D_2 ($2\cdot10^{-3}$ m^2/s) the ambipolar diffusion coefficient $\mu_i U_e$ is used, where μ_i denotes the ion mobility and U_e the Volt equivalent of the electron energy. Here, a value of 1 V is assumed. The life time τ is estimated by the ion drift velocity in the gap ($6 \cdot 10^{-7}$ s). The film density ρ equals $1 \cdot 10^3$ kg/m^3.

The deposition rate $3.7 \cdot 10^{-10}$ m/s is close to the experimental values for 2.1 W and 4.7 W. The ion concentration of $2 \cdot 10^{15}$ m^{-3} corresponds to a value averaged over the period. Experimental data[3] of electron concentrations in an atmospheric pressure glow discharge are two orders of magnitude higher ($3 \cdot 10^{17}$ m^{-3}), but these data are related to the active phase of this discharge which accounts only for a few hundredths of the period. The calculated value is therefore in the same order of magnitude. The decay of the monomer concentration by one order of magnitude at the end of the plasma region (15 mm) agrees quite well with the experimental value. The calculated decay of the deposition rate is smaller than the experimental result. The value $kn = 50$ can be verified. For a value of $E/p =$ 3V/cm torr a rate coefficient for HMDSO electron impact ionization of $3 \cdot 10^{-9}$ cm^3/s is valid[4]. The electron density may be taken from the experimental results[4] as $3 \cdot 10^{17}$ m^{-3}. With a time averaged value of $3 \cdot 10^{15}$ m^{-3} we receive $kn \sim 10$ s^{-1}. This value lies within the same order of magnitude as the kn value used. Moreover, it can not be excluded that there is also an additional ionization of HMDSO molecules by other mechanisms, like collisions with excited He atoms which would increase the effective kn value. As discussed in detail[3,4], the ion current which flows to the electrode surface is sufficient to account for the measured thin film deposition rate.

For the calculation of thin film formation by neutral radicals the diffusion coefficient D_2 ($2.39 \cdot 10^{-6}$ m^2/s) is determined with $c\lambda/3$ for the radical $(CH_3)_3SiOSi(CH_3)_2$. The lifetime can be estimated based on the assumption of a diffusion current from the center of the gap to the electrode surface and a negligible radical concentration on the surface:

$$\tau = \frac{d^2}{8 D_2} \tag{8}$$

The calculation according to equations (4)-(6) with kn 50 1/s and τ $3.3 \cdot 10^{-2}$s provides a maximum deposition rate of $1 \cdot 10^{-10}$ m/s. Its position is shifted 7 mm in flow direction. With kn 200 1/s the maximum rate equals $1.7 \cdot 10^{-10}$ m/s, and it is shifted only 3 mm. In this case the monomer concentration after leaving the discharge region is reduced by 4 orders of magnitude. The supposition of film formation by neutral particles leads to smaller deposition rates compared to the experimental results. The maximum is shifted to higher z values. The assumption of a higher kn value with reasonable shift, however, produces monomer decays that are too strong.

By comparing both cases with the experimental results the conclusion can be drawn, that the film formation in He DBD with small HMDSO admixtures is determined by ions.

5. CONCLUSIONS

The deposition profile of plasma polymers of the monomer hexamethyldisiloxane as small admixture in He in a homogeneous dielectric barrier discharge is compared with the results of a simple model of monomer activation and thin film formation by activated particles in a flowing gas. The maximum deposition rate and the decay of the monomer concentration in the gas flow support the assumption of the film formation by ions. This picture is be valid for low monomer concentrations. The dominance of neutral radicals is expected for higher concentrations according to the increased formation probability of these species.

6. ACKNOWLEDGEMENTS

The financial support by BMBF Project No. 13N7351 is gratefully acknowledged.

7. REFERENCES

1. U. Kogelschatz, Dielectric-barrier discharges: Their history, discharge physics, and industrial applications, *Plasma Chem. Plasma Processing* **23(1)**, 1-46 (2003).
2. M. Schmidt and J.F. Behnke, Surface treatment, in: *Non-equilibrium air plasmas at atmospheric pressure*, edited by K. Becker, R. Barker, and K. Schoenbach (IOP, Bristol, to be published 2004).
3. R. Foest, F. Sigeneger, and M. Schmidt, Deposition of plasma polymer films by an atmospheric pressure glow discharge, in: *Gaseous Dielectrics IX*, edited by L.G. Christophorou and D.R. James, (Kluwer Academic/Plenum Press, New York, 2001), pp. 113-118.
4. R. Foest, F. Adler, F. Sigeneger, and M. Schmidt, Study of an atmospheric pressure glow discharge (APG) for thin film deposition, *Surf. Coat. Technol.* **163-164**, 323-330 (2003).
5. M. Schmidt, K.D. Schulz, and M. Maas, Glow polymerisation of hexamethyldisiloxane in a flow system, *Proc. XIIIth Int. Conf. on Phenomena in Ionized Gases*, Berlin, Contributed Papers **1**, 381 (1977).
6. H. Deutsch and M. Schmidt, On the quantitative treatment of the growth rate of thin polymer films produced in glow discharges, *Contr. Plasma Phys.* **21**, 279-292 (1981).
7. R. Basner, R. Foest, M. Schmidt, K. Becker, and H. Deutsch, Absolute total and partial electron impact ionization cross sections of hexamethyldisiloxane, *Int. J. Mass Spectrom.* **176(3)**, 245-252 (1998).

ASSESSED TOTAL AND PARTIAL IONIZATION CROSS SECTIONS FOR CF_4, C_2F_6, C_3F_8, CHF_3, CF_3I, *c*-C_4F_8, Cl_2, CCl_2F_2, BCl_3, SF_6, AND FRAGMENTS OF CF_4 AND SF_6

Loucas G. Christophorou* and James K. Olthoff**

1. INTRODUCTION

We have recently completed an updated review, synthesis, assessment, and discussion of low-energy electron interactions with the plasma processing gases CF_4, C_2F_6, C_3F_8, CHF_3, CF_3I, *c*-C_4F_8, Cl_2, CCl_2F_2, BCl_3, SF_6. This work, along with fundamental knowledge on electron-molecule collision processes, the definition of the various electron collision cross sections and rate coefficients, and the experimental and theoretical methods used for their determination have been described in Christophorou and Olthoff.[1] It has been shown in this reference, that reliable data on the electron transport, electron attachment, and electron-impact ionization coefficients generally exist for most of these molecules, except for electron transport in strongly electronegative gases. Also, reliable data exist on the collision cross sections for most of these molecules, but this knowledge depends on the collision process and the molecule under consideration. The cross sections and coefficients that have been recommended or suggested[1] were based on experimental measurements. This work[1] has shown that while a great deal of quantitative data have become available, especially recently, for the molecules under consideration, there is still a pressing need for quantitative measurements of the cross sections for electron-impact dissociation into neutral fragments, direct and indirect vibrational (and electronic) excitation, and electron interaction processes involving excited and transient (radical) species.

---

* Academy of Athens, 28 Panepistimiou, Athens 108 79, Greece
** National Institute of Standards and Technology, Gaithersburg, MD 20899-8170, USA

Gaseous Dielectrics X.
Edited by Christophorou *et al.*, Springer, New York, 2004.

Substantial amounts of new data have recently become available for these ten plasma processing gases, especially on their total and partial ionization cross sections. These have been reviewed, synthesized, assessed and discussed and have been shown to generally be more accurate than earlier measurements allowing better recommended cross sections.[1] In this paper we present total and partial ionization cross sections for CF_4 and SF_6 as examples of these data. We also present a comparison of experimental data and calculated results on the electron-impact ionization cross sections for the fragments of CF_4 and SF_6.

2. EXAMPLES OF RECENTLY RECOMMENDED TOTAL IONIZATION CROSS SECTIONS OF MOLECULES

Substantial amounts of new data have recently become available for these cross sections which have been reviewed, synthesized and discussed in Ref. 1. Examples of the recently recommended $\sigma_{i,t}$ (ε) are shown in Figs. 1 and 2 for CF_4 and SF_6, respectively.

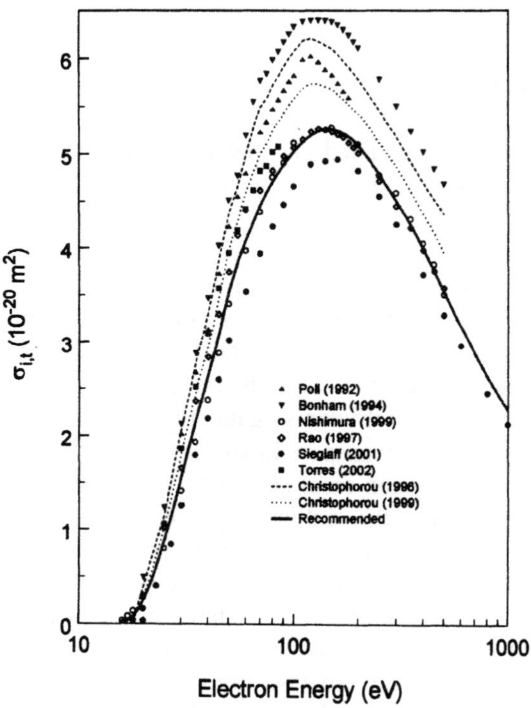

Figure 1. Total ionization cross section $\sigma_{i,t}$ (ε) as a function of electron energy for CF_4. ▲, Ref. 2; ▼, Refs. 3 and 4; ○, Ref. 5; ◇, Ref. 6; •, Ref.7; ■, Ref. 8; - - -, Ref. 9; . . ., Ref. 10; ___, recommended, Ref. 1.

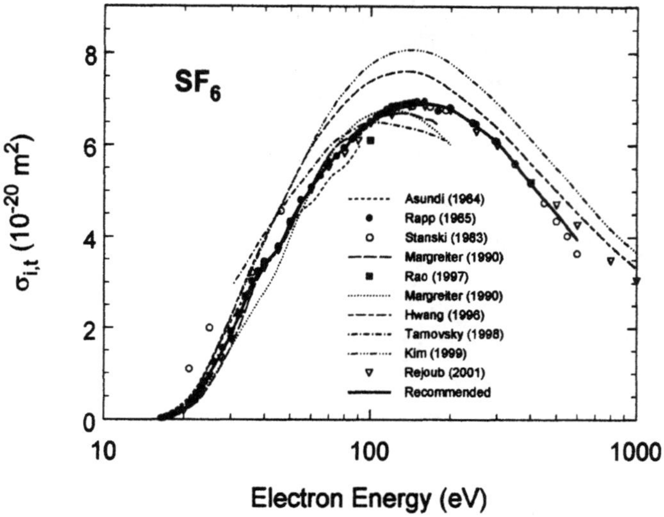

Figure 2. Total ionization cross section $\sigma_{i,t}$ (ε) as a function of electron energy for SF_6. - - -, Ref. 11; •, Ref. 12; ◦, Ref. 13; __ __, Ref. 14; ■, Ref. 15; ▿, Ref. 16; ..., Ref. 17; _ - _, Ref. 18; _ . _, Ref. 19; _ ... _, Ref. 20; ___, recommended.

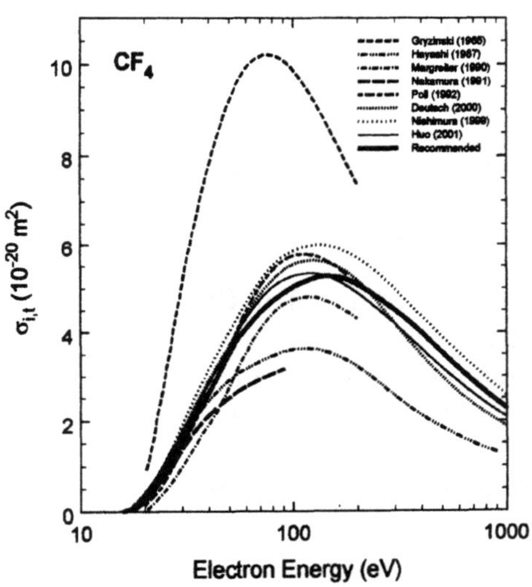

Figure 3. Comparison of the recommended $\sigma_{i,t}$ (ε) for CF_4 with the results of various calculations. __, from Fig. 1; - - -, Ref..21; _ - _, Ref. 2; - . -, Ref. 22; - ... - Ref. 23; _ _, Ref. 24 . . ., Ref. 5; ..., Ref. 25, __, Ref. 26.

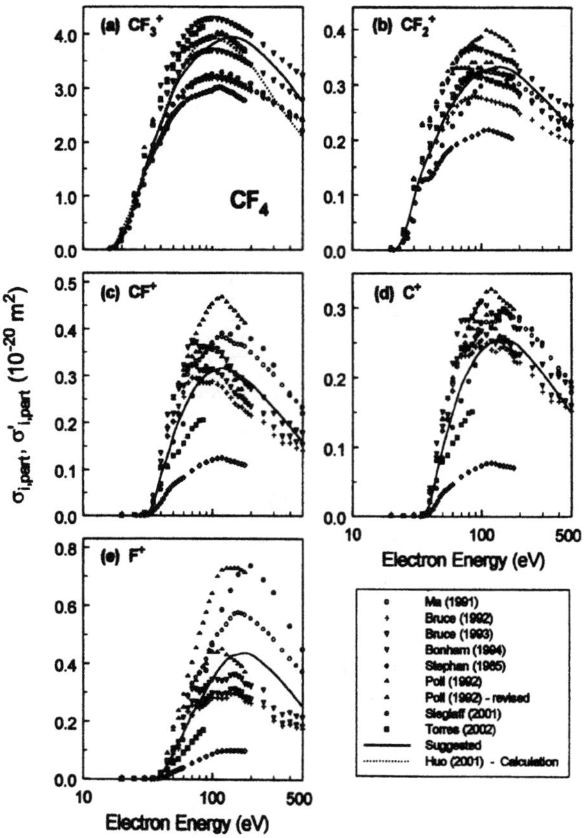

Figure 4. Partial ionization cross section for (a) CF_3^+, (b) CF_2^+, (c) CF^+, (d) C^+, and (e) F^+ by electron impact on CF_4: ○, Ref. 27; +, Ref. 28; ▽, Ref. 4; ▾, Ref. 3; ◇, Ref. 29; △, ▲ Refs. 2 and 1;• Ref. 7 ;■, Ref. 8; ____, suggested. Also shown (. . .) is a calculation of the cross section for CF_3^+ production by electron impact on CF_4 (Ref. 30).

In Fig. 3 the recommended $\sigma_{i,t}$ (ε) of CF_4 is compared with the results of various calculations. The results of the semiempirical (additivity-rule-based) calculations[5,25,26] are in reasonable agreement with experimental data, especially the results of the simple-Binary-Encounter-Dipole (sBED) model of Huo.[26]

3. PARTIAL IONIZATION CROSS SECTIONS FOR CF_4 AND SF_6

In Figs. 4 and 5 are shown the partial ionization cross sections, $\sigma_{i,part}$ (ε), for the formation of various ions by electron impact on CF_4 and SF_6 respectively. Although there is agreement among the measurements of various groups, the partial ionization cross sections for light fragments still seem to be uncertain due to ion discrimination problems.

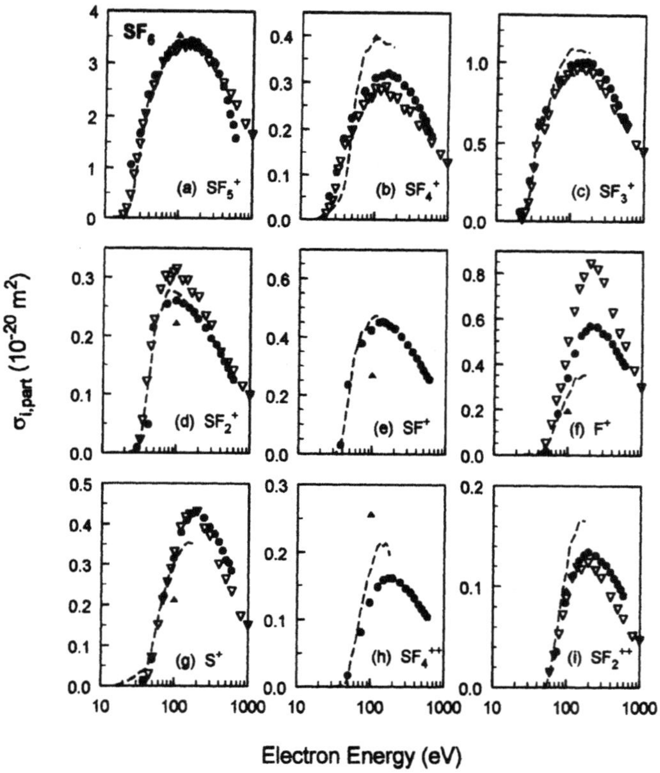

Figure 5. Partial ionization cross sections, $\sigma_{i,part}$ (ε), as a function of electron energy for SF_6. •, Ref. 13; - - -, Ref. 14; ▲, Ref. 15, ▽, Ref. 16.

4. TOTAL IONIZATION CROSS SECTIONS FOR RADICALS OF CF_4 AND SF_6

There are limited data on the partial and total ionization cross sections of radicals. In Fig. 6 are shown the available data on electron-impact ionization of fragments of CF_4. The data on the total ionization cross sections of the three radicals from CF_4 were obtained by summing up the partial ionization cross sections shown in the figure. They are in reasonable agreement with the results of semiempirical calculations.[25,31] This agreement is gratifying because it shows that these semiempirical calculations can be employed to provide values of the total ionization cross sections of radicals and reactive species, which are normally difficult to study experimentally.

Limited data are available for the ionization cross sections of SF_6 radicals. These are compared with the results of semiempirical calculations in Fig. 7.

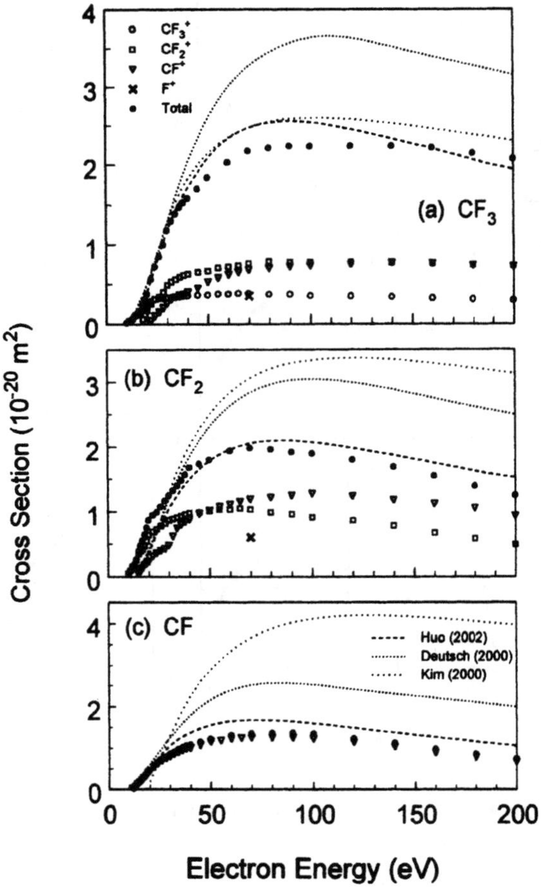

Figure 6. Electron-impact ionization cross section for the fragments (a) CF_3, (b) CF_2, and (c) CF: •, total ionization cross section (Ref. 31);∘, CF_3^+ (Refs. 32, 33);□, CF_2^+(Refs. 32, 33); ▿, CF^+ (Refs. 32, 33); ×, F^+ (Ref. 32). Also shown are calculated total ionization cross sections for each fragment; - - -, Ref. 31; ...., Ref. 25; . . ., Ref. 34.

5. CONCLUSIONS

Based on our comprehensive work[1] on fundamental electron interactions with plasma processing gases, we can conclude that:

- While much progress has been made in our understanding of low-energy electron-molecule interaction processes and reasonable progress has been made in our effort to build up a sound database on such processes, **much is still needed**.

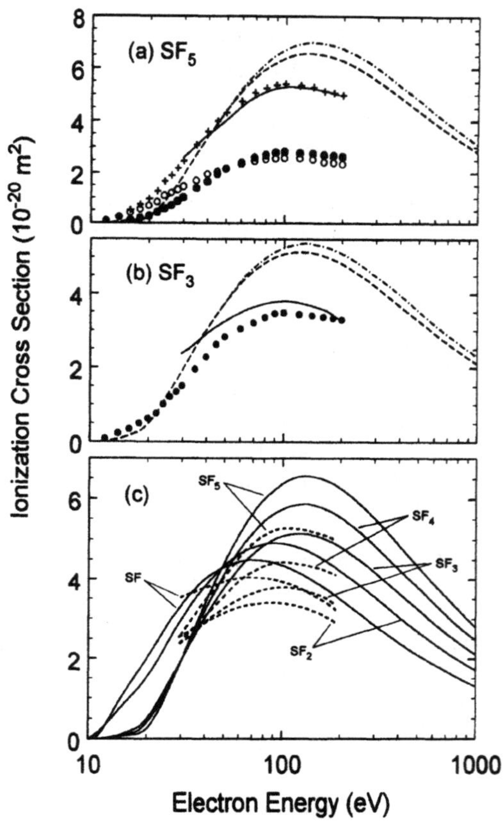

Figure 7. Electron-impact ionization of SF_6 radicals: (a) Cross section for SF_5^+ and SF_4^+ by electron impact on SF_5. Measurements (Ref. 19): ∘, SF_5^+; •, SF_4^+;+, sum. Calculations of the total single ionization cross section of SF_5: __, Ref. 19 - - -, __ . __, Ref. 35. (b) Cross section for SF_3^+ by electron impact on SF_3: •, measurements (Ref. 19); Calculations, __, Ref. 19, ---, _ ._ Ref. 35. (c) Calculated electron-impact total single ionization cross sections for the radicals SF_5, SF_4, SF_3, SF_2, and SF: - - -, Ref. 19, __, Ref. 35 (see Ref. 1)

- The fundamental generality of low-energy electron-molecule interactions, their generic significance to modern technology, and their basic role in life sciences, stresses the **need for further coordinated and quantitative studies in this area.**

5. References

1. L. G. Christophorou and J. K. Olthoff, *Fundamental Electron Interactions with Plasma Processing Gases*, Kluwer Academic/Plenum Publishers, New York, 2004.
2. H. U. Poll, C. Winlker, D. Margreiter, V. Grill, and T. D. Märk, Int. J. Mass Spectrom. Ion Processes **112**, 1(1992).

3. R. A. Bonham, Jpn. J. Appl. Phys. **33**, 4157 (1994).
4. M. R. Bruce and R. A. Bonham, Int. J. Mass Spectrom. Ion Processes **123**, 97 (1993).
5. H. Nishimura, W. M. Huo, M. A. Ali, and Y.-K. Kim, J. Chem. Phys. **110**, 3811 (1999).
6. M. V. V. S. Rao and S. K. Srivastava, in *Proceedings of the Twentieth International Conference on the Physics of Electronic and Atomic Collisions*, Vienna, Austria, July 1997, Vol. II, paper MO 150.
7. D. R. Sieglaff, R. Rejoub, B. G. Lindsay, and R. F. Stebbings, J. Phys. B **34**, 799 (2001).
8. I. Torres, R. Martinez, and F. Castaño, J. Phys. B **35**, 2423 (2002).
9. L. G. Christophorou, J. K. Olthoff, and M. V. V. Rao, J. Phys. Chem. Ref. Data **25**, 1341 (1996).
10. L. G. Christophorou and J. K. Olthoff, J. Phys. Chem. Ref. Data **28**, 967 (1999).
11. R. K. Asundi and J. D. Craggs, Proc. Phys. Soc. **83**, 611 (1964).
12. D. Rapp and P. Englander-Golden, J. Chem. Phys. **43**, 1464 (1965).
13. T. Stanski and B. Adamczyk, Int. J. Mass Spectrom. Ion Phys. **46**, 31 (1983).
14. D. Margreiter, G. Walder, H. Deutsch, H. U. Poll, C. Winkler, K. Stephan, and T. D. Märk, Int. J. Mass Spectrom. Ion Processes **100**, 143 (1990).
15. M. V. V. S. Rao and S. K. Srivastava, in *Proceedings of the Twentieth International Conference on the Physics of Electronic and Atomic Collisions*, Vienna, Austria, July 1997, Vol. II, paper MO 151.
16. R. Rejoub, D. R. Sieglaff, B. G. Lindsay, and R. F. Stebbings, J. Phys. B **34**, 1289 (2001).
17. D. Margreiter, H. Deutsch, M. Schmidt, and T. D. Märk, Int. J. Mass Spectrom. Ion Processes **100**, 157 (1990).
18. W. Hwang, Y.-K. Kim, and M. E. Rudd, J. Chem. Phys. **104**, 2956 (1996).
19. V. Tarnovsky, H. Deutsch, K. E. Martus, and K. Becker, J. Chem. Phys. **109**, 6596 (1998).
20. Y.-K. Kim and M. E. Rudd, Comments At. Mol. Phys. **34**, 309 (1999).
21. M. Gryzinski, Phys. Rev. A **138**, 305 (1965).
22. D. Margreiter, H. Deutsch, M. Schmidt, and T. D. Märk, Int. J. Mass Spectrom. Ion Processes **100**, 157 (1990).
23. M. Hayashi, in *Swarm Studies and Inelastic Electron-Molecule Collisions*, L. C. Pitchford, B. V. McKoy, A. Chutjian, and S. Trajmar (Eds.) (Springer, New York, 1987) p. 167.
24. Y. Nakamura, in *Gaseous Electronics and Their Applications*, R. W. Crompton, M. Hayashi, D. E. Boyd, and T. Makabe (Eds.), KTK Scientific, Tokyo, Japan, 1991, p. 178.
25. H. Deutsch, K. Becker, S. Matt, and T. D. Märk, Int. J. Mass Spectrom. Ion Processes **197**, 37 (2000).
26. W. M. Huo, Phs. Rev. A **64**, 042719 (2001).
27. C. Ma, M. R. Bruce, and R. A. Bonham, Phys. Rev. A **44**, 2921 (1991); Erratum: Phys. Rev. A **45**, 6932 (1992).
28. M. R. Bruce, L. Ma, and R. A. Bonham, Chem. Phys. Lett. **190**, 285 (1992).
29. K. Stephan, H. Deutsch, and T. D. Märk, J. Chem. Phys. **83**, 5712 (1985).
30. W. H. Huo, C. E. Dateo, and G. D. Fletcher, *in 22nd International Conference on Photonic, Electronic, and Atomic Collisions, Abstracts of Papers,* Santa Fe, NM, July 2001.
31. W. M. Huo, V. Tarnovsky, and K. H. Becker, Chem. Phys. Lett. **358**, 328 (2002).
32. V. Tarnovsky, P. Kurunczi, D. Rogozhnikov, and K. Becker, Int. J. Mass Spectrom. Ion Processes **128**, 181 (1993).
33. V. Tarnovsky and K. Becker, J. Chem. Phys. **98**, 7868 (1993).
34. Y.-K. Kim and K. K. Irikura, *Atomic and Molecular Data and Their Applications*, K. A. Berrington and K. L. Bell (Eds.) AIP 1-56396-971-8, Conference Proceedings CP543, 2000, p. 220.
35. M. A. Ali, K. K. Irikura, and Y.-K. Kim, Int. J. Mass Spectrom. **201**, 187 (2000).

SYSTEMATIC STUDY OF ELECTRON AND POSITRON COLLISIONS WITH PERFLUOROCARBONS C_4F_8, cyclo-C_4F_8 AND C_4F_{10}

Casten Makochekanwa, Osamu Sueoka and Mineo Kimura*

Abstract: Total cross section (TCS) measurements for 0.4-1000eV electron and 0.2-1000eV positron collisions have been studied for C_4F_8, cyclo-C_4F_8 and C_4F_{10} molecules in a linear time-of-flight apparatus. Electron TCSs for all molecules show strong enhancements, which may be due to the temporary-negative-ion resonances observed in another systematic study of perfluorocarbons by Sanabia *et al.* Positron TCSs, on the other hand, show no conspicuous structures but have TCS magnitudes that increase with increase in molecular size.

1. INTRIDUCTION

Both elastic and inelastic electron-molecule scattering processes have been shown to be useful for modeling various low-pressure cold plasmas. Accurate determination and understanding of a set of cross sections for electron scattering from perfluorocarbons is urgently needed in areas including plasma processing, material sciences, and earth sciences, since these molecules, owing to their shorter lifetime than that of CF_4 in the atmosphere, are considered environmentally friendly and hence constituting possible candidates for uses in these areas.[1] In view of this, suitable gases such as C_4F_8, cyclo-C_4F_8 and C_4F_{10} are such possible candidates which need to be studied. Besides the above uses, perfluorocarbons offer other attractive properties ranging from good thermal and chemical stability, low toxicity and high dielectric strengths to high electron attachment rate constants and drift velocities, making them also useful for applications as gaseous dielectrics. Whereas electron scattering processes have these wider applications, it is necessary to have an understanding of positron scattering processes too, for improved understanding of the electron scattering

* Corresponding author: Graduate School of Science and Engineering, Yamaguchi University, Tokiwadai 2-16-1, Ube, Yamaguchi 755-8611, Japan.

E-mail: mineo@yamaguchi-u.ac.jp TEL: (81) 836 85 9806, FAX: (81) 836 85 9801.

Gaseous Dielectrics X.
Edited by Christophorou *et al.*, Springer, New York, 2004.

phenomena in the comparative studies using these projectiles.

In this paper electron and positron TCSs are studied for a mixed form of octafluorobutane C_4F_8 gas (hereafter referred to as *m*-C_4F_8), octacyclofluorobutane (*c*-C_4F_8) and fluorobutane (C_4F_{10}) molecules. Comparative studies of these data with other groups' data is done only for the case of electron scattering from c-C_4F_8, where data is available in literature for TCSs by Sanabia *et al.*[2] and Nishimura[3], cross sections for modeling for plasma processing applications[4], partial dissociative ionization cross sections of Toyoda *et al.*,[5] and for C_4F_{10} where the electron attachment swarm experimental studies of Hunter *et al.*[6] are used. Otherwise, all the rest of the comparative studies are carried out using data from this laboratory, as no other data exists, to our knowledge. The results presented here have not been corrected for the forward scattering effect for lack of DCS data.

2. EXPERIMENTAL METHOD

The apparatus used for the total cross section (TCS) measurements investigated here is the linear transmission type retarding potential time-of-flight (RP-TOF), already discussed briefly elsewhere in this book of proceedings. Details about the apparatus and procedure can be found in our earlier report.[7]

The independence of TCS values on the cell pressure was ensured for electron collision. As Kennerly *et al.*[8] pointed out, this pressure gauge stability check and the pressure independence of the TCSs are an important aspect of measurements using a collision cell in the transmission RP-TOF setup.

3. RESULTS AND DISCUSSION

The TCS results for electron and positron impact are shown in Figure 1, for *c*-C_4F_8, and Figure 2, for *m*-C_4F_8 and C_4F_{10} molecules.

Electron impact. There are a number of interesting features observed from these data, which are summarized as follows. (i) Except for collision energies 16 – 25 eV, C_4F_{10} TCSs are greater than m-C_4F_8 and c-C_4F_8 TCSs. This is rather expected since C_4F_{10} molecules are the greatest in molecular size amongst these three, i.e. the molecular size effect resulting in spatially increased electron charge distribution leading to increased scattering events, as a result of the larger geometrical size. (ii) Below 2 eV, and C_4F_{10} TCSs decrease rather rapidly to produce a minimum at 0.6 eV before beginning to rise again below this energy, as suggested by our 0.4 eV data point. If proved to be true, then this minimum might be suggesting a Ramsauer-Townsend minimum, and the steep rise of the TCS towards 0 eV might be pointing to the existence of a very low-energy resonance or even the possibility of a stable bound-state ion. m-C_4F_8 TCSs increase down to 1 eV and remain at a plateau down to 0.4 eV. On the other hand, c-C_4F_8 TCSs continue to increase without showing any sign of a slowing down to any peak.

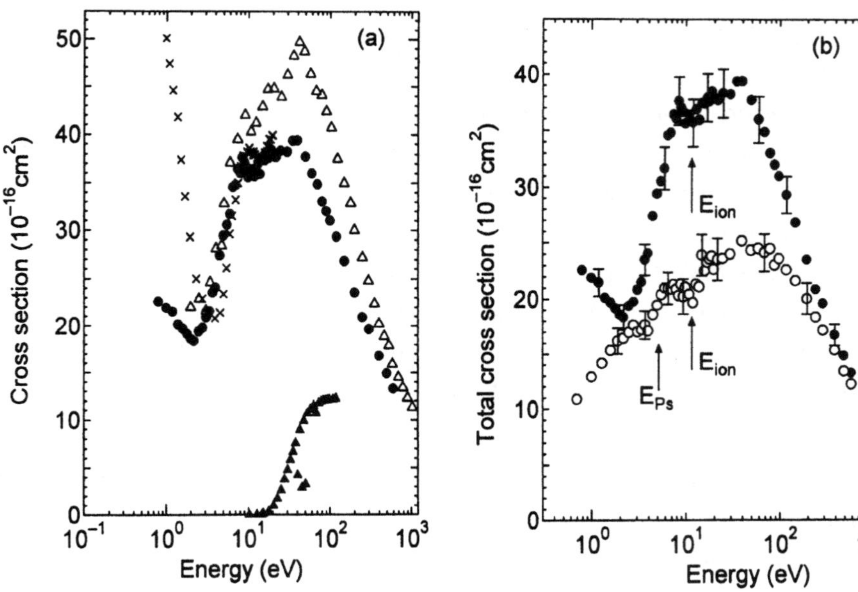

Figure 1. *c*-C_4F_8 (a) electron scattering present TCS (•); Sanabia *et al.*[2] TCS (×); Nishimura TCS (Δ); sum of partial ionisation cross sections of Toyoda *et al.*[5] (▲); (b) present electron TCS (•) and present positron TCS (○). The arrows labelled E_{ion} and E_{Ps} denote the thresholds for ionisation and positroniumu formation, respectively. Error bars show total uncertainties, consisting of statistical and errors due to gas density and estimation of collision cell effective length.

This behavior of c-C_4F_8 electron TCSs agrees with the electron attachment rates studies, below 1 eV, which have been found to rise steeply as the electron energy approaches 0 eV.[8-9] Besides, Miller *et al.*[10] have also estimated an electron affinity of +0.63 eV, for these molecules, from electron attachment experiments. This could in fact be presumably corresponding to electron capture into the LUMO.[11] Though Hunter *et al.*[6] report a C_4F_{10} electron attachment peak at 1.95

eV, no correspondence of this is observed in our TCSs, possibly because the cross section at this maximum is only 0.6×10^{-16} cm^2 compared to the TCS (our data) which is more than 30 times larger. (iii) At 2 – 3eV, C_4F_{10} TCSs show a peak (at 2.8 eV) which should be due to some rovibrational shape resonance. Contrarily, m-C_4F_8 and c-C_4F_8 TCSs show minima in this energy range. These minima should be suggesting Ramsauer-Townsend minima. Though c-C_4F_8 derivative electron transmission experiments have suggested a resonance peak at 5.8 eV,[11] this is not observed in our TCSs, possibly because its cross section is smaller than would be observed in TCSs which are more than 35×10^{-16} cm^2 at this energy range. Comparison of our c-C_4F_8 TCSs with other data, Figure 1(a), shows good qualitative agreement with both Sanabia *et al.*[2] and Nishimura[3] TCSs. However, quantitative differences are observed below 3 eV, where our data is comparable with the Nishimura data but becoming almost half the magnitudes of the Sanabia *et al.* data. In the 10-100 eV peak region, there is general agreement with the Sanabia *et al.* data though the Nishimura data is larger by about 25%. The peak appearing at about 40 eV should be consisting of strong contributions from the ionization channel. However, that the ionisation cross section, though peaking at about 100 eV, is only about 30% of the TCS (our data) is a property that makes this gas a possible candidate for the gaseous dielectric application, i.e. this minimal ionization means the gas will not produce many electrons by impact ionisation, and hence keeping the number of free electrons low.

Electron and positron impact compared. Though electron TCSs show a lot of resonance structures that have been discussed above, positron TCSs are rather smoother, but with few weak structure at energies corresponding to the opening up of the positronium formation channel, i.e. E_{Ps}, which is 5.45 and 6.25 eV for *c*-C_4F_8 and C_4F_{10} respectively. Small structures above 12 eV are due to the opening of ionization channels, and the changes of TCS slopes between E_{Ps} and E_{ion} should be due to electronic excitation. At low-to-intermediate energies, the TCSs for electron impact are about 1.5 times larger due to resonances over the 5-100 eV region. This characteristic is a clear reflection of differences in the interaction between the electron and positron with molecules. Electron and positron TCSs for all molecules tend toward merging above a few hundred eV. This should be because at these higher energies, only the first Born term dominates the scattering event.

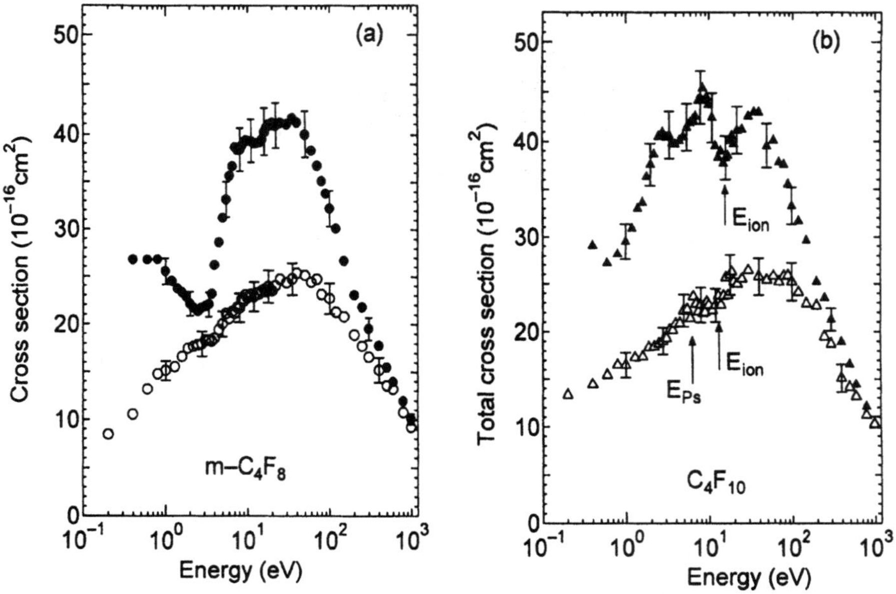

Figure 2. Present (a) m-C_4F_8 electron (•) and positron TCSs; (b) C_4F_{10} electron (▲) and positron (Δ) TCSs. The arrows labelled E_{ion} and E_{Ps} denote the thresholds for ionisation and positroniumu formation, respectively. Error bars show total uncertainties, consisting of statistical and errors due to gas density and estimation of collision cell effective length.

Summary

Except for collision energies 16 – 25 eV, C_4F_{10} TCSs are greater than m-C_4F_8 and c-C_4F_8 TCSs for both electron and positron impact, i.e. fingerprints of the molecular size effect. *c*- C_4F_8 has been observed to have small ionisation cross sections; a property that makes this gas a possible candidate for the gaseous dielectric application. At low-to-intermediate energies, the TCSs for electron impact are about 1.5 times larger due to resonances over the 5-100 eV region. Merging phenomena has been observed between electron and positron at energies beyond a few hundred eV and explained from the first Born approximation.

Acknowledgement

The work was supported in part by the Ministry of Education, Sport, Culture and Technology, Japan Society for Promotion of Science, and Cooperative grant from National Instutute for Fusion Science.

References

1. L. G. Christophorou, J. K. Olthoff and M. V. V. S. Rao, Electron interactions with CF_4, *Phys. Chem. Ref. Data* **25**(5), 1341-1388 (1996).
2. J. E. Sanabia, G. D. Cooper, J. A. Tossell, and J. H. Moore, Total electron-scattering cross sections for CHF_3, C_2F_6, C_3F_8, and c-C_4F_8, *J. Chem. Phys.* **108**(2), 389-391 (1998).
3. H. Nishimura, Total electron scattering cross sections for c-C_4F_8, in Book of Abstracts of the *International Symposium on Electron-Molecule Collisions and Swarms*, Tokyo, 18-20 July 1999, p. 103-104.
4. G. I. Font, W. L. Morgan, and G. Mennenga, Cross-section set and chemistry model for the simulation of *c*-C_4F_8 plasma discharges, *J. Appl. Phys.* **91**(6), 3530-3538 (2002).
5. H. Toyoda, M. Iio and, H. Sugai, Cross section measurements for electron-impact dissociation of C_4F_{10} into neutral and ionic radicals, Jpn. J. Appl. Phys. **36**(6A), 3730-3735 (1997).
5. S. R. Hunter and L. G. Christophorou, Electron attachment to the perfluoroalkanes n-C_NF_{2N+2}($N=1$-6) using high pressure swarm techniques, *J. Chem. Phys.* **80**(12), 6150-6164 (1984).
6. O. Sueoka, S. Mori and A. Hamada, Total cross sections for positrons and electrons colliding with molecules: I. SiH_4 and CF_4, *J. Phys. B* **27**, 1453-1465 (1994).
7. R. E. Kennerly and R. A. Bonham, Electron-helium absolute total scattering cross sections from 0.5 to 50 eV, *Phys. Rev. A* **17**(6), 1844-1854 (1978).
8. A. A. Christodoulides, L. G. Christophorou, R. Y. Pai and C. M. Tung, Electron attachment to perfluorocarbon compounds. I. c-C_4F_6, 2-C_4F_6, 1,3-C_4F_6, c-C_4F_8 and 2-C_4F_8, *J. Chem. Phys.* **70**(3), 1156-1168 (1979).
9. A. Chutjian and S. H. Alajajian, Measurements of electron attachment lineshapes and cross sections at ultra-low electron energies for CF_2Cl_2, c-C_4F_6, c-C_4F_8 and c-C_7F_{14}. *J. Phys. B* **20**(4), 839-846 (1987).
10. T. M. Miller, R. A. Morris, A. E. S. Miller, A. A. Viggiano and J. F. Paulson, Observation of thermal electron detachment from cyclo-$C_4F_8^-$ in FALP experiments, *Int. J. Mass Spectrom. Ion Processes* **135**(2-3), 195-205 (1994).
11. I. Ishii, R. McLaren, A. P. Hitchcock, K. D. Jordan, Y. Choi and M. B. Robin, *Can. J. Chem.* **66**, 2104 (1988).

TRANSPORT COEFFICIENTS FOR ELECTRONS IN CF_4 IN E(t)×B(t) FIELDS

S. Dujko, Z.M. Raspopović and Z.Lj. Petrović*

1. INTRODUCTION

Low pressure discharges sustained by radio-frequency (rf) electric fields, are widely used in microelectronic fabrication and manufacture of new materials[1]. In these applications it is desirable to generate high density, large volume non-equilibrium plasma in collision dominated regime. Most plasmas are driven by sources with frequencies ranging from several tens to several hundreds of MHz. In case of Inductively Coupled Plasmas (ICP) and magnetically enhanced plasmas, the magnetic field is present leading to spatial trapping of electrons. Investigation of electron transport and of the kinetic phenomena that may occur[2], in crossed dc and rf fields is highly desirable as the basis for plasma models under circumstances that are typical for plasma processing. On the other hand, modeling of gaseous dielectrics may benefit greatly from the studies of gases for plasma etching as often the same gases are used in both applications.

Modeling of rf plasmas usually relies on application of dc swarm data. At very high frequencies, effective field approximation gives reasonable results while at low frequencies instantaneous field approximation gives good results. However, for a wide range of frequencies and pressures, that are typical for applications of rf plasmas these approximations may fail. Therefore, the first critical step in modeling of rf plasmas is calculating the electron transport coefficient under the influence of time resolved fields.

The effect of magnetic field on electron transport is the second critical step. Namely, most models do not take into account the effect of time resolved magnetic field on electron transport. However, for example, in ICP the normalized magnetic field (B/N) may be very strong (of the order of 2000-5000 Hx in peak values).

* S. Dujko, Z.M. Raspopović and Z.Lj. Petrović, Institute of Physics P.O. Box 68, 11080 Zemun, Serbia and Montenegro.

Gaseous Dielectrics X.
Edited by Christophorou *et al.*, Springer, New York, 2004.

Finally, we want to emphasize the importance of the finite temporal and spatial relaxation of both electron transport coefficients and electron energy (velocity) distribution functions. Large time constants for energy and momentum relaxation as compared to the characteristic periods of the field variation may be the source of various kinetic phenomena such is anomalous longitudinal diffusion.

The aim of this work is to investigate behavior of the rate coefficients under the influence of crossed electric and magnetic fields in pure CF_4 (carbon tetra fluoride). We have applied our Monte Carlo technique[2] to calculate the transport coefficients, distribution functions, relaxation in space and time and obtain further understanding of electron transport in CF_4.

CF_4 is one of the most frequently used gases in plasma applications in plasma etching[3] for ultra large scale integrated (ULSI) circuits. As a result, a great effort has been made in order to produce complete sets of cross sections and related electron swarm data for modeling[4-9]. Rapidly rising cross section for vibrational excitation in the region of Ramsauer-Townsend minimum induces a strong anisotropy of the electron energy distribution function (EEDF) and requires exact calculation of transport data. One should bear in mind that the term *exact technique* implies any technique which is systematically tested in model and real gases against similar, well-established calculations.

Apart from calculation of transport coefficients in rf fields we show some results obtained for dc crossed electric and magnetic fields. These results should be the necessary input data for fluid models of magnetron[10] and ICP[11] plasma reactors. Comparisons against measurements in crossed fields is also an additional test of the validity of cross sections.

One should bear in mind that while transport data are the input for the fluid and also for hybrid codes, the cross sections verified by standard swarm procedures are input for Monte Carlo, PIC (particle in cell) and hybrid techniques (MC part).

2. MONTE CARLO TEHNIQUE

The Monte Carlo code applied in this work follows a large number of electrons (typically 10^4-10^6) over small time steps allowing them to relax to a quasi-stationary mode. The probability of scattering of electron with energy, ε in the time interval $(t, t+dt)$ may be written as:

$$p(t) = \nu_T(\varepsilon(t)) \exp\left[- \int_0^t \nu_T(\varepsilon(t))dt \right],$$

where $\nu_T(\varepsilon(t))$ is the time dependent total collision frequency. Instead of traditional null collision technique we have used integration technique for numerical solution where the small time steps are determined from the minimum of the three relevant time constants (mean collision time, period of the field and cyclotron period for **E**×**B**). The chosen time constant has been divided by a large number (20-100). At the same time the period of the field is always divided by 100 and these segments are used to sample the transport coefficients.

All calculations were performed for zero gas temperature. It is also assumed that the electron density is sufficiently small so Coulomb interactions between the particles as well as shielding of the field are negligible. The code has been tested, using Reid's ramp model gas and non-conservative models, against similar calculations and found to be in good agreement[12].

In our calculations, the gas number density is $3.54 \cdot 10^{22}$ [m^{-3}] (i.e. approximately 1 Torr). Initial electron energy distribution is Maxwellian with the mean energy of 1 eV. The phase between electric and magnetic field is 90°, which corresponds to the conditions found in ICP. We define our coordinate system according to the direction of two fields (**E** and **B**) and direction perpendicular to both (**E**×**B**). We chose a set of cross sections for CF_4 published by Kurihara[9] which is similar but not identical to that proposed by Christophorou[4].

3. RESULTS AND DISCUSSION

3.1. Results for crossed dc electric and magnetic fields

In Fig. 1 (a) we show the mean electron energy as a function of *E*/*N* for various values of *B*/*N* as indicated on the graph. The application of a magnetic field perpendicular to the electric field decreases the mean energy. As can be seen, there are three distinct regions of transport as *E*/*N* increases. The first region is at the lowest E/N (below 10 Td for 0 Hx) where the increase of the mean energy is very slow presumably due to the effect of vibrational excitation. This is followed by a sudden increase (10 Td- 50 Td) and then at the highest E/N the increase is again slow due to the effect ot electronic excitation and ionization. The onset of the sharp transition increases with increasing B/N since magnetic field decreases the mean energy. The fast variation of the mean energy ranging from 0.02 eV to 5 eV between 10 Td and 100 Td indicates that in this region plasma chemistry may be modified by changing the magnitude of an external magnetic field.

In Fig. 1 (b) we show drift velocity as a function of *E*/*N* for various *B*/*N* as indicated on the graph. In crossed fields, drift velocity has two components, in **E** and **E**×**B** directions. The effect of magnetic field on drift velocity is the most interesting in the region of negative differential conductivity (NDC). In the range of magnetic fields from 0 to 500 Hx, the NDC region is moving to the right to higher values of *E*/*N*. However, further increasing of *B*/*N* to 1000 Hx leads to the disappearance of the NDC. In addition, if we take a careful look at the range from 30 to 200 Td there are several points that show an increase of the drift velocity with increasing *B*/*N* values. This result appears to be unexpected.

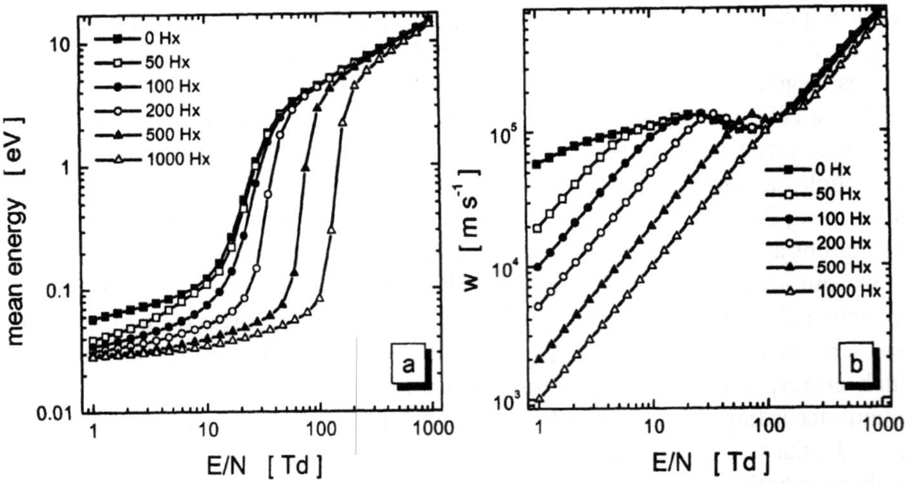

Figure 1. (a) Mean energy and (b) drift velocity as a function of *E/N* for a various magnetic field as indicated on the graphs

3.2. Results for crossed rf electric and magnetic fields

In Figs. 2 (a) and (b) we show time dependence of the drift velocity in **E** and **E**×**B** directions, respectively, for various magnetic fields as indicated on the graphs. Both components of the drift velocity lose symmetry and at the highest magnetic fields strong oscillation are induced due to the cyclotron rotation of the electron swarm. The **E**×**B** component is more asymmetric then component in **E** direction, thus having a mean value that is not zero. These graphs clearly show how dramatic influence of the magnetic field can be. At first, the magnetic field induces an effective force in **E**×**B** direction producing a macroscopic drift in that direction as well as complex, asymmetric behavior of drift velocity in **E** direction. Consequently, the magnetic field impedes the drift of electrons and reduces the mean energy of the electron swarm. However, most present-day plasma models do not take into account the effects of the magnetic field on the electron transport and **E**×**B** components of the drift velocity and diffusion tensor are neglected.

In Figs. 3 (a), (b), (c) and (d) we show the **E**, **E**×**B** and **B** components of the diffusion tensor. For zero value of magnetic field, the **E**×**B** and **B** components coincide. As *B/N* increases, **E** and **E**×**B** components start to differ from each other while the **E** component peaks at phases where magnetic field goes through zero as shown in Fig. 3 (b). As *B/N* further increases, the **E**×**B** component starts to behave the same as **E** component, as shown in Figs 3 (c) and (d). Also, we may observe that magnetic field does not affect the **B** component of the diffusion tensor. In addition, **B** component has a much smaller modulation then **E** and **E**×**B** components.

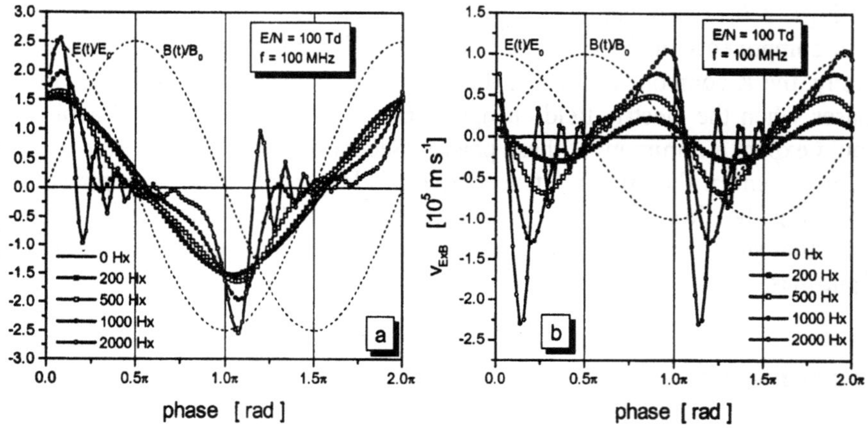

Figure 2. Time dependence of the drift velocity in (a) direction of the electric field and (b) in the **E**×**B** direction for various magnetic fields as indicated on the graphs. The frequency of the applied field is 100 MHz at 100 Td.

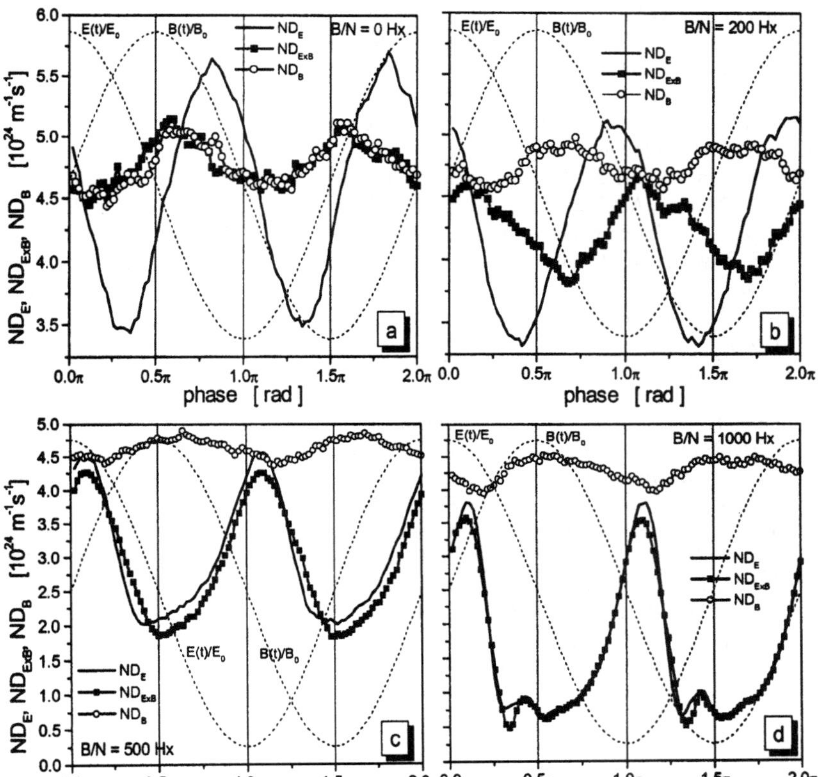

Figure 3. Time dependence of the **E**, **E**×**B** and **B** component of the diffusion tensor for (a) B/N=0 Hx, (b) B/N=200 Hx, (c) B/N=500 Hx and (d) B/N=1000 Hx. The frequency of the applied field is 100 MHz at 100 Td.

Figs. 3 (a)-(d) illustrate the effect of the anomalous behavior of the **E** and **E**×**B** components of the diffusion tensor[13-14]. The anomaly reflects two effects. Firstly, the **E** component of the diffusion tensor increases during or just after the phases when the electric field changes sign. In contrast, a quasi dc explanation would expect a minimum at these phases. Secondly, for a brief phase period, the **E** component of the diffusion tensor becomes larger then the **B** component, thus producing the flux of the electrons in direction of the electric field. One may claim that such flux of the electrons has to be taken into account in modeling of rf plasmas.

REFERENCES

1. M.A. Lieberman and A.J. Lichtenberg, *Principles of Plasma Discharges and Materials Processing* (Wiley, New York, 1994).
2. Z.Lj. Petrović, Z.M. Raspopović, S. Dujko and T. Makabe, in: *Advances in Low Temperature Plasmas*, edited by T. Makabe (Elsevier Science B.V., Amsterdam 2002), pp. 1-25.
3. H. Tanaka and O. Sueoka, in: Advances in Atomic, Molecular and Optical Physics, edited by M. Kimura and Y. Itikawa (Academic Press, San Diego 2001), vol. **44**, pp. 1-32.
4. L. G. Christophorou, J.K. Olthoff and M.V.V.S. Rao, Electron interactions with CF_4, *J. Phys. Chem. Ref. Data* **25**(5), 1341-1388 (1996).
5. M.C. Bordage, P. Segur and A. Chouki, Determination of a set of electron impact cross sections in tetrafluoromethane consistent with experimental determination of swarm parameters, *J. Appl. Phys.* **80**(3), 1325-1336 (1996).
6. M. Hayashi, in: *Swarm Studies and Inelastic Electron-Molecule Collisions*, edited by L.C. Pitchford, B.V. McKoy, A. Chutjian and S. Trajmar, (Springer, New York, 1987), pp. 167.
7. Y. Nakamura, in: *Gaseous Electronics and Their Applications*, edited by R.W. Crompton, M. Hayashi, D.E. Boyd and T. Makabe (KTK Scientific, Tokyo, 1991), pp. 178.
8. M.C. Bordage, P. Segur, L. G. Christophorou and J.K. Olthoff, Boltzmann analysis of electron swarm parameters in CF_4 using independently assessed electron-collision cross section, *J. Appl. Phys.* **86**, 355-363 (1999).
9. M. Kurihara, Z.Lj. Petrović and T. Makabe, Transport coefficients and scattering cross-sections for plasma modelling in CF_4-Ar mixtures: a swarm analysis, *J. Phys. D: Appl. Phys.* **33**, 2146-2153 (2000).
10. E. Shidoji, K. Ness and T. Makabe, Influence of gas pressure and magnetic field upon dc magnetron discharge, *Vacuum* **60**, 299-306 (2001).
11. T. Makabe, in: *Advances in Atomic, Molecular and Optical Physics*, edited by M. Kimura and Y. Itikawa, (Academic Press, San Diego, 2001) vol. **44**, pp. 127-154.
12. Z.M, Raspopović, S. Sakadžić, S. Bzenić and Z.Lj. Petrović, Benchmark Calculations for Monte Carlo Simulations of Electron Transport, *IEEE Trans. Plasma Sci.* **27**, 1241-1248 (1999).
13. R.D. White, K.F. Ness and R.E. Robson, Anomalous anisotropic diffusion in a.c. electric fields,", *Aust. J. Phys.* **48**, no. 6, pp. 925–938 (1995).
14. Z.R. Raspopović, S. Sakadžić, Z.Lj. Petrović and T. Makabe, Diffusion of electrons in time-dependent E(t)×B(t) fields, *J. Phys. D: Appl. Phys.* **33**, 1-5 (2000).

SECTION 5: HIGH PRESSURE GAS DIELECTRICS / MIXTURES

GASES AS A DIELECTRIC

Tadasu Takuma, Osamu Yamamoto, and Shoji Hamada

1. INTRODUCTION

Gas insulation plays a very important role in electric power systems, as GIS (gas-insulated switchgear), GIL (gas-insulated transmission line), GCB (gas circuit breaker), and other power equipment. In the present or conventional gas insulation, SF_6 (sulfur hexafluoride) is exclusively utilized as a principal insulating medium. It is needless to say that the use of solid insulating components is also inevitable there for supporting and fixing stressed parts because a gaseous medium alone cannot fulfill these tasks.

Until recently, SF_6 has been considered the best medium with few disadvantages among various gases for the gas insulation. Search for replacement gases made in 1970's and 1980's led to this conclusion. However, the global greenhouse effect, a serious environmental issue, has made SF_6 one of the accused gases, and the Kyoto-Conference of the Parties on climate change in 1997, so-called COP3, indeed designated it as one of the six gases to be regulated. Although SF_6 records by far a smaller amount of emission compared to CO_2, its GWP (global warming potential) is highest among most gases available. Since then, it has become an important and urgent task again in electric power industry worldwide to search for a new gas or gas mixture replacing SF_6, but this time with less environmental impact, that is, with a smaller GWP.

In the background of such environmental situations, a number of papers and reviews[1] have already been published. This paper reviews the fundamental aspects in selecting an appropriate gas under various conditions or constraints together with more recent data, and discusses the characteristics of some gases and gas mixtures concerned. The content of the paper roughly consists of the

Tadasu Takuma, CRIEPI, 2-11-1, Iwado-Kita, Komae-shi, Tokyo 201-8501, Japan, Osamu Yamamoto & Shoji Hamada, Dept. of Electrical Engineering, Kyoto University, Yoshida-Honmachi, Sakyo-ku, Kyoto 606-8501, Japan

Gaseous Dielectrics X.
Edited by Christophorou *et al.*, Springer, New York, 2004.

following items.
(a) Insulation in non-uniform field conditions
(b) Insulation in quasi-uniform field conditions
(c) Environmental aspects

2. SELECTION FOR NON-UNIFORM FIELD CONDITIONS

The selection of a gas for an insulating system makes a very important task for electrical engineers, although there exist not so many options compared to solid insulating materials.

Since the beginning of the commercial supply of electric power, transmission and distribution lines have been principally insulated by the atmospheric air. In this context, we here point out two important facts, which have been very favorable to electrical engineers up to now. First, any gas including air owns a fundamentally excellent insulating characteristic of low conductivity, or low leakage, compared to liquid or solid materials. This is an easily conceivable situation. If we had lived in water, for example, as "Wassermaenchen" appearing in German fairly tales, we could not have easily come to utilize electricity, or at least its use would have been so delayed.

Second, overhead lines comprise basically non-uniform field conditions. Although the atmospheric air shows much lower breakdown (sparkover) voltage in uniform conditions than (compressed) SF_6, it presents very good insulating ability for non-uniform long gaps. Figure 1 shows such characteristics compared to those of SF_6, where with increasing gap length (electrode separation), the sparkover voltage of SF_6 becomes lower than that of the atmospheric air[2]. Furthermore surprisingly, the sparkover voltage of SF_6 hardly increases even when the gas pressure is raised seven times to 0.7 MPa.

The low sparkover voltage of SF_6 in long gaps is attributed to the discharge channel with higher conductivity as a leader in lightning discharge, which extends to the opposite electrode (plane) with the effect of its own space charge. Time-resolved observation with an image-converter camera demonstrates that the discharge channel in a long-gap SF_6 propagates in intermittently illuminating steps[2], similarly to a stepped leader in natural lightning. In conclusion, the superior insulating property of the atmospheric air explained here has been so favorable to the power transmission with high voltage overhead lines, and also perhaps to the suppression of lightning strikes. Electrical engineers should thank it as the Grace of Nature.

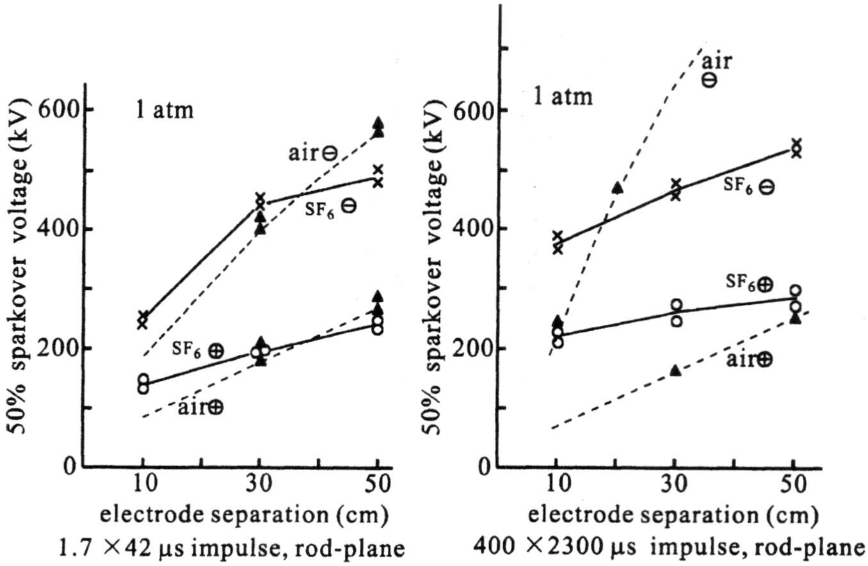

Figure 1. Comparison of sparkover voltage between air and SF_6 at an atmospheric pressure (0.1MPa) for long gaps

3. SELECTION FOR QUASI-UNIFORM FIELD CONDITIONS

Most of the gas-insulated systems form quasi-uniform field, consisting of relatively short gaps with rounded conductors, where discharge inception without preceding corona or partial discharge is a decisive factor in insulation design. However, discharge inception or sparkover voltage is not a unique characteristic to be considered. For example, CCl_4 exhibits discharge voltage roughly twice as high as SF_6 at an identical low pressure, but it is in a liquid state at room temperature. A gas must be compressed to a higher pressure to attain better insulation ability within the range of operating temperature.

In order to take into consideration the effect of gas temperature T of a gas, Devins has applied the following approximation[3] which relates vapor pressure P to T for a gas having a boiling point T_b,

$$P = \exp\{A\left(1 - T_b / T\right) / R\} \qquad (1)$$

where $A = 21$ cal/deg.mole, and R is the gas constant (= 2 cal/deg.mole). As the breakdown voltage of a gas is approximately proportional to Pd (d; gap length) for higher values of Pd, we can evaluate the dielectric

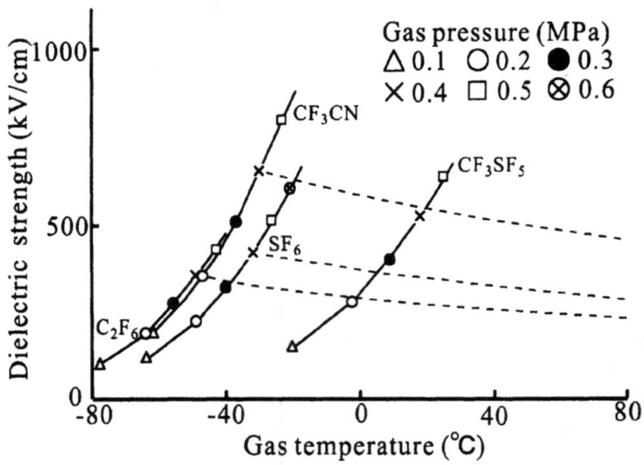

Figure 2. Dielectric strength relative to gas temperature

strength (kV/cm), that is insulating ability, dependent on T based on Eq. (1). Figure 2 compares such strength relative to gas temperature up to about 6 atms (0.6 MPa) for four gases, similarly as in ref. [4]. Solid lines in the Figure correspond to the values determined by vapor pressure, while dotted lines mean those by equipment pressure of 0.4 MPa. It is clear from this Figure that C_2F_6 is superior to SF_6 only at a very low temperature, while CF_3SF_5 is higher at a temperature above about 5 °C. The strength of CF_3CN is higher than that of SF_6 in all the operating temperature range, but this gas is highly toxic.

As already mentioned, the search for a gas or gas mixture replacing SF_6 was actively performed in 1970's and 1980's in the USA, Europe and also Japan. The main purpose was to improve some disadvantages of SF_6, such as relatively low breakdown (sparkover) voltage in non-uniform fields. The conclusion at the time was that SF_6 was the best medium for gas insulation considering the overall characteristics such as insulation ability (discharge voltage), boiling point and compressibility, chemical stability as well as non-toxicity.

4. SELECTION FROM ENVIRONMENTAL ASPECTS

A new environmental issue, greenhouse effect of gases, has again urged the search for a new gas or gas mixture replacing SF_6. A number of related papers have already been published. In one of them, Niemeyer listed the necessary selection criteria, and systematically surveyed all the possible candidate gases[5]. He concluded that the gas fulfilling all the criteria is only nitrogen (N_2) or air, but also suggested that substituting SF_6 by air might increase the total environmental impact of the equipment from the lifecycle assessment (LCA).

We here briefly review some of the related research and developments with more recent data.

4.1. Reduction of SF_6 Emission

We consider the best policy in the gas insulation is to use pure SF_6 and suppress its emission to the atmosphere to an allowable level. According to the estimate on the global warming effect of SF_6, the contribution of emitted SF_6 to the temperature rise is below 0.02 °C in 2100 even in the unrealistically severe case of the emission amounting to 10^4 ton per year and infinite lifetime[6]. This warming effect corresponds to less than 1 % of that due to CO_2.

A Task Committee (Chairman: T. Takuma), which was set up in the "Electric Technology Research Association" in Japan, proposed in 1998 the standards on the use of SF_6 gas in electric power equipment[7]. Its results have led to the voluntary action plans in handling and recycling SF_6 in the entire power sector in Japan. We here do not enter into the details of these actions, but present only Table 1, which gives the annual emission amount of SF_6 from 1995 to 2000 in various related sectors. 1 million GWP ton roughly corresponds to 42 actual tons. It can be seen in the Table that the annual emission in total decreased drastically to about one third in five years. This decrease resulted mainly in the gas producing and electric power sectors, while the emission increased considerably in the semi-conductor sector in the same interval. Anyhow, the emission of SF_6 in the order of a few hundreds of tons is considered allowable from the global warming effect as above mentioned.

4.2. Replacement Candidates of Gases or Gas Mixtures

Table 1. Annual emission amount of SF_6 in Japan (MGWPton)

	1995	1996	1997	1998	1999	2000
Gas manufacturers	4.7	4.2	2.6	2.1	1.5	0.9
Equipment manufacturers	9.6	10.0	8.5	7.7	4.2	2.3
Substations	1.4	1.7	1.8	1.3	0.8	0.5
Semi-conductors	1.5	1.4	1.6	1.6	1.9	2.1
Others	0	0	0	0	0.1	0.1
Total	17.2	17.4	14.4	12.8	8.4	5.7

We here discuss some possibilities of replacement gases; ordinary gas such as nitrogen (N_2) and carbon dioxide (CO_2), gas mixtures containing a fluorocarbon, a gas decomposable in the atmosphere, and vapor-mist dielectrics

4.2.1 Nitrogen or Carbon Dioxide

Discharge characteristics of N_2 and CO_2 were studied in considerable detail in the early days of gas insulation. Both gases are much more benign from the environmental aspects than SF_6. They exhibit little or no global warming effect compared to SF_6, and have less concern on harmful effect of discharge-decomposed components. Their disadvantage is well known, that is low insulating ability with discharge voltage roughly one third that of SF_6 under a same pressure. Therefore, a higher pressure over 1 MPa is needed to keep the dimension of gas-insulated equipment same as in the conventional one. We here do not further explain the characteristics of these gases.

4.2.2. Gas Mixtures Containing Fluorocarbon

Gas mixtures composed of an electronegative gas and an ordinary gas like N_2 have the following advantages; reduction of gas cost, drop of a boiling point or liquefaction temperature, and a nonlinear rise of discharge voltage for the low mixing ratio of the electronegative gas. The last characteristic can be well expressed by the following empirical formula, which one of us once proposed[8];

$$V_m = V_2 + \frac{k(V_1 - V_2)}{k + (1-k)C} \tag{2}$$

where V_1 and V_2 are the discharge voltage (sparkover voltage for a uniform field) of pure component gases 1 and 2 ($V_1 > V_2$), respectively, and V_m is the discharge voltage of the gas mixture. The variable k is the mixing ratio or partial pressure ratio of gas 1, and C is a constant depending only on the combination of the component gases where $0 < C < 1$. The lower value of C makes V_m the more nonlinear in the rise of discharge voltage relative to mixing ratio.

Most of gas mixtures recently studied for gas insulation are those consisting of SF_6 as one component and an ordinary gas (mostly N_2) as a principal one[1]. However, as most fluorocarbon gases have a GWP considerably lower than SF_6, gas mixtures with one of such gases can bring a larger decrease of the latent GWP concerned than with SF_6.

Various requirements must be fulfilled to apply a mixture in practical gas insulation. The fundamentally important requirements from the environmental aspects are non-toxicity, no ozone destruction

Table 2 Candidate component gases

	Diel. Str.	B.P.(°C)	GWP(100years)
C_2F_6	0.78-0.79	-78	6,200
C_3F_8	0.96-0.97	-37	7,000
C_4F_{10}	1.32-1.36	-2	
c-C_4F_8	1.25, 1.31	-6(-8)	8,700
CF_3SF_5	1.51-1.52	-24(-20)	
1,3-C_4F_6	1.3-1.5	-6.5	
CF_3OCF_3	0.84	-59	
SF_6	1	-64(subl.)	23,900

ability or no ozone depletion potential (ODP) and low GWP. From the requirement of zero ODP, we must exclude such gases as include chlorine or bromine, and thus limit our selection only to those containing fluorine such as HFC (hydro-fluorocarbon) or PFC (perfluorocarbon). Table 2 lists such candidate gases with no or unknown toxicity, which have discharge voltage (or dielectric strength) higher than, or comparable to SF_6. Among them, we have selected c-C_4F_8 (perfluorocyclobutan, C-318) as a possible component electronegative gas from electrical, physical and environmental aspects. Reference [9] explains various items related with gas mixtures containing c-C_4F_8, such as selection criteria of a component gas, environmental aspects, allowable mixing ratio, and sparkover characteristics under quasi-uniform field conditions. Figure 3 compares the GWP (100% for pure SF_6) of gas-insulated equipment (bus-bar) when we use a gas mixture having the insulating property corresponding to that of pure SF_6 at 0.4MPa[10].

The application of a mixture in gas insulation, however, entails some disadvantages. One serious problem is that the liquefied recovery of a component gas becomes very difficult, needing low temperature or high pressure, or both. We are also studying recovery or enrichment of a component gas using a polymer membrane[11].

4.2.3, Gases Decomposable in the Atmosphere

The greenhouse effect of an emitted gas, that is its GWP, depends on the amount of absorbing spectra in a molecule and the staying period (lifetime in the atmosphere) of the gas. This means that if a gas is easily decomposed when emitted into the atmosphere, such gas would exhibit a low GWP. One possibility is CF_3I (trifluoromethyl-iodide). Its GWP is as low as near 5 due to the photo-decomposition in the atmosphere. Figure 4 shows an example of sparkover voltage for a sphere-to-plane gap and AC voltage[12]. As the boiling point of CF_3I is not enough low, its mixture with

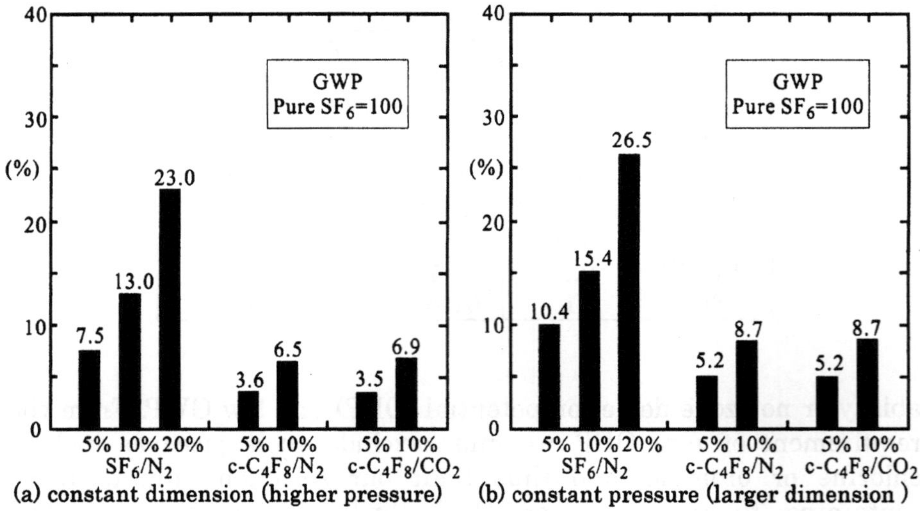

Figure 3. Decrease of latent GWP when a gas mixture is applied.

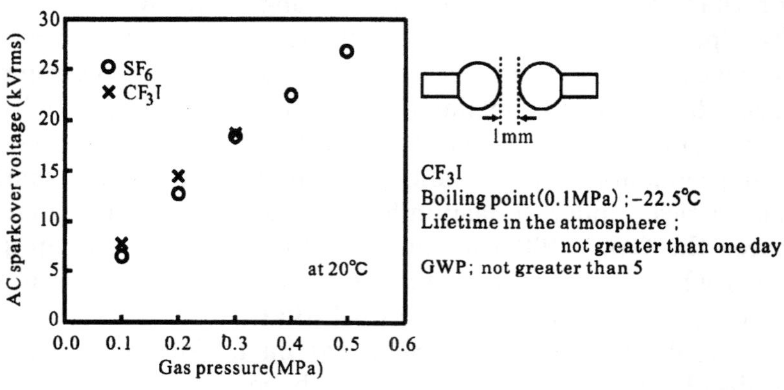

Figure 4. Comparison of AC sparkover voltage between CF_3I and SF

nitrogen is studied[13]. Furthermore, this gas suffers from slight toxicity as described in Table 3.

4.2.4 Vapor-Mist Dielectrics

If an additive can significantly improve insulation ability of gas-insulated equipment, the more compact design would contribute to the decrease of warming potential as a result of the diminished amount of the necessary component gas. It is needless to say that in this situation, the additive must have a negligibly small GWP.

Table 3 Some environmental properties of CF_3I

	CF_3I	SF_6
B.P.(atm. pressure)	-22.5°C	-63.8°C(subl.)
Lifetime	1-1.5 days	3,200 years
GWP (100 years)	below 5	23,900
ODP	below 0.0001	0
LC_{50} NOAEL LOAEL	274.000ppm/15min 2,000ppm 4,000ppm	

A unique insulation system called vapor-mist dielectric was proposed by Harrold in 1981[14]. It consists of a dense mist of micro-sized liquid droplets suspended in an (principal) insulating gas such as SF_6. He suggested it greatly increased sparkover voltage in nearly uniform fields, and possible applications as an insulating and cooling medium in power transformers. He also reported that vapor-mist dielectric effectively suppresses partial discharge inception under non-uniform fields[15].

According to the experiment conducted by a group of CRIEPI including one of us[16], mixing of C_2Cl_4 (tetrachloroethylene) into N_2 or SF_6 caused a remarkable or considerable increase in sparkover voltage in the case of impulse voltage application. However, such substantial improvement occurred only under lower gas pressures, and furthermore, sparkover voltage scarcely increased for DC and AC application except a slight rise at a pressure lower than 0.1 MPa.

The predominant factor for the rise of impulse sparkover voltage is considered due to the scarcity of initial or initiative electrons, which is caused by supplying C_2Cl_4 mist. This effect is practically significant for compact design of power equipment. as the impulse withstand level should be much higher than the corresponding AC value. However it is reported in ref. [16] that the mist suspended in the gas space was eliminated by applying AC or DC voltages. This suggests that it might be not so easy to control the mist condition in practical equipment.

5. CONCLUDING REMARKS

The atmospheric air is an excellent insulating medium in non-uniform field conditions, which has made it very suitable for insulating most of power transmission and distribution lines. For more compact designs of power equipment, we must resort to the usual gas-insulated systems with SF_6 in quasi-uniform arrangements. In regard to the greenhouse effect of SF_6, we consider the best policy is to use pure SF_6 and suppress its emission as low as possible. However, we have also briefly reviewed several possibilities of replacement gases. The gases mentioned are ordinary gases such as N_2 and CO_2, gas mixtures containing fluorocarbon, gases decomposable in the atmosphere like CF_3I, and vapor-mist dielectrics.

REFERENCES

1. For example, L. G. Christophorou, J. K. Olthoff, and D. S. Green, Gases for electrical insulation and arc interruption: possible present and future alternatives to pure SF_6, NIST Technical Note 1425, 1-44 (1997)
2. T. Takuma, Discharge characteristics of gaseous dielectrics, IEEE Trans. on Electrical Insulation, 21(6), 855-867 (1986)
3. J. C. Devins, Replacement gases for SF_6, ibid., 15(2), 81-86 (1980)
4. R. E. Wootton, S. J. Dale, and N. J. Zimmerman, Electric strength of some gases and gas mixtures, Gaseous Dielectrics II, 137-148 (1980)
5. L. Niemeyer, A systematic search for insulation gases and their environmental evaluation, Gaseous Dielectrics XIII, 459-464 (1998)
6. M. K. W. Ko, et al, Atmospheric sulfur hexafluoride: sources, sinks and greenhouse warming, J. Geophys. Research, 98(D6), 10,499-10,507 (1993)
7. Electric Technology Research Association, The handling standards on the use of SF_6 gas in electric power equipment (in Japanese), Electric Cooperative Research, 54(3), 1-155 (1998)
8. T. Takuma, T. Watanabe, and K. Kita, Breakdown characteristics of compressed gas mixtures in nearly uniform fields, Proc. IEE (Great Britain), 119(7), 927-928 (1972)
9. O. Yamamoto, T. Takuma, S. Hamada, and Y. Yamakawa, Applying a gas mixture containing c-C4F8 as an insulating medium, IEEE Trans. on Dielectrics and Electrical Insulation, 8(6),1075-1081 (2001)
10. T. Takuma, S. Hamada, and O. Yamamoto, A study on gas insulation applying gas mixtures (in Japanese), Report for the grant-in-aid from the Japanese Ministry of Education No.11450112 (2002)
11. O. Yamamoto, T. Takuma, and M. Kinouchi, Recovery of SF_6 from N_2/ SF_6 gas mixtures by using a polymer membrane, IEEE Electrical Insulation Magazine, 18(3), 32-37 (2002)
12. For example, Hitachi Ltd., Japanese Patent Application No. P2000-150253A (2000)
13. H. Toyoda, S. Matsuoka, K. Hidaka, Voltage-time characteristics of CF_3I-N_2 gas mixtures in nanosecond range (in Japanese), Joint Technical Meeting on Electrical Discharge and High Voltage, IEE Japan, ED-03-91, HV-03-62 (2003)
14. R. T. Harrold, Vapor-mist dielectrics, 1981 Annual Report of the IEEE Conference on Electrical Insulation and Dielectric Phenomena, 360-369 (1981), also R. T. Harrold, Physical aspects of vapor-mist dielectrics, IEEE Trans. on Industry Applications, 22(1), 63-69 (1986)
15. R. T. Harrold, Partial discharge suppression in vapor-mist dielectrics, 1984 Annual Report of the IEEE Conference on Electrical Insulation and Dielectric Phenomena, 364-373 (1984)
16. M. Yashima, H. Fujinami, and T. Takuma, Breakdown characteristics of gases mixed with tetrachloroethylene mist under nearly uniform fields, IEEE Trans. on Electrical Insulation, 25(2), 371-379 (1990)

INVESTIGATIONS ON SF_6-FREE GAS CIRCUIT BREAKER ADOPTING CO_2 GAS AS AN ALTERNATIVE ARC-QUENCHING AND INSULATING MEDIUM

T. Uchii, Y. Hoshina, T. Mori, H. Kawano, T. Nakamoto, and H. Mizoguchi

1. INTRODUCTION

SF_6 gas has widely been used for high-voltage and large-capacity electric power equipment such as gas insulated switchgears (GIS) and gas circuit breakers (GCB) due to its excellent insulating and arc-quenching capability. Although SF_6 gas strongly contributes to achieve compactness and high reliability of the equipment, it has been recognized as one of the potent greenhouse gases and was designated to reduce the emissions at COP3 at Kyoto in 1997. At present, strategic effort to reduce the emissions is being made, which actually pan out.[1] Over the long term, however, it is certainly preferable to reduce the consumption itself, because its atmospheric life time is observed to be quite long, thus the amount of SF_6 gas on the earth will inevitably get increasing in the future unless artificial destruction.

With the above background, a 72.5 kV-31.5 kA classCO_2 gas circuit breaker (hereinafter called CO_2-GCB) model, which contains neither SF_6 nor Freon at all, was designed and produced. As a result of current interruption and insulation tests, the CO_2-GCB model achieved satisfactory performance. In the present paper, the basic concept and technologies, the test results, and the estimation of its environmental impact will be described.

2. WHY CO_2 GAS?

The gases that are applicable to an environmentally-benign electric power equipment are required to have no or minimal toxicity, global warming effect and

* T. Uchii, Y. Hoshina, T. Mori, H. Kawano, T. Nakamoto, H. Mizoguchi, TMT&D Corporation, Kawasaki, JAPAN 210-0862.

ozone depletion effect, and should remain gaseous at low temperatures, for example, around –30 degrees C. When selecting the alternative gases widely from the above viewpoints, the possible candidates are narrowed down to air, N_2, O_2, H_2, CO_2, rare gases (He, Ar, etc), and their mixtures.[2] In practice, they are also required to have adequate insulating and arc-quenching capability, chemical stability, and have no flammability and explosiveness. Eventually, the possible candidates which can be applied as single gas or main gas of mixture could be limited only to N_2 and CO_2. (Here, air is regarded as an N_2-based mixture.)

Table 1 shows the comparison of fundamental gas properties of SF_6, CO_2 and N_2. As shown in Table 1, CO_2 meets the basic requirements for application to an environmentally-benign electric power equipment. In addition, CO_2 has a lower boiling temperature than SF_6, and it is known that CO_2 remains gaseous at low temperature ranges down to –40 degrees C even at a high gas pressure of 1.0 MPa-abs.

As for insulating capability, as also shown in Table 1, CO_2 is naturally lower than SF_6, but its 50% breakdown voltage is about 35% higher than that of N_2 at a high gas pressure of 0.9 MPa-abs.

In Table 1, arc-quenching capability is evaluated by arcing time constant as an index for the thermally interrupting capability of a gas. Qualitatively, smaller arcing time constant suggests better thermal interrupting capability. Table 1 shows that the arcing time constant of CO_2 is higher than that of SF_6, but is below one tenth of that of N_2.

In short, although CO_2 gas is inferior to SF_6 gas in insulating and arc-quenching capabilities, it surpasses N_2 gas which is regarded as a representative alternative gas in many previous works, particularly in arc-quenching capability. This suggests that CO_2 gas is a promising alternative gas, particularly for switching apparatus such as GCBs.

Table 1. Comparison of fundamental gas properties between SF_6, CO_2 and N_2.

Gas	SF_6	CO_2	N_2
Molecular mass	146.06	44.01	28.01
Density (kg/m^3)*1	5.9	1.8	1.1
GWP*2	23,900	1	~0
ODP*3	–	–	–
Toxicity*4	–	–	–
Chemical stability	Stable	Stable	Stable
Flammability / Explosibility	–	–	–
Boiling temperature (deg. C)*5	–51	–78	–198
Dielectric strength (%)*6	100 (–)	34 (–)	25 (+)
Arcing time constant (μs)*7	0.8	15	220

*1 At 300 K, 1 atm
*2 Global Warming Potential, Integrated period 100 years (IPCC, 1995)
*3 Ozone Depletion Potential
*4 As pure gas (Note that arced gas could be different)
*5 At 1 atm
*6 50% breakdown voltage measured by a full-scale coaxial cylindrical electrode[3] (The weak polarity value is shown), Lightning impulse, At 0.9 MPa-abs
*7 Measured for a free-burning arc at 1 atm

3. BASIC CONCEPT AND TECHNOLOGIES

3.1. Basic Concept for CO_2-GCB

As described above, CO_2 gas meets the basic requirements for application to an environmentally-benign power equipment and surpasses N_2 gas in terms of insulating and arc-quenching properties. However, when simply replacing SF_6 gas with CO_2 gas based on an existing SF_6 gas equipment, it must cause reduction of performance or increase in the size of the equipment. The authors, therefore, had the basic concept for a CO_2-GCB, namely (1) to understand the fundamental properties of CO_2 gas as an insulating and arc-quenching medium to draw its full potential, and (2) to develop new breakthrough technologies to compensate the performance gap between SF_6 and CO_2.

3.2. Key Technologies for Thermal Interruption

Focusing on Item(2) of the above, technologies particularly for enhancing thermal interruption performance of a CO_2-GCB are introduced below.

3.2.1. Application of Hybrid-puffer™ Interrupter

A Hybrid-puffer™ type interrupter[4] specially designed for CO_2 gas has been investigated. As known generally, the higher puffer pressure at current zero leads the higher thermal interruption performance. A Hybrid-puffer™ is one of the techniques enhancing the puffer pressure by utilizing arc energy effectively during a current interruption.[4] Figure 1 shows the analytical result of puffer pressure at current zero, compared between a conventional double-flow-type interrupter and Hybrid-puffer™-type for both CO_2 and SF_6. As noted in Figure 1, Hybrid-puffer™ technique is more effective for CO_2 than SF_6, which is because gas density and heat capacity of CO_2 are smaller than SF_6 and, in addition, arc voltage in CO_2 is larger than SF_6.

3.2.2. Application of Ablation Phenomena of Polymer Material

Application of ablation phenomena of polymer material has also been investigated as one of the breakthrough technologies improving interruption performance of CO_2-GCB. Figure 2 shows the principle of this technique, in which thermal(or radiative) energy from the arc brings about ablation of the polymer element during current interruption. This process causes both the puffer pressure and the arc-quenching capability of the blasting gas itself to be enhanced due to the generation of the ablated gas, including H or O elements, which have relatively satisfactory arc-quenching capability. It was observed in an experimental GCB model that the specific polymer element which was installed in the puffer cylinder enhanced the puffer pressure at the current zero by 16% and thermal interruption capability by 20~25% at the maximum.[5]

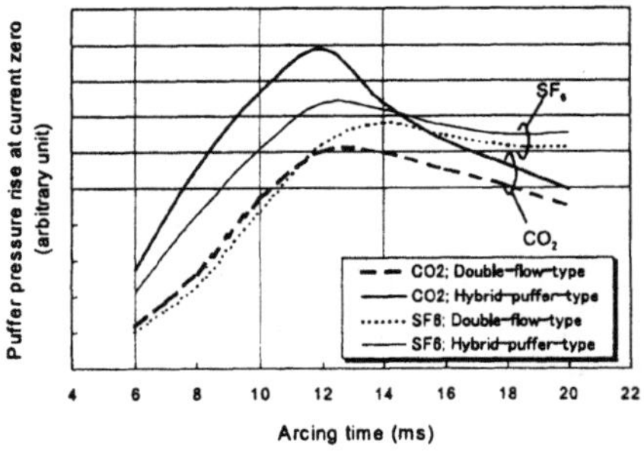

Figure 1. Comparison of puffer pressure enhancing effect by Hybrid-puffer™ technique between SF_6 and CO_2.

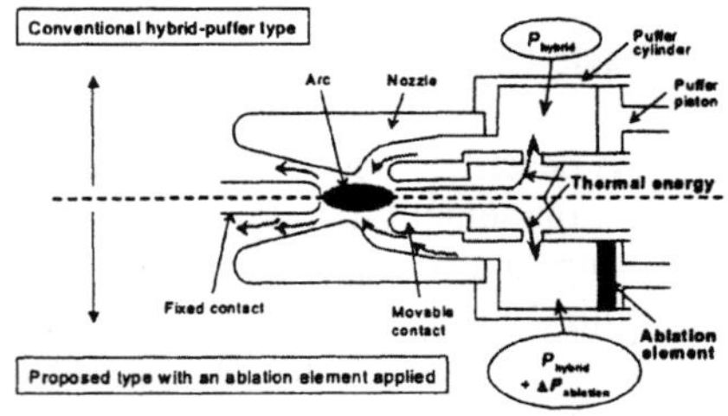

Figure 2. Principle of puffer pressure enhancing technique utilizing ablation phenomena of a polymer element.

4. INTERRUPTION AND INSULATION TEST OF CO_2-GCB MODEL

4.1. Specifications of the CO_2-GCB Model

Based on the basic investigations of CO_2 properties as an insulating and arc-quenching medium, a 72.5 kV-31.5 kA-class CO_2-GCB model, shown in Figure 3, was designed and produced. The specifications of the model are shown in Table 2. The main features of the model are as follows:

(1) A Hybrid-puffer™-type interrupter specially designed for CO_2 gas was adopted.
(2) An ablation element made of appropriate polymer material was applied.
(3) All the dimensions of the GCB components were determined based on fundamental dielectric data of CO_2 gas with sufficient margin.

Figure 3. View of the 72.5 kV-class CO_2-GCB model (for single phase).

Table 2. Specifications of the CO_2-GCB model.

Items	Value
Filling gas	CO_2 100%
Filling gas pressure	0.8 MPa-abs
Rated voltage	72.5 kV
Rated interrupting current	31.5 kA
Power frequency	50 Hz
Rated ACWV	160 kV_{rms}
Rated LIWV	375 kV

4.2. Test Result

Current interruption and insulation tests of the CO_2-GCB model were carried out based on the standard of IEC 62271-100. The test results are summarized in Table 3. As shown in Table 3, the CO_2-GCB model achieved satisfactory performance for major test-duties; namely, capacitive current switching, short-line fault 90%, terminal fault 100% (symmetrical and asymmetrical) interruption, power-frequency and lightning impulse insulation.

Table 3. Summary of the test results of the CO_2-GCB model. (o: Success, x: Failure)

RATINGS	72.5 kV-31.5kA-50Hz	
STANDARD	IEC 62271-100	
DUTY	RESULT	JUDGEMENT
Capacitive current switching (After 3 times 60% current interruptions)	NO Restrike during 16 times interruptions around the minimum arcing time with the recovery voltage of 1.7 x 72.5kV/1.73.	**Satisfactory**
Short line fault 90% (L90)	Arcing time: 	Interruption window: 9.5 ms secured **Satisfactory**
Breaker terminal fault 100%, Symmetrical (T100s(b))	Arcing time: 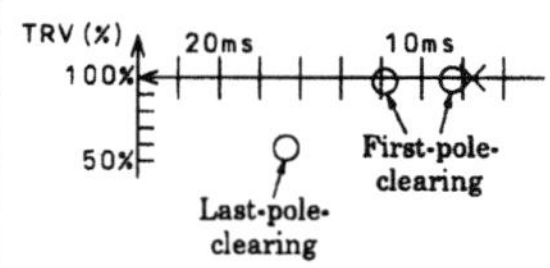	Interruption window: 3.3 ms for first-pole-clearing, 8.2 ms secured over all. **Satisfactory**
Breaker terminal fault 100%, Asymmetrical (T100a) (Major loop)	Arcing time: 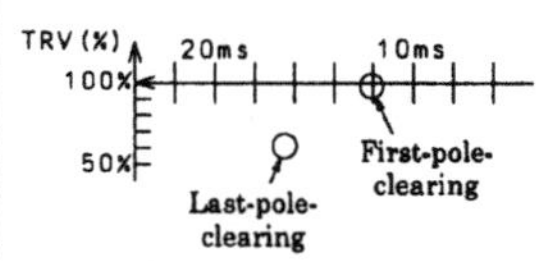	Interruption window*: 3.6 ms for first-pole-clearing, 8.0 ms secured over all. **Satisfactory**
LI insuration	375 kV No F.O.	**Satisfactory**
AC insulation	160 kV_{rms} No F.O.	**Satisfactory**

*Minimum arcing time for minor loop is assumed from T100s

4.3. Decomposed Products and Gases after Current Interruptions

Decomposed products and gases after current interruptions are one of the fundamental issues from the practical viewpoint. It was observed that no harmful arced gas which cannot be absorbed by Zeolite and that no controversial decomposed product for the practical use were detected even after more than 10 times large current interruptions.

5. ESTIMATION OF ENVIRONMENTAL IMPACT

A life cycle assessment (LCA) was carried out to evaluate the environmental contribution of the CO_2-GCB model. Figure 4 shows the considered life cycle flow and the conditions of the LCA, and the result is shown in Figure 5. It was noted from Figure 5 that the CO_2-GCB model could reduce the global warming impact by about 45% compared to the latest SF_6 gas circuit breaker in the same rating for 40 years operation including two maintenance.

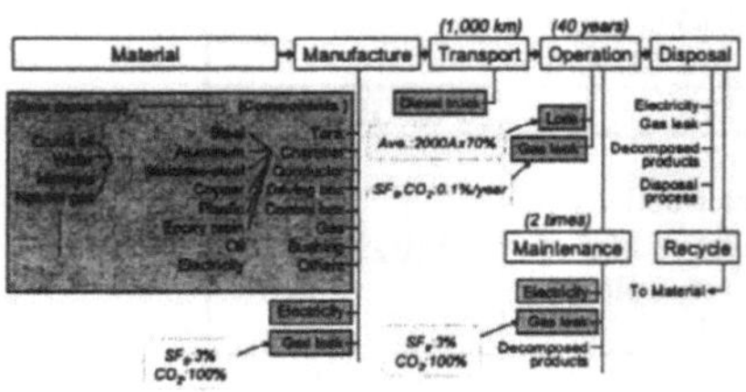

Figure 4. Considered life cycle flow and conditions of the LCA.

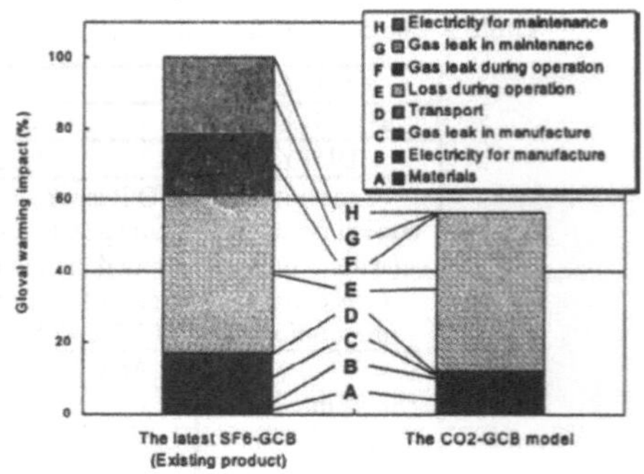

Figure 5. Reduction of global warming impact during 40 years operation by the CO_2-GCB model.

6. CONCLUSIONS

Based on the fundamental investigations of CO_2 properties as an insulating and arc-quenching medium, and adopting new technologies such as polymer ablation, a 72.5 kV-31.5 kA-class CO_2-GCB model was designed and produced. As a result of interruption and insulation tests, the model achieved satisfactory performance for major test-duties.

The present study was done as a part of work commissioned by the Agency for Natural Resources and Energy, METI to Institute of Applied Energy (IAE).

REFERENCES

1. U. S. Environmental Protection Agency, *International Conference on SF_6 and the Environment: Emission Reduction Strategies, San Diego*, URL: http://www.epa.gov/highgwp1/sf6/ (2002).
2. Toshiba Corporation, Development of Dielectric and Arc-quenching Gases and Their Applications to Power Equipment for Mitigating Global Warming, *Final R&D Report for FY 2001 Proposal-Basic R&D Program of New Energy and Industrial Technology Development Organization(NEDO)*,pp. 3-8 (2001) (in Japanese)
3. M. Sato et al., Dielectric Properties in High Pressure CO_2 and N_2 Gas Using Coaxial Cylindrical Electrode and on the Surface of Spacer, *IEEE/PES T&D Conference*, OR24-1 (2002).
4. S. Yanabu et al., Development of Novel Hybrid Puffer Interrupting Chamber for SF_6 Gas Circuit Breaker Utilizing Self-Pressure-Rise Phenomena by Arc, *IEEE Trans. on Power Delivery*, Vol. 4, No. 1, pp. 355-361 (1989).
5. T. Uchii et al., Development of 72 kV-Class Environmentally-Benign CO_2 Gas Circuit Breaker Model, *IEE of Japan Trans. PE*, Vol. 124, No. 3 (2004) (in Japanese)

N_2 AND N_2/CO_2 MIXTURE IN GAS INSULATED COMPARTMENTS UNDER HIGH PRESSURE

Edelhard Kynast and Karsten Juhre*

1. INTRODUCTION

SF_6 gas is widely used as insulating medium in high-voltage technology due to its nearly ideal insulating and switching behavior. Based on many years of experience SF_6 gas-insulated switchgear (GIS) is very compact and of highest reliability. Due to the discussions around the environmental potential of released SF_6 and also out of economic reasons alternatives to the insulating gas SF_6 had been looked for since long time in the past. According to today's consideration the most suitable insulating alternatives are nitrogen (N_2), carbon-dioxide (CO_2) and mixtures of both gases. For insulation <u>and</u> switching behavior no acceptable alternatives have been found[1]. In comparison to the SF_6 technique the practical use of N_2 and/or CO_2 means basically a specific reduction in insulating strength with the result of a much larger design or a strong increase in gas pressure. N_2 and CO_2 have been widely and intensively investigated in the past, but only few investigations dealt with N_2 and N_2/CO_2 under extreme pressures up to 2.0 MPa in real size GIS compartments. It has been found that there is a wide spread in literature data of the electrical field strength of e.g. N_2 under 2.0 MPa (between 12 and 40 kV/mm) and most investigations had been performed by using small gas gaps with AC and DC test voltage only.

Contributing to the various investigations breakdown and withstand voltage tests on N_2 and N_2/CO_2 insulated GIS components have been performed to

* Edelhard Kynast and Karsten Juhre, Siemens AG, Power Transmission and Distribution, High Voltage Technology, 13623 Berlin, Germany

Gaseous Dielectrics X.
Edited by Christophorou *et al.*, Springer, New York, 2004.

estimate their feasibility. Two different test setups were used:

- pure gas-gap configurations for basic tests,
- 245 kV GIS components including epoxy insulators for design tests.

2. BASIC INVESTIGATIONS BY USE OF GAS GAP

GIS components with pressure vessels applicable up to 2.0 MPa were used for basic investigations under N_2 and N_2/CO_2 (80%/20%) atmosphere in a cylindrical pure gas gap.

In the first basic tests the 50 Hz alternating (AC) withstand voltages and 1.2/50 μs lightning impulse (LI) withstand voltages at gas pressures 0.6, 0.9, 1.2, 1.5 and 1.8 MPa were determined. In addition the influences of surface roughness, coatings and imperfections were investigated. The test setup is shown in Figure 1. For a proper data basis the field distribution was calculated and the results are given as specific field strength values.

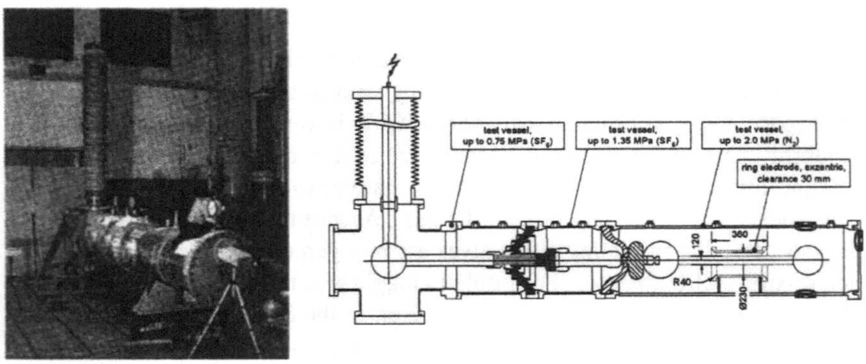

Figure 1: Test setup for basic investigations up to 1.8 MPa

The testing procedure was standardized to obtain comparable test results for all different test series which were repeated at least three times[2].

The results of the basic series with bare electrodes and without impurities are shown in Figure 2. For lightning impulse voltage, the withstand field strength increases generally with increasing pressure up to about 1.5 MPa. For N_2 LI withstand voltages with positive polarity are higher than those for negative polarity up to approx. 1.6 MPa. Within the investigated pressure range, the withstand field strengths of N_2/CO_2 gas mixture are remarkable higher than those of N_2. Up to 1.5 MPa the positive withstand values are higher than those of negative polarity, at higher pressures a reversal can be noticed. The lightning

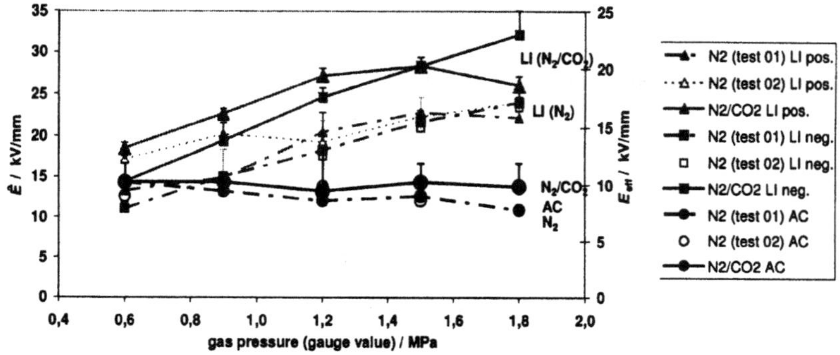

Figure 2: LI and AC withstand voltages (test results of basic series)

impulse withstand field strength of N_2/CO_2 at 1.5 MPa is comparable to this of 0.43 MPa SF_6 in practical applications.

For AC voltage the test results show that there is a slight decrease of AC withstand field strength from 0.6 MPa to 1.8 MPa with N_2. With N_2/CO_2 the AC withstand field strength is nearly constant within the investigated pressure range.

3. INFLUENCE OF THE SURFACE AND IMPERFECTIONS

With N_2 two different surface roughnesses were investigated. For test 01 an inner conductor with R_z = 1.5 ... 2.8 μm was used, the inner conductor for test 02 had surface roughnesses R_z = 10.8 ... 17.3 μm as commonly used for conventional GIS. No appreciable differences in test results have been observed (Figure 2).

In order to estimate the influences of dielectric coatings an epoxy coating of 0.5 mm was applied on the inner conductor. In comparison to the bare conductor a larger scatter of the LI and AC withstand voltages in N_2 has to be stated (Figure 3). In N_2 as well as in N_2/CO_2 a strong decrease of positive LI withstand voltage was observed. The behavior of AC withstand voltage shows a slight increase with pressure. The punctures of the epoxy coating were spread over the whole stressed area and not cumulated anywhere.

The feasibility of PD measurements for diagnostics in N_2 or N_2/CO_2 filled GIS components under high pressure was checked. For this an UHF PD monitoring system and a digital video camera system were installed during tests with AC voltage at pressures 1.2 to 1.8 MPa. Tests were performed with graphite powder on the grounded cylindrical electrode, fixed with a small amount of grease, and with a needle of 10 mm length on the inner conductor. The AC and LI withstand voltages in the test setup with the graphite powder decreased to about 50% of those under clean conditions. At the needle tip a small glowing point was visible without change up to breakdown voltage. No partial discharge activities were measured before breakdown.

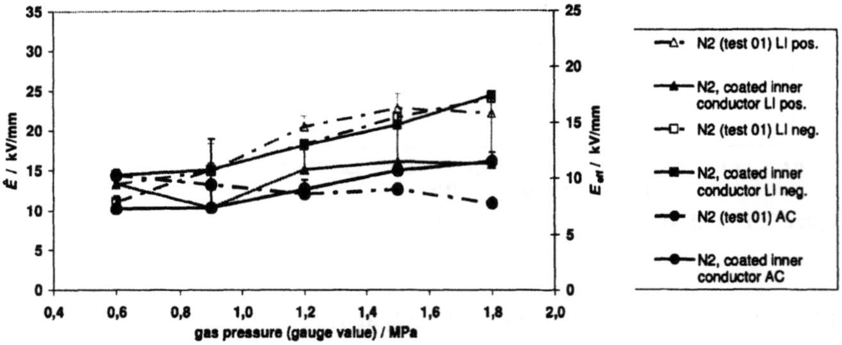

Figure 3: LI and AC withstand voltages with a coated inner conductor in N_2

4. TESTS ON GIS-COMPONENTS

The aim of the second part of the investigations was to verify as far as possible the results of the basic investigations from the gas gap model by testing GIS compartments from the standard production line of the nowadays used SF_6 technique. Compartments of a single phase enclosed switchgear type for U_m = 245 kV was chosen with a special selection of the test vessels for high pressure application of maximum 1.3 MPa. The selected compartments were current transformer, T-module, 30°-elbow connection and the necessary connecting parts including seven cast epoxy insulators. Field calculations and an estimated reduction of 40 % compared to the pure SF_6 insulation show that the withstand level of U_m = 145 kV should just be reached with N_2 under this high pressure.

The withstand and breakdown tests were carried out at 0.7, 1.0 and 1.3 MPa with pure nitrogen. In a second test series the gas mixture 80 % N_2 with 20 % CO_2 was used for 1.3 MPa. Figure 4 shows the results of the power-frequency respectively the lightning impulse voltage tests.

The power frequency breakdown and withstand test values show an increase from 0.7 MPa up to 1.3 MPa. Because of the contradiction to the basic tests verification tests at 1.3 MPa were carried out. Breakdowns occur at relatively low values compared with the fist tests (e.g. 71 % and 81 % of first withstand value). The breakdown and the withstand values show a large scatter.

The results with lightning impulse voltage show a significant difference for positive and negative polarity. The negative voltage values are generally higher than the positive ones. For positive impulses the difference between first breakdown and withstand value is quite small and both show a straight rise with increase of the pressure up to 566 kV withstand voltage.

During the first test series with negative impulses extreme high values for the first breakdown were reached for 1.0 MPa and 1.3 MPa, but the withstand voltages were far below mainly due to breakdown effects on the surface of the insulators. In repeated tests at 1.3 MPa breakdown and withstand voltages showed a normal correlation due to scatter of breakdown probability. The reason for the difference between the tests can only be residual contamination due to different manual handling of cleaning.

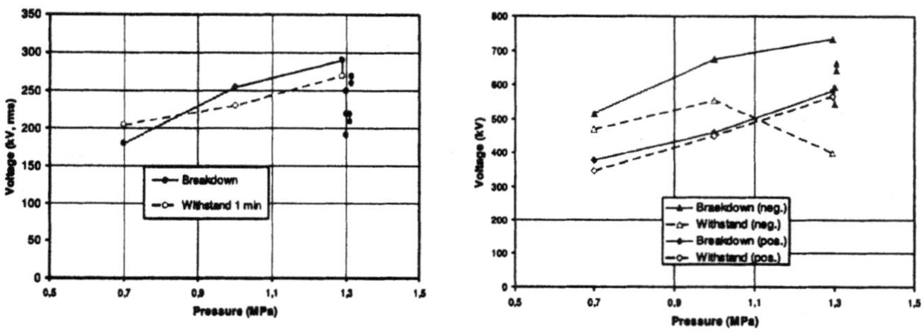

Power-frequency voltage

Lightning impulse voltage

Figure 4: Breakdown and withstand voltage values of GIS compartments with N_2

The inspection of the flashover locations showed that the insulator surfaces were involved very often. Furthermore breakdowns occurred at very small areas of field enhancements which do not harm with SF_6 insulation. An improvement of these areas led to an increase of the withstand voltages of some percents.

The tests with the gas mixture N_2/CO_2 (80 %/20 %) were carried out at 1.3 MPa only with power-frequency and lightning impulse voltage. Within a scatter of about 3 % the breakdown and withstand voltages values were the same compared with the results with pure N_2.

5. DISCUSSION

Considering the results out of both test setups the following conclusions can be drawn for a GIS insulation with pure N_2 and the mixture N_2/CO_2:

With the increase of pressure a nearly linear increase of the lightning impulse withstand capability is given up to the maximum investigated pressure of 1.8 MPa. For the negative polarity a high sensitivity concerning insulator surfaces and small irregularities gives no clear withstand results.

For power-frequency voltage no increase of the withstand voltage has been observed in the basic investigation setup. Due to an increase in scatter above 1.0 MPa no clear withstand capability can be given for the real size GIS compartments.

The high differences in results for repeated tests show that the insulation with high pressure N_2 is highly sensitive against residual contamination like dust. The sensitivity seems to be increased with increase of pressure. For practical use of N_2 in GIS, a much higher cleanliness would be necessary than with the well known SF_6 insulation, as well as the avoidance of small field enhancements, which can be tolerated in the conventional SF_6 technique.

The addition of 20 % CO_2 to nitrogen insulation leads to an remarkable increase of the dielectric withstand strength in pure gas gap configurations. The sensitivity concerning irregularities decreases [3]. But in practical application with insulator surfaces no remarkable differences have been found.

UHF partial discharge measurement is not suitable tool for detection of irregularities. In high pressured N_2 and N_2/CO_2 breakdown occurs without any measurable PD before [4].

6. SUMMARY

Basic investigations on gas gaps and first tests in real size GIS components with nitrogen and gas mixture N_2/CO_2 under high pressure showed that these insulating gases are not well appropriate for a reliable real application in GIS. The dimensions of a SF_6 insulated GIS type 245 kV are not sufficient for a N_2 insulated GIS for 145 kV at 1.3 MPa minimum insulating pressure. The required withstand voltages are just not met. The trend of the insulating capability and the scatter of the results lead to the assumption that also a further increase in pressure would not give sufficient results for a reliable GIS design.

7. REFERENCES

1. Knobloch, H.: The comparison of arc-extinguishing capability of SF_6 with alternative gases in high-voltage circuit-breakers; *Gaseous Dielectrics VIII (1998)*
2. Kynast, E.; Grund, A.; Finkel, M.; Diessner, A.: Dielectric properties of N_2/SF_6 mixtures for use in GIS or GIL; *International Symposium on High-Voltage Engineering ISH (1999)*
3. Goshima, H. et al.: Lightning impulse breakdown charcteristics of high-pressure CO_2 compared with those of N_2 as alternative insulation gases to SF_6; *ISH (2003)*
4. Graf, R.; Schoeffner, G.: Determination of inception and breakdown voltages of N_2-SF_6 gas mixtures in strong inhomogeneous fields; *Gaseous Dielectrics IX (2001)*

PARTIAL DISCHARGE AND BREAKDOWN CHARACTERISTICS OF CO_2–BASED GAS MIXTURES AS SF_6 SUBSTITUTES

N. Koshino[1], Y. Yoshitake[1], N. Hayakawa[1], and H. Okubo[2]

1. INTRODUCTION

SF_6 gas has been used as an excellent insulating and arc-quenching gas for electric power apparatus such as SF_6 gas insulated switchgear (GIS). However, since SF_6 gas was specified to be a greenhouse gas because of its high global warming potential (GWP), environmental-friendly gases or gas mixtures alternative to SF_6 gas are strongly required to be developed.

We focused our research on CO_2 gas and CO_2-based gas mixtures such as CO_2/O_2, CO_2/N_2 with small GWP as SF_6 substitutes, and investigated partial discharge (PD) and breakdown (BD) characteristics under lightning impulse and ac voltage applications. PD inception voltage (PDIV), BD voltage (BDV) and the impulse ratio were obtained in CO_2, N_2, SF_6 gases and CO_2-based gas mixtures. Experimental results and discussions revealed the advantages of CO_2 gas and CO_2-based gas mixtures as SF_6 substitutes.

2. EXPERIMENTAL SETUP

Figure 1 shows the experimental setup. We used sphere-plane electrode system for quasi-uniform electric field (sphere diameter ϕ=50mm, gap length g=10mm) and rod-plane electrode system for non-uniform electric field (rod diameter ϕ=5mm, g=40mm). The experimental tank was filled with CO_2, N_2, SF_6 gas or CO_2-based gas mixtures at the gas pressure P=0.1-0.4MPa. Standard lightning impulse voltage (1.2/50μs) or 60Hz ac high voltage was applied to sphere or rod electrode, and generated PD and BD. PD current pulses were detected through a matching resistance and PD light intensity pulses were detected through a photo multiplier tube (P.M.T.). The 50% probability PDIV

[1] Department of Electrical Engineering, Nagoya University, Nagoya, 464-8603, Japan
[2] Center for Integrated Research in Science and Engineering, Nagoya University, Nagoya, 464-8603, Japan

($PDIV_{50}$), 50% probability BDV (BDV_{50}) and their standard deviations were obtained by the up-and-down method under the impulse voltage application. All experiments in this paper were carried out under room temperature.

3. BREAKDOWN CHRACTERISTICS OF CO_2 GAS UNDER QUASI-UNIFORM ELECTRIC FIELD

3.1. IMPULSE RATIO

Figure 2 shows the impulse BDV_{50} and ac BDV in CO_2, N_2 and SF_6 gases at P=0.1MPa under sphere-plane electrode system (ϕ=50mm, g=10mm). The ac BDV in CO_2 gas was almost the same as that in N_2 gas, and the ac BDV in SF_6 gas was about three times of those in CO_2 and N_2 gases. On the other hands, the impulse BDV_{50} in SF_6 gas was twice of that in CO_2 gas. These results suggest the difference in the impulse ratio between in CO_2 and SF_6 gas.

Table 1 shows impulse ratio of positive and negative BDV (impulse BDV_{50}/ac BDV) in CO_2, N_2 and SF_6 gases. The impulse ratio of CO_2 gas was the highest in these gases, which suggested that the lack of initial electrons in CO_2 gas was more remarkable than those in N_2 and SF_6 gases. The initial electrons may be derived from the electron emission from the cathode surface, the electron detachment from negative ions and so on. According to the fact that the negative BDV_{50} was almost the same as the positive BDV_{50} in CO_2 and SF_6 gases, the electron detachment from negative ions could be dominant; if the electron emission from the cathode surface was dominant, the negative BDV_{50} would be lower than the positive BDV_{50}. Therefore, we focused on the electron detachment from negative ions.

Since the electron attachment coefficient in CO_2 gas is much lower than that in SF_6 gas, the negative ion density in CO_2 gas can be lower than that in SF_6 gas. Furthermore, the electron detachment coefficient in CO_2 gas is smaller than that in SF_6 gas[1]. As the result, the generation probability of initial electrons in CO_2 gas can be considered to be lower than that in SF_6 gas. In N_2 gas, the generation probability of initial electrons is higher than that in CO_2 gas, because the ionization coefficient of N_2 gas at low electric field increased the effective

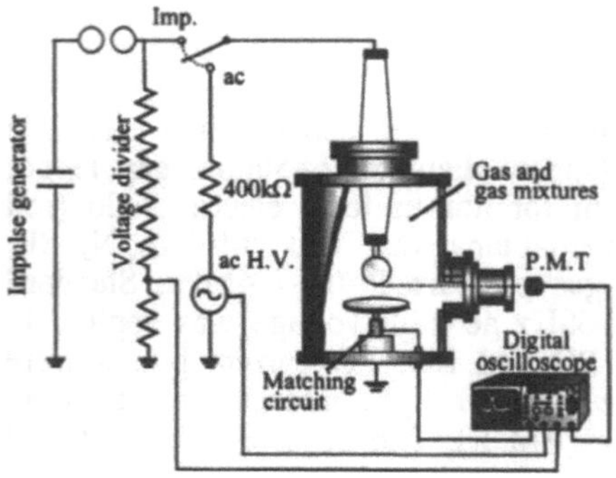

Figure 1. Experimental setup.

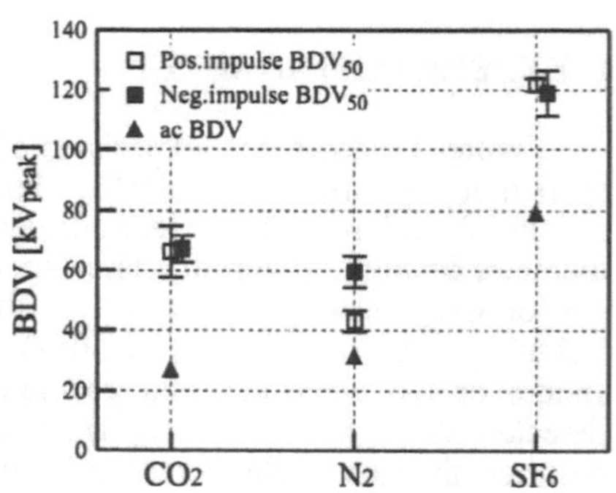

Figure 2. Impulse and ac BDV characteristics in CO_2, N_2 and SF_6 gases. (ϕ=50mm, g=10mm, P=0.1MPa)

Table 1. Impulse ratio of BDV in CO_2, N_2 and SF_6 gases for quasi-uniform electric field. (ϕ=50mm, g=10mm, P=0.1MPa, Non-uniformity factor η=1.28, $S_{90\%}$=221mm^2)

	CO_2	N_2	SF_6
Pos. impulse/ ac BDV	2.46	1.38	1.54
Neg. impulse/ ac BDV	2.49	1.90	1.50

ionizable volume. Consequently even under the quasi-uniform field, the generation probability of initial electrons in CO_2 gas would be lower than those in SF_6 and N_2 gases, i.e., the impulse ratio of CO_2 gas will be the highest.

3.2. BREAKDOWN ELECTRIC FIELD

Figure 3 shows the breakdown electric field (BDE) derived from BDV in Fig. 2, together with the theoretical value and the critical strength E_{cr}. BDE represents the maximum electric field strength at the tip of sphere electrode calculated by the charge simulation method. The theoretical BDE can be calculated by the streamer criterion as follows:

$$\int_0^x \bar{\alpha}\, dx = K \qquad \bar{\alpha}: \text{Effective ionization coefficient} \quad K: \text{constant}$$

Figure 4 shows the effective ionization coefficient as a function of electric field strength in CO_2, N_2 and SF_6 gases[2,3]. The constant K varies with gas (CO_2 : K=13-16, N_2 : K=4.5-5.5, SF_6 : K=10.5-18.6)[3,4]. Fig. 3 verifies that the theoretical BDE corresponds to ac BDE.

Figure 3 also shows that the ratio of the theoretical BDE to E_{cr} in CO_2 gas was 1.48, whereas that in SF_6 gas was 1.08. The higher ratio in CO_2 gas may be attributed to the slower increase in the effective ionization coefficient around E_{cr} in Fig. 4, where the electron avalanche in CO_2 gas would be hard to develop into the streamer discharge, compared with that in SF_6 gas.

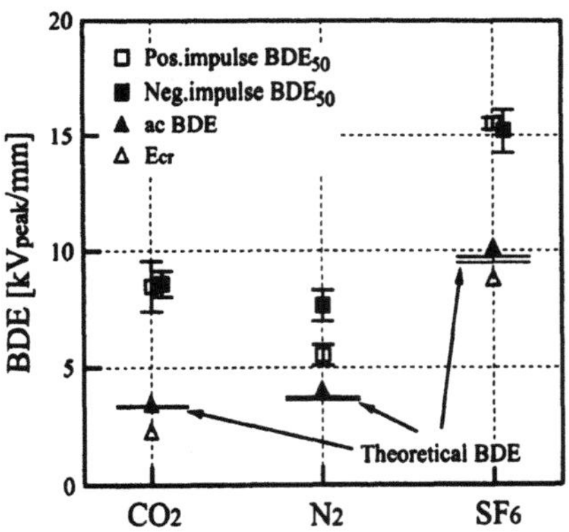

Figure 3. Impulse, ac and theoretical BDE characteristics in CO_2, N_2 and SF_6 gases. (ϕ=50mm, g=10mm, P=0.1MPa)

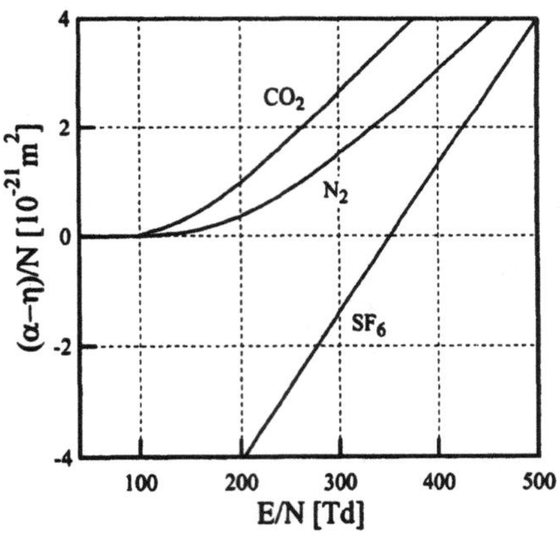

Figure 4. Effective ionization coefficient as a function of electric field strength in CO_2, N_2 and SF_6 gases[2,3].

4. PDIV AND BDV CHARACTERISTICS UNDER NON-UNIFORM ELECTRIC FIELD

4.1. PURE CO_2 AND SF_6 GAS

Positive BDV is usually crucial under non-uniform electric field in electronegative gases, because the positive discharge expands longer than the negative one. In this section, we discuss positive PD and BD characteristics under non-uniform electric field in CO_2, SF_6 gases and CO_2-based gas mixtures.

Figure 5 shows the positive impulse $PDIV_{50}$ and BDV_{50} as a function of gas pressure in CO_2 and SF_6 gases under rod-plane electrode system (ϕ=5mm, g=40mm). In addition, Fig. 6 shows the ac PDIV and BDV as a function of gas pressure in CO_2 and SF_6 gases. The impulse and ac PDIV in both gases were almost proportional to the gas pressure, whereas BDV exhibited nonlinear characteristics against the gas pressure due to the correlation between the corona stabilization effect and the streamer-to-leader transition[5].

Here, we defined the relative ratio R_{CO_2/SF_6} as the BDV of CO_2 gas normalized by that of SF_6 gas. Figure 7 shows R_{CO_2/SF_6} of the impulse BDV_{50} and ac BDV as a function of gas pressure. R_{CO_2/SF_6} tended to increase with the gas pressure, and reached 0.49 for ac BDV and 0.64 for impulse BDV_{50} at P=0.3MPa. These results suggest that CO_2 gas may have excellent dielectric characteristics as an SF_6 substitute, especially under impulse voltage condition.

4.2. CO_2-BASED GAS MIXTURES

Figure 8 shows the CO_2 content dependence of positive impulse $PDIV_{50}$ and BDV_{50} in CO_2/O_2 and CO_2/N_2 gas mixtures at P=0.1MPa. $PDIV_{50}$ in CO_2/O_2 and CO_2/N_2 gas mixtures slightly increased with the increase in CO_2 content. This result may be attributed to the lower generation of initial electrons in CO_2 gas, as described in section 3.1. BDV_{50} in CO_2/O_2 gas mixtures decreased with the increase in CO_2 content, whereas BDV_{50} in CO_2/N_2 gas mixtures slightly increased with the increase in CO_2 content. In other words, BDV_{50} in $CO_2$50%/$O_2$50% gas mixtures is higher than that in pure CO_2 gas as well as that

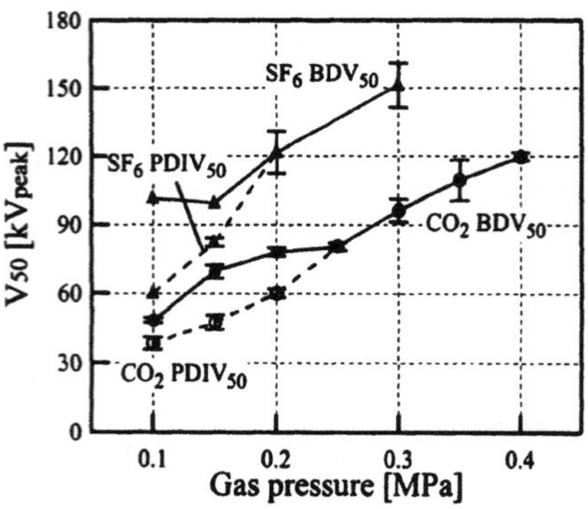

Figure 5. Positive impulse $PDIV_{50}$ and BDV_{50} as a function of gas pressure in CO_2 and SF_6 gases. (ϕ=5mm, g=40mm)

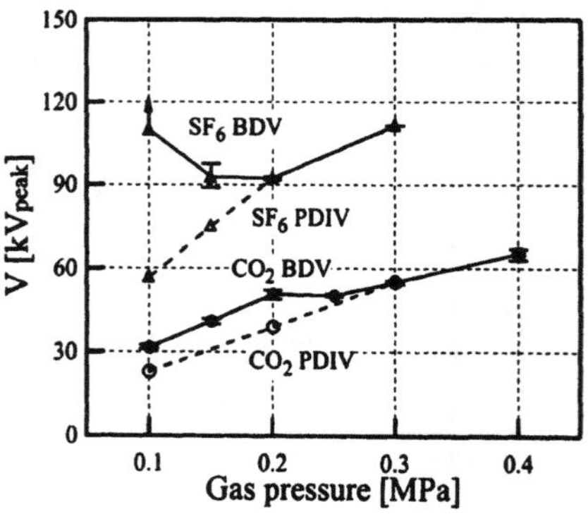

Figure 6. ac PDIV and BDV as a function of gas pressure in CO_2 and SF_6 gases. (ϕ=5mm, g=40mm)

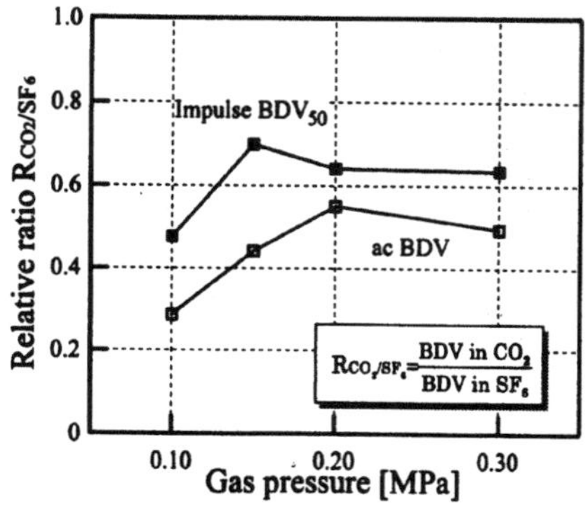

Figure 7. Relative ratio R_{CO_2/SF_6} as a function of gas pressure. (ϕ=5mm, g=40mm)

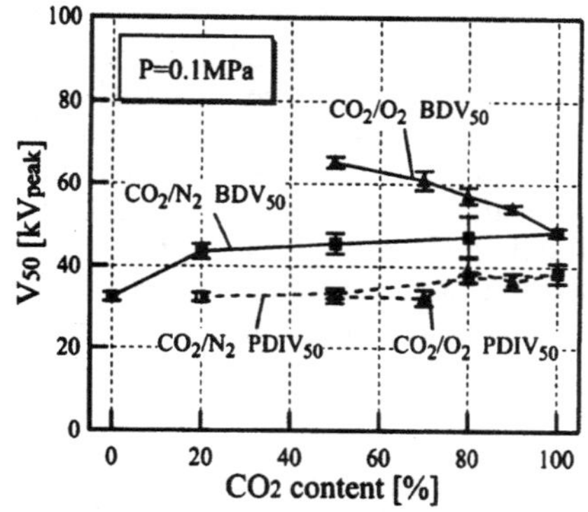

Figure 8. Positive impulse $PDIV_{50}$ and BDV_{50} as a function of CO_2 content in CO_2/O_2 and CO_2/N_2 gas mixtures. (ϕ=5mm, g=40mm)

in pure N_2 gas. These results can be derived from 4 times larger cross-sectional area of electron attachment in O_2 gas than that in CO_2 gas[6].

Figure 9 shows the positive impulse $PDIV_{50}$ and BDV_{50} as a function of gas pressure in $CO_2 50\%/O_2 50\%$ and $CO_2 50\%/N_2 50\%$ gas mixtures. $PDIV_{50}$ in both gas mixtures are almost the same at any gas pressure, whereas the local maximum in BDV_{50} and the corresponding gas pressure in both gas mixtures differ from each other. The higher BDV_{50} at the lower gas pressure in $CO_2 50\%/O_2 50\%$ gas mixtures can be attributed to the activated corona stabilization due to the combination of both electronegative gases. However, the stronger electronegativity in $CO_2 50\%/O_2 50\%$ gas mixtures also brought about the streamer-to-leader transition at the lower gas pressure, so that BDV_{50} at the higher gas pressure was lower than that in $CO_2 50\%/N_2 50\%$ gas mixtures.

The impulse ratio of discharge inception voltage under non-uniform electric field in CO_2, SF_6 gases and $CO_2 50\%/O_2 50\%$ gas mixtures is shown in Fig. 10. The impulse ratio in each gas increased with the gas pressure, which may be interpreted by that the lack of initial electrons become remarkable due to the

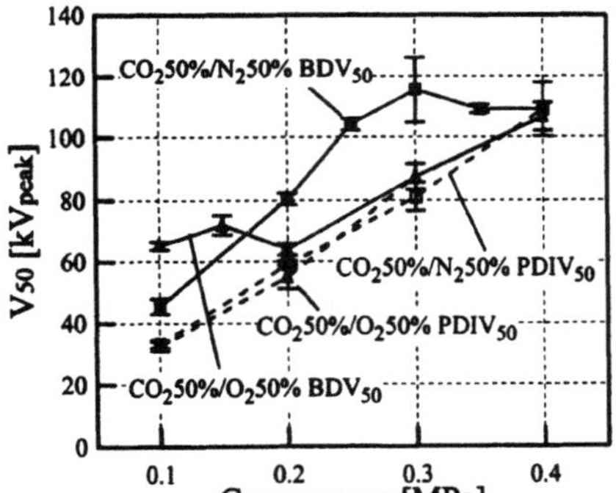

Figure 9. Positive impulse $PDIV_{50}$ and BDV_{50} as a function of gas pressure in $CO_2 50\%/O_2 50\%$ and $CO_2 50\%/N_2 50\%$ gas mixtures. (ϕ=5mm, g=40mm)

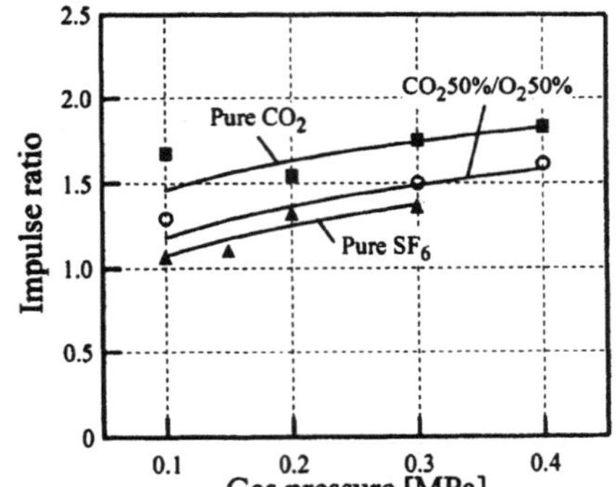

Figure 10. Impulse ratio as a function of gas pressure in pure CO_2, SF_6 gases and $CO_2 50\%/O_2 50\%$ gas mixtures. (ϕ=5mm, g=40mm)

shrinkage of critical volume around the tip of rod electrode with the increase in the gas pressure[7]. In addition, the impulse ratio in CO_2 gas and $CO_2$50%/$O_2$50% gas mixtures were higher than that in SF_6 gas.

From the results in Fig.10 and Table 1, the impulse ratio under non-uniform field was lower than that under quasi-uniform field. This would be caused by that concentrated stress under non-uniform field promotes the generation probability of initial electrons. The higher impulse ratio in CO_2 gas and CO_2-based gas mixtures will contribute to the efficient insulation design of power apparatus using such SF_6 substitutes.

5. CONCLUSIONS

In this paper, we obtained the impulse and ac PDIV, BDV and the impulse ratio in CO_2, N_2, SF_6 gases and CO_2-based gas mixtures under quasi-uniform and non-uniform electric field. The main results are summarized as follows:

1. The impulse ratio in CO_2 gas under quasi-uniform field was higher than those in that in N_2 and SF_6 gases, because the lower negative ion density and the smaller electron detachment coefficient in CO_2 gas could reduce the generation of initial electrons.
2. The impulse ratio in CO_2 gas and $CO_2$50%/$O_2$50% gas mixtures under non-uniform field was also higher than that in SF_6 gas, which will contribute to the efficient insulation design of power apparatus using SF_6 substitutes.
3. The ratio of the theoretical BDE to the critical strength E_{cr} in CO_2 gas was higher than that in SF_6 gas, because of the slower increase in the effective ionization coefficient around E_{cr}, where the electron avalanche would be hard to develop into the streamer discharge in CO_2 gas.
4. The relative ratio R_{CO_2/SF_6}(= BDV in CO_2/ BDV in SF_6) under non-uniform field tended to increase with the gas pressure, and reached 0.49 for ac BDV and 0.64 for impulse BDV_{50} at P=0.3MPa. These results suggest that CO_2 gas may have excellent dielectric characteristics as an SF_6 substitute, especially under impulse voltage condition.

REFERENCES

1. D. M. Xiao, X. G. Li and X Xu, Swarm parameters in SF_6 and CO_2 gas mixtures, *Journal of Physics D: Applied Physics*, Vol.34, No.24, L133-135, 2001.
2. J. L. Hernández-Ávila, E. Basurto, J. de Urquijo, Electron transport and swarm parameters in CO_2 and its mixtures with SF_6, *Journal of Physics D: Applied Physics*, Vol.35, No. 18, pp. 2264-2269, 2002.
3. Sayed A. Ward, Influence of Conducting Particles on The Breakdown Voltages of SF_6-N_2 Mixtures, *11th International Symposium on High Voltage Engineering*, Vol.3, pp.187-192, 1999.
4. N. H. Malik, A.H. Qureshi, Breakdown Gradients in SF_6-N_2, SF_6-Air and SF_6-CO_2 Mixtures, *IEEE Transactions on Electrical Insulation*, Vol. EI-15, No. 5, pp.413-418, 1980.
5. Y. Yoshitake, N. Hayakawa, T. Ueda, H. Okubo, Impulse partial discharge propagation mechanism under non-uniform electric field in SF_6/N_2 gas mixtures, *13th International Symposium on High Voltage Engineering*, p.471, 2003.
6. L. G. Christophorou and L. A. Pinnaduwage, Basic Physics of Gaseous Dielectrics, *IEEE Transactions on Electrical Insulation*, Vol.25, No. 1, pp.55-74, 1990.
7. H.Okubo, N. Koshino and N. Hayakawa, Electrical insulation characteristics of CO_2 and CO_2 gas mixtures under non-uniform electric field, Conference on Electrical Insulation and Dielectric Phenomena, pp.538-541, 2003.

COMPARATIVE MEASUREMENT OF CABLE GUN PLASMA DENSITY WITH BIASED AND AIR-EMBEDDED CHARGE COLLECTORS

Zhengzhong Zeng, Yulan Chen, Yuchang Qiu, and Edmund Kuffel*

1. INTRODUCTION

Low-temperature (~eV) plasmas produced with cable guns or flash boards are the indispensable switching media for plasma opening switch (POS) used in pulsed power apparatus.[1, 2] The determination of the plasma density is of interest to theoretical evaluation and experimental adjustment of the switch performance. Several diagnostic techniques have been developed to characterize the plasmas, including electrical probes,[2-4] charge collectors,[3, 4] microwave measurements,[5] interferometry[6, 7] and spectroscopy,[8, 9] with charge collector being among simple and effective ones.

Charge collectors are practically used with bias as shown in Figure 1. The collection of charged particles of plasma originates from the coulombian force on the particles by the bias electric field through the plasma sheath. Due to the difference between the drift energies of the electrons and ions, the repelling of the field to the ions with positive bias may be less effective than that to the electrons with negative bias for a certain collector and bias voltage amplitude. This implicates the necessity to crosscheck the results for different bias polarities.

Low-pressure air-embedded charge collector is an idea for comparative measurement. At appropriate low air pressure (~Pa), the discharge of the plasma guns will not be affected significantly compared with that in high vacuum,[10] while the mean-free-path for the plasma ions under this pressure can be com-

*Zhengzhong Zeng and Yulan Chen, Northwest Institute of Nuclear Technology, P.O. Box 69-1, Xi'an, 710024, P. R. China. Yuchang Qiu, Xi'an Jiaotong University, Xi'an, 710049, P. R. China. Edmund Kuffel, Department of Electrical and Computer Engineering, University of Manitoba, Winnipeg R3T 5V6, Canada.

Gaseous Dielectrics X.
Edited by Christophorou *et al.*, Springer, New York, 2004.

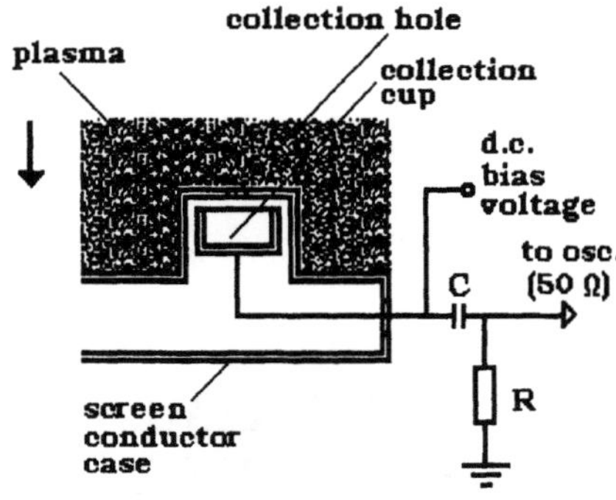

Figure 1. Schematic of biased charge collector.

parable with the POS gap and much smaller than that of electrons.[11, 12] This makes it possible to utilize non-biased low-pressure air-embedded charge collectors to measure the plasma electron density.

2. EXPERIMENTAL SETUP

In the experimental setup shown in Figure 2, 92 cable plasma guns mounted on the outer coaxial electrode of the POS (0.4 μs conduction time, 600 to 800 kA peak current) are used to prefill the switch coupled to an electron diode. 36 charge collectors are arranged in between the POS gap and on the inner electrode of the POS in proper positions to measure radically drifting plasma particles. Each brass collection cup is placed inside an individual brass screen case with a 1-mm-diameter and 0.5-mm-depth collection hole. For the convenience of description, the coordinates is defined as z=0 at the axial center of the POS active region, r=0 on the POS anode surface and r=20 mm on the POS cathode surface.

Figure 3 shows the scheme of the plasma source. A single 3 μF capacitor charged to 22 to 30 kV drives the cable plasma guns with about 2.5 μs of quarter period. When the switched voltage reaches the end of the cable guns, the guns flash over the insulator between the inner and outer conductors and thus plasma is produced. Some fraction of the plasma is injected into the POS region at a typical speed of several cm/μs, primarily owing to the self magnetic force $\boldsymbol{J}\times\boldsymbol{B}$ (the gun discharge current interacting with its own magnetic field). The sensitivity of the collectors is designed to be 1.3×10^{13} $V^{-1}\cdot cm^{-3}$ for single charge state (the case for cable plasma gun[1]) and drift velocity 1 cm/μs.

3. RESULTS AND DISCUSSION

Shown in Figure 4 are the plasma density waveforms measured at two positions with radial space of 0.9 cm. The time delay between the two waveforms is read

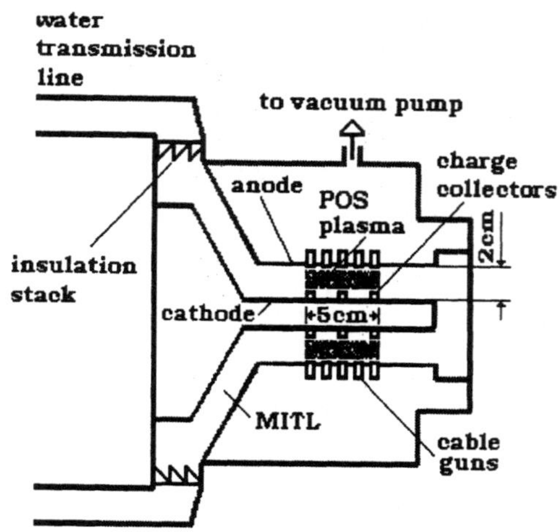

Figure 2. Diagram of the experimental setup.

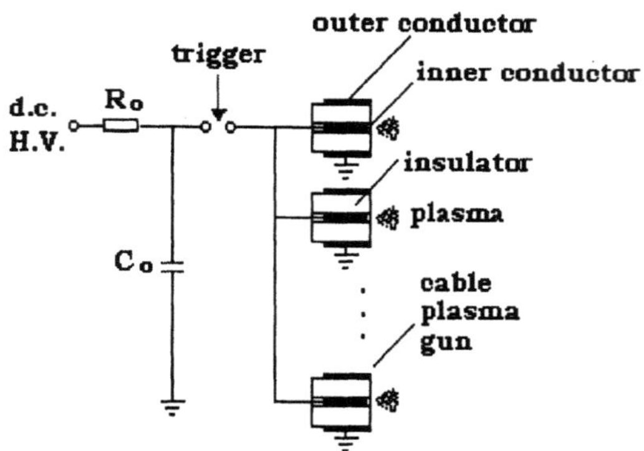

Figure 3. Schematic of the plasma gun source.

to be 2.75 μs. The drift velocity of plasma is thus given to be 2.4 cm/μs. This value is nearly independent of axial and radial positions as well as bias polarity. Figure 4 also demonstrates a significant difference of plasma density in the median region of POS gap and near the POS cathode surface due to plasma cumulation[13] on the surface.

Figure 5 shows the plasma densities obtained with negative and positive bias as well as low-pressure air embedment (0.5 Pa) at the same position. The plasma electron density with positive bias is about 30% lower than the ion density obtained with negative bias while the electron density with air embedment is about

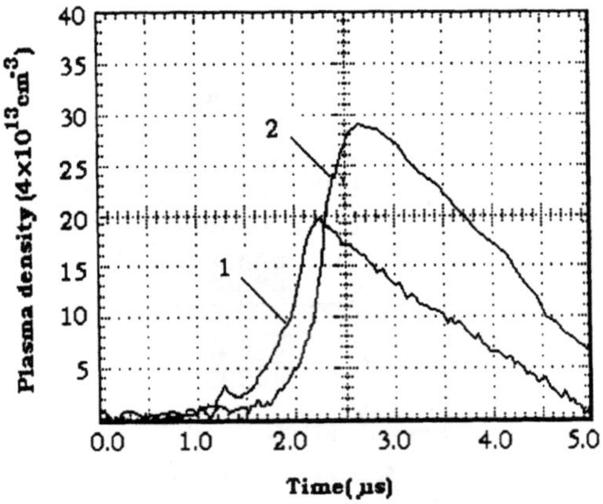

1- r=10 mm, Z=0; 2- r=19 mm, Z=0. Bias: -117 V

Figure 4. Measured plasma density waveforms for determination of plasma drift velocity.

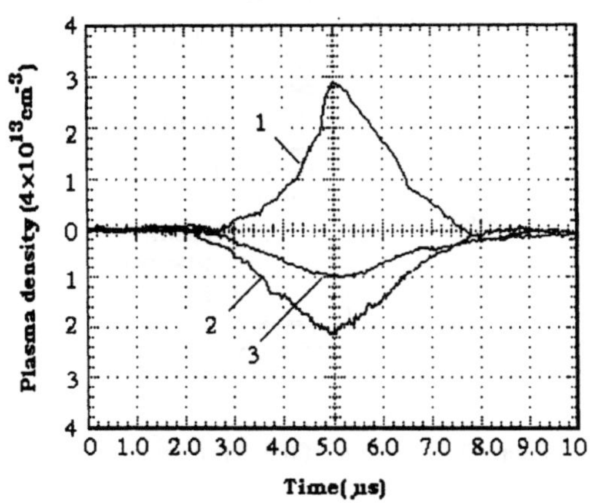

1- bias: -117 V; 2- bias: 135 V; 3- no bias, 0.5 Pa air embedment. Position: r=19 mm, Z=0

Figure 5. Plasma density waveforms obtained with negative and positive bias as well as low-pressure air embedment (0.5 Pa) at the same position.

50% lower than that with positive bias. The measurement shows weak dependence of the data deviation on axial and radial positions.

The deviation of measured plasma density can be explained as follows. At a drift velocity of several cm/μs, the kinetic energy of the ions is several electron volts that is three orders greater than that of the electrons. This means that the electrons are easy to repel, while the ions have more probability to get into the screen case, giving rise to a reduction of recorded plasma density. In the case of collection with air embedment, the mean-free-path of ions is about 2 cm[11, 12] allowing only one time of collision with background air molecule. This indicates that about 37% of the ions will enter the case resulting in a reduction of recorded plasma density.

In regard to the secondary charged particle emission on the collector cup surface that may influence the measurement, the kinetic energy for an electron to induce secondary electron emission from brass surface is about 377 eV while that for an ion is ranging from 10 eV to 30 eV.[14, 15] With these facts one can see that at bias voltage -117 V or 135 V the ions or electrons are difficult to induce secondary electron emission.

Shown in Table 1 is a comparison of measured flashboard plasma density using charge collector with that using microwave or spectroscopy method. With these data it can be convincing that the deviation using charge collector method in this work is reasonable.

Table 1. Comparison of measured flashboard plasma density using different method.

Data Sources	plasma density (1×10^{13} cm^{-3}) (charge collector)	(microwave)	(spectroscopy)
Experiment I[5]	15 (negative bias)	6	-
Experiment II[16]	1 (negative bias)	-	1 to 2

4. CONCLUSION

The measured plasma electron density with positive bias is lower than the ion density obtained with negative bias due to the difference of drift energy of the particles, while the electron density with air embedment is smaller than that with positive bias because of the POS gap limitation. These deviations are comparable to those given by previous experiments in which charge collection, microwave and spectroscopy methods were used for comparison, indicating the reasonableness of the deviations and the effectiveness of the charge collection method, with negative bias being preferably acceptable for POS plasma density measurement.

ACKNOWLEDGEMENTS

The authors wish to thank Sun Fengju, Kuai Bin, Yin Jiahui, Cong Peitian and Liang Tianxue for their help during the work.

1. S. Kohno, Y. Teramoto, I. V. Lisitsyn, S. Katsuki, and H. Akiyama, *IEEE Trans. Plas. Sci.* **27**, 778-785 (1999).

2. T. J. Renk, *J. Appl Phys.* **65**, 2652-2663 (1989).

3. Y. E. Krasik and A. Weingarten, *IEEE Trans. Plas. Sci.* **26**, 208-219 (1998).

4. L. Veron, R. Boivinet, C. Rouille, B. Etlicher, C. Peugner, and B. Dufour, *J. Appl. Phys.* **71**, 3002-3009 (1992).

5. B. V. Weber, R. J. Commisso, G. Cooperstein, J. M. Grossmann, D. D. Hinshelwood, D. Mosher, J. M. Neri, P. F. Ottinger, and S. J. Stephanakis, *IEEE Trans. Plas. Sci.* **15**, 635-648 (1987).

6. B. V. Weber and S. F. Fulghum, *Rev. Sci. Instrum.* **68**, 1227-1232 (1997).

7. D. D. Hinshelwood, B. V. Weber, J. M. Grossmann, and R. J. Commisso, *Phys. Rev. Lett.* **68**, 3567-3570 (1992).

8. R. Arad, K. Tsigutkin, Y. V. Ralchenko, and Y. Maron, *Phys. Plas.* **7**, 3797-3807 (2000).

9. A. Weingarten, V. A. Bernshtam, A. Fruchtman, C. Grabowski, Y. E. Krasik, and Y. Maron, *IEEE Trans. Plas. Sci.* **27**, 1596-1605 (1999).

10. A. S. Pillai and R. Hackam, *J. Appl. Phys.* **58**, 146-153 (1985).

11. E. Nasser, *Fundamentals of Gaseous Ionization and Plasma Electronics* (Wiley-Interscience, New York, 1971).

12. M. N. Hirsh and H. J. Oskam, *Gaseous Electronics, Vol. 1, Electrical Discharges* (Academic Press, New York, 1978).

13. L. Veron, B. Etlicher, A. S. Chuvatin, C. Rouille and J. P. Stephan, *IEEE Trans. Plas. Sci.* **21**, 529-536 (1993).

14. J. S. Blakemore, *Solid State Physics* (Cambridge University Press, New York, 1985).

15. A. Van der Ziel, *Solid State Physical Electronics* (Prentice-Hall, Englewood Cliffs, 1976).

16. A. B. Baranga, N. Qi and D. A. Hammer, *IEEE Trans. Plas. Sci.* **20**, 562-567 (1992).

NEGATIVE DC PRE-STRESSING ON CONDUCTOR-ROD GAPS UNDER POSITIVE IMPULSE VOLTAGES

P. N. Mikropoulos, C. A. Stassinopoulos, and C. G. Yakinthos*

ABSTRACT

The influence of the superimposition of positive impulses on negative DC pre-stressed conductor-rod gaps with spacing of 50 cm is investigated. Previous experiments have shown that when a conductor of 2 cm in diameter is used a marked increase in the dielectric strength of the gap is observed when negative DC voltages between 100 kV and 125 kV in value are applied at the rod; this is not the case for a 3 cm conductor. This paper provides new data concerning the above phenomenon; by applying higher values of pre-stressing voltage a similar behaviour as with the 2 cm conductor is observed for the 3 cm conductor. The discharge mechanism is discussed taking into consideration parameters affecting the above behaviour, such as the value of the pre-stressing DC voltage, the waveshape of the applied impulse voltage and the variation of absolute humidity.

1. INTRODUCTION

The conductor-rod gap exhibits higher dielectric strength compared with the rod-plane gap under impulse voltages at the conductor. The high gap factor, that is the ratio between the minimum breakdown voltages of the two gap configurations, makes the conductor-rod gap important when studying the subject of overhead line compaction.

For the conductor-rod gap configuration, as the one investigated in this study, the breakdown mechanism starts with a negative corona at the rod that influences

* P.N. Mikropoulos, Department of Electrical and Computer Engineering, Aristotle University of Thessaloniki, Thessaloniki, 541 24 Greece

Gaseous Dielectrics X.
Edited by Christophorou *et al.*, Springer, New York, 2004.

the emergence and development of the much more complex and, in most cases, longer positive discharge[1, 2]. For the same gap the effect of pre-stressing with negative DC voltage at the rod is for long front impulses to increase considerably the dielectric strength of gap whereas for lightning impulses it has minimal influence upon the breakdown voltage of the gap[3].

This behaviour has already been discussed in terms of the associated space charges of the DC coronas occurring at the electrodes which tend to create a more 'uniform' field distribution in the gap impeding the development of the positive impulse discharge from the conductor[3]. Pre-stressing reduces the streamer corona from the conductor and suppresses the growth of the leader[4].

Evidence has been presented to show that the effect of pre-stressing is determined by several parameters. For an energised conductor 2 cm in diameter and DC voltages up to 125kV pre-stressing resulted in a marked increase in the dielectric strength whereas for a conductor 3 cm in diameter had no effect on the breakdown voltage of the gap[3]. This study extends our previous work by applying higher values of pre-stressing DC voltage at the rod. The discharge mechanism is discussed taking into consideration the gap geometry, the value of the pre-stressing DC voltage, the waveshape of the applied impulse and the variation of absolute humidity.

2. EXPERIMENTAL LAYOUT

A conductor-rod gap, vertically mounted, with its' axis perpendicular to the ground has been used (Figure 1). The conductor 3 cm in diameter and 4 m in length was energised through a lead at mid point. The earthed electrode was a cylindrical brass rod 80cm long and 2 cm in diameter with hemispherical end. Two gap configurations have been used, axial geometry with the rod set beneath the mid point of the conductor (point A, Fig. 1) and non-axial geometry with the rod set at 1m from the central axis of the gap (point B, Fig. 1).

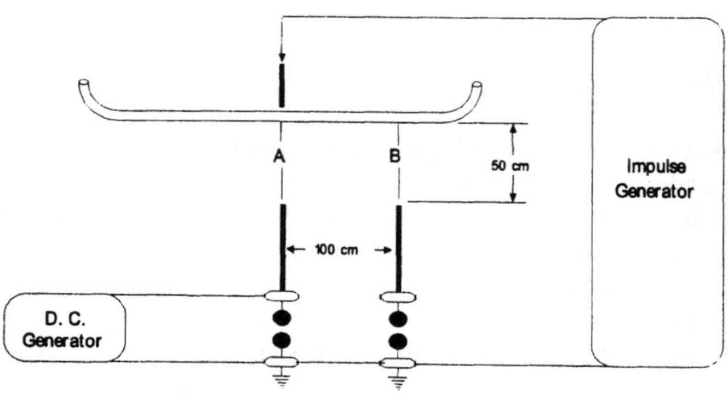

Figure 1. Schematic experimental layout

A ten-stage, 1 MV, 7 kJ Marx generator produced positive impulse voltages with the following waveshapes: 1.2/50μs (from now on referred to as *LI*), 2/1700μs, 50/1950μs, and 250/2500μs (from now on referred to as *SI*). The gap was pre-stressed by negative DC voltages U_{DC} of 150, 200, 250 kV in amplitude produced by a three-stage voltage multiplier circuit using the Cockroft-Walton principle with a maximum output approximately 420kV. An auxiliary sphere gap set to break down at a voltage slightly above the U_{DC} was used for the protection of the D.C. generating unit in cases of breakdown of the main gap (Fig. 1).

Breakdown probability curves were obtained through the multiple level test method[6] for breakdown probabilities ranging from 0 to 100%. Each voltage level was higher 2 ÷ 3% from the previous one and consisted of 20 impulses applied at time intervals of around 30 s.

The breakdown probability curves were fitted with the cumulative Normal (Gauss) distribution and a good fit was observed in their vast majority. From each curve the value of 50% breakdown voltage U_{50} was calculated. For the pre-stressed gaps U_{50} is considered as the algebraic sum of the negative and positive voltage components. The voltages were corrected for air density according to IEC. Absolute humidity was found to vary naturally between 7 and 20 gm^{-3}.

3. EXPERIMENTAL RESULTS

3.1. Breakdown Voltages

A marked increase in the dielectric strength of the gap is observed when negative DC voltages higher than 150 kV in value are applied at the rod especially for impulses with long wavefront durations. This is demonstrated in Figure 2 for the non-axial geometry; the values of U_{50} for U_{DC}= 0 and 125 kV are taken from previous studies[2, 3].

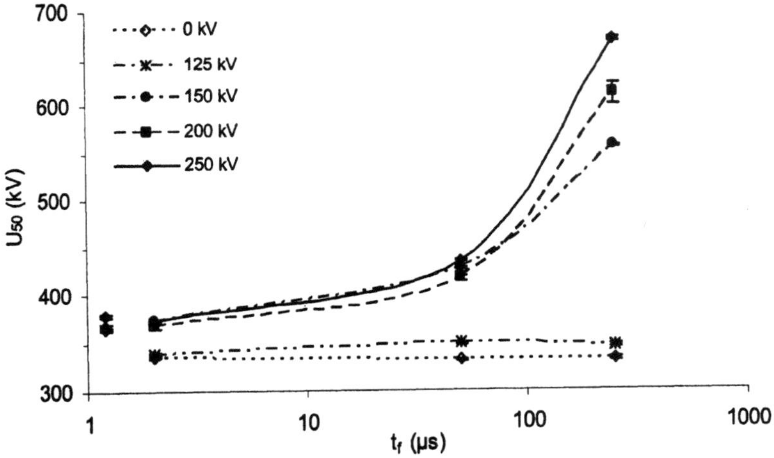

Figure 2. U_{50} versus t_f with U_{DC} as parameter. Non-axial geometry; vertical bars represent the 90% confidence limits of U_{50}; average humidity 12.5 gm^{-3}

The general trend is U_{50} to increase with U_{DC}, on the other hand, as is evident from Figure 2, pre-stressing has almost no effect on the values of U_{50} for *LI*. This behaviour is observed for axial geometry as well, however in a less pronounced way. The most considerable case is for the non-axial geometry under *SI* and U_{DC}= 250 kV where U_{50} is almost 100% higher than its value for U_{DC}= 0 kV.

In general, U_{50} is higher for axial than non-axial geometry for any U_{DC} value applied at the rod. This is shown in Figures 3 and 4 for *LI* and *SI* respectively. For both gap configurations pre-stressing for *LI* has no significant effect on U_{50} whereas for *SI* it results in a marked increase in U_{50}.

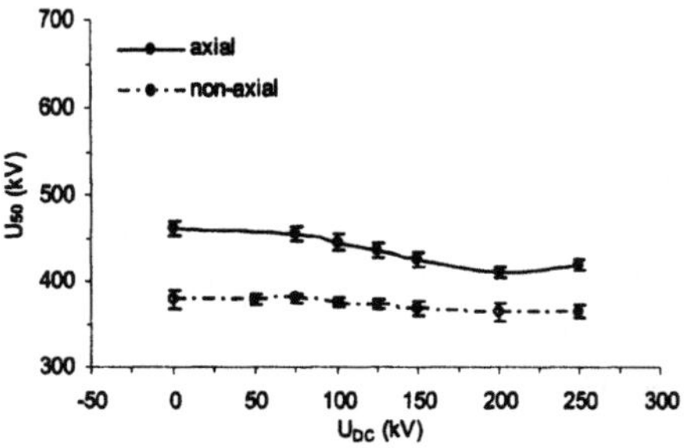

Figure 3. U_{50} versus U_{DC} with the gap geometry as parameter; *LI*, average humidity 12.5 gm^{-3}

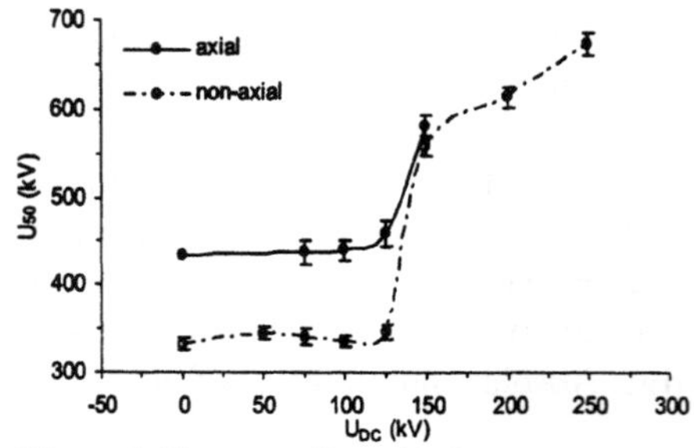

Figure 4. U_{50} versus U_{DC} with the gap geometry as parameter; *SI*, average humidity 12.5 gm^{-3}

Figures 5 and 6 show the effect of absolute humidity on U_{50}. In general, U_{50} increases with humidity. This effect was found stronger for impulses with long wavefront durations (Fig. 5), for axial geometry and with increasing U_{DC} (Fig. 6). The only cases where U_{50} was found to decrease with humidity are for non-axial geometry under U_{DC}= 150 and 200 kV.

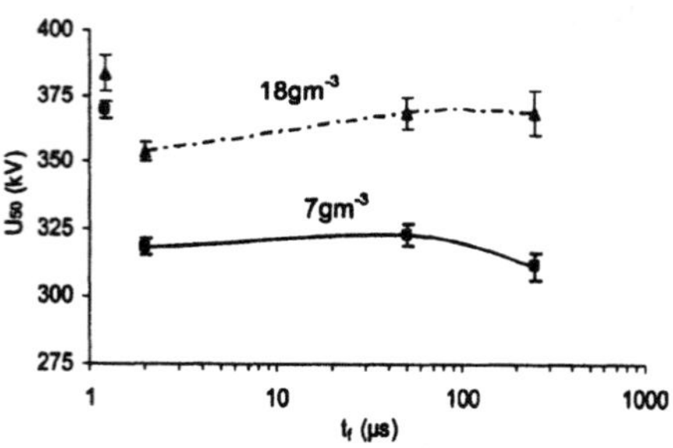

Figure 5. U_{50} versus t_f with humidity as parameter; U_{DC}=125 kV, non-axial geometry

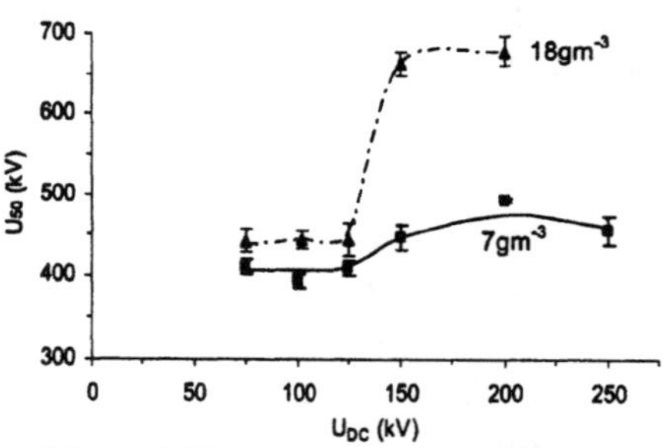

Figure 6. U_{50} versus U_{DC} with humidity as parameter; *SI*, axial geometry

3.2. Gap Factor

Gap factors have been calculated using the ratio between the U_{50} of each gap configuration to the U_{50} value of the positive rod-plane gap of the same spacing under critical t_f, 226 kV for a conical tip of the rod[5]. This method has the advantage of a single reference voltage per gap length, for all the examined configurations[1]. The result is shown in Figure 7 for both gap configurations.

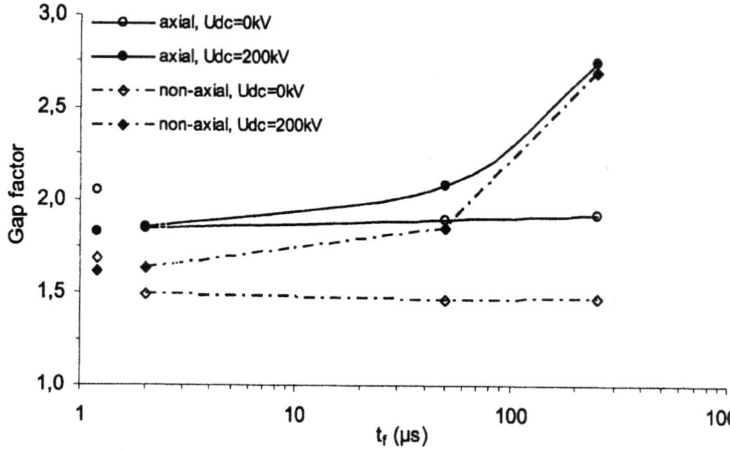

Figure 7. Gap factor versus t_f with gap geometry as parameter; average humidity 12.5 gm^{-3}

The gap factors are higher for axial than non-axial geometry. For both gap configurations and for U_{DC} = 0 the highest gap factors are for *LI* but with pre-stressing they become highest for *SI*. The gap factors were found to increase with increasing U_{DC} under impulse voltages with long front durations.

4. DISCUSSION

For the conductor-rod gap, the effect of pre-stressing with DC voltages at the rod is to give rise to DC coronas at both electrodes. The associated space charges decrease the divergence of the field, the field at the gap becomes more 'uniform', thus impeding the development of the positive impulse discharge from the conductor cause an increase in U_{50}. This increase in U_{50} was attributed to the higher applied voltages required so as to achieve field values at the electrodes sufficient enough to overcome the inhibitory effect of the space charges[3].

Comparison with earlier work (Figure 1) shows that when the conductor is 3 cm in diameter and U_{DC} takes values up to 125 kV pre-stressing has minimal influence upon the breakdown voltage of the gap. For U_{DC} higher than 150 kV a marked increase in the dielectric strength of the gap is observed especially for impulses with long wavefront durations (Figure 1). It is suggested that for such U_{DC} values the growth of the positive DC corona at the conductor is significant thus higher applied voltages are required to overcome the inhibitory effect of associated space charges. The relation between DC corona development and U_{DC} was demonstrated by photomultiplier measurements before; the light output of the DC corona increases with U_{DC} indicating a greater corona growth with increasing U_{DC}[3]. For the same gap under composite impulse/direct voltages pre-stressing at the conductor with similar U_{DC} values caused a marked increase in U_{50} when *SI* was applied at the conductor[6].

Axial geometry exhibits higher breakdown voltages and is less influenced by pre-stressing than non-axial geometry (Figurers 3 and 4). Also, the effect of humidity was found to be stronger for axial than non-axial geometry. These findings are in accordance with previous work, conductor 2 cm in diameter, and were explained in terms of the field values at the conductor[2, 3].

For axial geometry U_{50} increases with humidity. This effect becomes stronger with increasing U_{DC} especially for impulses with long front durations where breakdown occurs with the development of a small positive leader. Pre-stressing by suppressing the growth of the leader[4] results in a breakdown mechanism which is more influenced by humidity since the bigger the part of the gap occupied by streamers the stronger the influence of humidity on U_{50}[7, 8].

5. CONCLUSIONS

In the conductor-rod gap under impulse voltages, negative DC voltage pre-stressing at the rod results in a significant increase in the breakdown voltage whenever breakdown occurs through a leader breakdown mechanism. There is a threshold value of the DC voltage below which the so produced DC coronas at the electrodes have minimal effect upon the dielectric strength of the gap.

6. REFERENCES

1. N. L. Allen, M. Boutlendj, and C. A. Stassinopoulos, Gap Factor and Breakdown Mechanism in the Conductor - Rod Geometry'. *Proc. 9th I.S.H.*, Graz, Paper 20.96 (1995)
2. D. E. Gourgoulis, P. N. Mikropoulos, C. A. Stassinopoulos, and C. G Yakinthos, Behaviour of Positive Conductor – Rod Gaps stressed by Impulse Voltages in Atmospheric Air, *IEE Proc.-Sci. Meas. Technol.*, **144**(5), 209-214 (1997)
3. N. L. Allen, D. E. Gourgoulis, P. N. Mikropoulos, C. A. Stassinopoulos, and C. G Yakinthos, Effects of Negative Direct Voltage Pre-stressing on the Breakdown of Conductor – Rod Gaps under Positive Impulse Voltages, *IEE Proc.-Sci. Meas. Technol.*, **145**(3), 105-109 (1998)
4. N. L. Allen, C. F. Huang, K. J. Cornick, and D. A. Greaves, Sparkover in the Conductor-rod and Conductor-plane Test Gaps under Composite Slow Front Impulse/direct Voltages, *IEE Proc.-Sci. Meas. Technol.*, **146**(3), 135-141 (1999)
5. P. N. Mikropoulos, and C. A. Stassinopoulos, Impulse Breakdown of Short Rod - Plane Gaps and the Influence of Humidity, *IEE Proc.-Sci. Meas. Technol.*, **145**(4), 141-146 (1998)
6. N. L. Allen, C. F. Huang, K. J. Cornick, and D. A. Greaves, The Sparkover of Air Gaps under Composite Impulse/direct Voltages, *Proc. 10th I.S.H.*, Montreal, Vol. 3, 157-160 (1997)
7. K. Feser, and J. Schmid, Influence of Atmospheric Conditions on the Impulse Breakdown of Rod-plane Gaps, *Proc. 5th I.S.H.*, Braunschweig, Paper 11.01 (1987)
8. P. N. Mikropoulos, and C. A. Stassinopoulos, Influence of humidity on the Breakdown Mechanism of Medium Length Rod-plane Gaps stressed by Positive Impulse Voltages, *IEE Proc.-Sci. Meas. Technol.*, **141**(5), 407-417 (1994)

EXPERIMENTAL INVESTIGATION OF ARC IN FUSE ELEMENTS DURING THE INTERRUPTION PROCESS

C.G. Karagiannopoulos*

1. INTRODUCTION

Electric fuses usually operate under either less than their nominal current or momentarily excessive or short-circuit currents. In the case of operation under excessive or heavy fault currents, the design of fuses is based on the principle that they must interrupt those currents in a very short time period [1-4]. In this case, an increase in the fuse element's resistance causes the temperature to rise rapidly until the melting point is reached. The latent heat of fusion is produced gradually by the current during the melting process until the material is completely vaporized [1,4,5]. When the material of the element vaporizes, electrical arcs strike between the remaining solid parts of the element. This dynamic process gives rise to a rapid temperature increase, while the current decreases rapidly until its flow is interrupted [6,7]. The time period from the instant when the current exceeds the critical value until the melting and vaporization of the elements is taking place is known as the pre-arcing period and the following time until the arc interruption is achieved is called the arcing period. During the pre-arcing period the current varies with time in a way mainly determined by the source and the impendances in the circuit up to the point of fault. The fundamental operation of fuses has been described extensively, among others, by P.G. Newbery, A. Wright, E. Jacks, R. Rüdenberg, D.R. Barrow and T. Chicata [1–7]. This research focuses on the arcing period. The fundamental operation is well understood in general, but not all of the physicochemical processes that occur during the arcing period seem to be completely understood.
This paper attempts to explore the behavior of fusible elements under DC short circuit heavy fault currents. DC currents are considered as an unfavourable case

* C.G. Karagiannopoulos, National Technical University of Athens, Department of Electrical and Computer Engineering, 9 Iroon Polytechniou Str., GR 157-80, Athens, Greece.

in arcs extinguishing, therefore being appropriate for experimental investigation. Measurements of the current through, and the voltage drop across fuse elements are presented, which allow investigation of the physicochemical process that occurs during the pre-arcing and arcing periods.

2. EXPERIMENTAL SET-UP

Fig. 1 shows the simplified schematic diagram of the experimental set-up, used for the estimation of the current and the voltage drop across fuse elements, under short-circuit conditions. The measurements were performed through an analog/digital converter with high sampling frequency. The sampling system was designed and built specifically for this application.

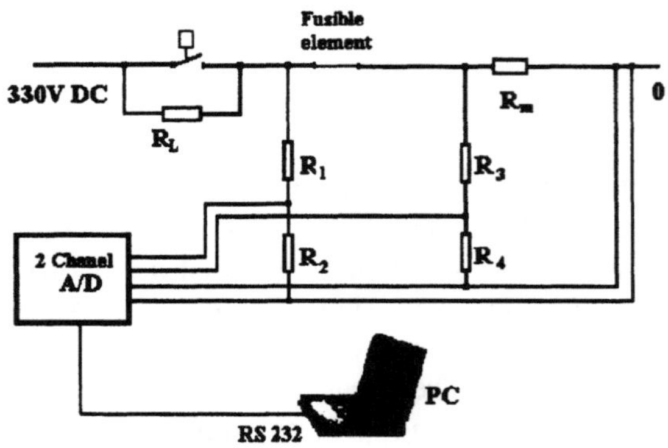

Figure 1. Simplified schematic diagram of the measuring setup.
R_1, R_2 voltage divider for the voltage measurement across the element, R_3, R_4 and R_m voltage divider and measuring resistance for the current measurement through the element

The 12-bit, 20 MSPS A/D converters employed, were connected with the computer via RS-232 communication port. Signals varied between ±2,5V and the sampling frequency was 0,5MHz. Two voltage dividers were used to reduce the maximum voltage value to ±2,5V.

The experimental investigation of the arcing period was based on measurements of the voltage drop and the current in fusible elements. Thin cylindrical wires simulating the fuse elements, were connected to the circuit using appropriate terminals. A typical ohmic load connected in parallel with an open switch was used to simulate the nominal current operation and short-circuit conditions respectively. A measuring resistance R_m was used in order to estimate the current through the fuse element. The measuring resistance had zero value thermal coefficient ($\alpha \approx 0$). The maximum absolute measuring error was 2.5mV as the measurements were performed in ±2,5V scale.

Measurements were performed on conductors (fusible elements) stretched in air. The conductors were thin cylindrical copper wires (99,9% Cu). Figures 2I(a) and 2II(a) show the voltage drop and the current waveforms during the current interruption process in fuse elements with length of 5cm and 4cm respectively while Figures 2I(b) and 2II(b) show the estimated corresponding time dependent resistance.

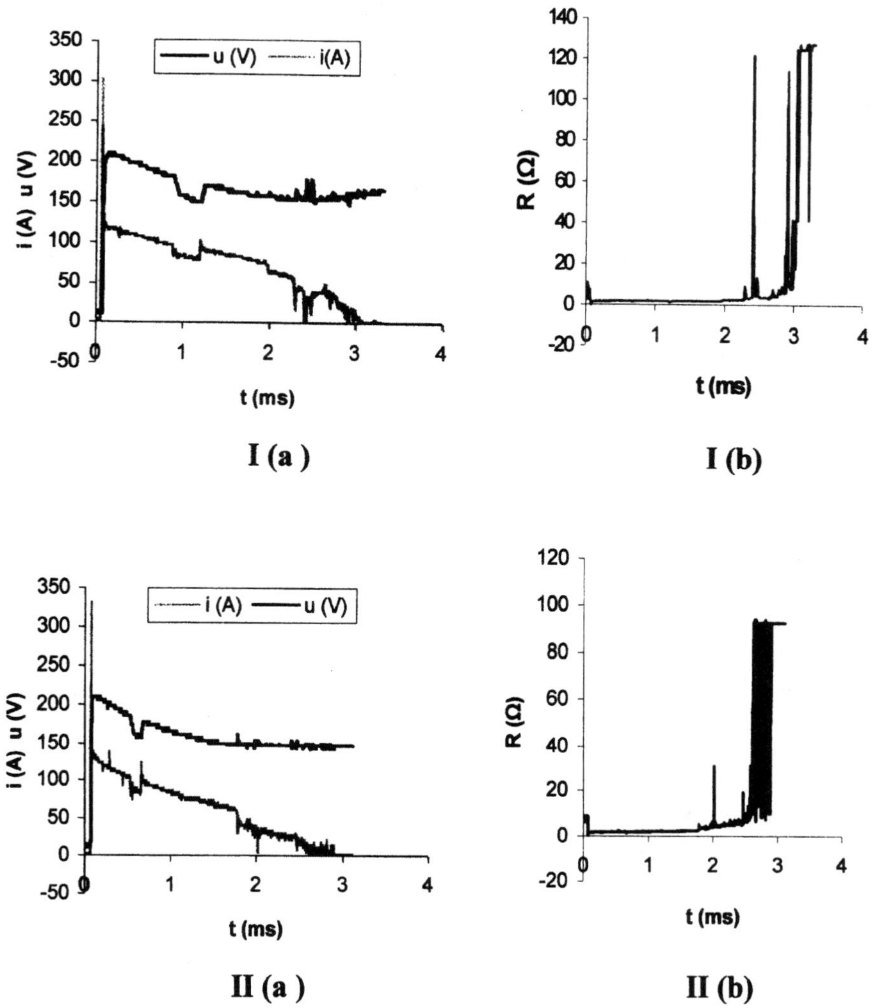

Figure 2. I) Typical waveforms of the voltage drop and the current during short-circuit, in a copper fusible element (wire).
II)Fusible 's element resistance (R) during the interruption process.
Geometric characteristics of the fusible element:
Length a) 5 cm , b) 4 cm, diameter 0.15mm, material cooper.

3. MEASUREMENTS RESULTS AND DISCUSSION.

Figure 2 present indicative measurements of voltage drop and current waveforms across and through the fusible element, as well as corresponding plots of the resulting resistance as a function of short circuit duration. The initial time period where impulse sort circuit current is in progress, is presented to the plots by an small area where the current presents picks. This time period, according to the measurements performed, seems to have a duration of 100 – 300μsec. After this period the current is almost stabilised (steady state sort circuit current), with a decreasing tense. This may be explained by the increase of the specific resistance due to overheating of the thin fusible element. If we consider the values provided by the plots, the well-known relation $\rho = \rho_0(1 + a\Delta\theta + b\Delta\theta^2)$ seems to be valid. The end of the pre-arcing period is marked by instabilities of the resistance presented to a time instant 0.5–1 msec after sort circuit's initiation. Instabilities result from the sinking of the voltage drop, which is a consequence of the progressive and finally complete liquefaction of the element. The phenomena presented could be as follows : the increase of the specific resistance (above mentioned) causes an accelerated temperature rise until the melting point of the fuse wire's material. After reaching the melting point, the wire starts to liquefy. In general the inner parts of the cross section are hotter than the outer parts and fuse first. Then the melting zone will spread out until it reaches the periphery. At melting temperature the resistivity of the fused material is much higher than that of solid material, and increases with progressive melting. The data for the specific resistivity of the solid cooper in melting point, liquid cooper in melting point and during the melting, support the above mentioned (the values are 10.2×10^{-6} Ohm-cm, 21.3×10^{-6} Ohm-cm and 13.8×10^{-6} Ohm-cm respectively) [8]. During the sort times here in question, the surface tension and the gravitational forces seem that do not play any important role and thus the now liquid wire material will remain in place and be heated further to the temperature of vaporization. The complete volatilization of the fuse metal also needs some time. At an instant before the material is completely vaporized, the metallic coherence within the fuse wire is interrupted and an arc is formed. [1,9,10]. However, it is doubtful whether this phenomenon will occur in a regular way or whether the liquid cylinder will explode under the mechanical action of the vapor pressure [11,12].

From the investigation in the field of exploding wires it is known that fuse elements after reaching evaporation temperature, under atmospheric conditions, can become highly superheated in the liquid state. Evaporation takes place at temperatures at which thermal ionisation can be expected [11-14]. Careful investigation of different factors, including the energy input during the liquid phase of fusible elements, have shown that in case of fuses heated by low and moderate currents a considerable retardation of the evaporation is presented. At evaporation temperatures the metal vapor column presents a resistance which is considerably higher than that of the fuse element at the same temperature but in the liquid state [1]. Thus the transition from molten to evaporated state, involves a considerable increase of the fuse resistance. So the electrical behavior of a fuse element can be characterized by an increase of the resistance at the instant of evaporation.

Considering Figure 2, the resistance during arcing period seems to present two distinguished behaviors (areas). In the first area, the resistance has a rather moderate increasing tense. In the second area that lasts until the interruption of the current occurs (respectively marked by infinite resistance), fast oscillations groups with corresponding peaks are presented. Those oscillations present approximately a duration (period) of 2–10μs each one and cannot be explained by the classical arc's theory as is established till now. According to this theory a metal vapor column sustained electric discharges in which plasma occurs at high pressure and temperature in local thermodynamic equilibrium. It would expected, as the burn – back procedure is in progress, that the vapor column extends lengthwise (arc's elongation) with respective reduction of it's cross–section. This normally results to a progressive rapid increase of the arc's resistance. The rapid changes in arc's dimensions and the expansion of the metal vapor column (which increases rapidly the arc resistance), as well as the possible decrease of the voltage drop across the fuse edges (caused by the source impedance) leads to the eventual arc's extinction.

Interpretations regarding the fast oscillations of the resistance presented on Figure 2 seem to involve fluctuation of the arc's current density due to the collisional process involving heavy ionised particles and free electrons. It's unlikely to presume that the phenomenon of multiple arcing caused by globules' formation, participates to the oscillations of the resistance value. A possible explanation of the phenomena could be as follows.

The inductance presented in any circuit in practice in general dictates a variation of current. The time constant τ=L/R of the circuit and the rise time of the fuse resistance during current interruption, precise the voltage variation to the end-cups of the fuse. At the moment of the current interruption, overvoltage appears between the still remaining solid parts of the fuse element. This is reasonable as the current interruption process is fast. Therefore it can be assumed that there is ignition of a new arc which cannot be sustained for long, as the overvoltage to the solid parts of the fuse element doesn't still exists. If this procedure if repeated leads to ignition and arcs extinguishment. Figure 2 support the aforementioned. The final extinguishment of arcs could be owed to an incidental event but mainly to the elongated distance of the two remaining solid parts of the fuse element.

Finally, observations concerning the short – circuit duration and the measured resistance values in relation the different geometric characteristics of the fusible elements, seems to present a normality since the different wire length implies different resistance and consequently current densities. The corresponding energy to melt is obviously analogous to the diameter and length of the fusible element.

4. CONCLUSIONS

To the present work measurements obtained using high sampling frequency analogue /digital (A/D) converters are presented, regarding the voltage drop across and the current through fusible elements (thin cylindrical copper wires) stressed by heavy DC fault current, in air. Existing knowledge of the process is demonstrated and further interpretation of the phenomena during short-circuit is attempted. Multiple arcing and arcing period phenomena are investigated considering the arc resistance. The present work aspire to be useful in technoeconomic selections.

5. REFERENCES

[1] A. Wright and P. G. Newbery, *Electric fuses*, (IEE-Peter Peregrinus, London, 1995, 2nd edition).

[2] J. G. Leach, P. G. Newbery and A. Wright, Analysis of high-rupturing-capacity fuse links prearcing phenomena by a finite differential method, *IEE Proc.* **120**(9), 987-993 (1973).

[3] D. R. Barrow, A.F. Howe and N. Cook, The chemistry of electric fuse arcing, *IEE Proc.* A**138**(1), 83-88 (1991).

[4] L. A. V. Cheim and A. F. Howe, Spectroscopic observation of high breaking capacity fuse arcs, *IEE Proc. Sci. Meas. Tech.* **141**(2), 123-128 (1994).

[5] V. N. Narancic and G. Fecteau, Arc energy and critical tests for HV current-limiting fuse, *Ibid.* 236-251.

[6] T. Chicata, Y. Ueda, Y. Murai and T. Miyamoto, Spectroscopic observations of arcs in current limiting fuse through sand, *Proc. of Int. Conference on Electric Fuses and their Applications,* Liverpool, UK, 114-121 (1976).

[7] D. R. Barrow, A. F. Howe and A. Wrigth, Methods of determining fuse arc parameters, *Int. Conference on Electrical Contacts, Arcs, Apparatus and their Applications*, Xi'an Jiaotong Univercity, Xi'an, China, (1989).

[8] Reinhold Rüdenberg, *Transient Performance of Electric Power Systems*,(M.I.T. Press, 1970).

[9] L. Vermij, The voltage across a fuse during the current interruption process, *IEEE Trans. Plasma Sc.*, PS-**8**(4), 460-468 (1980).

[10] V. T. Morgan, Rating of conductors for sort-duration currents, *IEE Proc.* **118** (3/4), 555-570 (1971).

[11] D. V. Lebedev, Explosion of a metal by an electric current, *Sov. Phys.–JETP* **5**(2) 243-252 (1957).

[12] I. M. Vitkovitsky and V. E. Scherrer, Recovery characteristic of exploding wire fuses in air and vacuum, *J. Appl. Phys.*, **52**(4), 3012-3015 (1981).

[13] T. L. Berger, Effects of surrounding medium on electrically exploded aluminum foil fuses, *IEEE Trans. Plasma Sc.*, PS-**8**(3), 213-216 (1980).

[14] Y. P. Raizer, *Gas Discharge Physics*, (Springer-Verlang, Heidelberg, 1991).

SECTION 6: GAS MIXTURES AS ALTERNATIVES TO SF_6

DIELECTRIC CHARACTERISTICS AND ELECTRICAL INSULATION DESIGN TECHNIQUES OF GASES AND GAS MIXTURES AS ALTERNATIVES TO SF_6

H.Okubo[1] and N.Hayakawa[2]

1. INTRODUCTION

SF_6 gas has been widely used and contributed to the downsizing and cost reduction of gas-insulated power apparatus such as switchgears (GIS) and circuit breakers (GCB). However, since SF_6 gas was specified as one of greenhouse gases with a high global warming potential (GWP) at COP3 in 1997[1], research and development has been accelerated to find new insulation gases alternative to SF_6 gas. N_2/SF_6 gas mixtures have been regarded as one of the promising SF_6 substitutes and utilized for an actual power apparatus such as transmission lines (GIL)[2], owing to low cost, unlimited resources, low liquefaction point, reduction of SF_6 gas amount, and remarkable synergistic effect on dielectric characteristics. Furthermore, other insulating gases and gas mixtures without SF_6 gas have also been investigated to find better solutions for future environment-friendly power apparatus and systems. Such research and development of SF_6 substitutes are to be strongly anticipated in the technical field of gaseous dielectrics.

From the above background, we have been engaged in the research on dielectric characteristics of various insulating gases and gas mixtures as alternatives to SF_6 gas for electrical insulation design of next-generation power apparatus. The electrical insulation design techniques should be based on the understanding of discharge mechanisms in insulating gases, so that we discussed the partial discharge (PD) inception and propagation characteristics leading to breakdown (BD). This paper summarizes our up-to-date activities on dielectric characteristics and discharge mechanisms of N_2/SF_6 gas mixtures, N_2 gas, CO_2 gas and CO_2-based gas mixtures, together with their comparison with those of SF_6 gas. Fundamental concepts for the electrical insulation design of environment-benign power apparatus are also introduced.

[1] Center for Integrated Research in Science and Engineering, Nagoya University, Nagoya 464-8603, Japan
[2] Department of Electrical Engineering, Nagoya University, Nagoya 464-8603, Japan

2. DATABASE OF DIELECTRIC CHATACTERISTICS IN GASES

Fundamental dielectric characteristics of insulating gases are gas pressure dependence of PD inception voltage (PDIV) and BD voltage (BDV). Figures 1 and 2 show typical gas pressure dependence of (a) PDIV and (b) BDV of different gases and gas mixtures under standard lightning impulse and ac voltage applications, respectively. The electrode arrangement in Fig.1 is rod-plane (diameter ϕ=5mm, gap length g=40mm), and that in Fig.2 is needle-plane (ϕ=1mm, g=10mm). Basic properties of SF_6, N_2, O_2 and CO_2 gases are listed in Table 1. In each condition, PDIV increased linearly with the gas pressure, whereas BDV exhibited non-linear characteristics against the gas pressure. The non-linearity of BDV is peculiar to electronegative gases, and attributed to the correlation between corona stabilization effect at the lower gas pressure and the streamer/leader transition at the higher gas pressure[3]. The gas pressure dependence of other gases and gas mixtures has also been acquired as a database to discuss the discharge characteristics and mechanisms.

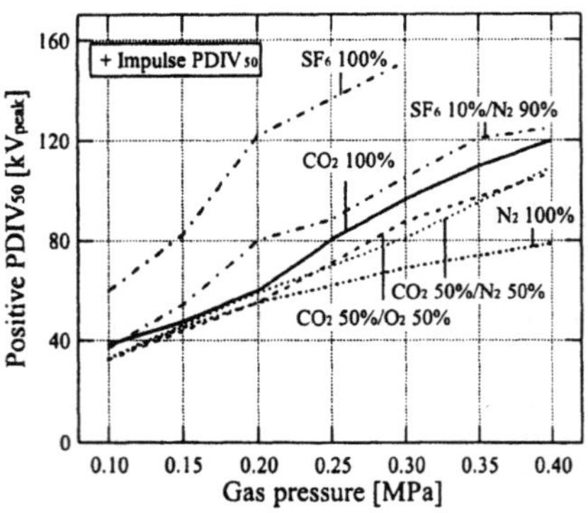

(a) PDIV

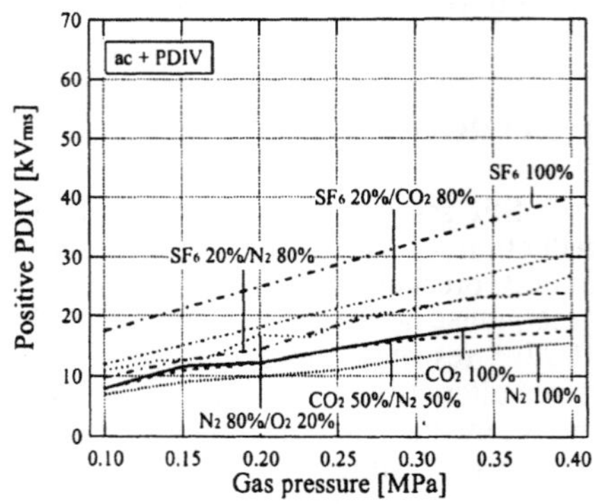

(a) PDIV

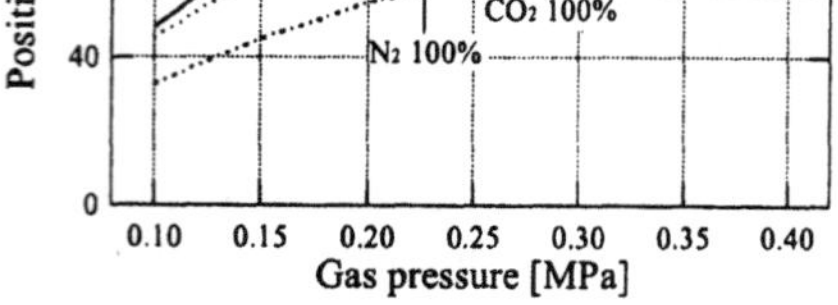

(b) BDV

Fig.1 Gas pressure dependence of PDIV and BDV under lightning impulse voltage application. (ϕ=5mm g=40mm)

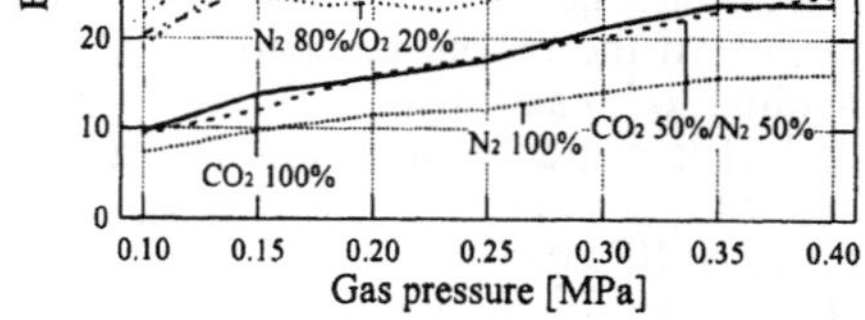

(b) BDV

Fig.2 Gas pressure dependence of PDIV and BDV under ac voltage application. (ϕ=1mm, g=10mm)

Table 1 Basic properties of insulating gases.

Gas	Molecular weight	Critical strength Ecr [kV(mm bar)$^{-1}$]	Boiling point [K]	Ionization potential [eV]	GWP
SF_6	146.06	8.9	209	15.8	23,900
N_2	28.01	3.29	77	15.6	0
O_2	32.00	3.26	90	12.2	0
CO_2	44.01	3.01	194	13.7	1

3. DIELECTRIC CHATACTERISTICS OF N_2/SF_6 GAS MIXTURES

As described in Introduction, N_2/SF_6 gas mixtures have been regarded as one of the promising SF_6 substitutes. This section deals with the dielectric characteristics and mechanisms of N_2/SF_6 gas mixtures. The other SF_6 substitutes without SF_6 gas such as CO_2 gas and CO_2-based gas mixtures will be described in the next section.

3.1 Synergistic effect and critical content of electronegative gas

N_2/SF_6 gas mixtures are characterized by the synergistic effect on BDV by a small addition of SF_6 gas in N_2 gas. This synergistic effect has been derived from the strong electron attachment of electronegative SF_6 gas. In order to discuss the electronegativity and the contribution of the additive gas to the dielectric characteristics in more detail, we investigated PD and BD characteristics of ultra-dilute SF_6/N_2 and PFC/N_2 (C_2F_6/N_2, C_3F_8/N_2) gas mixtures.

Figure 3 shows ac PDIV and BDV of a needle-plane electrode system (ϕ=1mm, g=50mm, P=0.1MPa) as a function of electronegative gas content (K) from the order of ppm to several %[4]. PDIV in all gas mixtures increased with the electronegative gas content, which suggested that PDIV was dominated by N_2 gas. On the other hand, BDV greatly depended on the additive gas, i.e. BDV started to increase with the electronegative gas content at K(SF_6)>10ppm, K(C_2F_6)>0.1% and K(C_3F_8)>1%, respectively, which could be referred to as the critical content of electronegative gas to activate the electron attachment in N_2 gas. Furthermore, the pre-breakdown streamer inception voltage (PSIV) was defined as the minimum voltage where the pre-breakdown streamer discharge[5] occurred beyond the glow discharge region around the needle-tip. PSIV decreased with the increase in the electronegative gas content in the range where the return stroke was suppressed and BDV increased.

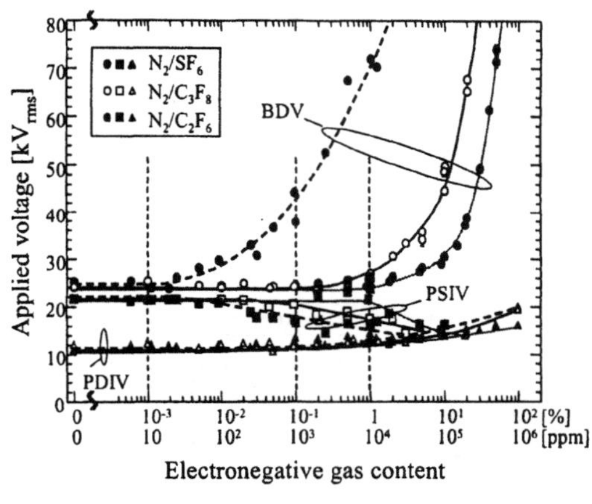

Fig.3 PDIV, PSIV and BDV of ultra-dilute SF_6/N_2, C_2F_6/N_2, C_3F_8/N_2 gas mixtures. (ϕ=1mm, g=50mm, P=0.1MPa)

3.2 Impulse partial discharge

Impulse PD propagation characteristics have been obtained in N_2/SF_6 gas mixtures for a needle-plane electrode system (ϕ=1mm, g=40mm). Figure 4 shows still images of the impulse PD in $N_2$90%/$SF_6$10% gas mixtures for different gas pressures. The brush-like PD around the needle-tip in Fig.4 (a) is the streamer discharge, whereas the stepwise PD in Fig.4 (b) is the leader discharge with the longer propagation length than that of the streamer. Each step length and the total length of the impulse PD propagation in Fig.4 (c) become shorter with the increase in the gas pressure. Figure 5 shows the applied impulse voltage, PD current and light intensity waveforms and streak light emission image at P=0.25MPa[6], where 3-peak pulses within the gate width of 600ns and the streak image verified the stepwise propagation of the impulse PD. The first step length is shown in Fig.6 as a function of gas pressure for different SF_6 contents in N_2

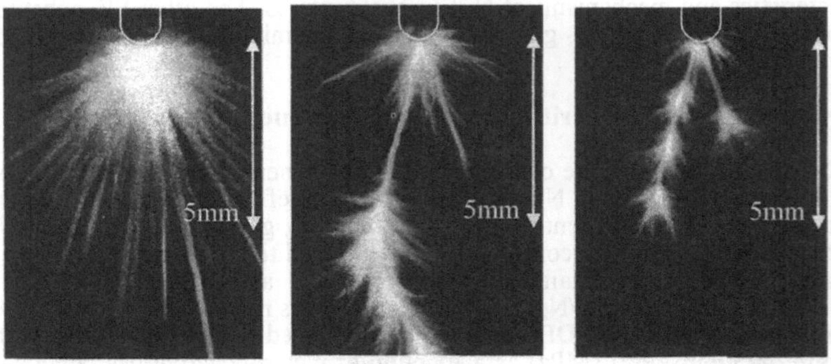

(a) P=0.1 MPa (b) P=0.2 MPa (c) P=0.3 MPa

Fig.4 Impulse PD propagation in $N_2$90%/$SF_6$10% gas mixtures. (ϕ=1mm, g=40mm)

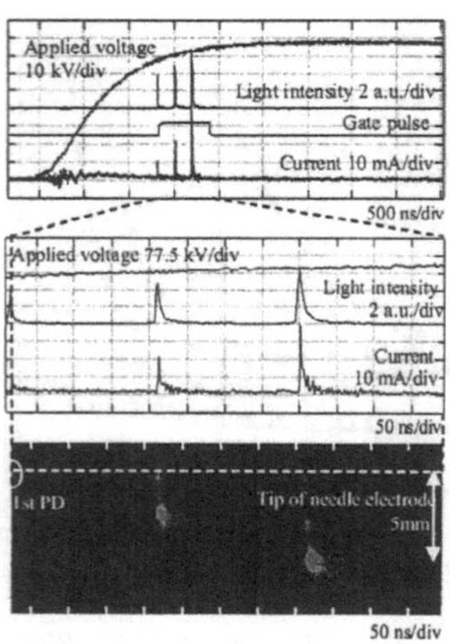

Fig.5 Stepwise impulse PD propagation in $N_2$90%/$SF_6$10% gas mixtures. (ϕ=1mm, g=40mm, P=0.25MPa)

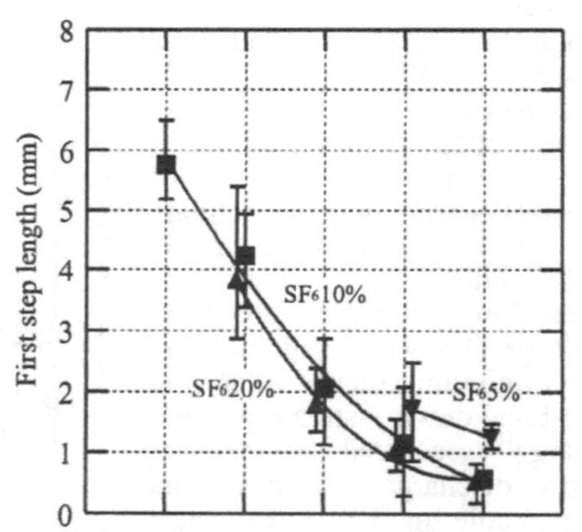

Fig.6 First step length as a function of gas pressure for different SF_6 contents in N_2 gas. (ϕ=1mm, g=40mm)

gas. The dependence of the step length on the gas pressure was greater than that of the SF_6 content, which would be derived from the difference in dependences of the critical strength of SF_6 gas on the gas pressure and the SF_6 content[3].

Figure 7 shows $PDIV_{50}$, BDV_{50} and leader discharge onset voltage (LOV) as a function of gas pressure in $N_2$90%/$SF_6$10% gas mixtures[7]. LOV was defined as the minimum voltage where the stepwise PD propagation with multi-peak pulses was observed, as shown in Fig.5. These results enabled us to classify the impulse PD type in N_2/SF_6 gas mixtures into streamer discharge or leader discharge regimes. Similar diagrams as that in Fig.7 have been obtained for different SF_6 contents. As the result, the corona stabilization was still effective even under the impulse voltage application. The leader discharge regime became larger with the gas pressure and the SF_6 content, whereas the streamer discharge regime became smaller. Moreover, leader discharge regime also enlarged with the SF_6 content at the lower gas pressure. These are attributed to the enhancement of electronegativity in N_2/SF_6 gas mixtures with the higher gas pressure and SF_6 content, which could induce the shrinkage and heating of leader column and the activation of the streamer/leader transition. Such classification of impulse PD type has been reported in pure SF_6 gas[8], and in this paper, the one in N_2/SF_6 gas mixtures was able to be shown.

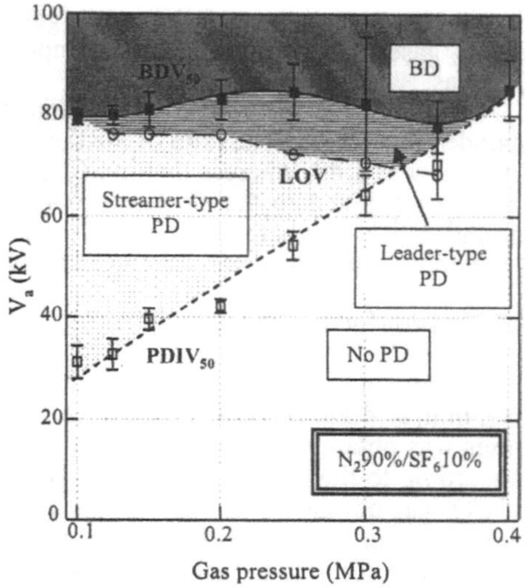

Fig.7 Classification of impulse PD type in $N_2$90%/$SF_6$10% gas mixtures. (ϕ=1mm, g=40mm)

4. DIELECTRIC CHATACTERISTICS OF CO_2 GAS AND CO_2-BASED GAS MIXTURES

This section presents the dielectric characteristics and mechanisms of CO_2 gas and CO_2-based gas mixtures as SF_6-free substitutes. The results will also be compared with those of SF_6 gas.

4.1 Quasi-uniform electric field

Under quasi-uniform electric field composed of sphere-plane electrode system (ϕ=50mm, g=10mm), Fig.8 shows breakdown electric field BDE at the tip of sphere electrode in SF_6, CO_2 and N_2 gases at P=0.1MPa under ac and impulse voltage applications. The ac BDE in CO_2 gas was almost the same as

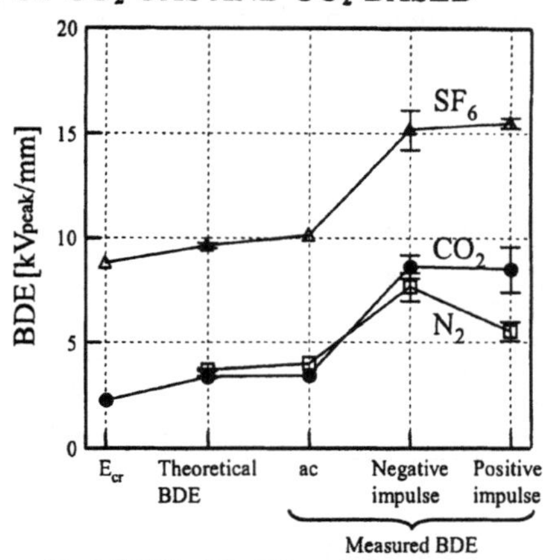

Fig.8 BDE in SF_6, CO_2 and N_2 gases. (ϕ=50mm, g=10mm, P=0.1MPa)

that in N_2 gas, and the ac BDE in SF_6 gas was about three times of those in CO_2 and N_2 gases. On the other hand, the impulse BDE in SF_6 gas was twice of that in CO_2 gas.

In Fig.8, the theoretical BDE calculated by the streamer criterion[9] corresponded to ac BDE. The ratio of the theoretical BDE to the critical strength E_{cr} in CO_2 gas was 1.48, whereas that in SF_6 gas was 1.08. The higher ratio in CO_2 gas may be attributed to the slower increase in the effective ionization coefficient around E_{cr}[10,11], where the electron avalanche in CO_2 gas would be hard to develop into the streamer discharge, compared with that in SF_6 gas.

4.2 Non-uniform electric field

Under non-uniform electric field composed of rod-plane electrode system (ϕ=5mm, g=40mm), gas pressure dependence of PDIV and BDV in CO_2 and SF_6 gases was obtained under positive lightning impulse and ac voltage applications. According to the results, the relative BDV ratio R_{CO_2/SF_6} was defined as the BDV of CO_2 gas normalized by that of SF_6 gas. Figure 9 shows R_{CO_2/SF_6} of the impulse BDV_{50} and ac BDV as a function of gas pressure. R_{CO_2/SF_6} tended to increase with the gas pressure, and reached 0.49 for ac BDV and 0.64 for impulse BDV_{50} at P=0.3MPa. These results suggest that CO_2 gas may have excellent dielectric characteristics as an SF_6 substitute, especially under impulse voltage application.

Figure 10 shows CO_2 content dependence of positive impulse $PDIV_{50}$ and BDV_{50} in CO_2/O_2 and CO_2/N_2 gas mixtures at P=0.1MPa[12]. $PDIV_{50}$ in CO_2/O_2 and CO_2/N_2 gas mixtures slightly increased with the CO_2 content, which may be attributed to the lower generation of initial electrons in CO_2 gas, as described in section 4.1. BDV_{50} in CO_2/O_2 gas mixtures decreased with the increase in CO_2 content, whereas BDV_{50} in CO_2/N_2 gas mixtures slightly increased with the CO_2 content. In other words, BDV_{50} in $CO_2$50%/$O_2$50% gas mixtures was higher than that in pure CO_2 and N_2 gases. These results can be derived from 4 times larger cross-sectional area of electron attachment in O_2 gas than that in CO_2 gas[13].

4.3 Impulse ratio

Figure 11 shows the impulse ratio under quasi-uniform and non-uniform electric fields in CO_2 and SF_6 gases at P=0.1MPa. The impulse ratio of CO_2 gas

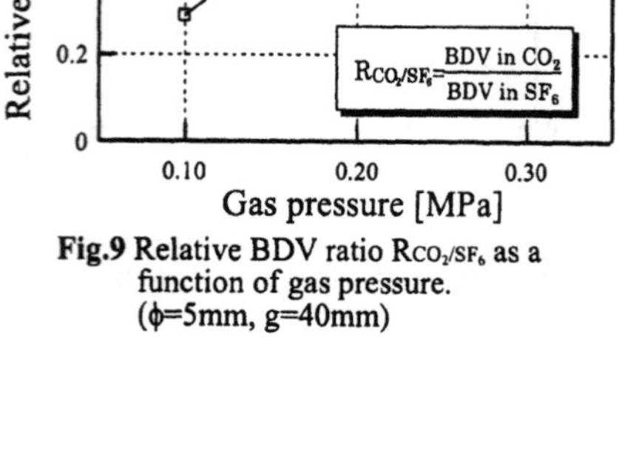

Fig.9 Relative BDV ratio R_{CO_2/SF_6} as a function of gas pressure. (ϕ=5mm, g=40mm)

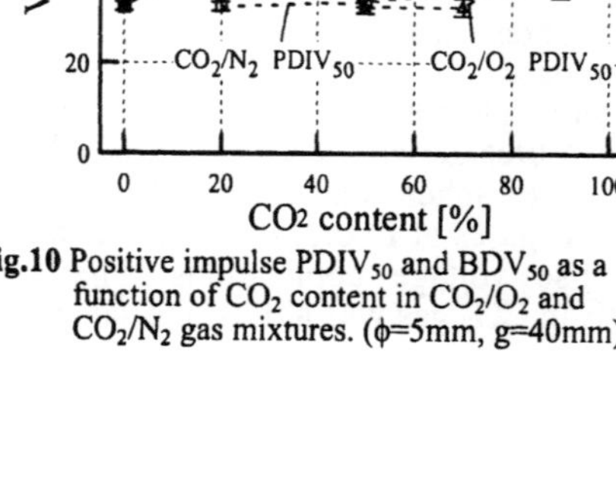

Fig.10 Positive impulse $PDIV_{50}$ and BDV_{50} as a function of CO_2 content in CO_2/O_2 and CO_2/N_2 gas mixtures. (ϕ=5mm, g=40mm)

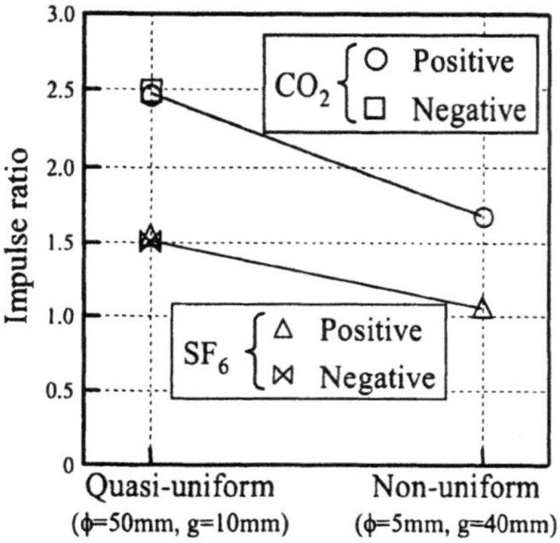

Fig.11 Impulse ratio under quasi-uniform and non-uniform electric fields in CO_2 and SF_6 gases. (P=0.1MPa)

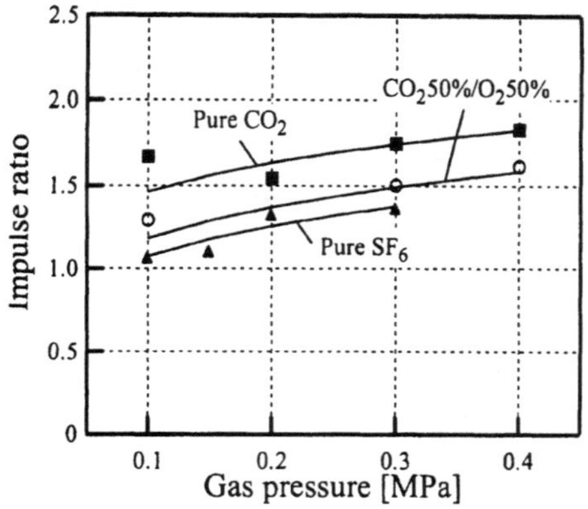

Fig.12 Impulse ratio as a function of gas pressure in CO_2, SF_6 gases and $CO_2$50% /$O_2$50% gas mixtures. (ϕ=5mm, g=40mm)

was higher than that of SF_6 gas, i.e. the lack of initial electrons in CO_2 gas was more remarkable than that in SF_6 gas, which can be interpreted as follows: Since the electron attachment coefficient in CO_2 gas is much lower than that in SF_6 gas, the negative ion density in CO_2 gas can be lower than that in SF_6 gas. Furthermore, the electron detachment coefficient in CO_2 gas is smaller than that in SF_6 gas[14]. As the result, the generation probability of initial electrons in CO_2 gas can be considered to be smaller than that in SF_6 gas, i.e. the impulse ratio of CO_2 gas would be higher than that of SF_6 gas. Furthermore, the impulse ratio under non-uniform field was lower than that under quasi-uniform field. This would be caused by that the concentrated stress under non-uniform field promotes the generation probability of initial electrons.

The impulse ratio of discharge inception voltage as a function of gas pressure under non-uniform electric field in CO_2, SF_6 gases and $CO_2$50%/$O_2$50% gas mixtures is shown in Fig.12. The impulse ratio in each gas increased with the gas pressure, which may be interpreted by that the lack of initial electrons become remarkable due to the shrinkage of critical volume around the tip of rod electrode with the increase in the gas pressure[12]. In addition, the impulse ratio in CO_2 gas and $CO_2$50%/$O_2$50% gas mixtures were higher than that in SF_6 gas.

The higher impulse ratio in CO_2 gas and CO_2-based gas mixtures will contribute to the efficient insulation design of power apparatus using such SF_6 substitutes.

5. FUNDAMENTAL CONCEPTS OF ELECTRICAL INSULATION DESIGN OF POWER APPARATUS USING SF_6 SUBSTITUTES

Figure 13 summarizes the factors related to the insulation design of gas-insulated power apparatus composed of high-voltage conductor, solid spacers, insulating gases and metallic enclosure. Taking account of the dielectric characteristics and mechanisms in the previous sections, fundamental concepts for electrical insulation design of N_2 and/or CO_2 gas-insulated switchgears will be described in this section.

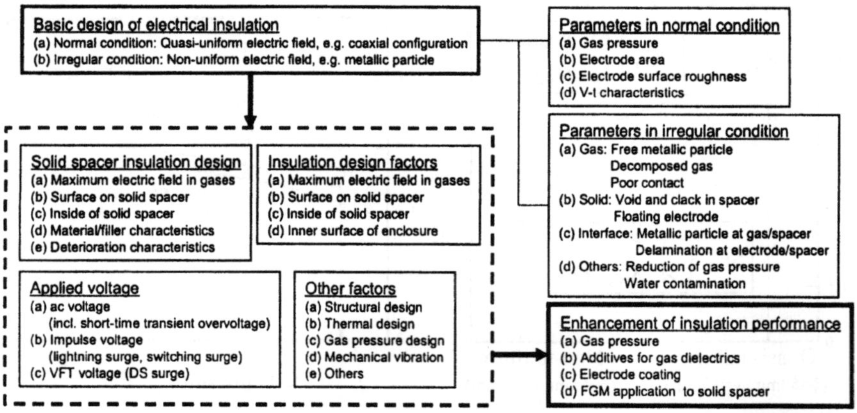

Fig.13 Factors related to insulation design of gas-insulated power apparatus by SF_6 substitutes.

5.1 Maximum electric field in gases

For the electrical insulation design of the power equipment, the maximum electric field E in gases can be given as follows:

$$E = E_{cr} \times P^{\alpha} \times S^{\beta} \times m^{\gamma} \times (1-k\delta)$$

where E_{cr} is critical strength [kV/cm] of the gas, P and α are gas pressure and its index, S and β are highly-stressed electrode area and its index, m and γ are electrode surface roughness and its index, δ is standard deviation, k is constant.

In the case of N_2/SF_6 gas mixtures with SF_6 content higher than 10%, the insulation design of N_2/SF_6-GIS can be based on that of SF_6-GIS, because the discharge characteristics are similar to those of pure SF_6 gas. The difference between the impulse ratio of discharge inception under quasi-uniform electric field and that of testing voltage is inevitable to be clarified, in order to identify the dominant factors to the electrical insulation design, where the polarity of the applied impulse voltage will also depend on the gas pressure to be adopted. These points should be discussed in detail with experimental verification.

Impulse and ac breakdown characteristics at the higher gas pressure up to 1.0-1.5MPa are necessary with consideration of area effect and surface roughness effect. When CO_2 gas with the pressure higher than 1.0-1.5MPa is expected, liquefaction and sublimation points should also be taken into account. V-t characteristics for the basic insulation design are quite few and needed to be obtained together with the standard deviation of breakdown stress. Considerations on dilute gas addition, gas mixture, electrode coating and their synergism are also crucial to enhance the dielectric characteristics of CO_2-based power apparatus.

5.2 Solid spacer and creepage insulation

Insulation design of solid spacers can be categorized as follows:

(A) Electric field relaxation at the triple junction with the high-voltage conductor
(B) Electric field relaxation at the triple junction with the grounded enclosure
(C) Electric field within the solid spacer under operating ac stress
(D) Electric field on the solid spacer surface

(A) and (B) can be determined by the maximum stress in gases described in section 5.1, together with the configurations of spacer, embedded electrodes and shielding electrodes. Experiences of SF_6-GIS will contribute to (C).

The concept for insulation design in SF_6 gas can be applied to (D). Generally, impulse voltage will be dominant due to the higher stress level and the less effective corona stabilization effect than those under ac voltage application, which is also inevitable to be discussed in detail with experimental verification. The surface configuration of the solid spacer can be designed according to the relationship between surface flashover stress and attached metallic particle length under impulse voltage application. Since the less effective corona stabilization effect in N_2 and/or CO_2 gases than that in SF_6 gas may reduce ac breakdown stress level, the impulse ratio of surface flashover voltage in N_2 and/or CO_2 gases will be required.

In addition, the electrical insulation design should be coordinated with the mechanical design against the gas pressure difference between both sides of the spacer. Especially, since the rated gas pressure of N_2 and/or CO_2 gas is expected to be higher than that of SF_6 gas, the gas pressure dependence of mechanical characteristics will become important. Thermal characteristics are also claimed to be balanced with the electrical and mechanical characteristics.

5.3 Enclosure stress design

Insulation design on enclosure surface in SF_6-GIS has been determined by the lift-off, crossing and breakdown characteristics of free metallic particles in SF_6 gas [15]. According to Fig.14[16,17,18], the stress margin between the particle lift-off and breakdown in N_2 gas and N_2/SF_6 gas mixtures was much smaller than that in SF_6 gas. This means that the particle-induced breakdown in N_2 gas and N_2/SF_6 gas mixtures may happen at the moment of particle lift-off even under ac voltage application. On the other hand, in CO_2 gas, breakdown did not occur even at the applied voltage of 1.8 times higher than the particle lift-off voltage, which suggested that CO_2 gas might be superior to N_2 gas and N_2/SF_6 gas mixtures in terms of the particle contamination. The crossing characteristics in CO_2 gas and CO_2-based gas mixtures as well as the effect of electrode coating on enclosure surface should be investigated in more detail.

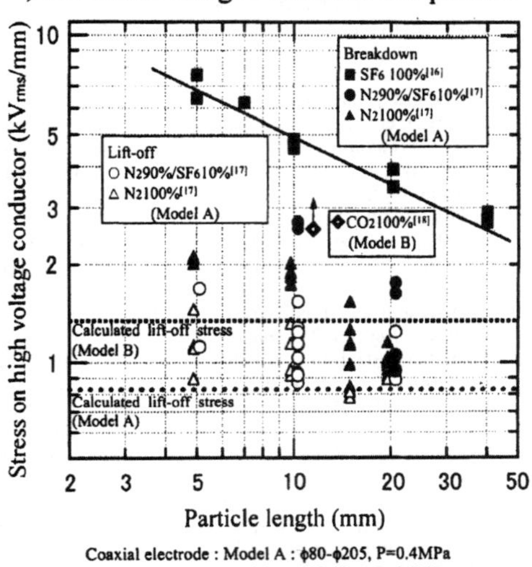

Coaxial electrode : Model A : ϕ80-ϕ205, P=0.4MPa
Model B : ϕ90-ϕ340, P=0.5MPa

Fig.14 Particle lift-off and breakdown stresses as a function of particle length.

6. CONCLUSIONS

In this paper, dielectric characteristics of N_2/SF_6 gas mixtures, CO_2 gas and CO_2-based gas mixtures as SF_6 substitutes were presented and compared with those of SF_6 gas on the basis of discharge mechanisms of electronegative gases. The main results are summarized as follows:

1. A database of PD and BD characteristics of insulating gases and gas mixtures has been acquired, expecting to make the most use of it.
2. Synergistic effect on BDV was discussed in N_2/SF_6 gas mixtures, and the critical SF_6 content to activate the electron attachment in N_2 gas was evaluated as 10ppm.
3. Stepwise PD propagation due to streamer/leader transition under impulse voltage application was clarified, and PD type in N_2/SF_6 gas mixtures was classified into streamer discharge or leader discharge regimes.
4. Ratio of theoretical BDE to critical strength in CO_2 gas was higher than that in SF_6 gas, because the electron avalanche in CO_2 gas would be hard to develop into the streamer discharge, compared with that in SF_6 gas.
5. Impulse ratio of CO_2 gas and CO_2-based gas mixtures under quasi-uniform and non-uniform electric field was higher than that in SF_6 gas, because of the lower generation of initial electrons in CO_2 gas with small electron attachment and detachment coefficients.
6. Fundamental concepts for electrical insulation design of CO_2 and N_2/SF_6 gas-insulated switchgears were presented, and technical problems to be solved were clarified.

REFERENCES

1. Intergovernmental Panel on Climate Change (IPCC): "The Science of Climate Change – Contribution of Working Group I to the Second Assessment Report of the IPCC", (1996).
2. A.Sabot: "Insulation Co-ordination Procedure for 420kV Gas Insulated Lines (GIL)", *11th International Symposium on High Voltage Engineering*, Vol.3, pp.1-10 (1999).
3. I.Gallimberti and N.Wiegart: "Breakdown Mechanisms in Electronegative Gases", *Gaseous Dielectrics V*, pp.61-79 (1987).
4. H.Okubo, et al.: "Partial discharge and breakdown mechanisms in ultra-dilute SF_6 and PFC gases mixed with N_2 gas", *Journal of Physics D: Applied Physics*, Vol.35, No.21, pp.2760-2765 (2002).
5. T.Yamada, et al.: "Partial Discharge and Breakdown Mechanisms in Ultra-dilute SF_6/N_2 Gas Mixtures", *IEEE Transactions on Dielectrics and Electrical Insulation*, Vol.8, No.1, pp.137-142 (2001).
6. Y.Yoshitake, et al.: "Impulse Partial Discharge and Breakdown Mechanism under Non-uniform Electric Field in N_2/SF_6 Gas Mixtures", *Conference on Electrical Insulation and Dielectric Phenomena*, pp.793-796 (2002).
7. Y.Yoshitake, et al.: "Impulse Partial Discharge Propagation Mechanism under Non-uniform Electric Field in N_2/SF_6 Gas Mixtures", *13th International Symposium on High Voltage Engineering*, p.471 (2003).
8. F.Pinnekamp and L.Niemeyer: "Qualitative Model of Breakdown in SF_6 in Inhomogeneous Gaps", *Journal of Physics D: Applied Physics*, Vol.16, pp.1293-1302 (1983).
9. H.Raether: "Electron avalanches and breakdown in gases", London, Butterworths (1964).
10. J.L.Hernández-Ávila, et al.: "Electron transport and swarm parameters in CO_2 and its mixtures with SF_6", *Journal of Physics D: Applied Physics*, Vol.35, No. 18, pp. 2264-2269 (2002).
11. Sayed A.Ward: "Influence of Conducting Particles on The Breakdown Voltages of SF_6-N_2 Mixtures", *11th International Symposium on High Voltage Engineering*, Vol.3, pp.187-192 (1999).
12. H.Okubo, et al.: "Electrical insulation characteristics of CO_2 and CO_2 gas mixtures under non-uniform electric field", *Conference on Electrical Insulation and Dielectric Phenomena*, pp.538-541 (2003).
13. L.G.Christophorou and L.A.Pinnaduwage: "Basic Physics of Gaseous Dielectrics", *IEEE Transactions on Electrical Insulation*, Vol.25, No. 1, pp.55-74 (1990).
14. D.M.Xiao, et al.: "Swarm parameters in SF_6 and CO_2 gas mixtures", *Journal of Physics D: Applied Physics*, Vol.34, No.24, L133-135 (2001).
15. S.Kobayashi, et al.: "Particle-Initiated Flash-over Caused by Disconnector Restriking Surges in GIS", *5th International Symposium on High Voltage Engineering*, No.1203 (1987).
16. T.Hattori, et al.: "A Study on Effects of Conducting Particles in SF_6 Gas and Test Methods for GIS", *IEEE Transactions on Power Delivery*, Vol.3, No. 1, pp.197-204 (1988).
17. H.Okubo, et al.: "Metallic Particle Behavior and Breakdown Characteristics in N_2/SF_6 Gas Mixture GIS", *International Conference on Gas Discharges and Their Applications*, Vol.1, pp.446-449 (2000).
18. M.Onomoto, et al.: "Study on metallic particle motion and partial discharge characteristics in GIS tank filled with CO_2", *2003 Japan-Korea Joint Symposium on ED & HVE*, pp.57-60 (2003).

ESTIMATION OF CROSS-SECTIONAL SIZE OF GAS-INSULATED APPARATUS USING HYBRID INSULATION SYSTEM WITH SF_6 SUBSTITUTE

H. Goshima, H. Shinkai, T. Kawamoto, M. Yashima, and H. Fujinami

1. INTRODUCTION

SF_6 has been identified as a greenhouse gas with a long atmospheric lifetime[1]. Therefore, recycling guidelines for SF_6 in electric power apparatus have been studied, and the reduction in the amount of SF_6 released into the atmosphere has progressed[2]. In the long term, it is preferable to reduce the amount of SF_6 used. Therefore, it may be important to discuss the possibility of using environmentally friendly gases as an alternative insulation gas for practical gas-insulated apparatus.

In the selection of an alternative gas, it is thought that high-pressure natural gases, such as air, nitrogen (N_2), and carbon dioxide (CO_2) are promising environmentally friendly candidates. However, some strategies for enhancing insulation performance are necessary for their practical applications to apparatus because the dielectric strengths of these gases are about 1/3 that of SF_6. In this study, we investigated the possibility of gas-insulated apparatus using a hybrid gas insulation system which has the combined features of a solid insulator and a dielectric gas. Because the hybrid gas insulation system has a structure which the high-electric-field part is insulated using a solid insulator[3], there is a possibility that the same degree of compactness as that of the present apparatus can be achieved in spite of the application of a natural gas (N_2 or CO_2). Therefore, it may be useful for one of the alternative equipment using the hybrid gas insulation system when a present compact gas-insulated apparatus is replaced. In this paper, we describe the estimation of the cross-sectional size of a gas-insulated apparatus using a hybrid gas insulation system based on insulation and thermal designs.

2. STRUCTURE OF HYBRID GAS INSULATION SYSTEM

Figures 1 show the basic structures of the hybrid gas insulation system for single-phase model. Figure 1(a) shows the single-core-type conductor which has a coaxial cylindrical structure. The central conductor is covered by a solid insulator

---

Electrical Insulation Department, Komae Research Laboratory, Central Research Institute of Electric Power Industry (CRIEPI), 2-11-1 Iwado-kita, Komae-shi, Tokyo, 201-8511, Japan

Gaseous Dielectrics X.
Edited by Christophorou *et al.*, Springer, New York, 2004.

which we assumed to be polyethylene (PE). In the case of this single-core-type structure, the maximum electric field in the gas can be reduced in the presence of PE and the field distribution tends to become uniform[3].

Figure 1(b) shows the triple-core-type conductor structure. The central conductor is divided into three subconductors[4] individually covered by PE. The Y-shaped spacer shown as a dotted line supports the subconductors. In the case of the triple-core-type structure, there are some merits such as the simple method of conductor support (shielding effect at the support points of the subconductors), flexibility of the subconductors, and so on[3]. In this paper, we estimated the cross-sectional size of the two types of hybrid gas insulation system.

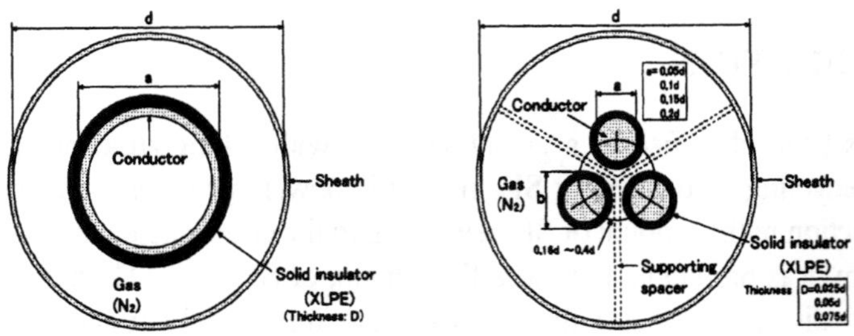

(a) Single-core type (b) Triple-core type

Figure 1. Basic structures of hybrid insulation system for single-phase operation.

3. INSULATION AND THERMAL DESIGNS

3.1 Insulation design

The electric field design was basically the same as that of conventional gas-insulated transmission line (GIL). We determined the allowable electric field E_L from the maximum electric field in the gas because the dielectric strength of a solid insulator was generally much higher than that of a gas. Moreover, it is assumed that the severest conditions are achieved by applying a lightning impulse voltage to an apparatus using the hybrid gas insulation system as well as a conventional SF_6 apparatus. Hence, E_L was determined from the maximum electric field at the surface of PE when a lightning impulse withstand voltage (LIWV) was applied.

3.1.1 Breakdown characteristics of N_2 and CO_2

Figure 2 shows breakdown strengths of N_2, CO_2 and SF_6 as functions of gas pressure. The data shown by black markers were acquired in the sphere-to-plane electrode configuration[5,6]. It is seen from this figure that the saturation tendency for the gas pressure dependency of the breakdown strength of CO_2 is smaller than that of N_2. Therefore, the breakdown strength of CO_2 is higher than that of N_2 in the high-pressure region. However, the breakdown strength of CO_2 is nearly equal to that of N_2 up to 1.5MPa.

Recently, the breakdown characteristics of high-pressure N_2 under several electrode conditions have been investigated[7-10]. We obtained the breakdown strength of N_2 under coaxial cylindrical electrodes from the reference data shown in Fig.2. It is seen from this figure that the breakdown strength of the coaxial cylindrical electrode is about 20% lower than that of the sphere-to-plane electrode because of the area effect. We basically determine E_L using the insulation design from the data of the coaxial cylinder as follows.

In this study, N_2 is used as the insulation gas, and the gas pressure is set to be 0.8MPa(abs). The breakdown strength E_B of N_2 at 0.8MPa is estimated at 15.8kV/mm from Fig.2. E_L is determined based on E_B using[11]

$$E_L = E_B(1-3\sigma)\frac{\eta}{s}, \qquad (1)$$

where σ represents the standard deviation which is set to be 5% from the results of the scattering of the breakdown voltage in high-pressure N_2[5], η is the spacer efficiency which is set to be 0.7 from the results using a simple cylinder spacer model, and s is the margin at 1.2 which may be experimentally used with SF_6 gas-insulated equipment. Thus, E_L of N_2 at 0.8MPa is estimated at 8kV/mm.

3.1.2 Improvement of breakdown strength using hybrid gas insulation system

Figure 3 shows the schematic of the breakdown process for the hybrid gas insulation system when a conducting particle exists in the gas. The breakdown for the hybrid gas insulation system may be caused by three types of discharge process: discharge in the gas, creeping discharge at the surface of PE and surface discharge at the PE interface. The larger the creeping discharge length, the larger the proportion of the creeping discharge voltage in the breakdown voltage. Hence, it can be expected that breakdown strength is increased by covering the center conductor with a solid insulator. Therefore, the breakdown strength for the hybrid gas insulation system may be higher than that for the conventional gas-insulated system[12], for which discharge in the gas causes sudden breakdown. Hence, the allowable electric field E_L of the hybrid gas insulation system may be raised compared with that of the conventional system. Therefore, we also estimate the fundamental specifications when the E_L is increased. Concretely, from the above estimation in section 3.1.1, the E_L is set to be 10kV/mm (level of no margin) or 13kV/mm (level of breakdown strength of the gas).

3.1.3 Insulation performance in the presence of a conducting particle

It was reported that a long conducting particle with roughly double length is permissible in the hybrid gas insulation system as compared with that in the normal gas insulation system from the viewpoint of insulation reliability[12]. The harmful effect of a conducting particle may be reduced because discharge in the gas does not cause sudden breakdown (short-circuit or earth fault), as shown in Fig.3. Therefore, we assume that we do not set an allowable electric field on the inner surface of the sheath. As a result, it is noted that the electric field of the sheath is below that of the conventional gas-insulated apparatus.

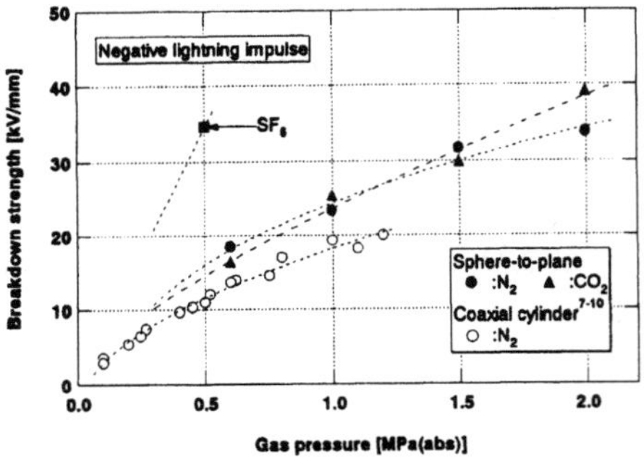

Figure 2. Breakdown strength as a function of gas pressure.

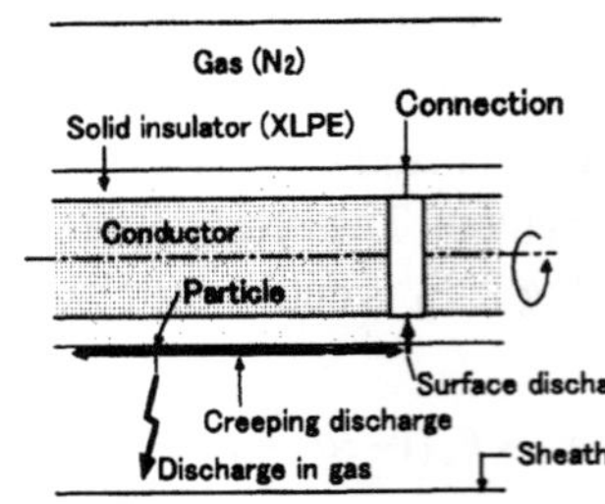

Figure 3. Breakdown process of hybrid insulation system.

3.2 Thermal design

The applied calculation method for the resistances of the conductor and sheath, dielectric loss of PE and heat transfer through PE was basically the same as that for an XLPE cable, based mainly on the Japanese Cable Standard (JCS 168-D). As to conductor (single-core type) and sheath loss, heat radiation between PE and the sheath and heat radiation between the sheath and ambient air, the calculation method for a normal GIL was applied. Current capacity in the steady state was obtained under the condition in which heat generation from the conductor, PE and sheath was balanced by the heat transfer through PE and heat radiation in both N_2 and ambient air[3].

3.3 Principal conditions assumed

Table 1 shows the principal conditions assumed such as structure and some coefficients using current capacity. We targeted 275kV class as the voltage class of apparatus. The LIWV is set at 950kV. The allowable electric field E_L and PE thickness are assumed. For the triple-core type, the conductor structure of the XLPE cable (without sheath) is assumed because the diameter of a subconductor can be reduced, and the flexibility of the subconductors and long distance of the subconductors can be achieved.

4. ESTIMATION OF BASIC SPECIFICATIONS

Table 2 shows examples of the fundamental specifications for the hybrid gas insulation system with an alternative gas to SF_6 viewed from the same specifications (e.g., cross-sectional size, rated voltage and current) as those of the present gas-insulated apparatus. If the allowable electric field is set at about 13kV/mm, the cross-sectional size of the sheath can be the same as that of the present apparatus.

Table 1. Principal conditions assumed.

Conductor	Structure	Single-core type	Aluminum alloy pipe (thickness t=15mm)
		Triple-core type	Conductor structure of XLPE cable is assumed. Bundle of stranded copper wire
	Resistance	Single-core type	Consideration of skin effect, No consideration of proximity effect
		Triple-core type	Consideration of skin effect and proximity effect of other conductor
	Allowable electric field	E_L=13kV/mm, 10kV/mm, 8kV/mm	
	LIWV	950kV (275kV class)	
	Allowable temperature	T_c=90 deg.	
Sheath	Structure	Aluminum alloy pipe (thickness t=10mm)	
	Resistance	Consideration of skin effect. No consideration of proximity effect.	
	Radiation coefficient	Inner: 0.1(aluminum), outer: 0.8 (with N7-Munsellpaint)	
	Bonding of sheath	Bonded at both ends (sheath current is nearly equal to total current of conductor)	
Solid insulator	Material	Cross-linked polyethylene, relative permittivity ε_r=2.3, tan δ =0.001	
	Radiation coefficient	0.1 (equal to aluminum)	
	Intrinsic thermal resistance	450 [deg. cm / W]	
Laying condition	Atmospheric temperature	T_0=40 deg. (installed in tunnel)	
	Separation of sheath	400mm (inner distance between neighboring sheaths)	

Table 2. Examples of fundamental specification for hybrid gas insulation system.

Item	Example-1	Example-2
Conductor type	Single-core type	Triple-core type
Rated voltage/ current	275kV／4000A	275kV／6300A
Gas, Pressure	N_2, 0.8MPa(abs)	
Solid insulator	XLPE (relative permittivity: 2.3)	
LIWV / E_L	V_L=950kV, E_L=13kV/mm	
T_c	90 deg	
Cross-sectional size (Single phase)	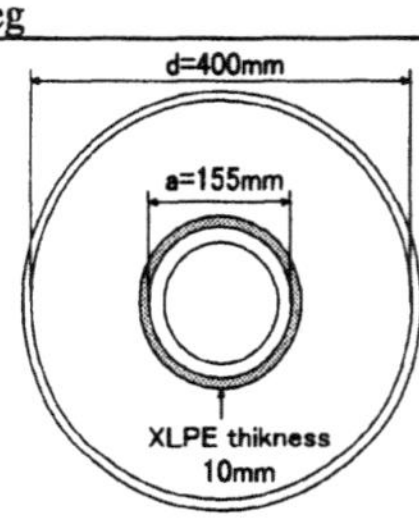	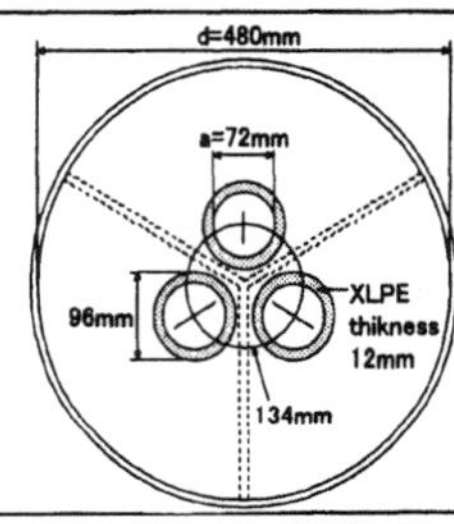
Applicable equipment	Gas-insulated bus in substation	Large-capacity transmission line

For the single-core type, the structure is simple, and its advantage is large in terms of the electric field distribution (e.g., reduction in the maximum electric field of the gas). Therefore, the hybrid gas insulation system of the single-core type may be expected for application as an alternative gas-insulated bus. However, there are important issues to be resolved prior to its practical applications, such as the breakdown mechanism of the composite insulation system, a method of supporting and connecting the conductor, and practical construction.

On the other hand, for the triple-core type, its advantage is large in terms of the method of conductor support, flexibility of the subconductor, and long distance of the subconductors. Thus, the hybrid gas insulation system of the triple-core type may be promising as a large-capacity transmission line.

5. CONCLUSIONS

In this paper, we estimated the cross-sectional size of a gas-insulated apparatus using a hybrid gas insulation system with a natural gas (N_2) as an alternative gas to SF_6 based on insulation and thermal designs. By applying the hybrid gas insulation system, a conventional SF_6-gas-insulated apparatus may be replaced with a natural-gas-insulated one without a considerable increase in cross-sectional size. In the case of applying gases such as N_2 and CO_2, which have lower insulation performance than SF_6, the use of the hybrid gas insulation system may be effective as a method of achieving a higher insulation performance.

This research was executed as one of the feasibility studies in the electric power technique development project of the strategic electric power technique development investigation consigned by the Ministry of Economy, Trade and Industry Agency for Natural Resources and Energy to the Institute of Applied Energy.

References

1. IPCC, "Radiative Forcing of Climate Change", p172 (1994)
2. CIGRE TF 01 of WG 23.10, "SF_6 Recycling Guide, Re-use of SF_6 Gas in Electric Power Equipment and Final Disposal", *Electra*, **173**, pp.43-71 (1997)
3. M.Yashima, T. Takuma and T. Kawamoto, "Basic Study on Hybrid-Gas-Insulated Transmission Line (H-GIL)", *Proc. of 9th ISH*, No.7880, Graz (1995)
4. B.F. Hampton, D.N. Browning, and R.M. Mayers, "Outline of a flexible, SF6 insulated e.h.v. cable", *Proc. IEE*, 123 (2), pp.159-164 (1976)
5. H. Goshima, H. Shinkai, M. Yashima, "Lightning Impulse Breakdown Characteristics of High-Pressure N_2 As an Alternative Insulation Gas to SF_6", *Gaseous Dielectrics IX*, pp.359-364 (2001)
6. H. Goshima, H. Shinkai, H. Fujinami, "Lightning impulse breakdown characteristics of high-pressure CO_2 compared with those of N_2 as alternative insulation gases to SF_6", *Proc. of 13th ISH*, p.443, Netherlands (2003)
7. A.H. Cookson, and B.O. Pedersen, "Analysis of The High Voltage Breakdown Results for mixtures of SF_6 with CO_2, N_2 and air", *Proc. of 3rd ISH*, No.31.10, Milan (1979)
8. H.I. Marsden, S.J. Dale, M.D. Hopkins, and C.R. Eck III, "High Voltage Performance of a Gas Insulated Cable with N_2 and N_2/SF_6 Mixtures", *Proc. of 10th ISH*, Vol.2, pp.9-12, Montreal (1997)
9. L.G Christophorou, J.K. Olthoff, and D.S. Green, "Gaseous for Electrical Insulation and Arc Interruption: Possible Present and Future Alternatives to Pure SF_6", *NIST Technical Note*, No.1425 (1997)
10. M. Shiiki, M. Sato, M. Hanai, and K. Suzuki, "Dielectric Performance of CO_2 Gas Compared with N_2 Gas", *Gaseous Dielectrics IX*, pp.365-370, (2001)
11. H. Saito, K. Morita, T. Kikkawa, N. Hayakawa, and H. Okubo, "Fundamental Study on Electrical Insulation Design of N_2/SF_6 Gas Insulated Electric Power Apparatus", *T.IEE Japan*, **121** (4), pp.449-454, (2001) (in Japanese)
12. M. Yashima, T. Kawamoto, H. Fujinami and T. Takuma, "A Study on Breakdown Characteristics of Hybrid Gas Insulated Transmission Line (H-GIL)", *Proc. of 10th ISH*, Vol.2, pp.5-8, Montreal (1997)

INFLUENCE OF TRANSIENTS ON BREAKDOWN PROBABILITY OF FIXED DEFECTS IN SF_6

Sander Meijer, Robert G.A. Zoetmulder, Paul V.M. van Nes, Johan J. Smit[1]

1. INTRODUCTION

SF_6 gas has nowadays established itself as a reliable insulating medium for gas-insulated switchgear (GIS). In GIS however different kind of defects may occur, such as protrusions fixed to the conductors. Such defects locally enhance the electric field, which can result in partial discharge (PD) activity or even breakdown.

For the application of condition based maintenance (CBM) on GIS, it is important to be able to assess its insulation condition. Aspects which have to be taken into account are sensitive detection of partial discharge activity, identification and location of the discharging source and probability for breakdown.

In the last decades different PD detection methods such as VHF, UHF and acoustic detection techniques have been successfully applied. Computer based classification of the different defects which can occur in GIS is well established. Using time-of-flight measurements, the location of the defect can be determined. However the criticality of the defects is also of importance for CBM purposes.

In this paper, the criticality of fixed defects has been investigated. For this purpose, AC voltages to study the breakdown probabilities under pure AC condition have been applied. Secondly, the defects have been tested with standard lightning impulses (LI) and switching impulses (SI).

[1] Sander Meijer, Robert G.A. Zoetmulder, Paul V.M. van Nes and Johan J. Smit, Delft University of Technology, High Voltage Technology and Management, Delft, The Netherlands.

Gaseous Dielectrics X.
Edited by Christophorou *et al.*, Springer, New York, 2004.

2. MEASURING SETUP

For the tests, a small test vessel has been used, see figure 1. During AC testing partial discharge measurements have been performed using a detection circuit according to the IEC 60270 recommendations with a 1000 pF coupling capacitor. The noise level was below 0.5 pC. Moreover, a UHF PD detection circuit was used consisting of an internal disk antenna, pre-amplifier and spectrum analyser. Results have been measured up to breakdown.

During lightning and switching impulse testing only two stages of the impulse generator were used allowing small voltage steps.

Figure 1: Photograph of the measuring set-up: a) LI testing and b) during SI testing.

Inside the test-vessel, protrusion of 1, 2 and 3 mm length have been tested in a plane-plane configuration, see figure 2a. Figure 2b shows a detail of the investigated protrusion, which is 1 mm thick and has a tip-radius of 70-90 μm. The distance between both electrodes is 5 mm. The voltage is applied to the upper electrode; the lower is connected to the earthed enclosure. The gas pressure was 4.5 bar pure SF_6.

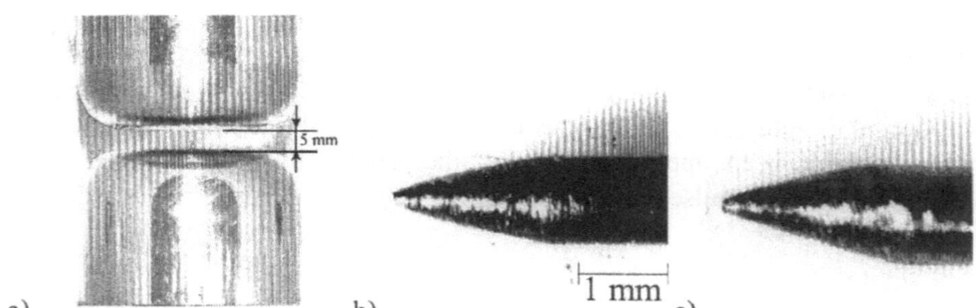

Figure 2: Detail of the test arrangement: a) electrode configuration with protrusion, b) protrusion before testing and c) protrusion after testing.

3. MEASURING RESULTS

3.1 AC testing

Table 1: The breakdown voltage and its standard deviation as measured for different protrusion lengths.

	$\hat{U}_{bd}$ [kV]	St.Deviation [kV]
1 mm	91.7	1.75
2 mm	54.3	0.70
3 mm	40.7	1.59

The 50 Hz AC breakdown voltage level was determined by increasing the voltage level slowly until breakdown occurred. This procedure was repeated 5 times. Table 1 presents the results for the mentioned protrusion lengths. As expected, the breakdown voltage increases with decreasing protrusion length.

Figure 3a shows the breakdown probability for the three different protrusions as function of the electric field strength. The nominal operating field strengths of coaxial GIS systems at the conductor (E_{inner}) and at the enclosure (E_{outer}) are also indicated. It is clear that the breakdown field of these protrusions are well above the nominal operating field strengths. However, as indicated in figure 3b, the PD inception for the 2 and 3 mm protrusions are in the region of the nominal field strength at the inner conductor, which means PD activity under normal operating conditions.

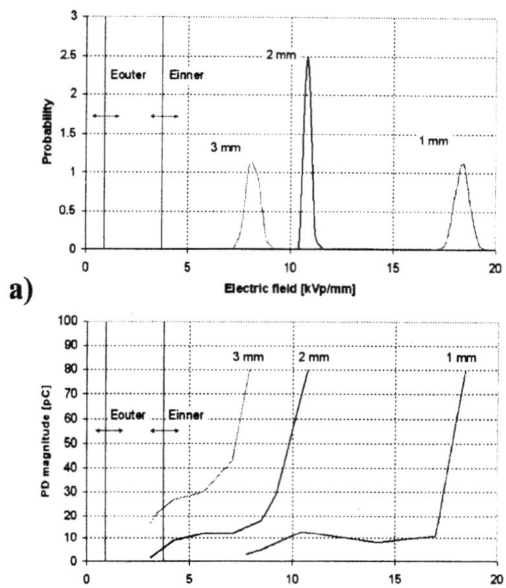

Figure 3: Measuring results obtain with AC voltage: a) Influence of the protrusion length on the breakdown probability; b) the PD magnitude as function of the electric field strength.

Breakdown, however, is not expected unless the operating voltage is increased to 2.2 p.u. in case of the 3 mm protrusion and even more for the 1 and 2 mm protrusions. It is known that in case of a single-line-to-ground fault a maximum temporary over voltage of 1.8 p.u. can be expected[1]. So for AC voltage application and temporary AC over voltages no breakdown is expected for the defects investigated.

As shown, for protrusions of 2 mm and longer, the PD inception voltage is in the same order of the operating voltage. It can be concluded that such protrusions can be detected using PD detection. In laboratory conditions, where a noise level of less than 0.5 pC can be reached, this can easily be done. However in the field, a PD detection circuit according to the IEC 60270 is more difficult, due to noise. To solve this problem and to reach sufficient sensitivity, the ultra-high frequency (UHF) detection technique was developed and introduced some 15 years ago[2].

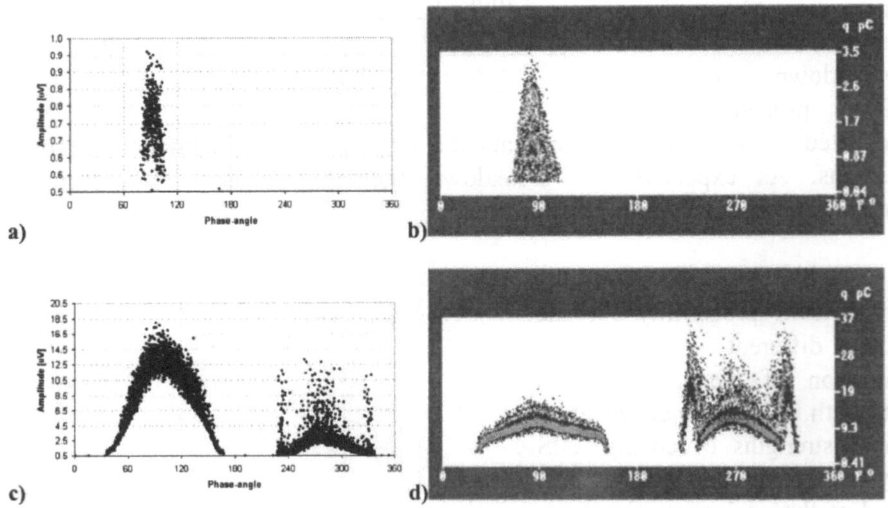

Figure 4: Examples of measuring results for a 2 mm protrusion

- obtained using the UHF technique with a centre frequency of 568 MHz at
  a) 11 kV and
  c) 30 kV
- obtained using a detection circuit according to the IEC 60270 recommendations at
  b) 11 kV and
  d) 30 kV

During the laboratory experiments, simultaneous measurements have been performed using both techniques. Examples of measurement results are shown in figure 4.

Although the shape of the patterns looks similar, a difference in amplitude is clearly visible for figures 4c and 4d. According to the calibrated IEC 60270 the PD magnitude around the negative peak is in the order of 28 pC and around the positive peak around 15 pC. However, detected with the UHF technique, the magnitude around the positive peak is bigger (16 μV) than around the negative peak (13 μV). This can be explained by the fact that the pulse-shape of positive and negative PD pulses differs[3]:

A: if the protrusion is negative, a short steep pulse occurs;

B: if the protrusion is positive, a longer less steep pulse occurs.

In case A, the steep pulse results in more UHF signal compared to case B. However, the pulse in case B is longer compared to case A that gives a higher PD magnitude in pC.

3.2 Lightning Impulse Testing

The U_{50} breakdown values for lightning impulse were determined by using the up-and-down method. Table 2 presents the results for different protrusion lengths and for positive and negative voltage application.

Table 2: The U_{50} breakdown voltage and its standard deviation as measured for different protrusion lengths.

	U_{50+} [kV]	St.Dev. [kV]	U_{50-} [kV]	St.Dev. [kV]
1 mm	103	2.8	91	17.0
2 mm	62	2.1	46	2.7
3 mm	49	1.4	35	1.7

3.3 Switching Impulse Testing

The U_{50} breakdown values for switching impulse were determined by using the up-and-down method. Table 3 presents the results for different protrusion lengths and for positive and negative voltage application.

Table 3: The U_{50} breakdown voltage and its standard deviation as measured for different protrusion lengths.

	U_{50+} [kV]	St.Dev. [kV]	U_{50-} [kV]	St.Dev. [kV]
1 mm	108	2.2	99	6.0
2 mm	61	1.3	55	1.4
3 mm	46	1.4	43	1.4

4. SUMMARY

In the previous sections small protrusions have been investigated under different voltage application. Table 4 shows the breakdown voltage level for AC, LI and SI in p.u.-values.

Table 4: The 100% breakdown voltage for AC, LI and SI in p.u for different protrusion lengths..

	AC [p.u.]	LI+[p.u.]	LI-[p.u.]	SI+[p.u.]	SI-[p.u.]
1 mm	4.9	5.9	7.4	6.4	6.4
2 mm	2.9	3.8	2.9	3.5	3.2
3 mm	2.2	2.9	2.2	2.9	2.6

Table 5 shows over voltage levels that can occur under different operating conditions such as switching or under the influence of lightning or faults.

Table 5: Possible over voltage levels in p.u. under different conditions.

	Over voltage [p.u.]
Single-line-to-ground fault (AC)	1.1-1.8
Switching	1.3-2.7
Lightning	3.4

In case of AC overvoltages, maximum levels of 1.8 p.u. have been reported. Comparison with table 4 shows that these levels are well below the 2.2-4.9 p.u. breakdown voltage for the investigated protrusions of 1-3 mm lengths.

In case of switching a maximum level of 2.7 p.u. has been reported. This means that 3 mm protrusions will not result in breakdown during pure AC application, but switching will trigger the breakdown.

For lightning impulses, levels around 3.4 p.u. can be expected. It can be concluded that in such cases, protrusions of 2-3 mm length will trigger breakdown[4].

5. CONCLUSIONS

From the experiments described in this paper, the following conclusions can be drawn:

1) Protrusions fixed to the conductor of a GIS of 1-3 mm length are not critical under nominal voltage application or temporary AC overvoltages;
2) Protrusions of 3 mm and longer are critical in case of switching overvoltages, protrusions of 2 mm and longer are critical in case of lightning overvoltages;
3) Critical defects longer than 2 mm give partial discharge activity under nominal voltage operation.
4) Protrusions which may cause breakdown by impulse overvoltages can be detected under normal operating voltage using PD detection techniques.

6. REFERENCES

1. F.H. Kreuger, Industrial High Voltage, Vol. 2, Delft University Press, 1991.
2. B.F. Hampton, R.J. Meats, Diagnostic measurements at UHF in gas insulated substations, IEE Proceedings, Vol. 135, Pt. C, No. 2, pp. 137-144, 1988.
3. M. Albiez, Partial Discharge measurements on SF_6 insulated switchgear, PhD Thesis, 1992 (In German)
4. K. Feser, R. Kurrer, Insulation Monitoring of Gas-Insulated Substations with Respect to Life Cycle, Tutorial In-Service Diagnostics of High Voltage Power Apparatus, ISH 1997.

EVALUATION OF INTERRUPTION CAPABILITY ON VARIOUS GASES

K.Mochizuki[1], T.Ueno[1], H.Mizoguchi[1], S.Yanabu[1], S.Yuasa[2],and S.Okabe[2]

1.INTRODUCTION

SF_6 gas is widely used as interrupting medium for disconnecting switch and circuit breaker in power system whose voltage is above 72kV or medium voltage. Global warming effect of SF_6, however, is well known and emission of this gas is strictly limited by regulation. As for research to find out substitute for SF_6 gas, many works have worked out. Gleitzes [1] compared with SF_6 and N_2 and found out characteristic of dv/dt that SF_6 was about three times larger than that of N_2 Mr. Grant [2] examined dv/dt to the mixture ratio of N_2-SF_6 at the total pressure 5bar, 6bar, and 7bar of di/dt=9 A/μ s for SLF interruption performance. It was shown that interruptible dv/dt became N_2-SF_6 gas mixture of 65% to 90% had higher than pure SF_6 gas, and this will be explained by synergistic effect. Also for He-SF_6 gas mixture its rate varied from 25%,50% to 75% and interruptible dv/dt characteristic were observed higher for gas mixture compared with pure SF_6. Lee[3] also carried out experiment and showed that the higher the mixture ratio of SF_6 with N_2 and He higher the interruptible current and that of He mixture was always inferior to that of N_2 Mr. Nakagawa [4] measured di/dt, which could interrupt the mixed gas when increasing the pressure of N_2 with a constant SF_6 pressure of 3 bars. As a result, if the pressure of SF_6 was the same, it was shown that increasing of total pressure became higher, interruption performance would be smaller. As mentioned before interruption and insulation abilities with various gas and their mixture have been studied by many researchers, but research especially for interruption was not done sufficiently. SF_6 gas including fluorine or chlorine gases was studied sufficiently but all gases have global warming effect or ozone problem. Natural gases such as H_2, He, Air, authors believe, must studied more in details and in our paper we study natural gas and their mixture of SF_6 using same interrupting chamber for standard. Interrupting performances

[1]Tokyo Denki University, Tokyo, Japan [2]Tokyo Electric Company, Yokohama, Japan

compared each other. In this experiment, CO_2 , Air , He and SF_6 and their mixture with SF_6 are used. And Mayr arc time constant and power loss coefficient are obtained from observation results and with EMTP-Models [5] we can simulate interruption ability of short line fault and comparisons are made.

2.MEASURING CONCEPT TO OBTAIN ARC PARAMETERS

The calculating method using vibration of a conductance was used. The equation of the arc by Mayr is expressed with formula (1).

$$\frac{1}{g}\frac{dg}{dt} = \frac{1}{\theta}\left\{\frac{vi}{N_0} - 1\right\} \cdots\cdots\cdots\cdots (1)$$

v : Arc voltage, i : Arc current, g : Arc conductance, θ : Time constant,
N_0 : Power loss coefficient

High frequency oscillation is generated just before current zero on power frequency current. To make the high frequency oscillation, a circuit composed of series connected reactance L and capacitance C. Current and arc voltage were measured and it found that the waveform of the arc conductance corresponding to high frequency vibration. It found that time constant and power loss coefficient in a formula (1), and they were expressed as a function of an arc conductance. [6]

3.EXPERIMENT EQUIPMENT AND CIRCUIT

Configuration of experimental equipment was shown in Fig.1. The value of the pressure gauge of upstream vessel was kept at 0.2MPa-a, which was supplied from reserve tank. The movable contact opened by tension spring travel length was about 20mm and the mean speed was adjusted to 1.0m/s. In addition, the distance between movable contact and nozzle was set to be 10mm at the final position of opening operation. The material of nozzle was PTFE thickness was 10mm and whole diameter was 5mm. The movable contact and fixed contact were made from Cu (20%)-W(80%) and these diameters were 4mm and 7mm respectively. Experiment circuit was shown in Fig.2. Current was supplied by C, L resonance circuit composed of C=0.0815F and L=6.0mH.

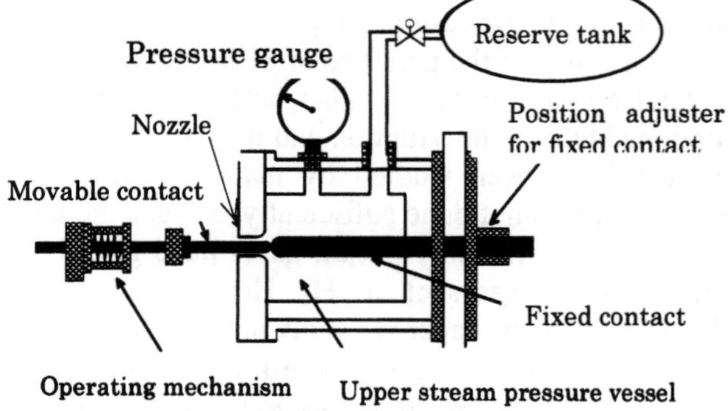

Fig.1 Experiment equipment

In this experiment, C_2 and L_2 were connected to equipment in parallel in order to generate oscillating current. The values of C_1 and C_2 were fixed to 0.5 μ F and 0.025 μ F respectively. And the frequency of oscillation current was set to be 45kHz, 100kHz and 180kHz by adjusting value of L2. The expansion waveform of arc current and arc voltage near current zero were shown in Fig. 3.

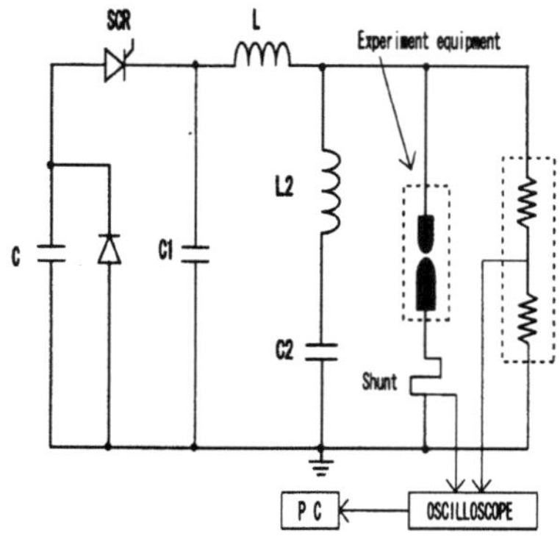

Fig.2 Experiment circuit

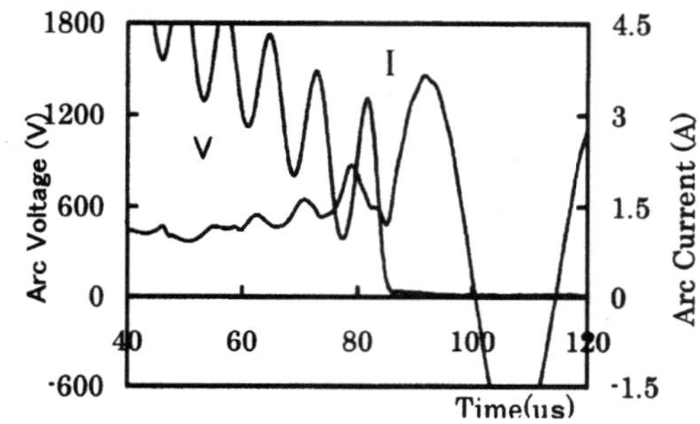

Fig.3 Arc Current, Voltage waveform nearby Current Zero

4.EXPERIMENT RESULT

The interruption performances were first evaluated by this research for Air, CO_2, He, SF_6, and the mixed gas (pressure ratio of SF_6: 50%) of them. In this experiment, the blasted pressure was fixed to be 0.2MPa-a.This Fig.3 showed the arc time constant of these gases vs. conductance. Fig.4 was calculated the vibration near current zero. And the oscillation frequency was 45 kHz. From the result of this figure, the arc time constant SF_6 was smaller and power loss coefficient was bigger than pure CO_2, Air He gases. And interruption capability of SF_6 was better than SF_6 mixture gases. However, even if it changed oscillating frequency to 100kHz and 180kHz, the relation of the conductance vs time constant and loss coefficient of various gas did not change a lot. When pure gases were compared, the result showed that CO_2 was better than Air and He gases. But when pure gas was compared, SF_6-He gas became good for the next of SF_6. In the case of He gas, if the pressure ratio of He at 50% changed into a mass ratio, this ratio of SF_6 to He was 97.5%:2.5%. However, SF_6-CO_2 reversed the loss coefficient highly rather than SF_6-He. Also oscillating frequency was changed to 100kHz and 180kHz, time constant and power loss coefficient against conductance did not change significantly. Next, about CO_2 and Air, the interruption performance was evaluated for the mixture ratio of SF_6 as 100%, 50%, 20%, 10%, and 0%. The relation between a time constant and power loss coefficient were shown in Fig. 5 and 6 at the various gases.Arc time constant was decreasing and power loss coefficient was increasing with increasing SF_6 ratio.

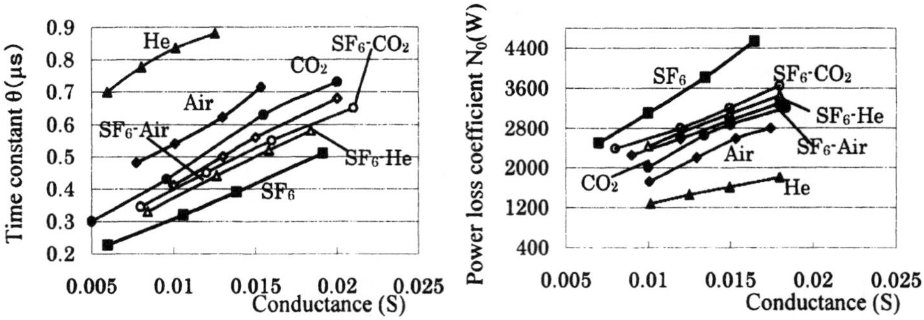

Fig.4 Conductance vs time constant and power loss coefficient on various gases (45kHz、0.2MPa-a)

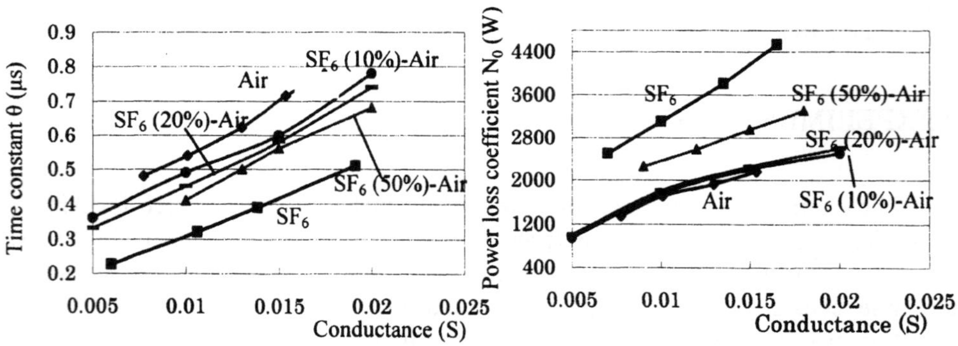

Fig.5 Conductance vs time constant and power loss coefficient with SF_6- Air mixing ratio (45kHz、0.2MPa-a)

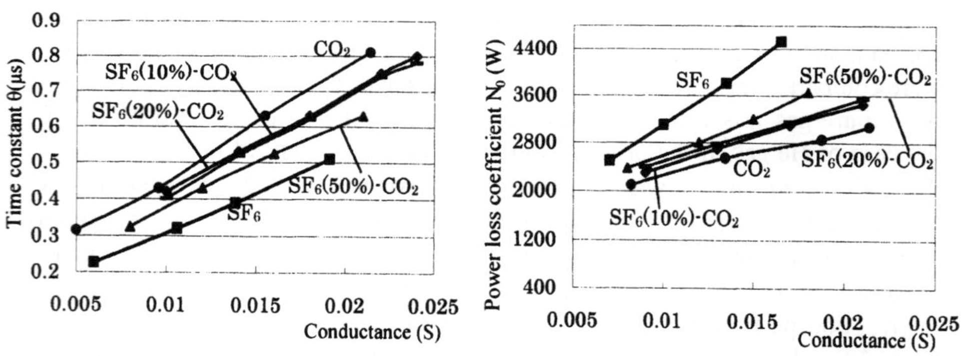

Fig.6 Conductance vs. arc time constant and power loss coefficient with SF_6- CO_2 mixing ratio (45kHz、0.2MPa-a)

5. SIMULATION USING BY EMTP

In Fig.7, simulation circuit was illustrated for short line fault of breaker. Surge impedance was taken as parameter and in order to obtain di/dt, the voltages of capacitor 30 μ F were changed, the success interruption of dv/dt could be obtained. In order to get di/dt and dv/dt, transient recovery voltage wave form at current zero point (several tens of micro second) were obtained and measured. EMTP-Models was used for simulation and wave shape was calculated by combining with arc parameters. When the charge voltage of a capacitor was raised, it could interrupt in the beginning. Furthermore, if it rises, post arc current appêar. After that, it would be interruption failure. Just before resulting interruption failure, di/dt,dv/dt of an interruption success limit were calculated.

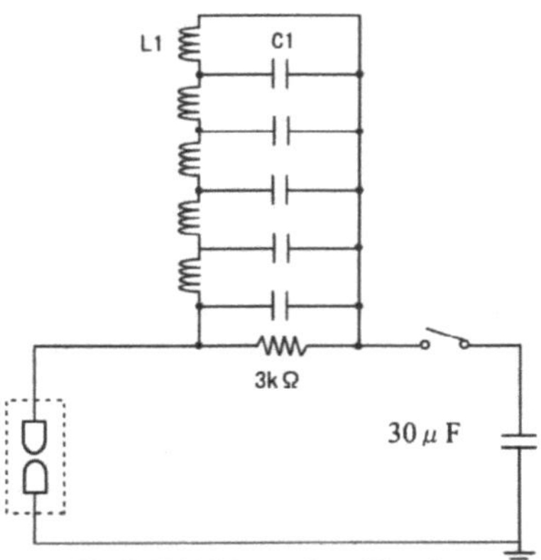

Fig.7 SLF Measuring Circuit

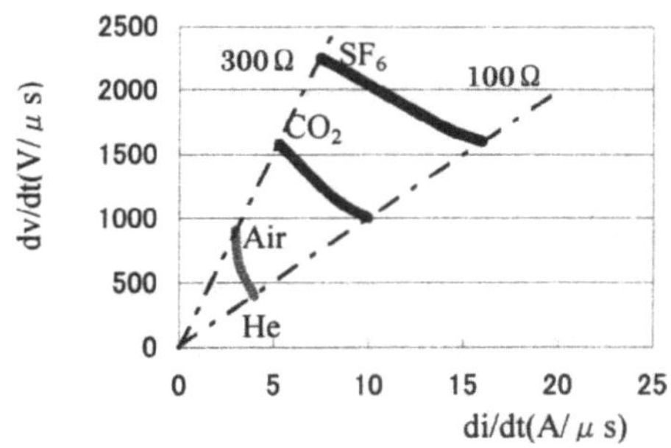

Fig. 8 di/dt-dv/dt characteristic

di/dt value was taken around from the current zero point and consequently dv/dt was also obtained from the voltage appeared at current zero.

Surge impedance was set to 450, 300, and 100Ω, and the result which calculated di/dt, dv/dt of an interruption limit about Air, CO_2, and helium and SF_6 is shown in Fig. 8.

6.DISCUSSION

(1) As a result of asking which gases were good for current interruption with a current vibration type method, it turned out that an extinction performance was good in the order of SF_6>CO_2>Air>He. When mixed gases were compared, the result with the performance was much better for large volume component of SF_6 rather than other mixed gases. The performance of SF_6-He by the pressure ratio was compared, He component was smaller from the mass ratio, and it turned out SF_6 had more influential.

(2) In the experiment, which changed the mixed ratio of mixed gas with SF_6, Air, and CO_2, and compared the extinction performance, the arc time constant was small at the time of SF_6 quantity was large, and loss coefficient became small. In the case of the specific rate of mixture, it was said some times, the destructive voltage came out by a discontinuous rise (Synergy effect), but the extinction performance in an experiment was changing continuously. It is the result of the ability saying that there was no synergy effect.

(3) By performing an EMTP simulation using an arc parameter, comparison of the SLF interruption performance of each gas was possible. It was appropriate for the same tendency might be appeared as gas constants. For the current vibrating method, interruption ability became SF_6>CO_2>Air>He

(4) When the performance of SLF interruption of CO_2 and SF_6 was compared with the simulation from calculation, it turned out that CO_2 served as with one half of dv/dt of SF_6 in the same di/dt, and served as one third of di/dt in the same dv/dt. Moreover, about Air and helium, they also turned out that their extinction performances were quite lower than CO_2.

Reference

[1] A.Gleizes, A.M.Rahal, H.Delacroix, P.V.Doan, "Study of a Circuit Breaker Arc with Self-Generated Flow:Part I -Energy Transfer in the High-Current Phase" IEEE Transactions on Plasma Science,Vol16,pp.606-614,1988

[2] D.M.Grant, J.F.Perkins, L.C.Campbell, O.E.Ibrahim, and O.Farish, "Comparative Interruption Studies of Gas-Blasted Arcsin SF_6/N_2 and SF_6/Hemixtures,"Proc.4thIntern.Conf.Gas Discharges,IEEConf.Publ.No143,pp48-51,1976

[3] A.Lee and L.S.Frost, "Interruption Capability of Gased and GasMixtures in a Puffer-Type Interrupter, "IEEE Trans.Plasma Science,Vol.PS-8,pp.362-367,1980

[4] Y.Nakagawa,M.Tsukushi,K.Hirasawa,andY.Yoshioka "Interruption Ability of SF_6/N_2 Mixtures and SF_6/CF_4 Mixtures in Puffer Type Gas Blast Circuit Breaker," Proxeedings4th International Symposium on Switching Arc Phenomena,September,22-24,pp210-217,1996

[5] T.Koshizuka, T.Shinkai, S.Nishiwaki, S.Yanabu, Y.Kawaguchi, "EMTP Simulation of SLF Interrupting Performance for Serially Connected Two Arcs with Different Arc Parameters", T.IEE Japan, Vol.121-B, No7, pp845-852 (2001)

[6] Okabe, Zaima, Yorozuya, Nishiwaki, Ishikawa, Nakamoto, Kosakada, "Simulation of High Frequency Arc Extinction on SF_6 Gas Circuit Breaker"T.IEE Japan, Vol11-B, No.11, pp.1233-1242 (1991)

SECTION 7: SF_6 INSULATED APPARATUS: DIAGNOSTICS

GASEOUS ELECTRONICS APPLIED TO THE DEVELOPMENT AND DIAGNOSTICS OF INSULATION SYSTEMS

Michel F. Fréchette[a]

1. INTRODUCTION

The author proposes a personal assessment of the progress in gaseous electronics and its impact on the development and diagnostics of insulating systems. The scope of the discussion is delimited by the major preoccupations experienced by a utility company over the last 20 years. These concerns shaped research activities that pertained to the selection and safe long-term use of high-voltage gas-insulated equipments in distribution and transmission networks. The commentary is developed along the author's own scientific work during the period 1980-2000. The Fleece[b] thread has been woven along the following four topics: collisional processes and electron swarms under uniform fields, avalanches under nonuniform-field conditions, gas breakdown phenomena, and partial discharges.

2. COLLISIONAL PROCESSES AND ELECTRON SWARMS

At the end of the seventies, in the midst of the first few symposia on gaseous dielectrics,[1] detailed knowledge of the dielectric behavior and macroscopic transport coefficients of SF_6 was still of practical interest to the electrical industry. A considerable amount of experimental data was being collected on sulfur hexafluoride.[2, 3] However, the volume of experimental work called for was ever-increasing in face of the wide scope of operating conditions and in view of the substantial interest in gas mixtures.[4] The SF_6 technology had just been introduced at the 800-kV class. At Hydro-Québec, the implementation of SF_6 technology progressed at a slow pace involving mainly the installation of stand-alone breakers in the transmission network. It was realized that this technology

[a] Institut de recherche d'Hydro-Québec, Varennes, Québec, Canada, J3X 1S1

[b] It was meant as a pun; the invited paper being presented in Greece, the word refers to Jason, the Argonauts and the Golden Fleece.

Gaseous Dielectrics X.
Edited by Christophorou *et al.*, Springer, New York, 2004.

was inadequate for outdoor applications at low temperatures while it was decided that the whole territory[5] would remain with a single specification relative to the temperature requirement of the equipments. The limit of $-50\ ^{\circ}C$ was about to be set, opening a field for the application of SF_6 gas mixtures. IREQ got involved around this time. With a limited effort, the assignment was to simulate and predict the breakdown characteristics of various gas and gas mixtures of interest.

Electron transport phenomena in complex molecular electronegative gases were currently subjects of growing interest to applied research workers in the field of high-voltage insulation. The relevance of macroscopic transport coefficients to dielectric strength had long been recognized but the description and prediction of dielectric behavior had traditionally been conceived in terms of empirical and semi-empirical formulae and/or simple criteria. Although these had proved useful for practical purposes, a more accurate and more general description was needed in the context of new gaseous dielectrics. Such a description would be based on microphysical material properties, i.e. collisional cross sections.

The primary advantage of this approach is to provide a basis for qualitative prediction of the insulating properties of gases and gaseous mixtures. Furthermore, knowledge of the representative set of cross sections, i.e. ionization, attachment, momentum transfer and the principal inelastic cross sections, enables quantitative calculations of the integral coefficients using a Boltzmann equation or the Monte Carlo method. On the other hand, some unknown cross sections can be determined by fitting of the theoretical and experimental macroscopic coefficients. Of the seminal pathway along these lines, the pioneering works of Phelps and co-workers[6, 7] and Nighan[8] should be mentioned. The aim of our study was to evaluate the sets of effective cross sections for multiple gases using the Boltzmann approach. After a while, it appeared a too formidable task, and our contribution limited itself to the following cases: SF_6 (sulfur hexafluoride), CCl_2F_2 (dichlorofluoromethane), c-C_4F_8 (perfluorocyclobutane) and to use the representative sets to calculate the macroscopic coefficients of insulating mixtures: SF_6/N_2, CCl_2F_2/N_2, c-C_4F_8/N_2, CCl_2F_2/SF_6, $CCl_2F_2/SF_6/N_2$.

Detailed knowledge of the dielectric and transport properties of SF_6 and its mixture with N_2 was of substantial practical interest and had been treated previously by Kline[9] and Yoshizawa[10] in the year 1979. However, some opened questions remained. What was a representative set of cross sections of SF_6? What readjustments, if any, of the experimental cross sections were required to obtain agreement between calculated and measured macroscopic coefficients? In our study related to SF_6, it was shown for the first time that reasonable agreement could be obtained between experiment and theory without *ad hoc* readjustment of the cross sections.[11, 12] Macroscopic coefficients for the SF_6/N_2 were predicted and reproduced with satisfactory agreement.[12, 13] The amalgam of theory and experiment was realized in 1987 with the subtle explanation[14] of the limit-field behaviour for the mixtures SF_6/N_2, CCl_2F_2, CCl_2F_2/SF_6, and $CCl_2F_2/$ SF_6/N_2 in the framework of a Boltzmann-equation analysis.

Dichlorofluoromethane (CCl_2F_2) was an other strong electronegative gas and had been identified as a potential component for insulating gas mixtures. There was a certain quantity of experimental data pertaining to its transport properties but, not much was known about its electron-molecule scattering cross sections. Our theoretical results were obtained concurrently to the work of Okabe and Kouno[15] who then had proposed an effective set of cross sections for CCl_2F_2. This task was of great complexity since only ionization and attachment cross sections were known, leaving both the momentum transfer and the inelastic collisional cross sections to be determined. Consequently, the only possible test consisted in the prediction of the attachment coefficient.[16] Prediction of the transport coefficients of the CCl_2F_2/N_2 mixtures constituted a further test. Our measurements were found to be quite consistent with theoretical calculations.[17]

Since considerable progress was achieved, modeling was fully developed, the theoretical approach was mastered, the steady-state Townsend experimental method was fully operational, in particular prediction of synergistic effects had been forecast and verified, the research program was then extended to an exotic gas, i.e. cyclo-C_44F_8, to demonstrate the general applicability of the method of analysis. But lack of data concerning both collisional processes and transport coefficients had rendered this task somewhat arbitrary. At least, the knowledge of ionization, attachment and momentum transfer cross sections extending over a reasonable range of energy, and the availability of a representative set of macroscopic transport data is not only necessary but barely sufficient for achieving an effective set of collisional cross sections. Useless to say that the study involving c-C_4F_8 was highly exploratory.[18]

It soon became clear that SF_6 was and was to remain the optimal dielectric for electrotechnical applications.[19, 20] This and the fact that no limitation on space was constraining Hydro-Québec to accelerate the implementation of SF_6 high-voltage substations have put a term to the rationale sustaining this research program. Some particular applications required the introduction of gas mixtures. In that case, the buffer gas was N_2 or a slightly electronegative gas CF_4. The work of Phelps and Van Brunt[21] put a close to the debate, as being the last published work on SF_6 in this period. Further details can be found in the author's thesis.[22] The study of electron-molecule interactions has continued for other applications and needs, other than those of the electrical industry.

3. AVALANCHES UNDER NONUNIFORM-FIELD CONDITIONS

For about 5 years and reported[23] on as early as 1993, we studied avalanche phenomena in various conditions. Risk of restriking in distribution interrupters was under scrutiny. In order to grasp some of the physical mechanisms involved, a doubly difficult approach was conceived. An avalanche was generated by the release of a large amount of photoelectrons that grew under a geometrically nonuniform field. Both experimental conditions would contribute to reproduce the reality of the situation associated with the equipment. This approach which provided the ability to experimentally produce rapid space-charge growth in a stressed gaseous dielectrics was of interest to simulate residual-charge effects during the recovery phase of an interrupting medium or to help in the assessment of a mixed-voltage situation, e.g. when both DC and time-varying voltages are superposed.

Basically, 3 situations were investigated for N_2 and SF_6: low-pressure conditions (e.g. 5 Torr),[24-28] atmospheric-pressure conditions (e.g. 760 Torr), and moisture conditions (up to 15%). It is clear that the resulting field-distribution has helped to evidence the avalanche growth in its early stage of development. At both low and atmospheric conditions, it was demonstrated that the breakdown voltage could be reduced under static fields, upon the photo-release of a large amount of initial electrons. Reduction in voltage withstand provides supporting evidence in favor of an active role of space charge during discharge development. Conditions are that of an under-voltage breakdown development.[29]

At low pressures, the time-resolved signals have exhibited particular features over the first transit time. In the early stage of development, the avalanche growth varied exponentially with time, showing a lower rate than that observed under uniform-field conditions. A Pulse-Townsend experiment may be given as a reference situation. This observation, plus the particular time-location of the delayed secondaries, concur in the supporting of an early buildup of space charge. Finally, the presence of depressed field-conditions towards the anode affected the registered waveform: Collisional attachment dominates in this region, resulting into a decrease of electron density. The rapid fall of the time-resolved signal might not be fully associated with neutralization at the electrode, the electron growth being quenched.

Study of electron swarms is usually performed at low pressure and uses a low intensity light pulse to avoid the build up of a too large avalanche size. A large avalanche would perturb the field uniformity. However, when this occurs, applied voltage and pressure permitting, it is usually contended that space-charge effects intervene in the late stage of electron arrival at the anode. For an intense source, space charge thus created close to the cathode may elicit the early manifestation of a locally perturbed field, and may upon circumstances lead shortly to breakdown.[29-31] It is the discharge characteristics prevailing to this temporal transition in atmospheric N_2 that is presented next. Increase of pressure would possibly help in generating substantial space charge at lower pressure-reduced field and would at the same time bring the situation closer to the reality of the HV equipment.

Figure 1 illustrates some results pertaining to atmospheric N_2.[32] The temporal varia-

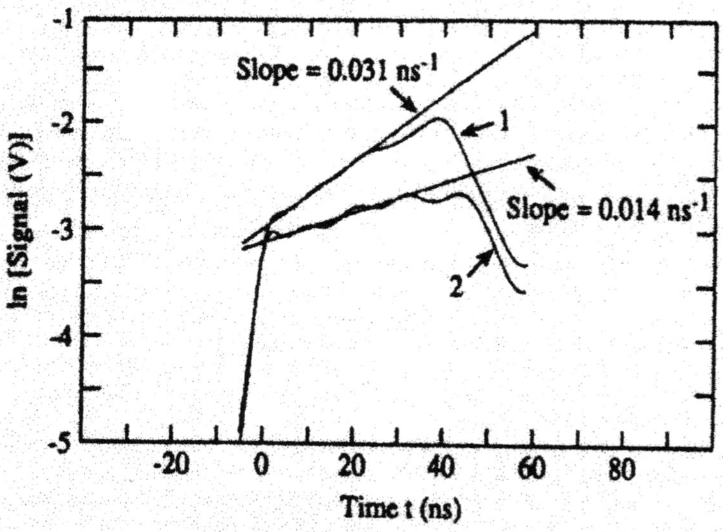

Figure 1. Electron temporal growth over the transit time; ln (signal) vs time; 1- 40.8 $Vcm^{-1}Torr^{-1}$, 2- 36.8 $Vcm^{-1}Torr^{-1}$, gap 0.5 cm, atmospheric N_2.

tion over the transit time is rather steep, steeper at the larger E/p-value. Despite growing in a geometrically decreasing field, the swarms exhibited an exponential growth, at least up to half the transit time. When displayed on a log-scale, the time-resolved signals varied linearly over the transit. This would indicate that the field close to the cathode has been enhanced due to space-charge effects, this occurring in the early stage of discharge development. A simple model, assuming a space-charge free environment, permits to interpret the constant slope as the product of the ionization coefficient times the drift velocity. The products derived from these exponential portions were found to lay considerably lower than those obtained under uniform-field conditions.

In the usual conditions, the avalanche growth in atmospheric SF_6 was found to be completely quenched. We had to recourse to artifices to finally obtained detectable signals.[33] Even with 15% water contamination, it was possible to extend the measurement pressure limit only from about 30 to 50 Torr.[34] It was shown in the end that the risk at atmospheric pressure in SF_6 for the discharge to restart once the contacts of the interrupters are opened would necessitate a highly nonuniform geometrical field where the onset would occur.

The scientific target was attained but a few other aspects could have been dealt with. This program lasted for about 1 more year, while it was documented and technical implications were drawn and discussed. This research activity stopped giving way to the concept of "necessary but sufficient information". International interest in the field diminished as well, Henk Gaxiola's thesis[35] was published in 1999 putting a term to the efforts of the other involved research group. It is worth underlying that the geometrical characteristics of the gap may serve in controlling both space-charge magnitude and location, illustrating the concept of the space-charge structural design. When quenched, the avalanche becomes a "partial avalanche", analogous to a partial discharge!

4. GAS BREAKDOWN PHENOMENA

Québec has a 4-season climate. During winter, especially over January and February, temperature may be quite low. January 8th of this year (2004), the temperature in Montreal reached –38 °C with the wind factor. In the eighties, Hydro-Québec used –50 °C as the downer temperature limit in its technical specifications for outdoor gas-insulated equipments. It was based on the observation of the temperature in the region of Fermont in the northern Québec, on the frontier to Labrador. This is the area of Manic V, the world's largest hollow body arch and buttress dam. Many generation-center 315-kV power lines gather for up transformation at 735 kV to reach the major consumption areas. Outdoor use at 735 kV and disengagement of the manufacturers towards air technology were the main reasons why SF_6-mixture technology was implemented for circuit breaking in the transmission network.

Late in the 80s, engineers and management at HQ were still questioning the situation. Only limited experience with outdoor SF_6/N_2 breakers was gained at Hydro-Québec. At the time (circa 1987), about 24 breakers were in service in the class 69 kV to 315 kV. Cost for low temperature testing at full scale was considerable; the necessary use of several chambers was also expensive. What was it really that prevented the development of a pure SF_6 breaker for low temperature application? Eventually, Hydro-Québec and GEC Alsthom would develop jointly a low-temperature 735-kV breaker. Early 1988, precise questions regarding the breakdown behavior of SF_6 were asked to us. After a brief review[36] and based on some physical considerations,[37] force was to conclude, due to conflicting experimental evidence, that a precise and definite answer could not be given.

We got support to solve this imbroglio and embarked on a program to resolve the situation. Much effort was devoted to the control[38-40] of experimental parameters so that it would be allowed in the end to draw a definite conclusion on the matter. At the core of this experimental approach was the concept of an insulation system consisting of geometrical fields (originating from the electrode profile), interfaces and a gaseous dielectric. The system forms a whole with the dielectric, i.e. an effect due to lowering the surrounding temperature may not be distinguishable from that of the insulating system. Early recognition of the potential role played by the degree of nonuniformity of the geometrical field provided guidelines for the definition of the experimental framework. Preference was given to an electrode arrangement that provided uniform-field conditions while allowing the interface to be active under certain conditions. This was achieved using high fields under negative DC polarity. The study thus undertaken, essentially experimental, relied mainly on the two following hypotheses: That the insulating system, not the dielectric, exhibits a temperature-dependent breakdown behavior, and that the cathode-gas interface comes into play.

The results[41, 42] have strongly evidenced the variance of the breakdown behavior of an SF_6-insulated system at subnormal temperatures. Despite a constant molecular density and under preferably uniform-field conditions, lowering the environmental temperature was found to bring about a reduction in strength of the withstand capability of the system. This decrease amounts to approximately 10%, appears at a temperature threshold around -25 to -30 °C, and remains constant down to -50 °C. No definite trend with regards to a temperature dependence was unraveled under high-field conditions. This leads to believe that the variance observed under highly uniform-field conditions may have to do with a triggered perturbation (field-related) linked to the modification of the cathode-surface state. Since the breakdown behavior of SF_6-insulated systems was shown to depend on the environmental temperature, the conclusion may be that the HV equipment design should take this into account because of the implicit reduction in the operational safety factor. In this regard, the findings set the order of magnitude for the withstand drop below -25°C, i.e. 10%, which provides a basis for deciding what measures to take. The fact that the effect was observed under negative DC voltage is not restrictive, whereas AC voltage includes a negative-voltage phase. These results were presented in the fall of 1992 and discussed along the ones obtained by Beroual and Buret.[43]

Low-temperature negative DC breakdown of an SF_6/N_2 gas-insulated system was investigated next.[44] The gas mixture studied had a 40% partial molecular density in SF_6. Low-temperature data sets showed as good statistical characteristics as the sets obtained at room temperature. The data sets were taken respectively at room and subnormal temperatures, in a consecutive order. Breakdown values were measured as a function of the same gap lengths while keeping the gas molecular density constant. This allowed a direct one-on-one comparison of the experimental breakdown values. Such an example is given in Fig. 2, where low-temperature breakdown voltage (here -50 °C) is compared with its room-temperature counterpart at a constant density of 3.252×10^{19} cm^{-3}. Conclusion is

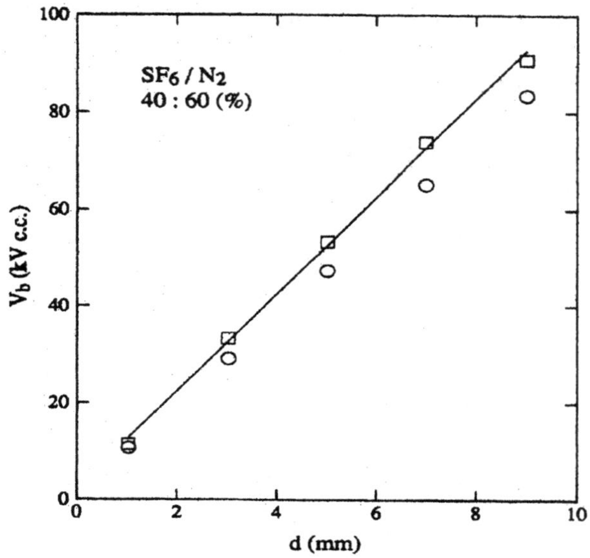

Figure 2. Breakdown voltage (V_b) as a function of gap length (d) for constant molecular density; the density is 3.252×10^{19} cm^{-3}, squares - breakdown data at 1.33 bar and 24 °C, and circles - values for 1.00 bar at -50 °C.

self-evident: Lowering the temperature results into a systematic reduction in breakdown strength of the insulated system, in spite of the fact that the gas density remained the same. Moreover, the effect is clearly present at each particular length. The magnitude of the drop varies slightly, averaging to a more or less constant percentage of approximately 10%. It is worth noticing that the absolute voltage deviations with respect to room-temperature data exceed substantially the corresponding standard deviations, underlying the unfortuitous nature of the observation.

5. PARTIAL-DISCHARGE PHENOMENOLOGY

Before a gas-insulated system ends up to breakdown, a series of events precedes. For a utility company, it is important to gain experience of these events, in order to understand the various pathways leading to the deterioration and ageing of the insulation conditions. The study of these phenomena is also essential if there is to be involvement into the development of a diagnostic system. Although these phenomena exhibit electrical, optical,[45, 46] ultrasonical[47] properties, only the former ones are referred to here. Furthermore, the pre-breakdown characteristics may be the results of various interplays.

In our earlier studies, it appeared clearly that one of the major drawbacks with recording the electrical signals associated with partial discharges was the sampling. Indeed, statistical analysis based on randomly acquisitioned numerous single pulses does not reflect the total complexity of the phenomenon.[48, 49] But at the time, circa 1991, other concerns prevailed. A study was started with the aim of determining the effect of water vapor on the partial-discharge activity. Planning of an increased implementation of SF_6 interrupters in the distribution network (25 kV) was effective, and standardization was under way. One of the selling aspects of the technology was its maintenance-free status, and this was questioned by the distribution management. Furthermore, some of the equipments would be operated in the underground network where they could be submerged. Besides, in spite of the use of molecular sieve, the gas was found sometimes to be rather humid, 800 ppm_v. An experiment was conceived using the single-pulse sampling and, in spite of its limitations, allowed to conclude[50, 51] that the insulation integrity was maintained at least up to a moisture content reaching 2000 ppm_v.

While monitoring of the condition-state of various equipments was becoming of interest for the utilities,[52] it warranted the introduction of a different measuring technique that consists of acquisitioning continuous large time frames (typically of the order of ms) of digitized information. These time frames feature high-resolution (typically of the order of ns) information on the signal, offering at the same time an access window on the pulse-to-pulse relations. This was achieved in the mid-90s[53, 54] and applied to the perturbation of the Trichel regime in air with the adjunct of small relative concentation of SF_6. The effort to circumscribe an adequate methodology of work permitting sensible interpretation of partial-discharge phenomenology was pursued. This period corresponds to a time where benchmarking of exploratory research became a must. The situation developed into a collaboration with a Japanese manufacturer, Toshiba Corporation, and a sharpened focus of the program research.

The recognition of the detrimental potentiality of partial discharges has elicited much effort along various relevant aspects of the research field.[55] Of the many areas, the present contribution sides with the preoccupation of interpreting data with an aim of predicting the time-evolution outcome of PD phenomena once set in and concerns on-line monitoring of the insulation state. Assuming the presence of a minor fault, assuming further that the resulting PD activity exhibits measurable distinctive patterns, how can these signals be used to predict and/or interpret their time evolution ? When to decide on carrying out maintenance or some intervention to avoid a major fault? The present developed methodology[56] connects with complementary experimental approaches[57-59] with a marked different emphasis on interpulse characteristics via a high time-resolution. At the core of the original methodology is the seeking of the definition of a discharge regime: A set of discharge parameters or an unraveling of some significant parameters associated with the erratic and random behavior of the discharge. It is contended here that the stochastic nature of the phenomena involved[60] can be exploited by providing a gain of information contributory to the definition of a discharge regime. Van Brunt's introductory lead into the subject[55, 61, 62] is considered, though here the concept of stochascity is taken loosely referring to memory effects diverse in origins. Pulse characteristics are in part affected by their past history. The intermittent shielding due to space charge and moving charged clouds plays certainly a role. But, often blurred by the measuring technique, the stochastic nature of the whole phenomenon, including the condition state of the various constituents of the insulation system (e.g. emission surfaces) and deviation from initial conditions (e.g. erosion of active electrodes) shows more complexity that can be handled. However, it is believed that the wealth of information hidden may be of great help.

The choice of particular experimental conditions has helped greatly in the demonstration of the present methodology. We have in mind here the fact that recordings of the electrical signals are basically taken at corona threshold. This context brings about the production of erratic discharges of low intensity, featuring the discharge conditions of a minor defect. Equally, this context produces low-repetition rate discharging, well defining individual discharge events. It also allows to observe time-confined sequences of pulses and facilitates the unraveling of stochastic effects. So far we have made substantial progress in the essentials of the present methodology. Despite the recording of a wealth of information, we have managed to keep in the ballpark of a few indicative or key parameters to describe the discharge regime. The concept of a discharge regime, that is the existence of a representative ensemble of parameters associated with the discharge, was demonstrated successfully.[56, 63, 64] It is essential that the parameters of such a regime be established, if monitoring of change over time is sought eventually. Situations involving gases of different nature were resolved, whereas the methodology unraveled and quantified respective phenomenological patterns and specific parameters.[65-67] A discharge situation involving the interface between solid and gaseous insulation was also explored in which air and SF_6 were selected as to offer direct comparison with results from the earlier studies involving gaseous insulation alone.[68, 69]

In the present example,[69] measurements were also taken at corona threshold. Erratic discharges of low intensity were produced in compressed air at 500 kPa using an electrode configuration simulating a minor defect. Alongside a point electrode and partially in contact with it, is the solid bulk surface (epoxy). The whole gap length fixed at 5 mm, defined by the distance between the point electrode end and facing plane, is constituted by the interface between the gas medium and the solid material surface. DC voltage was used, and the system run under negative corona conditions. Applied voltage was set approximately at corona threshold and maintained constant for the duration of the measurements. When discharges were produced along the gap, a discharge regime could be described based on the recording and analysis of digitized real-time sequences. It was established that the discharge regime thus observed could be greatly modified, once the insulation system was submitted to partial discharge over time. Indeed our methodology of work allowed to detect and describe the new discharge regime. The occurrence of the change was observed to be more drastic than progressive. The new regime showed a marked presence of discharge events consisting of isolated single pulses and short pulse sequences. These short sequences would be non-equilibrium discharge events as shown in Fig. 3.

6. CLOSING COMMENTS

We have presented several studies of applied gaseous electronics contributing to the development and/or diagnostics of HV equipments. It has been recognized that sometimes the borderline between development and diagnostic is thin. Peculiar items or measures taken care of during development may preclude the need for a diagnostic, or indicate the dominant parameters to be monitored, or fix the maximum levels or limits for fault to become likely. Understanding may also limit the need for a diagnostic, or at least limit the extent of it, or facilitate its interpretation. Albeit the concerns at the source may be different, it remains a business of developing knowledge to assist decision-making processes. This precious knowledge can be used (A utility is a user!) to elicit the eclosion of a particular equipment (It may be developed jointly), to assess a risk (e.g. fault, health,

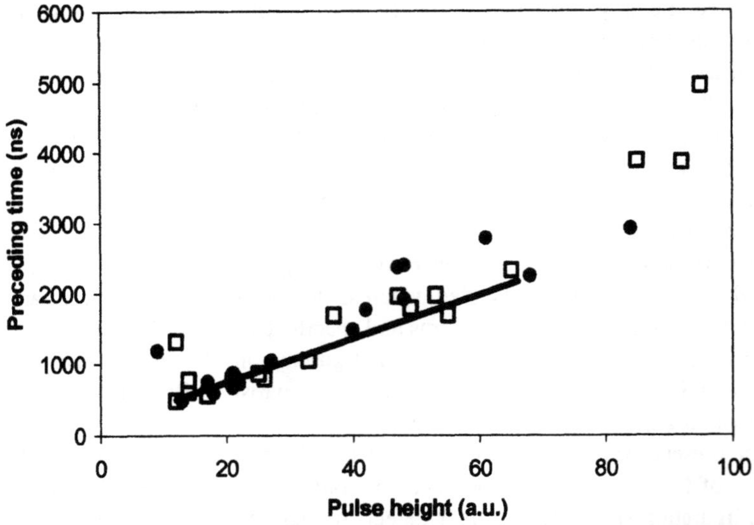

Figure 3. Preceding time versus pulse height, in air (5 bars) with epoxy; Full line - Long sequence, circles and squares – short sequences.

environment), to develop a new or more adapted standard, to provide guidelines, to recommend equipments suitable to the network operations, to set limit values of parameters relevant to insulation integrity and maintenance. Support for the research activities is eventually and theoretically driven by the state of "necessary but sufficient information", a state-of-situation reflecting the perception that the minimum limit of information has been reached. The follow-up steps will depend greatly on the technical gains and profitability and how it is assessed. This nowadays exercise would have killed 20th century progress as a whole.[70]

Although no recent management decision was taken to implement new diagnostics to assess the insulation state and integrity (It is a trait of research to be advanced compared to the need ...), much work has been achieved so far. Besides partial discharges[71] and avalanches under moist conditions, SF_6 by-products formation and toxicity were investigated with emphasis on the mitigation aspects.[72-78] The administrative worries behind these studies stemmed from an increased use of SF_6 in the distribution network causing a dispersion and penetration in the public. The only research activity remaining in the field of gaseous-insulation condition state concerns the development of a general metrology[79-81] to detect and quantify polar contaminants in a lesser polar gas background. For the moment, it is applied to moisture in SF_6 and seeks to produce an absolute measurement.

Finally, a gas-insulated system forms a whole and comes with solid insulation as well. Introducing new gases, as it was the case at Hydro-Québec when SF_6/CF_4 mixtures were selected as an interrupting medium in the mid 90s or when N_2/SF_6 mixtures with low content of SF_6 are chosen for GIL applications, often requires that material compatibility be established. Considering the nature of the gas mixtures, the heavy ions could have deleterious effects, and these were investigated.[82, 83]

The SF_6 technology has reached maturity. The environmental situation has been stabilized.[19, 20] It is our strong feeling that the ongoing developments based on gas mixtures will not require as much effort. Much has been understood, and these are variants. However, there are promising approaches that mix solid and gaseous insulation for current interruption.[84]

ACKNOWLEDGMENTS

The author is very much grateful to the researchers, graduate students, and technicians who have accompanied him throughout these research activities. Contributions from the senior technologist R.Y. Larocque were remarkable and are specially acknowledged. Special thanks to G. Bélanger, Manager Chimie et Matériaux, for encouragement and support. Thank you all !

REFERENCES

1. Proceedings of the International Symposium on Gaseous Dielectrics, Knoxville, Tennessee, USA, March 6-8, 1978, Ed. L. G. Christophorou, 454 pages. A total of 5 were held between 1980 and 1990.
2. C. M. Cooke and A.H. Cookson, The nature and practice of gases as electrical insulators, IEEE Trans. Electric. Insul. EI-13, 239 (1978).
3. N.H. Malik and A.H. Qureshi, Breakdown mechanisms in sulfur-hexafluoride, IEEE Trans. Electr. Insul., Vol. EI-13, No. 3, pp. 135-145 (1978).
4. M.O. Pace, L.G. Christophorou, D.R. James, R.Y. Pai, R.A. Mathis, and D.W. Bouldin, Improved unitary and multicomponent gaseous insulators, IEEE Trans. Electr. Insul., Vol. EI-13, No. 1, pp. 31-36 (1978).
5. The province of Québec is located in the northeast of the North American continent, Canada's largest province covers an area of 1,667,926 km^2. See www.bonjourquebec.com.
6. L.S. Frost and A.V. Phelps, Rotational excitation and momentum transfer cross sections for electrons in H_2 and N_2 from transport coefficients, Phys. Rev. 127, 1621 (1962).
7. A.G. Engelhardt and A.V. Phelps, Elastic and inelastic collision cross sections in hydrogen and deuterium from transport coefficients, Phys. Rev., Vol. 131, pp. 2115-2128 (1963).
8. W. L. Nighan, Electron energy distributions and collision rates in electrically excited N_2, CO, and CO_2, Phys. Rev. A, Vol. 2, pp. 1989-2000 (1970).
9. L.E. Kline, D.K. Davies, C.L. Chen and P.J. Chantry, Dielectric properties of SF_6 and SF_6 mixtures predicted from basic data, J. Appl. Phys. 50, 6789 (1979).
10. T. Yoshizawa, Y. Sakai, H. Tagashira and S. Sakamoto, Boltzmann equation analysis of the electron swarm development in SF_6, J. Phys. D.: Appl. Phys., 12, 1839 (1979).
11. J.P. Novak and M. Fréchette, Calculation of SF_6 transport coefficients from revised data, J. Phys. D: Appl.Phys., 15 L105-110 (1982).
12. J.P. Novak and M.F. Fréchette, Transport coefficients of SF_6 and SF_6-N_2 mixtures from revised data, J. Appl. Phys. 55, 107 (1984).
13. M.F. Fréchette, Experimental study of SF_6/N_2 and SF_6/CCl_2F_2 mixtures by the steady-state Townsend method, J. Appl. Phys. 59, 3684 (1986).
14. M.F. Fréchette and J.P. Novak, Limit-field behavior of various gas mixtures discussed in the framework of a Boltzmann-equation analysis, IEEE Trans. on Electrical Insulation, EI-22, 691 (1987).
15. S. Okabe and T. Kouno, Analysis of discharge properties in CCl_2F_2 by the Boltzmann equation method, Jpn. J. Appl. Phys., Vol. 24, 1335 (1985).
16. J.P. Novak and M.F. Fréchette, Collisional cross sections of CCl_2F_2 and transport coefficients of CCl_2F_2 and N_2-CCl_2F_2 mixtures, J. Appl. Phys. 57, 4368 (1985).
17. M.F. Fréchette, Electron swarm properties in CCl_2F_2 and mixtures with N_2 under steady-state conditions, J. Appl. Phys. 61, 5254 (1987).
18. J.P. Novak and M.F. Fréchette, Collisional cross sections of c-C_4F_8 and transport coefficients of c-C_4F_8 and N_2-(c-C_4F_8) mixtures, J. Appl. Phys. 63, 2570 (1988).

19. W. Boeck, Situation and treatment of environmental SF_6 problems in Germany, Meeting CIGRÉ D1 WG03; Future electric power equipment applying dielectric gases, Workshop on Future Electric Power Equipment and Electrical Insulation Techniques, Nagoya University, Japan, June 6th 2003.
20. SF_6 Recycling Guide – Revision 2003, Task Force B3.02.01, Electra, No. 209, pp. 34-47, August 2003.
21. A.V. Phelps and R.J. Van Brunt, Electron-transport, ionization, attachment, and dissociation coefficients in SF_6 and its mixtures, J. Appl. Phys. Vol. 64, 9, p. 4269 (1988)
22. M. F. Fréchette, Swarm-derived scattering cross sections of N_2, SF_6, CCl_2F_2, c-C_4F_8 and transport coefficients of electron swarms in these gases and various gas mixtures, Ph. D. thesis, Physics Dpt., Université de Montréal, 1987.
23. M.F. Fréchette, N. Bouchelouh and R.Y. Larocque, Time-resolved avalanches in SF_6 under sligthly nonuniform field, 46th Gaseous Electronics Conference, Montréal CAN, Oct. 19-22 1993.
24. M.F. Fréchette, N. Bouchelouh, and R.Y. Larocque, Time-resolved avalanches in SF_6 under geometrically-perturbed uniform field, Proceedings International Symposium on Gaseous Dielectrics, Knoxville TN USA, April 24-28 1994.
25. N. Bouchelouh, M.F. Fréchette, and R.Y. Larocque, Avalanche study in low-pressure N_2 under geometrically decreasing field, Annual Report Conference on Electrical Insulation and Dielectric Phenomena, Arlington TX, Oct. 23-26 1994.
26. M. F. Fréchette, N. Bouchelouh, and R. Y. Larocque, Laser-induced SF_6 undervoltage-break-down characteristics in a geometrically decreasing field, Proceedings International Symposium on High Voltage Engineering, Graz Austria, Aug. 28/Sept. 1 1995.
27. N. Bouchelouh, M.F. Fréchette, and R.Y. Larocque, Avalanche study in low-pressure SF_6 under geometrically decreasing field, Proceedings International Conference on Gas Discharges and their Applications, Tokyo Japan, Sept. 11-15 1995.
28. N. Bouchelouh, M. F. Fréchette, and R. Y. Larocque, Large avalanche buildup in low-pressure SF_6 under nonuniform-field conditions, Annual Report Conference on Electrical Insulation and Dielectric Phenomena, Virginia USA, Oct. 22-25 1995.
29. H. Raether, *Electron avalanche and breakdown in gases*, Butterworths, London, (1964).
30. N. Sato and S. Sakamoto, Undervoltage breakdown between parallel plates in air, J. Phys. D: Appl. Phys., Vol. 12, pp. 875-85, 1979.
31. J. T. Kennedy, J. M. Wetzer, and P. C .T. van der Laan, Space charge induced breakdown in undervolted gaps in nitrogen and air, Proc. 8th Int. Symposium on High Voltage Engineering, Yokohama, Japan, Aug. 23-27, paper 34.07, 1993.
32. M.F. Fréchette, N. Bouchelouh, and R.Y. Larocque, Laser-induced undervoltage breakdown in atmospheric N_2 correlated with time-resolved avalanches, Proceedings International Symposium on Electrical Insulation, Pittsburgh Penn. USA, June 5-8 1994.
33. N. Bouchelouh, M. F. Fréchette, S. Kamel, and R. Y. Larocque, Avalanche current measurements at high SF_6 pressure, Annual Report Conference on Electrical Insulation and Dielectric Phenomena, San Francisco CA USA, October 20-23th 1996.
34. N. Bouchelouh, M. F. Fréchette, S. Kamel, and R. Y. Larocque, Avalanche development in SF_6 contaminated with water vapor, Proceedings International Symposium on Electrical Insulation, Montréal CAN, June16-19 1996.
35. Henk Gaxiola, Dielectric breakdown in insulating gases: Space charge effects and non-uniform fields, Ph. D. thesis, Eindhoven University, 1999.
36. M.F. Fréchette, SF_6 gas breakdown at low temperature, Proceedings International Symposium on Electrical Insulation, Toronto CAN, June 3-6 1990.
37. M.F. Fréchette and D. Roberge, Discussion of some space-charge effects in the context of low- temperature and high-field steady-state gas discharges, Proceedings Sixth Inernational Symposium on Gaseous Dielectrics, Knoxville TE, USA, Sept. 23-27 1990.
38. M.F. Fréchette and D. Roberge, Experimental setup for the investigation of gas discharge phenomena at low temperature, Proceedings Third International Conference on Properties and Applications of Dielectrics Materials, Waseda University, Tokyo Japan, July 8-12 1991.
39. M.F. Fréchette, D. Roberge, and R.Y. Larocque, High-voltage photocathode for gas breakdown studies, Proceedings Seventh International Symposium on High Voltage Engineering, Dresden Germany, August 26-30 1991.

40. M.F. Fréchette, SF_6 breakdown characteristics with a specially-designed cathode, Proceedings Symposium on Electrical Insulation, Baltimore MD USA, June 7-10 1992.
41. M.F. Fréchette, D. Roberge and R.Y. Larocque, Breakdown behavior of SF_6 gas-insulated systems down to -50 °C, Proceedings Tenth International Conference on Gas Discharges and their Applications, Swansea UK, Sept. 13-18 1992.
42. M.F. Fréchette, D. Roberge, and R.Y. Larocque, Breakdown behavior of SF_6 gas-insulated systems at low temperature, IEEE Trans. on Dielectrics and EI, Digest Ed., 2, 925 (1995).
43. A. Beroual and F. Buret, SF_6 dielectric strength at low temperatures under DC voltages, Proc. 10th Int. Conf. on Gas Discharges and their Applications, Swansea, United Kingdom, pp. 406-409, Sept. 1992.
44. M.F. Fréchette, D. Roberge and R. Y. Larocque, Breakdown performance of SF_6/N_2 gas-insulated systems at low temperature, Proceedings Eight International Symposium on High Voltage Engineering, Yokohama Japan, August 23-27 1993.
45. M.F. Fréchette, G.R. Mitchel, and J.L. Parpal, Negative DC corona discharge in SF_6 for highly divergent field, Proc. Fifth International Symposium on Gaseous Dielectrics, Knoxville, TE USA, May 3-7 1987.
46. M. Côté, A. Croes, M.F. Fréchette, and R.Y. Larocque, Spectroscopic study of positive corona discharge in SF_6, Proc. International Conf. on Gas Discharges and their applications, Tokyo Japan, Sept.11-15 1995.
47. M. F. Fréchette and R. Y. Larocque, Partial discharge activity at the intercept between solid and air insulation: Qualitative interpretation of the ultrasonic signatures, IEEE Conference on electrical insualtion and dielectric phenomena, Albuquerque NM USA, October 19-22 2003.
48. M.F. Fréchette and S. I. Kamel, DC corona discharge in SF_6: Evidence of an electronic component for the negative polarity, Proc. International Symposium on Electrical Insulation, Boston USA, June 5-8 1988.
49. M.F. Fréchette and S.I. Kamel, Negative DC corona in SF_6 at low pressure and short gap length, Proc. IX International Conference on Gas Discharges and their Applications, Venice, Italy, Sept. 19-23 1988.
50. M.F. Fréchette, M. Côté, N.G. Trinh, and R.Y. Larocque, Partial-discharge activity for a negative DC corona in atmospheric SF_6, Proceedings International Symposium on Electrical Insulation, Pittsburgh Penn. USA, June 5-8 1994.
51. M. Côté, M.F. Fréchette, and R.Y. Larocque, Negative DC corona characteristics in SF_6 under moisture contamination, Annual Report Conference on Electrical Insulation and Dielectric Phenomena, Arlington TX USA, Oct. 23-26 1994.
52. CIGRE Symposium on diagnostics and maintenance techniques, Berlin, Germany, 19-21 April 1993.
53. M.F. Fréchette, R. Bartnikas, S. Kamel, and R.Y. Larocque, Negative corona discharge in air with small SF_6 content, Proceedings International Symposium on Electrical Insulation, Montréal CAN, June 16-19 1996.
54. M.F. Fréchette, S. Kamel, R. Bartnikas, and R.Y. Larocque, Change in the Pulse Regime of Negative Corona in Air Due to a Small Adjunct of SF_6, Annual Report Conference on Electrical Insulation and Dielectric Phenomena, San Francisco, October 20-23th 1996.
55. R.J. Van Brunt, Physics and chemistry of partial discharges and corona, IEEE Trans. on Diel. and Electr. Insul., Vol 1, No 5, Oct. 1994, pp 761-784.
56. M.F. Fréchette, C. Hudon, M. Germain, R. Larocque, S. Matsumoto, and T. Umemura, On the stochastic nature of SF_6 partial discharges close to onset conditions, Int. Symp. on Electr. Insul., Anaheim CA, April 2-5 2000.
57. P. Von Glahn and R.J. Van Brunt, Continuous recording and stochastic analysis of PD, IEEE Trans. on Diel. and Electr. Insul., Vol 2, No 4, Aug. 1995, pp.590-601.
58. R. Patsch and F. Berton, Pulse-sequence-analysis – Chances to characterize defects, Conf. on Electr. Insul. and Diel. Phen., Texas, Oct. 1999, pp 243-248.
59. T. Takahashi, T. Yamada, N. Hayakawa, and H. Okubo, Space charge behavior in SF_6 gas and sequential generation of PD pulses, IEEE Trans. on Diel. and Electr. Insul., Vol 7, No 1, Feb. 2000, pp 141-145.
60. C. Heitz, A generalized model for partial discharge processes based on a stochastic process approach, J. Phys. D, Sept. 1999, pp 1012-1023.
61. R.J. Van Brunt and S.V. Kulkarni, Stochastic properties of partial-discharge phenomena, IEEE Trans. on Diel. and Elec.Insul., Vol 26, 1991, pp 902-948.
62. R.J. Van Brunt, Stochastic properties of Trichel-pulse corona: A non-Markovian random point process, Phys. Rev. A, Vol 42, No 8, Oct. 1990, pp 4908-4932.
63. M.F. Fréchette, C. Hudon, M. Germain, R.Y. Larocque, S. Matsumoto, and T. Umemura, Partial-discharge phenomenology under stochastic conditions, Annual Report Conference on Electrical Insulation and Dielectric Phenomena, Victoria CA, 15-18 Oct. 2000.

64. C. Hudon, M.F. Fréchette, M. Germain, R.Y. Larocque, S. Matsumoto, and T. Umemura, Evaluation of distinctive PD parameters in SF_6 under ac voltage, Annual Report Conference on Electrical Insulation and Dielectric Phenomena, Victoria CA, Oct. 15-18 2000.
65. M.F. Fréchette, R.Y. Larocque, C. Hudon, T. Umemura, S. Matsumoto and T. Takahashi, Corona discharge phenomenology at onset in weakly electronegative gas mixtures, Proceedings Ninth International Symposium on Gaseous Dielectrics, Baltimore USA, May 21-25, 2001.
66. T. Takahashi, M.F. Fréchette, R.Y. Larocque, C. Hudon, T. Umemura, S. Matsumoto, and H. Okubo, Long-term sequential characteristics of negative partial discharge in pressurized SF_6 gas, Proceedings Ninth International Symposium on Gaseous Dielectrics, Baltimore USA, May 21-25, 2001.
67. M.F. Fréchette, T. Takahashi, H. Okubo and R.Y. Larocque, Instability of self-sustained gas discharges at corona onset for various experimental conditions, IEEE International Conference on properties and Applications of Dielectric Materials, June –15, 2003, Nagoya Japan.
68. M.F. Fréchette and R.Y. Larocque, Partial discharge behavior at the interface between solid and gaseous insulation, International Symposium on Electrical Insulation, Boston, USA, pp. 395-398, Avril 2002.
69. M.F. Fréchette and R.Y Larocque, Negative partial discharge in compressed air: Sequential characteristics evolving over time, IEEE Conference on Electrical Insulation and Dielectric Phenomena, Quinta Roo, Mexico, October 20-24 2002.
70. I. Flatlow, *They All Laughed...From Light Bulbs to Lasers: The Fascinating Stories Behind the Great Inventions That Have Changed Our Lives.*, (Published by HarperCollins, New York), 1993.
71. M. F. Fréchette, State-of-situation with gaseous dielectrics and emerging insulation materials, Workshop on Future Electric Power Equipment and Electrical Insulation Techniques, Nagoya University, Japan, June 6th 2003.
72. C. Laverdure, M. Massé and M.F. Fréchette, Gas integrity and adsorbent performance in an SF_6 circuit breaker after test duty, Annual Report Conference on Electrical Insulation and Dielectric Phenomena, Pocono Manor, Penn. USA, Oct. 17-20 1993.
73. M.F. Fréchette, Mitigation studies, Proceedings Workshop on Investigation of S_2F_{10} Production and Mitigation in Compressed SF_6-Insulated Power Systems, Pittsburgh, June 9th 1994.
74. T. Lussier, M.F. Fréchette, and R.Y. Larocque, Gas/adsorbent interaction dynamics: An experimental approach, Annual Report Conference on Electrical Insulation and Dielectric Phenomena, Virginia USA, Oct. 22-25 1995.
75. T. Lussier, M.F. Fréchette, and R.Y. Larocque, Gaseous contaminants in SF_6 interacting with alumina, Proceedings International Symposium on Electrical Insulation, Montréal CAN, June16-19 1996.
76. M. Trudeau, A. Rochefort, and M.F. Fréchette, Determination of adsorbed species in the surface hydrolysis of SOF_2 on activated alumina, Annual Report CEIDP, San Francisco, October 20-23th 1996.
77. T. Lussier, M.F. Fréchette, A. Rochefort, and R.Y. Larocque, Interaction of SOF_2 with alumina and zeolite 4A: A comparative study, Annual Report CEIDP, San Francisco, October 20-23th 1996.
78. T. Lussier, M.F. Fréchette, and R.Y. Larocque, Interactions of SOF_2 with molecular sieve 13X, Annual Report CEIDP, Minneapolis, MN USA, Oct. 19-22, 1997.
79. J.F. Rouleau, J. Goyette, T.K. Bose, and M.F. Fréchette, Investigation of a microwave differential cavity resonator device for the measurement of humidity in gases, Rev. Scientific Instruments, Vol. 70, No 9, pp 3590-3594, Sept. 1999.
80. J.F. Rouleau, J. Goyette, T.K. Bose, and M.F. Fréchette, Performance of a microwave sensor for the precise measurement of water vapor in gases", IEEE Trans. on Diel. and Electr. Insul., Vol. 7, No. 6, pp. 825-831 (2000).
81. J. F. Rouleau, J. Goyette, T.K. Bose, and M.F. Fréchette, Optimization study of a microwave differential technique for humidity measurements in gases, IEEE Trans. on Instrum. and Meas., Vol. 50, No. 3, pp. 839-845 (2001).
82. M. F. Fréchette, M. Trudeau, and R. Y. Larocque, Solid materials subjected to partial discharge activity in pure SF_6 and mixtures with CF_4, Proceedings ISEI, Washington USA, June 7-10 1998.
83. I. Paun, M. F. Fréchette, M. R. Wertheimer, and R.Y. Larocque, Characterization of epoxy resin surfaces exposed to partial discharges in SF_6 and N_2/SF_6 mixtures, Annual Report Conference on Electrical Insulation and Dielectric Phenomena, Atlanta USA, Oct. 25-28 1998.
84. T. Uchii, T. Shinkai, and K. Suzuki, Thermal interruption capability of carbon dioxide in a puffer-type circuit breaker utilizing po;ymer ablation, IEEE PES T&D Conference and exhibition 2002: Asia Pacific, October 6-10 2002, Yokohama, Japan, Submission No. 388.

INTRODUCTION OF A NEWLY DEVELOPED PURIFICATION PROCESS FOR USED SF_6 FROM ELECTRICAL EQUIPMENT

Peter Jannick, Michael Pittroff*

1. Introduction

Sulphur hexafluoride (SF_6) is a physically and chemically stable, non-toxic and inert gas with a high dielectric strength and thermal stability[1]. It is generally used in electrical equipment for power transmission and distribution such as gas-insulated circuit breakers, gas-insulated transmission lines, gas-insulated transformers and gas-insulated substations, as an insulating and arc quenching gas.

Besides its unique electro-technical properties, SF_6 has a long atmospheric life and a high GWP compared to CO_2 when released into the atmosphere. A closed loop treatment of SF_6 is therefore required to minimize potential emissions wherever possible, which also enables its reuse, such as that implemented with the SF_6 ReUse Concept[2, 3].

Solvay Fluor has been treating SF_6 throughout the world since 1991, and thus enabling used SF_6 to be utilized for the production of new SF_6 - fulfilling and even exceeding the IEC376 specification[3].

However, the used SF_6 collected from different fields of application such as GIS and test fields contains various gaseous impurities. The removal of these impurities, especially S_2F_{10}, has been investigated and is discussed here in order to fulfil the ReUse specification in regard to avoiding process contamination after reintroduction into the SF_6 production process[3, 4].

Other contaminations such as dust, moisture, oil etc. that may occur during the handling and use of SF_6 are removed using state of the art technology, suitable filters and adsorption materials[5].

2. Impurities in SF_6 after use in different electrical applications

A representative composition of used SF_6 exceeding the IEC480 specification, shown in table 1, is considered on the basis of the total amount of used SF_6 that was collected by Solvay Fluor during 2001.

* Peter Jannick, author for correspondence, Solvay Fluor und Derivate GmbH und Co. KG, Technical Service SF_6, Hans-Böckler-Allee 20, 30173 Hanover, Germany,
Tel.: 0049/511/857-2441, Fax: 0049/511/857-2166, e-mail: peter.jannick@solvay.com.
Michael Pittroff, Solvay Fluor und Derivate GmbH und Co. KG - Marketing, Tel.: 0049/511/857-3448, Fax: 0049/511/857-2146, e-mail: michael.pittroff@solvay.com

Table 1: Average impurities in relation to volume in used SF_6 exceeding the IEC480 specification in 2001 that was reused by Solvay

Impurity	IEC480[5]	Average value /max. value	Behaviour / Property
SF_6	96.99	97.17 / 99.99%	Inert
Air (O_2, N_2)	3.0%	1.59 / 59.74%	Inert
CF_4		1.15 / 2.70%	
SO_2	50ppm	300ppm / 1.4%	Hydrolysable
SO_2F_2		200ppm / 1.5%	
S_2F_{10}	Not mentioned	174ppm / 1.55%	Stable / Non hydrolysable

3. Purification processes

The impurities in table 1 can be subdivided into inert, hydrolysable and stable/non hydrolysable components which can be removed from SF_6 by the following purification processes.

3.1 Membrane separation

Inert gas impurities such as oxygen, O_2, and nitrogen, N_2, can be removed by the membrane technique. The main gas flow is divided into a permeate stream and a product stream. Membrane separation units are used to separate gases with different molecule sizes such as SF_6 from N_2 and O_2[6].

3.2 Gas scrubber

Soluble gaseous components such as sulphur-oxy-fluorides, sulphur-dioxide (SO_2) and sulfuryl-di-fluoride (SO_2F_2) are removed from SF_6 by gas scrubbers. These components have a high solubility in water compared to SF_6, as shown in table 2, which can even be improved by using alkaline solutions. This allows them to be separated from SF_6, while the sulphur-oxy-fluorides remain in the alkaline solution inside the gas scrubbers.

Table 2: Solubility of different gases at 10°C and 1.013bar[7-11]

Component	SF_6	CF_4	N_2	O_2	S_2F_{10}	SO_2	SO_2F_2
Solubility (g/kg H_2O)	0.06	0.04	0.02	0.05	-	164.1	1.00

3.3 Pyrolysis and photolysis

Due to its properties, di-sulphur-decafluoride, S_2F_{10} is the most challenging part for removal. When separated, S_2F_{10} is obtained in a concentrated form, which creates a waste disposal problem due to its toxicological profile[12, 13].

Therefore this component must be destroyed. The most promising destruction methods for S_2F_{10} are pyrolysis[14, 15] and photolysis[1, 16].

Figure 1 shows different wavelength windows for the destruction of gases utilizing photolysis[1, 16].

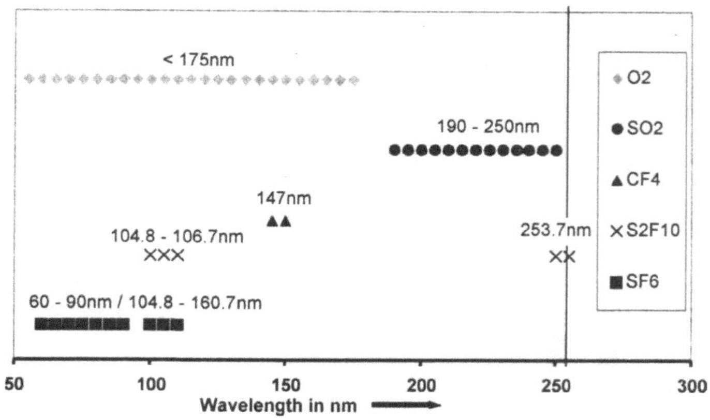

Figure 1: Wavelength for gas destruction using photolysis

The use of gas scrubbers and membranes during the manufacture and purification of SF_6 is common practice. However the destruction of S_2F_{10} with pyrolysis and photolysis has not yet been tried and tested on an industrial scale. Because of this fact this study was addressed to investigate the pyrolysis and photolysis process and to develop a pre-treatment pilot plant for used SF_6.

4. Test facility for the investigation of S_2F_{10} destruction via pyrolysis and photolysis

The designed facility for the investigation of the different destruction processes for S_2F_{10} is illustrated in figure 2.

Used SF_6 from a SF_6 ReUse cylinder[3, 4] (1) is evaporated in an evaporator (2) and the gas pressure is adjusted via a pressure reducer (3). The SF_6 gas stream is regulated with a fine regulating valve (4) and fed either into the photolysis (5) – or pyrolysis reactor (6).

The photolysis reactor comprises 150W and 700W water-cooled 253.7nm UV lamps with 180 and 520ml volume operating at 90°C.

A 20m spiral shaped copper tube with an inner diameter of 2mm, heated with 750W heat coils was used for the pyrolysis reactor. Temperatures up to 350°C can be achieved.

After the gas treatment the SF_6 is captured in a cooling trap (7) where it is liquefied.

GC analysis is carried out before and after each test series.

The volumetric flow, treatment time, temperature and pressure are continuously measured, controlled and recorded with a data acquisition system.

In order to avoid any gas release, the system is kept under a slight vacuum.

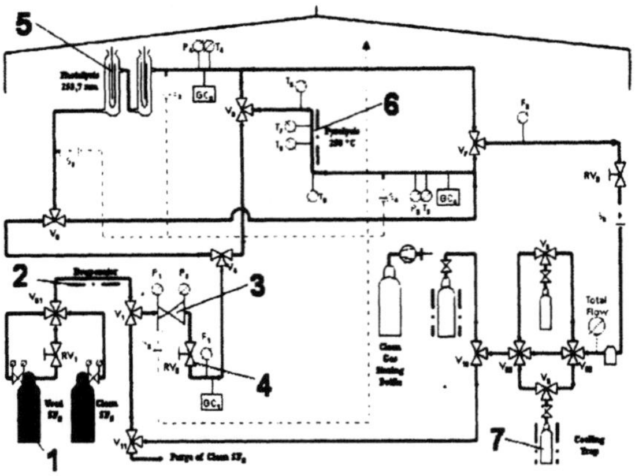

Figure 2: Test facility for the investigation of the destruction of S_2F_{10} via photolysis and pyrolysis

4.1 Results of pyrolysis

Table 3 shows the results of the pyrolysis trials at 250°C and 0.8bar. The heating capacity of the reactor was 750W. A gas treatment time in the reactor of 7.6s was achieved at a volumetric flow of 0.5 l/min.

Table 3: Impurities related to volume before (base) and after the pyrolysis treatment

Experiment	CF_4 (%)	Air (%)	SF_6 (%)	SO_2F_2 (ppm)	S_2F_{10} (ppm)	SO_2 (ppm)
Base	**1.15**	**1.59**	**97.17**	**200**	**174**	**300**
1	0.95	1.64	95.8	-	136	610
2	0.91	1.69	97.7	-	135	1740
3	1.25	1.58	97.02	-	136	653

The amount of S_2F_{10} was slightly reduced using pyrolysis. A complete destruction of the S_2F_{10} present at 174ppm was achieved by changing the temperature and treatment time[14, 15]. SF_6 and air exhibited inert behaviour in this process. The detected concentration of SO_2 is higher compared to the base concentration. This becomes clear because SO_2 is formed from SO_2F_2, S_2F_{10} and O_2 during pyrolysis at 250°C[14, 15].

4.2 Results of the photolysis

All photolysis trials were performed at 253.7nm, 90°C and 0.8bar. The adjusted capacity of the two UV lamps was 700W. A gas treatment time of 1.26 min. was achieved at a volumetric flow of 0.5l/min. using a reactor volume of 700ml. The photolysis results are listed in table 4.

Table 4: Impurities related to volume before (base) and after the photolysis treatment

Experiment	CF_4 (%)	Air (%)	SF_6 (%)	SO_2F_2 (ppm)	S_2F_{10} (ppm)	SO_2 (ppm)
Base	**1.15**	**1.59**	**97.17**	**200**	**174**	**300**
1	1.02	1.52	97.2	< LOD	<LOD	-
2	0.92	1.55	97.2	< LOD	<LOD	<LOD
3	1.23	1.60	97.8	161	<LOD	<LOD

The detected amounts of S_2F_{10} and even SO_2F_2, with the exception of one value, and SO_2 were below the limit of detection (LOD = 5ppm) after the UV treatment. SF_6 and air were not influenced by photolysis under the test conditions[16]. Therefore photolysis is probably the most appropriate destruction method for S_2F_{10}. The results should be further verified.

5. Pilot plant for the pre-treatment of used SF_6

The different purification processes for removal of S_2F_{10}, SO_2F_2, SO_2, O_2 and N_2 were arranged in a pilot plant to ensure adequate pre-treatment of the used SF_6 in order to comply with the ReUse specification[4]. The pilot plant in figure 3 for the pre-treatment of used SF_6 mainly consist of a photolysis reactor, an alkaline gas scrubber and membrane separation units.

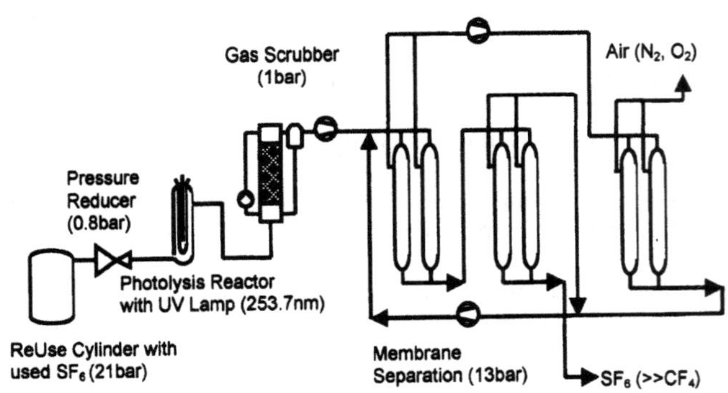

Figure 3: Pilot plant for the pre-treatment of used SF_6 before reintroduction to the SF_6 production process

The contaminated SF_6 flows through a pressure reducer to the different purification processes. S_2F_{10} is efficiently destroyed in the photolysis reactor. The reaction products and other hydrolysable components are separated from the SF_6 gas stream with an alkaline scrubber (NaOH). Inert gases, e.g. N_2, O_2, are separated from the SF_6 in a subsequent step via membranes[6].

6. Conclusion

Photolysis is the most appropriate process for the destruction of S_2F_{10}. Therefore that process was implemented into a pilot plant.

The pilot plant comprises a photolysis system to destroy S_2F_{10}, an alkaline scrubber to remove SO_2 and SO_2F_2 and membrane units to separate O_2 and N_2 from SF_6 utilizing gas pre-treatment for the production of new SF_6.

This completes Solvay's SF_6 ReUse Concept comprising the collection after use, pre-treatment, purification and reintroduction into the SF_6 production process.

7. References

1. L. Gmelin: Handbuch der Anorganischen Chemie, Springer-Verlag, Vol. 9, 8th Edition, 1978, pp. 33-201, Vol. 14, 1974, pp. 131-132, Vol. 3, 1958, pp. 639-643
2. M. Pittroff, H. Krähling, E. Preisegger: Product Stewardship for SF_6, Proceedings of 2001 Int. Symp. on Electr. Insult. Materials, pp.254-257, Himeji, November 2001
3. P. Jannick: Das SF_6-ReUse-Konzept als Beispiel weltweiter Produktverantwortlichkeit, ETG Kongress, Hamburg, September 2003
4. Solvay Fluor und Derivate GmbH: Concept for the ReUse of Used SF_6, 08/2000
5. CIGRE 234: SF_6 Recycling Guide, 2003
6. M. Pittroff: Separation of SF_6/N_2 Mixtures, MEIE'2000, 2nd European Conference on Industrial Electrical Equipment and Environment, pp. 61-68, January 2000
7. J.T. Asthon et al.: The solubility of Certain Gaseous Fluorine Compounds in Water, J. Chem. Soc. (A) 1968, pp. 1793-1796.
8. M. Trautz, K. Ehrmann: Beiträge zur Chemie des Schwefels. II.- Zur Kenntnis der Fluoride und Oxyfluoride des Schwefels, J. prakt. Chem. Vol. 142, 1935, pp. 79-124
9. Kirk-Othmer: Encyclopedia of Chemical Technology, Vol. 11-16, 4th Ed, John Wiley & Sons, 1994-1996
10. G.H Cady, S. Misra: Hydrolysis of Sulfuryl Fluoride, Inorganic Chemistry, Vol. 13, No. 4, 1974, pp.837-841
11. E.E. Kenaga: Some Biological, Chemical and Physical Properties of Sulfuryl Fluoride as an Insecticidal Fumigant, Journal of Economic Entomology, Vol. 50, No. 1, February 1957, pp. 1-6
12. TLW/TWA (S_2F_{10}) = 25ppb, Niosh/Osha: US Department of Labour and Health
13. M. Piemontesi et. al: Sorption of SF_6 and SF_6 decomposition products by activated alumina and molecular sieve 13X, IEEE International Symposium on Electrical Insulation, Montreal, Quebec, Canada, June 16-19, 1996
14. I. Sauers, et al.: On S_2F_{10} Formation in Spark Breakdown of SF_6, Conference Record, 1988, IEEE Symposium on Electrical Insulation, pp. 112-115, June 5-8 1988
15. J.T. Herron: A Critical Review of the Chemical Kinetics of SF_4, SF_5, and S_2F_{10} in Gas Phase, Int. J. Chem. Kinet., pp 129-142, Vol 19, 1987
16. H.J. Emeléus, K.J. Packer: The photochemical reactions of some pentafluorosulphur derivatives with sulphur dioxide, Journal of the Chemical Society, pp. 771-775, 1962

RESPONSIBLE HANDLING OF SULFUR HEXAFLUORIDE (SF_6) BY MANUFACTURERS OF HIGH-VOLTAGE SWITCHGEAR

Hartmut Knobloch and Klaus Schuler•

1. INTRODUCTION

The electricity industry has been using sulphur hexafluoride (SF_6) for more than 40 years. SF_6 is an excellent insulating and arc-quenching medium and allows systems to be more powerful and more compact than any other available medium. As well as handling the highest voltages and make/break ratings, SF_6 switchgear is also very compact and has been responsible for many resource-saving developments, superseding both oil and air for arc-quenching and insulation[1]. Over the last few years, however, criticism of SF_6 gas has become ever stronger due to its potential effect on the world's climate. Although the contribution of accumulated SF_6 to manmade greenhouse gases in the atmosphere is, at 0.2%, relatively minimal, the Kyoto Conference in 1997 nevertheless still included it on the list of the six human-made greenhouse gases for monitoring[2].

SF_6 is extremely long-lived in the atmosphere (for approximately 3,200 years), and its greenhouse potential is 22,200 times greater than that of CO_2[3]. However, the results of new studies based on actual measurements point to a life of 650 years and a greenhouse potential 20,900 times greater than that of CO_2[4]. For insulation purposes at pressures between 0.45 and 0.7 MPa, the attainable field strengths for SF_6 are greater than those of air at an atmospheric pressure of 0.1 MPa by a factor of approximately 10. In quenching arcs, SF_6 outperforms air

• Hartmut Knobloch and Klaus Schuler, Siemens AG, Power Transmission and Distribution, Technology, 13623 Berlin, Germany

Gaseous Dielectrics X.
Edited by Christophorou *et al.*, Springer, New York, 2004.

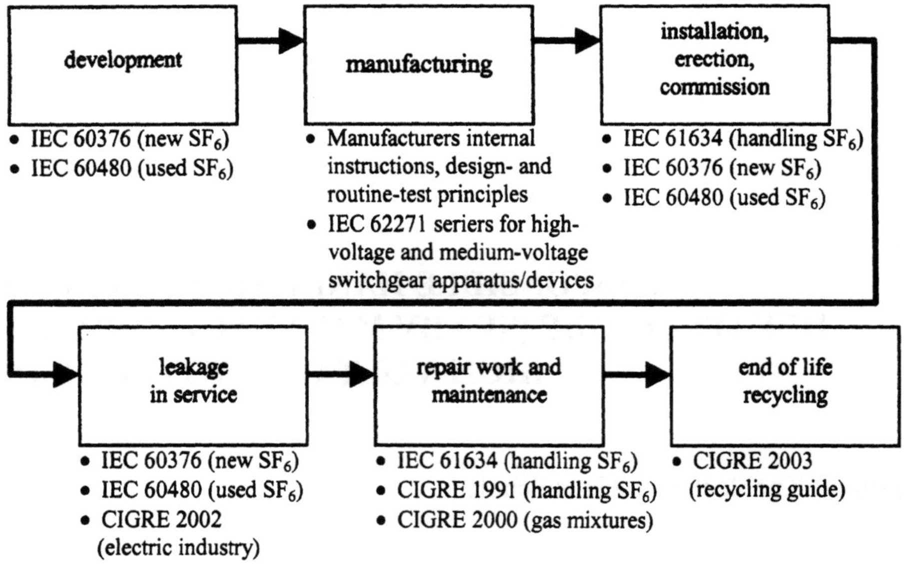

Figure 1. Switchgear standards referring to SF_6-issues.

by a factor of several times ten, assuming the same conditions. So far no other substances suitable for insulating and arc-quenching have been found that come anywhere near these values, while at the same time possessing a lower greenhouse potential[5, 6].

2. LIFE CYCLE AND SF_6 ISSUES

Figure 1 shows the SF_6 issues "from cradle to grave", i.e. from the development stage through manufacturing, assembly, installation and commissioning to operation (involving leakage, handling losses during maintenance and repair work) and ultimate end-of-life recycling. The whole cycle is covered by IEC (International Electrotechnical Commission) standards and reports, by manufacturers' instructions and also by CIGRE (Conseil International des Grands Reseaux Electriques) reports. Based on the particular life cycle the global problem of SF_6 emissions is combatted. E.g. the actual maintenance work is carried out exclusively by trained personnel who have been well instructed in the handling of SF_6. In order to minimize handling losses, all the relevant processes involved in assembly, commissioning, maintenance and repair (including disposal), are described in full detail in the operating instructions provided by the switchgear manufacturers.

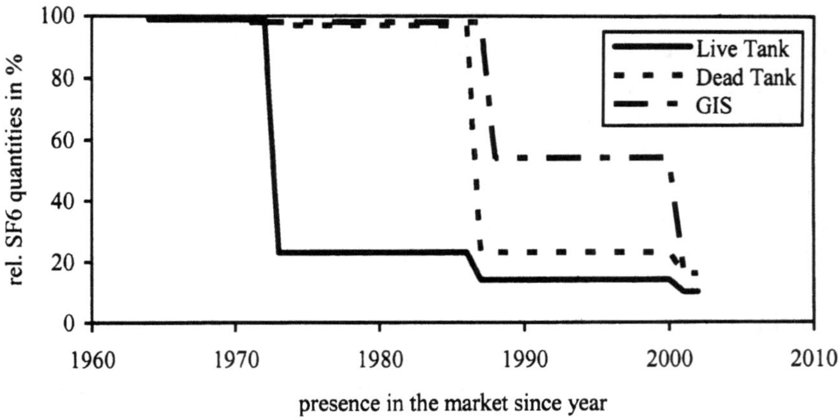

Figure 2. SF_6 quantities per functional unit in switchgear.

3. METHODS FOR MANUFACTURERS TO REDUCE SF_6 EMISSIONS

3.1 Design and development stage

One efficient way of reducing relative SF_6 emissions from electrical equipment is to reduce the amount of SF_6 required in the equipment, without placing its functional performance at risk. Figure 2 shows how the quantities of SF_6 used in various switchgear units (from the late 1960s/early 1970s up to the present time) have developed. For live-tank, dead-tank and GIS circuit-breakers, the quantities of SF_6 required have been reduced to some 20% of their original levels. One reason is optimization of the dielectric field aspect, which has resulted in smaller, more compact designs. Another reason is the increased breaking capacity of each interrupter unit, which has resulted in fewer breaker gaps needed to control the current and voltage. Furthermore, the development of integrated units with combined functions such as earthing switches and disconnectors in the same gas compartment has also helped reduce the amount of SF_6 needed.

From a design viewpoint, care must be taken with the sealing of the joints between the individual enclosures and fittings that comprise a closed system. Therefore, the number of such joints should be kept to a minimum and the length of the seal should be as short as possible. The most popular materials for the enclosures are welded steel or aluminium, and cast iron or cast aluminium. Corrosion protection is unnecessary. It is essential for aluminium castings to be gas-tight. It is not necessary to "impregnate" the material afterwards in order to achieve the required seal. When selecting the materials for sealing, it is important to note the following criteria and parameters: the diffusion rate of SF_6

through the sealing material, the service temperature range, the residual elasticity at low temperatures, the hardness, the tearing elongation, the ageing characteristics and the resistance to the decomposition products of SF_6. With older designs of switchgear, leakage rates of 3% p.a. and more were quite usual. Developments over the last 20 years, however, have reduced those rates to guaranteed values of less than 0.5% per annum per gas compartment. These values also have to be verified in the context of type tests. In actual practice especially in medium-voltage systems below 52 kV, leakage rates of less than 0.1% are being achieved.

3.2 Testing stage

Development work also involves the testing of equipment. Dielectric tests, power tests, mechanical tests and temperature-rise tests are necessary in order to evaluate different designs. The chart in Figure 3 shows the SF_6 handling procedure to be adopted during these tests, to ensure that the maximum amount of gas can be re-used in a closed cycle. While equipment is being developed and tested it has to be opened up repeatedly in order to change certain internal components. During such work it is obviously important not to allow any SF_6 to escape into the atmosphere. Used gas is removed from the switchgear with the help of service units which, in addition to a compressor, also incorporate tanks for storage and filters for treating the gas prior to its re-use. The hoses used for removing the SF_6 and for refilling the switchgear, and which are disconnected from the gas compartments afterwards, must have self-sealing couplings at the ends. Gas contaminated with by-products must be either recycled or returned to the manufac-

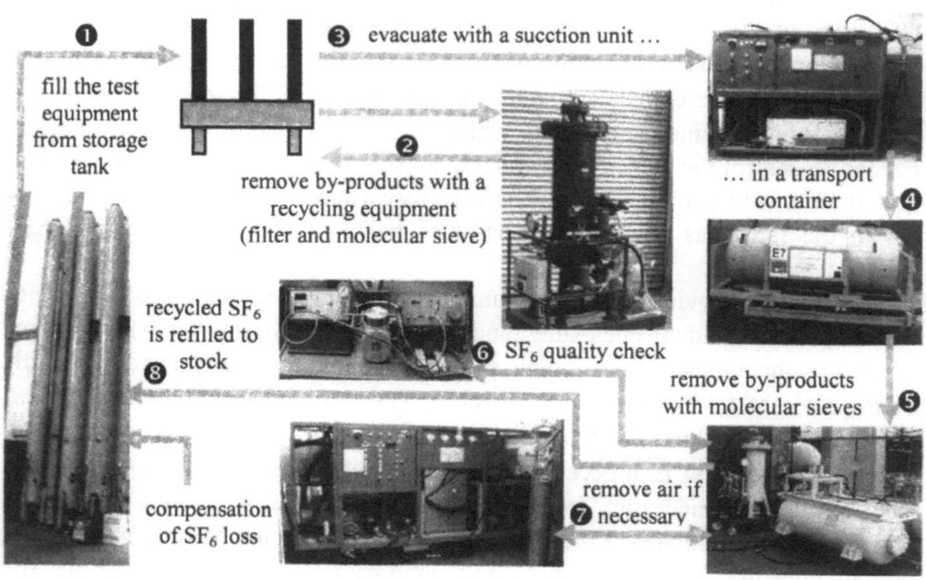

Figure 3. Equipment for using SF_6 in the development.

turer for disposal. After recycling, the guaranteed quality of the gas is sufficient for further testing and also conforms to the IEC standard for used gas[7]. Naturally, these basic rules are also applicable to routine testing and to switchgear in normal service undergoing maintenance.

3.3 Production stage

When the parts of the enclosures that accommodate the static seals are produced it is very important to ensure that the quality of surface finish of the sealing faces is as set down in the specification. In the case of castings it is essential for the sealing faces not to show even the slightest signs of porosity after they have been machined. At the Siemens High-Voltage Switchgear Factory every gas compartment is subjected to a 100% pressure and leak test before the enclosures are assembled. The leak test is carried out with helium gas. Since helium molecules are only about half the size of SF_6 molecules, the test will expose even the very smallest leak in the enclosures. Using helium instead of SF_6 for the leak test not only reduces the testing time and increases the sensitivity but also avoids unnecessary use of SF_6. The sensitivity of the routine test employed for leak detection is better than 10^{-6} bar cm^3/s. Theoretically, this means that it would take approximately 4.8 years to lose one gram of SF_6. Fully assembled equipment and subassemblies are examined at their sealed joints with high-sensitivity leak detectors.

3.4 Assembly and installation stage

Shipping the largest possible preassembled and prefilled modules not only improves the quality of assembly over that achieved when switchgear is put together on site, but also reduces the amount of SF_6 lost during commissioning. These days, switchbay units with ratings up to 245 kV are shipped fully assembled. SF_6 handling losses during the assembly, erection and commissioning of larger equipment, and also during maintenance and repair work, can be avoided completely - or at least be significantly reduced - by using handling gear similar to that shown in Figure 3.

4. MEASURING THE EMISSIONS AND RESULTS

So far the reductions in greenhouse gases demanded by the Kyoto protocol are being implemented through declarations of intent and specific legislation. In certain countries such as Switzerland and Germany, trade associations, gas suppliers, switchgear manufacturers and switchgear users have made voluntary agreements[8]. These declarations require an SF_6 balance sheet to be drawn up every year (using inventory methodology) as suggested by CAPIEL[9]. The results are then made available to the various government ministries responsible for environmental protection through the respective national associations.

Efforts aimed at reducing SF_6 emissions are already producing good results. Enquiries among power utilities and switchgear manufacturers e.g. in Germany[8] showed that the manufacturer's figures for development, manufacture and installation in 2001 were only 2% of the quantity of SF_6 used (4.4% in 1997), whereas SF_6 losses due to leaks and handling by operators were only 0.8% for high-voltage equipment (0.9% in 1997). For modern medium-voltage equipment no handling by operators is necessary due to "sealed-for-life" technology. The leakage rates for this equipment is below 0.1%.

5. SUMMARY

There are currently no alternatives to SF_6 for high-voltage switchgear that do not involve significant losses in performance. Although the SF_6 used in high-voltage equipment only makes a very small contribution to the greenhouse effect, it is still very important to adopt a responsible attitude towards the handling of this important arc-quenching and insulating gas. The technology required is already available to guarantee a closed SF_6 cycle from "cradle to grave". A joint declaration of responsibility for the reduction of SF_6 emissions has provided an excellent starting point.

6. REFERENCES

1. Project Group ABB, Preussen Elektra, RWE, Siemens, Solvay, Electrical supply using SF_6 technology - *Life Cycle Assessment*, Id. No. 80/801/06.99/007/5000
2. Kyoto-Protocol to the United Nations Framework Convention on Climate Change, *FCCC/CP/1997/L.7/Add.1*, Dec. 10 (1997)
3. Carbon Dioxide Information Analysis Center (CDIAC). *Current greenhouse gas concentrations*, (Sept.), (2001); http://cdiac.esd.ornl.gov/pns/current_ghg.html
4. Owens, J.G., Calculation of the Global Warming Potential for Sulfur Hexafluoride Using the Updated Atmospheric Lifetime from Morre et al., *Gaseous Dielectrics IX, edited by Christophorou and Olthoff / Plenum Publishers, New York,* 91-92, (2001)
5. Niemeyer, L., A systematic search for insulation gases and their environmental evaluation, *Gaseous Dielectrics VIII, edited by Christophorou and Olthoff / Plenum Publishers, New York,* 459-464, (1998)
6. Knobloch, H., The comparison of arc-extinguishing capability of sulfur hexafluoride (SF_6) with alternative gases in high-voltage circuit-breakers, *Gaseous Dielectrics VIII, edited by Christophorou and Olthoff / Plenum Publishers, New York,* 565-570 (1998)
7. International Electrotechnical Commission, Guide to the checking and treatment of SF_6 taken from electrical equipment, *IEC 60480,* under revision
8. ZVEI, Freiwillige Selbstverpflichtung, *Solvay, ZVEI,VIK, VDN,* Juli 2003,.http://www.zvei.org
9. Capiel, SF_6 Inventory Methology 2001, *CAPIEL HV-ESDD1-R1-1.02,* Paris, http://www.capiel-electric.com

ANALYSIS OF BY-PRODUCTS OF N_2–SF_6 GAS MIXTURES SPARKED UNDER INHOMOGENOUS FIELD CONDITIONS

J. Sundara Rajan, K. Dwarakanath [1] and N. Srinivasan[2]

1. INTRODUCTION

A vast data on formation of by-products due to arc and spark decomposition of SF_6 is available in literature [1-10]. But experimental work relating to N_2 – SF_6 gas mixtures is limited. In view of increasing application of gas mixtures, it is essential to understand the nature of by-products formed in gas mixtures. Many recent studies have highlighted the advantages of using N_2-SF_6 gas mixtures as a replacement to SF_6 gas for High Voltage application, with a view to reduce emission of SF_6 gas into the atmosphere. Majority of these studies are aimed at determining the basic characteristics of N_2-SF_6 gas mixtures. However, this study attempts to understand the nature and quantum of different species formed in N_2-SF_6 mixtures in the presence of insulating spacers when sparked under inhomogeneous field conditions.

---

[1] J. Sundara Rajan and K. Dwarakanath, DMD, Central Power Research Institute Bangalore 560080, India
[2] N. Srinivasan, Electrical Engineering Department, UVCE, Bangalore University, Bangalore 560001, India

Gaseous Dielectrics X.
Edited by Christophorou *et al.*, Springer, New York, 2004.

2. GC-MS APPARATUS AND THE METHOD

For the determination of N_2 - SF_6 gas ratio, a gas chromatograph (GC) was used and for identification of the species formed during the discharge processes, the Mass Spectrometer (MS) was used. The GCMS analysis was carried out with HP model 5890 series II equipment. Throughout the studies, Thermal Conductivity Detector (TCD) and Mass Selective Detector (MSD) were used. Analysis of sparked SF_6, SF_6-N_2 gas mixture and SF_6-N_2 gas mixture with six different spacer materials have been carried out and reported in this paper. A needle - plane electrode configuration was used and the by-products formed due to spark discharges under these conditions have been compared. The duration of sparking was extended beyond 24 hours depending upon the nature and concentration of by-products formed and their time-dependence. The electrode gap distance for gas gaps was 10 mm and the discs of spacer material of this thickness were inserted into the gap for studying the influence of different materials like PTFE, Polyester, Polypropylene (PP), Nylon, Cycloaliphatic resin and Polymethyl-methacrylate (PMMA).

3. BY-PRODUCTS WITH SF_6 AND N_2/SF_6 GAS MIXTURES

In case of pure SF_6 gas the abundance of SF was 2.4% while SF_4 accounted for 1.75%. Even the percentage of SF_2 was comparatively less, about 2.33. Further, SF_3 is seen to the extent of 14% much higher than all other species like SF, SF_4 and SF_2. On the other hand, SF_4 is also very significant by its presence and its abundance is 6.2%. These results are repeatable and the variation in relative abundance (%) is within ±3%. Thus under non-uniform field conditions the most significant by-product formed is SF_3.

Table 1. Relative abundance of by-products formed after different durations of sparking

Species	Duration of sparking		
	4 hours	6 hours	10 hours
SF	2.4	7	4.7
SF4	1.75	1	4.4
SF2	2.33	6.4	4.4
SF3	14	30	18
SF4	6.2	11	7

With the continuation of sparking at 30 kV for 6 hours it will be interesting to see if there is any increase or otherwise in the concentration of the constituents formed. This result is shown in table 1. SF is observed to the extent of 7% while SF_4 is present marginally at 1%. SF_2 is significant and has a relative abundance of 6.4%. The presence of SF_3 is significant since its relative abundance is as high as

30%. On the other hand, SF_4 is seen to the extent of 11% and SF_3 complex has a relative abundance of 65.4%.

With the continuation of sparking for 10 hours, the presence of SF_5 is seen in larger proportion and the presence of SF_3 is significant in addition to SF_2. The abundance of SF is 4.7% and of SF_2 is 4.4%. On the other hand, SF_3 is significant with the relative abundance percentage of 18.11%. In addition to these ions, complexes involving SF_4 account for 0.8% and even SF_4 is also present and its percentage is 6.8. SF_4 complex is observed in a very low concentration of 0.9%.

4. BY-PRODUCTS IN 80:20 N_2 - SF_6 GAS MIXTURES

In this case, SF and SF_2 have a relative abundance of 4% each after 10 hours of sparking at 30 kV under point-plane field configuration. In fact, there was no difference in the species and their relative concentrations after 6 and 10 hours of sparking at 30 kV. As in the earlier cases, SF_3 has the highest relative abundance of 17.45% and that of SF_4 is 7%. Complexes involving SF_2, SF_3 and SF_4 or similar species are observed in relative concentrations varying from 1 to 4.75 %. Thus the comparison of by-products of SF_6 and SF_6 - N_2 gas mixtures show similar trends with SF_3 being the major constituent.

5. BY-PRODUCTS WITH CYCLOALIPHATIC RESIN

In this case, SF has a relative abundance of 4.3%, SF_4: 3.1% and SF_2: 4.09%. However, SF_3 is a major constituent with a relative abundance of 22%, followed by SF_4 with 8%. Complexes involving SF_4 are seen to the extent of 0.81 %; those involving SF_3 are about 4%. These complexes are also significant when sparking is continued for 8 hours. After 12 hours of sparking there is a slight increase in abundance of SF_5 (5.5%), SF4 (4.7%), SF_2:(4.38%), SF_3 (23.25%), and SF_4 (8%). Complexes of SF_3 and SF_5 are less than 1%. Thus the presence of CAR does not result in formation of new species as compared to the results in SF_6 and SF_6/N_2 gas mixture (20:80). Thus CAR is observed to be stable and does not contribute any new species when sparked in N_2-SF_6 gas mixture.

6. BY-PRODUCTS WITH POLYESTER

After 6 hours of sparking with polyester, it is observed that all the species reported earlier are formed are in a very low concentration. With 12 hours of sparking, there is a significant change in the concentration of various constituents formed. After 16 hours of sparking, there is a further decrease in relative concentration of different species. Complexes involving SF_5 (lower), SF_5 (higher), SF_4+SO_2F and SF_5NF_3 are also observed in a very low percentage.

Though these complexes are present, they do not seem to increase in concentration with duration of sparking.

7. BY-PRODUCTS WITH NYLON

In presence of nylon, in 80:20 N_2 - SF_6 gas mixture the relative abundance of SF is 7.6%, SF_4 6%, SF_2 6%, SF_3 27%, SF_4 9.35% respectively. The complex ions involving SF_3 has a relative abundance of 1.53%, SF_5 (lower): 1% and SF_5 (higher): 4.6%. Thus the trend seen in N_2 - SF_6 mixture is also evident in nylon. The presence of SF_3 to the extent of 27% is a significant factor, which is observed in other cases, except polyester.

The sparking was continued for 12 hours and the relative abundance of SF was 5%, SF_4^{++} is 3.8%, SF_2 is 4.36%, SF_3 is 21.8% and SF_4^+ 8%. Though SF complexes are observed, their concentrations are significantly low (less than 1%) except SF_5, which is 3.8%. Also present in the gas mixture, though to a very low extent is SF_3NF [8], which was not observed in presence of cycloaliphatic resin spacer or for that matter in SF_6 and N_2SF_6 gas mixture.

8. BY-PRODUCTS WITH POLYPROPYLENE

Even in case of PP, SF has a relative abundance of 3%, SF_4: 2.7%, SF_2: 3.12%, SF_3: 19%, SF_4^{++}: 7.75%. Though several complexes involving SF_2, SF_3, SO_2F_{10}, SO_2+SO_2F, $S_2O_3F_6$, SF_5+SF_5 etc. are formed, their concentration is insignificant. Thus PP appears to be quite stable and major by-products formed are almost identical.

9. BY-PRODUCTS WITH PMMA

With PMMA material after 6 hours of sparking in 80:20 N_2/SF_6 gas mixtures, the relative abundance of SF in this case is 2.7%, and SF complex is 2.45, SF_4 is 2.42, SF_2^+ is 2.77, SF_3 is 17 % and SF_4^+ is 7%. Species like $S_2O_3F_6$ and SF_4+SOF_2 though observed are in very low concentration.

The sparking was continued for 10 hours and SF content is observed to increase to 3.88, SF_4 to 3.55, SF_2 to 3.5. Thus on an average SF, SF_4 and SF_2 have a relative abundance of 3.7%. Even SF_3 content increases to 20% while SF_4 increases to 8%. SF_5 complexes also increase to 4.5%, which was not observed earlier.

With further increase in the sparking duration to 16 hours, SF concentration goes up further to 4.35%, SF_4 to 3.9% and SF_2 to 3.8%. Thus all the species show increase in relative abundance as compared to the results of sparking for 8 hours. In case of SF_3, increase in relative abundance (%) is from 20 to 21.65% when sparking duration is increased from 10 to 16 hours and SF_4 remains almost constant at 8%. Even concentration of SF_5 complex is almost constant at 4.3%.

10. BY-PRODUCTS WITH PTFE

SF_6 - N_2 (20:80) gas mixture was sparked in presence of PTFE disc at 30 kV initially for a period of two hours. Because of the popularity of this material for many applications, it was decided to monitor closely the by-products formed from the initial stages. In this case, SF has a relative abundance to 2.6%, SF_4 0.9 %, SF_2 2.8 %, SF_3 15.64 %, and SF_4^+ 6.5 %. In addition HN_3O_3, C_4NO_3, $C_4H_2NO_4$, $C_3H_2N_2O_4$, and SF_5NF_3 are also formed. Some of these fragments are peculiar to PTFE. Thus SF_3 is a major constituent, in addition to SF_2, SF, and SF_4. Similar results are also obtained after 6 hours of sparking. After 8 hours of sparking, species like HN_3O_3, C_4NO_3, $C_4H_2NO_4$ and $C_3H_2N_2O_4$ are absent but SF_5NF_3 is present in a very low concentration.

With the sparking being continued for 10 hours, the relative abundance of SF was 4.37%, SF_4: 3.88%, SF_2: 2.7 % and SF_4: 7.58%.

11. CONCLUSIONS

The following are the important conclusions of this study:

(1) In pure SF_6 gas, SF percentage increases gradually with sparking from 4 hours to 6 hours, but from 6 hours to 10 hours, there is an indication of recombination and its percentage goes down. On the other hand even SF_4 seems to show the same trend and its percentage increases initially but tends to decrease with continuous sparking. SF_3 is always formed with the highest relative abundance and a decrease in its concentration is evident as the duration of sparking is increased. Similar results are also seen with SF_6– N_2 gas mixture.

(2) With the CAR, SF_3 is a major constituent of the spark by-products followed by SF_4. With sparking of longer duration, the abundance of SF_5 increases.

(3) With PE, SF_3 is again a major constituent of the by-products followed by SF_4. The presence of SF_4+ SOF_2 points towards formation of more such complexes or fragments with further increase in duration of sparking.

(4) With the increase in duration of sparking, N_2 has a tendency to inhibit the formation of certain by-products, seen in SF_6 and SF_6 - N_2 gas mixtures.

12. ACKNOWLEDGEMENTS

The authorities of CPRI, Bangalore, India are thanked for their support and permission to publish this work

13. REFERENCES

[1] F.Y.Chu, "SF_6 Decomposition in Gas Insulated Equipment", IEEE Trans. On Electrical Insulation, vol.E1-21, pp 693-725, 1986.

[2] CIGRE Working Group 23-03, "Handling of SF_6 and its Decomposition Products in Gas Insulated Switchgear", Parts 1 & 2, Electra, Vol.136 & 137, 1991.

[3] CIGRE Working Group 23-10, "SF_6 and Global Atmosphere", Electra, Vol. 154, February 1996.

[4] L.G. Christophorou, J.K. Olthoff and D S"Green Gases for Electrical Insulation and Arc Interruption: Possible Present and Future Alternatives to pure SF_6, NIST Technical Note 1425, National Institute of Standards and Technology", Gaithersburg, USA, 1997.

[5] A.M.Casanovas, L. Vial, I. Coll, M. Storer, J. Casanovas and R. Claveul, "Decomposition of SF_6 under ac and dc Corona Discharges in High-Pressure SF_6 and SF_6/N_2 (10-90%) Mixtures", pp 379-384, Gaseous Dielectrics VIII, edited by Loucas G Christophorou and James K Olthoff, Kluwer Academic/Plenum Publishers, NewYork, 1998.

[6] I. Coll, A.M. Casanovas, C. Pradayrol and J Casanovas, "Influence of a Solid Insulator on the Sark Decomposition of SF_6 and SF_6 + 50% CF_4 Mixtures", pp387-393, Gaseous Dielectrics VIII, edited by Loucas G Christophorou and James K Olthoff, Kluwer Academic/Plenum Publishers, New York, 1998.

[7] J. Castonguay, "Chemical Reactions and Kinetics of Mixtures of SF_6 and Fluorocarbon Dielectric Gases in Electrical Discharges", Gaseous Dielectrics VIII, pp 395-401, edited by Loucas G Christophorou and James K Olthoff Kluwer Academic/Plenum Publishers, New York, 1998.

[8] C.T. Dervos and P. Vassiliou, "SF_6 Handling and Maintenance Processes Offered by Quadrupole Mass Spectrometry", Gaseous Dielectrics VIII, pp 417-422, edited by Loucas G Christophorou and James K Olthoff, Kluwer Academic/Plenum Publishers, New York, 1998.

[9] Constantine T. Dervos and Panayota Vassiliou "Byproducts in the insulating gaseous matrix of a GIS', Gaseous Dielectrics IX, edited by Loucas G Christophorou and James K Olthoff, Kluwer Academic/Plenum Publishers, New York, 2001.

[10] E.J. Dolin, "The United States Environmental Protection Agency's SF_6 Emissions Reduction Partnership for Electric Power Systems: An Opportunity for Industry", pp 425-430, Gaseous Dielectrics VIII, edited by Loucas G Christophorou and James K Olthoff, Kluwer Academic/Plenum Publishers, New York, 1998.

THERMALLY STRESSED SF_6 EFFECTS ON CONSTRUCTION METALS

Constantine T. Dervos, Panayota Vassiliou and John A. Mergos*

1. INTRODUCTION

Sulfur hexaflouride (SF_6) is a man-made gas[1], offering excellent electrical, thermal, and acoustical insulation properties, and as such, it has been extensively employed for a variety of applications, i.e. power distribution (mainly), semiconductor foundries, metallurgy, tracer gas, sound sealing (double glazing, torpedo noise silencer), thermal sealing, diving equipment, aeronautical, wind tunnels, telecommunications, relays, antennas, medical (ultrasounds, ophthalmology) etc [2]. SF_6 is primarily employed worldwide (approximately 80% of its total production) as a gaseous insulator for high-voltage Gas Insulated Switchgear (GIS) systems, including gas insulated transmission lines, and power transformers.

The chemical stability of SF_6 and of its mixtures with inert gases such as N_2, or/and CO_2 is taken for granted by the producers and the users. In this work an attempt is made to investigate the effect of SF_6 on construction materials after heating at temperatures as high as 450°C. The possible decomposition of the insulating gas and the long term byproduct formation, caused either by heating to this temperature, or/and by accelerated decomposition effects due to corrosion products formed on the metallic surfaces, has also been examined. The employed metals are Cu and Al, which are commonly encountered in GIS equipment either as components in bus bars, connectors, field smoothing devices, windings and tap changers of transformers.

* Constantine T. Dervos, J. A. Mergos, School of Electrical and Computer Engineering, Panayota Vassiliou, School of Chemical Engineering, National Technical University of Athens, 9, Iroon Polytechniou Str., Zografou 157 73, Athens, Greece.
emails: cdervos@central.ntua.gr , pvas@orfeas.chemeng.ntua.gr

Gaseous Dielectrics X.
Edited by Christophorou *et al.*, Springer, New York, 2004.

The electrical stressed SF_6 gas byproduct formation has been described elsewhere[3]. Owing to the reaction tendency of the fluorine and the sulfur atoms, any impurity (e.g. metallic vapors, organic vapors) that may be present in the gas matrix, will prevent SF_6 from recombining, thus producing fragmentation and/or chemical dissociation leading to deterioration of its dielectric properties. In the presence of an arc or a spark, the metallic electrodes of a GIS vaporize, thus reacting readily with the fluorine or sulfur atoms. After chemical reaction of SF_6 with metallic surfaces the possible byproducts are for the studied metals:

$Cu + 2F \rightarrow CuF_2$ $\quad\quad$ $2Al + 3S \rightarrow Al_2S_3$

$Cu + S \rightarrow CuS$ $\quad\quad$ $Al + 3F \rightarrow AlF_3$

The use of SF_6/N_2 (10-90) mixture causes (independently of the presence of oxygen, water or organic insulator) a drop in the formation of the major decomposition products (SF_4+SOF_2 and S_2F_{10}) of SF_6 under the effect of sparking. This drop is due to the lower initial proportion of SF_6 in the mixture. But when this production is expressed per unit SF_6, the higher levels of degradation product in the mixture, compared to SF_6 become apparent. Thus, since the recombination of SF_x fragments diminishes, other reactions of oxidation and/or hydrolysis are favored leading to the formation of oxyfluorides, as in the case of room temperature corona discharges[4].

In a relevant experimental work[5] it was investigated by differential thermal analysis (DTA) the effect of SF_6 on metals and on their oxides. The work concludes that SF_6 reacts with metals at about 500-600°C to produce sulfides of the metals. Specifically for Cu, there were no thermal effects registered by DTA but it was found a 30% change in weight. For Al, there was a thermal effect at 560°C but no change in weight.

In this work the presence of dry O_2 in a small concentration in thermally stressed SF_6 matrix is also examined, since in real conditions practically no high vacuum is attained when power company employees replenish the SF_6 containers of the power units. Vacuum practically attained in GIS installation is of the order of 10mbar before gas insertion.

2. EXPERIMENTAL

The reactor employed was a cylinder of stainless steel (DIN 17440) of a volume of about 2 l with an entrance (3.5cm diameter) for a solid specimen to be inserted. The reactor was connected to a gas analysis unit by a system of valves and at the same time to gas cylinders either of SF_6 or O_2. A heating mantle controlled externally was strapped to the reactor body. The reactor through piping and vacuum valves was connected to high vacuum pumps to empty or fill the gas. The gas analysis system was a quadrupole mass spectrometer located in a near by Ultra High Vacuum environment. The employed quadrupole mass analyzer has been extensively described elsewhere[3]. The experimental setup of the reactor is shown in Fig. 1. The solid byproducts on the exposed coupons surface were analyzed by Scanning electron micrography (SEM) coupled with Electron diffraction spectroscopy analysis (EDS), to identify the surface compounds.

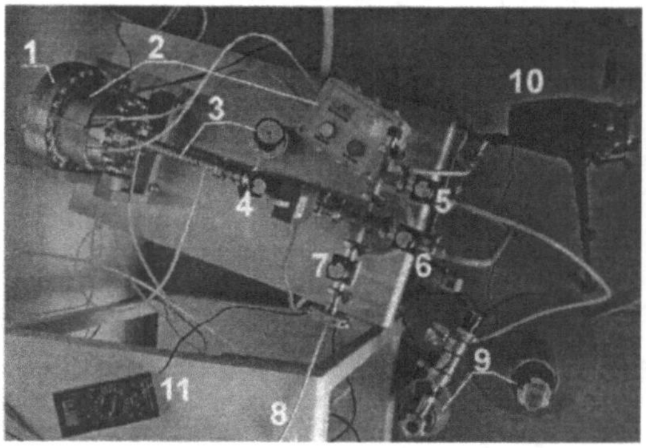

Figure 1. The experimental setup with the reactor and the measuring system.

1. Stainless steel reactor
2. Heating jacket and controller
3. Cooling system and thermocouple
4. Gas chamber isolation valve
5. Gas dosing valve
6. Vacuum pump connector valve
7. Gas connection valve to quadrupole mass analysis unit (QMA)
8. Line to QMA unit
9. Gas cylinders (SF_6 and O_2)
10. Infrared exterior thermometer
11. Thermocouple for reactor temperature monitoring

Coupons of metals of deoxidized electrolytic copper quality (OFHC), annealed before usage and of aluminum-magnesium alloy 5054 were inserted in separate runs in the reactor chamber and after sealing, the reactor, was vacuum emptied and then filled with SF_6 either pure or with oxygen (0.5% v). To fill the system for the experiments with the gas mixture, it was at first emptied, then flushed with oxygen, then emptied to a specified vacuum level and filled with SF_6 to 1.2atm. The system was heated to 450°C and kept to this temperature for 4 hours. The gas, before entering the mass spectrometer, was cooled to near room temperature by a water cooling coil attached to the outgoing gas line from the reactor. Thus, the long term formed byproducts could be monitored at room temperature. During each experimental run there were several gas spectra recordings performed to check on the possible changes of concentrations of gas by-products during the heating process. At the end of the 4h exposure the system was emptied of gases and was opened to extract the metal coupons for surface analysis. The microstructure was also examined for heat effects by measuring the Vickers microhardness of the metals. The corrosion products are examined as previously mentioned.

The Vickers microhardness measurements (nine tests) under an optical microscope of both types of metals tested gave the following results: Copper before heating: 138HV and after heating 118HV and for aluminum before heating 101HV and after heating 102HV. These results signify that for aluminum there was not any structure change affected by the thermal process. but for copper there was a change caused by a recrystallization process in the mass of the metal.

Figure 2. SEM micrograph x50 of the copper metal surface after reaction to SF_6 and O_2 for 4h at 450°C. The scales of copper sulfide are fragile and detach from the surface.

Figure 3. SEM micrograph x1000 of the black gelatinous scale of copper sulfide formed on the copper surface.

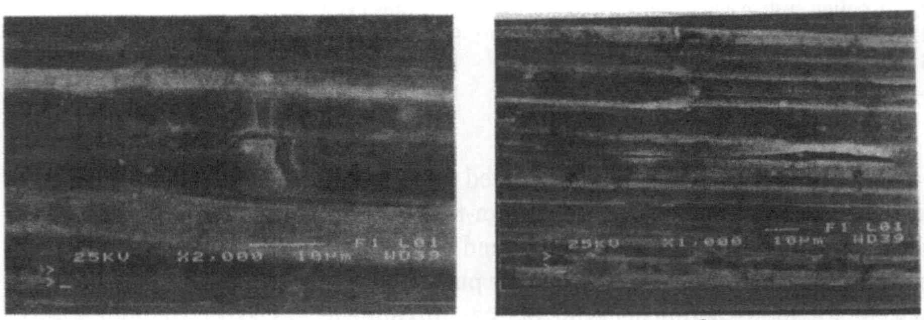

a b

Figure 4. SEM micrograph x2000 (a) and x1000 (b) of the aluminum surface after reacting with SF_6 after 4h at 450°C. The aluminum has been affected by fluorine and forms a thin layer of AlF_3.

Copper coupons after their reaction to the mixture of SF_6 and O_2 for 4h at 450°C form scales of copper sulfide which are fragile and detach easily from the substrate (Fig.2). The scale is made of Cu_2S (Fig. 3) and also underneath this scale an oxide of copper is observed with small pits created on the surface of the metal. The copper flanges of the rector vacuum system were also affected after heating forming a very thin oxide layer with optical fringes, as well as sulfur compounds. Aluminum, on the other hand, reacts with SF_6 and forms a thin AlF_3 overlayer, which is very adherent to the substrate (Fig. 4). No reaction with sulfur has been observed on the aluminum surface, contrary to what was observed for copper. The aluminum metal has a thin oxide layer on its surface, spontaneously formed by its reaction to air. This oxide might react with SF_6 and enhance its decomposition rate. Thus, the fluorine produced reacts with aluminum to form the observed aluminum fluoride. The presence of the different compounds on the surfaces has been identified by EDS analysis.

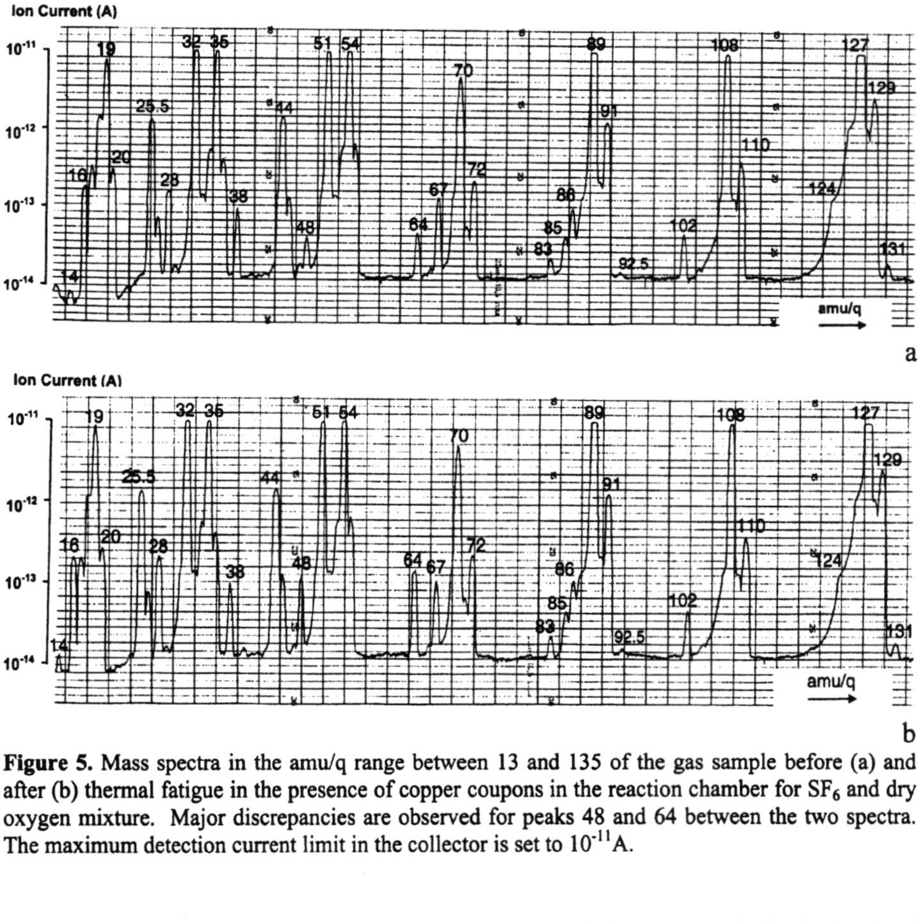

Figure 5. Mass spectra in the amu/q range between 13 and 135 of the gas sample before (a) and after (b) thermal fatigue in the presence of copper coupons in the reaction chamber for SF_6 and dry oxygen mixture. Major discrepancies are observed for peaks 48 and 64 between the two spectra. The maximum detection current limit in the collector is set to 10^{-11}A.

The quadrupole mass spectrometry characteristic peaks of singly ionized fragments of pure dry SF_6 are encountered at the following amu/q positions: $F^+ = 19$, $^{32}S^+ = 32$, $^{32}SF^+ = 51$, $^{32}SF_2^+ = 70$, $^{32}SF_3^+ = 89$, $^{32}SF_4^+ = 108$, and $^{32}SF_5^+ = 127$. Doubly ionized SF_6 fragments may be detected at the following amu/q positions: $^{32}SF^{++} = 25.5$, $^{32}SF_2^{++} = 35$, $^{32}SF_3^{++} = 44.5$, $^{32}SF_4^{++} = 54$, and $^{32}SF_5^{++} = 63.5$. Singly ionized fragments formed by the ^{34}S isotope, having natural abundance 4.2%, are shifted by two mass units from the corresponding ^{32}S ions. For the doubly ionized fragments this difference turns out to be only one mass unit.

$^{32}SF_3^+$ and $^{32}SF_5^+$ provide large characteristic peaks and therefore, suitable mass tracking sites for SF_6 detection in air. The additional peaks recorded in the gas matrix were detected at the following amu/q positions: $O^+ = 16$, $OH^+ = 17$, $H_2O^+ = 18$, $N_2^+ = 28$, $HF^+ = 20$, $H_2S^+ = 34$, $CO_2^+ = 44$, $SO^+ = 48$, $SO_2^+ = 64$, $SO_3^+ = 80$, $SOF^+ = 67$, $NF_3^+ = 71$, $SOF_2^+ = 86$, $SOF_4^+ = 124$, $SO_2F^+ = 83$, $SO_2F_2^+ = 102$, $SO_5F^+ = 131$, $SO_2F_4^+ = 140$, and $S_2F_{10}^+ = 254$.

Typical SF_6-O_2 spectra before and after thermal fatigue for the case of copper coupons in the reactor are shown in Fig. 5. It has been observed that SO^+ and SO_2^+ concentrations increased by a factor of five after the heating process, while all other peaks remained practically unaltered. Also, SO_3^+ becomes evident in the gas matrix after heating at sub-ppm levels.

In the case of the aluminum, enhanced sulfur oxides concentration after heating were observed as well. Additionally, at position 85 the ion current peak was intensified by a factor of six. This mass number could be possibly attributed to an sp3 hybridization of sulfur generating the unstable radical: $H_2SO_2F^+$.

The experimentally observed increase of sulfur oxides after heating. could be possibly attributed to the catalytic action of the metal surface that under thermal excitation dissociates the SF_6 gas at much lower temperatures than frequently referred in the literature. This has to lead to reconsideration of the chemical stability issue of the SF_6 when heated in the presence of GIS construction metals.

3. CONCLUSIONS

The metals copper and aluminum in the presence of SF_6 with O_2 and heated to 450°C react with the gas and produce corrosion products, either with sulfur or with fluorine. Aluminum reacts with fluorine and copper with sulfur. The gas byproducts are also changed towards sulfur oxide formation. The issue of the inertness and chemical stability of heated SF_6 has to be reconsidered in all relevant practical applications.

4. REFERENCES

1. H. Moissan and P. Lebau, *Compt. Rend. Acad. Scin. Paris*, 130, 865-984 (1900).
2. C.T. Dervos and P. Vassiliou, Sulfur Hexafluoride (SF_6): Global environmental effects and toxic byproduct formation, *J. Air & Waste Manage. Assoc.* **50**, 137-141, (2000).
3. C.T. Dervos and P. Vassiliou, Byproducts in the insulating gaseous matrix of GIS, In *Gaseous Dielectrics IX*, Christophorou L.G. and Olthoff J.K. Eds; (Kluwer Academic/Plenum Publishers, New York, 2001), pp.403-412.
4. R.J. Van Brunt, Production rates for oxyfluorides SOF_2, SO_2F_2 and SOF_4 in SF_6 corona discharges, *J. Res. Nat. Bur. Stand.*, **90** (3), 229-253 (1985).
5. A.A.Opalesky and E.U. Lobcov, Interaction of Sulphur Hexafluoride with Metals and Oxides, *Journal of Fluorine Chemistry*, 2, (5), 349-359 (1973).

CBM OF HV SWITCHGEAR USING ON-LINE PD MEASUREMENTS

Robert G.A. Zoetmulder, Sander Meijer, Johan J. Smit[1]

1. INTRODUCTION

Condition Based Maintenance (CBM) is a concept already applied to HV switchgear. The principle of CBM is that the condition of the high voltage insulation of the equipment is assessed by inspection and diagnosis. For this purpose, techniques that detect partial discharge activity can be used. The maintenance actions are only performed when necessary. The CBM process consists mainly of five items[1]:

1) Detection of PD activity
2) Defect identification
3) Defect location
4) Risk assessment
5) Economical aspects

In this paper the ultra high frequency (UHF) technique has been used to detect partial discharges (PD) activity inside the equipment. To identify the origin of the measured signals, raw data such as single pulse, patterns, shape, crest factors, etc, has to be analyzed. Also physical defect models and system-knowledge is necessary for a well-defined basis for the CBM process. The location of the PD source can for example be found by means of flight time measurements[2].

Risk analysis is needed to obtain the criticality of the defect. Although it is applied to HV switchgear for several years, the part of risk-analysis needs more investigation. For example the risk of different moving particles should be mapped. When the economical risk is reasonably high, it can be decided to maintain the installation.

---

[1] Robert G.A. Zoetmulder, Sander Meijer and Johan J. Smit, Delft University of Technology, High Voltage Technology and Management, Delft, The Netherlands.

In this paper the applicability of UHF PD measurements in HV switchgear has been investigated during laboratory, see figure 1, and on-site testing[3].

Figure 1: PD-monitoring set-up used in the laboratory existing of the following components:

1a- internal UHF-coupler
1b- external UHF-coupler
2- preamplifier
3- spectrum analyzer
4- computer

2. PD DETECTION: UHF MEASURING TECHNIQUE

Partial discharges in switchgear are caused by defects of the insulating system such as protrusions and particles. The discharge currents at the defects have rise-times that can be less than one hundred pico-seconds[4]. As a result, electromagnetic (EM) wave transients with frequency content of more than 2 GHz are emitted[5]. These signals propagate within the coaxial or rectangular bays of a switchgear system not only in the basic mode (TEM) but also in many higher order modes (TE_{mn},TM_{mn}).

At locations where the enclosure is interrupted by a dielectric material such as: epoxy resin, glass inspection window, gas rupture, etc., these high frequency signals will couple out of the switchgear. These "dielectric windows" offer the possibility to pick-up the signals by means of external couplers, which is often applied in older types of installations. In new types of GIS's usually internal sensors are installed.

The amplitude of the signals reduces due to damping but mainly due to reflections at discontinuities such as circuit-breakers and disconnectors. As consequence of UHF signal reduction, several couplers have to be installed or external couplers can be used to monitor at more locations.

In case of UHF measurements with internal couplers, the enclosure of the installation will function as a shield for most of the extern signals. When external couplers are used, more background noise signals can be measured. Often noise peaks come from navigation air-traffic, TV broadcast and GSM.

The advantage of using external UHF sensors is the cost effectiveness and no invasion into the initial switchgear design. When the sensor is placed at the "dielectric window" its detection sensitivity varies widely depending on the type of out-coupling of the EM-waves and in general to the geometries of the bay or installation-section[6]. This results in the need for an on-site sensitivity check, which can be performed during a scheduled outage. A procedure for the on-site sensitivity check of the UHF monitoring system with the use of external couplers is reported in[6].

The following measuring procedure was followed to assess the insulation condition of HV installations:

- A background noise measurement was performed. Here the external UHF coupler is usually placed in front of the HV bay to obtain all EM-wave frequency components at the outside of the installation.
- Secondly a signal spectrum was measured at a dielectric window of the installation. The measuring locations were e.g. gas rupture disks, spacers and cable terminations.
- A signal to noise ratio was made, to determine frequency peaks probably due to PD activity
- A center frequency is chosen to make a phase resolved PD pattern
- Then the PD level is estimated with the data from the sensitivity check performed on the different installations.

3. TYPICAL PD SOURCES

Although the GIS installation is very reliable, still defects can occur. The behavior of occurrence of the defects can be random and are not only present due to aging effects.

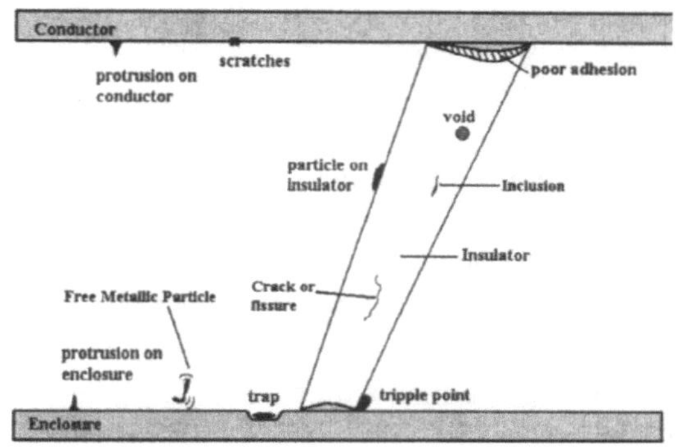

Figure 2: Typical defects which can occur in GIS installations.

Examples of typical defects which can occur in GIS are, see figure 2:

- Protrusions on conductor,
- free (conducting) particle,
- (conducting) particle stuck on spacer
- defects in spacers of support insulators (inclusions, voids, cracks)
- electrically floating shields

4. LOCALIZATION

The location of the PD source can for example be found by means of time-of-flight measurements[2]. A PD pulse will be picked up at two couplers. The Δt, which is measured between the two couplers, is a measure for the location of the defect.

5. FIELD EXPERIENCE

After exploring the measurement technique in the high voltage laboratory, field measurements are needed to confirm a proper 'working' of the measurement system. As an example, a 150 kV and a 380 kV GIS have been investigated.

During earlier on-site/on-line testing at the 150 kV and the 380 kV GIS installations, an adapted sensitivity check has been carried out [3, 6]. Based on this adapted sensitivity check it is possible to estimate the PD amplitude of the defect in pC.

5.1 Example 1: 150 kV GIS

Measurements have been performed at a 150 kV GIS installation in the Dutch utility grid. For the measurement external UHF couplers were used. The external UHF coupler is fixed to the gas rupture disk, see figure 3. At a few locations at this installation PD activity was measured.

One and a half year ago, at one bay, which was connected to a power plant, a flashover occurred. After the outage and repair of that specific bay of the 150kV GIS, a test was performed with a classical PD test set-up according IEC 60270. With the classical PD detection no PD activity was recorded. During later online PD inspections this specific location of the generator-connecting bay, was also inspected on of PD activity. The measurement procedure was followed according the earlier described procedure. In this case the coupler was fixed to the gas rupture disk. The background- and signal-spectrum can be seen in figure 4a.

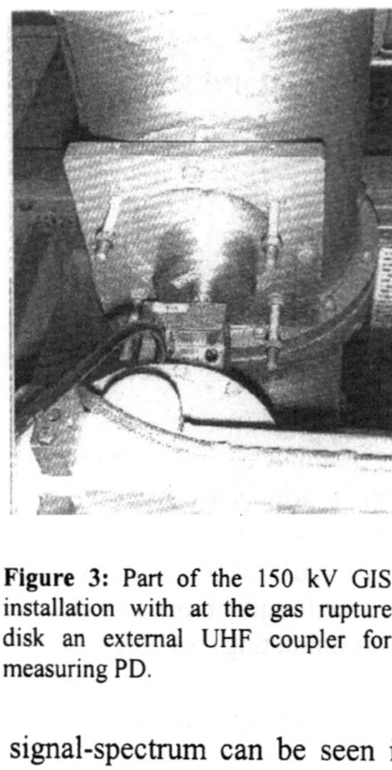

Figure 3: Part of the 150 kV GIS installation with at the gas rupture disk an external UHF coupler for measuring PD.

Figure 4b shows the signal-to-noise ratio. Around 260 MHz a positive frequency peak can be seen in this signal-to-noise ratio. The reason for the negative results in the signal-to-noise ratio can be explained by the fact that the noise spectrum is recorded with the sensor in front of the GIS and the signal spectrum against an aperture in the enclosure. Due to the screening of the enclosure and sensor, the background noise is reduced by about 20 dB. As a result, all peaks in the frequency spectrum coming from noise become negative after a signal-to-noise ratio. The phase resolved pattern can be seen in figure 4C.

According to the sensitivity-check 22μV corresponds to about 8 pC. The defect at the found location had an estimated PD amplitude between the 10 and 20 pC, which is higher than the allowed limit of 5 pC. Although the found location is monitored several times, the PD activity was not always detectable. Thus in this case continuous online PD measurements are more likely preferred above periodical online/onsite PD measurements. Moreover because of this irregular pattern, the fact that the type of defect is yet unknown and thus its criticality, it was decided to leave the installation as it is and continue with periodic inspections.

5.2 Example 2: 380 kV GIS

A second investigated installation was a 380 kV GIS installation of 30 years old. Originally this installation was designed for the application as an air insulated installation used for a voltage level of 150 kV. Using SF_6 gas as an insulating medium and using different circuit breakers, it was upgraded to a 380 kV installation.

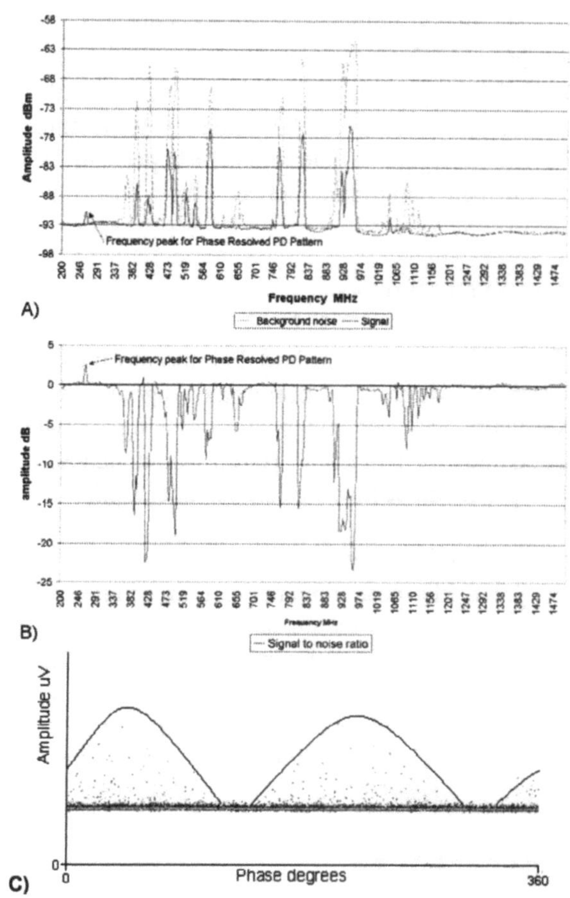

Figure 4: A) Frequency spectra and the signal to noise ratio of the 150 kV GIS installation measured at the suspicious location. B) Signal to Noise ratio of the background noise and signal spectra at the suspicious location. C) Phase Resolved PD Pattern.

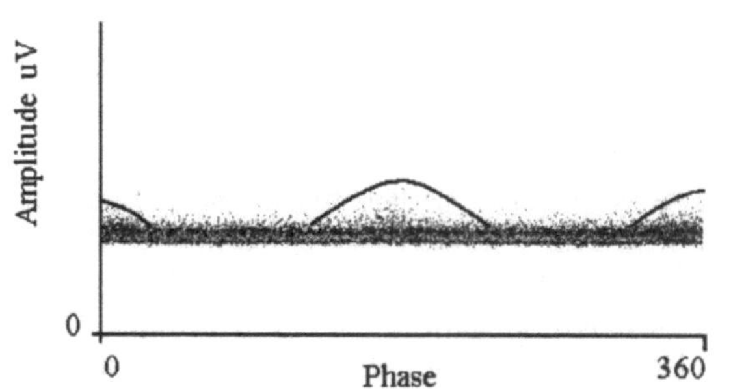

Figure 5: Phase Resolved PD Pattern of the defect measured at the 380 kV GIS installation bay.

At this installation measurements with external UHF couplers have been performed. These couplers could be fixed to the spacers of the installation. In our case the coupler was fixed to the spacer near the busbar-disconnector.

A phase resolved PD pattern was obtained at a center frequency of 816,3 MHz, see figure 5. The highest PD peak in the pattern was 14,6 μV resulting in an estimated amplitude between the 5 and 10 pC.

6. CONCLUSIONS

A portable on-site/on-line PD monitoring system has been developed using external couplers. Appropriate signal conditioning, data-acquisition by a spectrum analyser and post-processing software makes it possible to create a good tool for Condition Based Maintenance. Based on the results described in this paper, the following can be concluded:

- Measurements with the external UHF coupler are applicable to different types of HV switchgear systems.
- UHF PD measurements can contribute to CBM procedures, because on-line assessment of the insulation condition is possible.
- Using the adapted sensitivity check, PD magnitudes in older GIS can be estimated when measuring with an external UHF coupler.
- In the given examples continuous PD monitoring is preferred above periodical PD monitoring.

7. REFERENCES

1. CIGRE 15/23-0, 1992 session, 30 August- 5 September, "Diagnostic Methods for GIS Insulating Systems", Working group 15.03
2. L.E. Lundgaard, M. Runde, B. Skyberg, October 1990, "Acoustic diagnosis of SF_6 gas insulated substations: a theoretical and experimental basis ", IEEE Transaction on Power Delivery, Vol.5, No.4, pp.1751-1759,.
3. Sander Meijer, Robert G.A. Zoetmulder, Johan J. Smit, Edwin R.S. Groot, Martin D. Judd, 2000, "Development and application of a portable multipurpose on-site/on-line PD measuring system", DMMA conference publication, IEE, No. 473.
4. M.D. Judd, O. Farish, June 1998, "High Bandwidth Measurement of Partial Discharge Current Pulses", ISEI paper of IEEE, Vol. 2, pp. 436-439,.
5. B.F. Hampton, R.J. Meats, 1988, "Diagnostic measurements at UHF in gas insulated substations", IEE Proceedings, Vol. 135, Pt. C, No. 2, pp137-144,.
6. Robert G.A. Zoetmulder, July 11th 2000, "Partial Discharge Diagnostics by External Monitoring with UHF couplers", Masters Thesis, Delft University of Technology, Delft, The Netherlands, 2000.

SECTION 8: GAS-INSULATED TRANSMISSION LINES

IMPROVEMENT OF SF_6 GAS INSULATION SYSTEM BY APPLYING THICK DIELECTRIC-LAYERED ELECTRODE AND ITS APPLICATION TO HIGH VOLTAGE EQUIPMENT

H. Hama, K. Inami, T. Ootsuka, Y. Shimizu, E. Nagao, K. Oosumi and M. Yoshimura*

1. INTRODUCTION

SF_6 has widely been used as insulating and arc-quenching media for high voltage (HV) power equipment such as gas insulated switchgear (GIS), gas circuit breaker (GCB), gas insulated bus (GIB), gas insulated line (GIL) and gas insulated transformer (GIT) which contribute to achieve compactness and high reliability of the equipment. Improvement of SF_6 insulation system in the HV power equipment is however required, if we want a further downsizing of the equipment from a view point of saving the investment cost or reducing SF_6 usage that is identified as one of the potent green house gases.

A thin dielectric coating on bare electrodes[1] is a well-known method to increase breakdown voltages for the HV equipment. However, area effect of breakdown cannot be controlled completely when effective electrode area is infinitely large.

The concept on hybrid insulation system featuring compressed SF_6 gas and a thick dielectric layer[2] is proposed for GIL. A conductor covered with thick solid insulator of polyethylene (PE) is located at the center of the enclosure. The PE with higher insulation performance than SF_6 insulates the HV conductor surface, and the other areas are by SF_6.

The breakdown characteristics of thick dielectric layer made of cast epoxy on electrode under low SF_6 pressure up to 0.2MPa were reported.[3] However, the fundamental characteristics of the insulation system and the data at high pressure, which are important for insulation design of the HV equipment, are not

* H. Hama, K. Inami, T. Ootsuka, Y. Shimizu, E. Nagao and K. Oosumi, TMT&D Corporation, Hyogo, JAPAN 661-8661. M. Yoshimura, Mitsubishi Electric Corporation, Hyogo, JAPAN 661-8661.

Gaseous Dielectrics X.
Edited by Christophorou *et al.*, Springer, New York, 2004.

sufficiently reported so far.

The paper deals with the improvement of a current SF_6 gas insulation system of the HV equipment by using thick dielectric-layered electrode (TDLE). The goal of the study is to clarify the fundamental properties of TDLE and to apply the TDLE techniques to downsize the equipment. Firstly, we investigated the breakdown properties of TDLE in SF_6 gas at lightning impulse (LI) voltages at high pressure up to 0.6MPa, comparing with the properties of bare electrodes. Secondly, the properties of surface degradation and puncture of the dielectric layer by repetitive discharge in SF_6 were also studied to use the TDLE as the substitutes of metallic electrodes. Finally, the examples of applying the TDLE to the HV equipment are introduced in view of downsizing the components of the equipment.

2. BREAKDOWN PROPERTIES OF TDLE IN SF_6

2.1. Experiments

We used two kinds of TDLE test samples for the experiments. A coaxial electrode configuration (Type 1) using an actual bus of 5,200mm in length and 340mm in diameter[4,5] and a rod to plane electrode configuration[6] (Type 2) are shown in Figs. 1(a) and (b), respectively. The embedded electrode was molded in

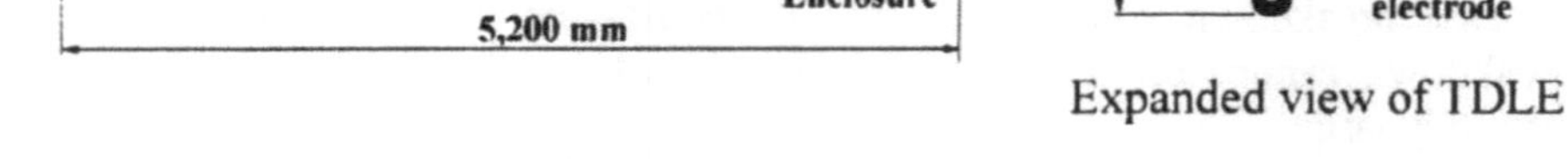

(a) Type1 TDLE of coaxial electrode configuration using actual bus.

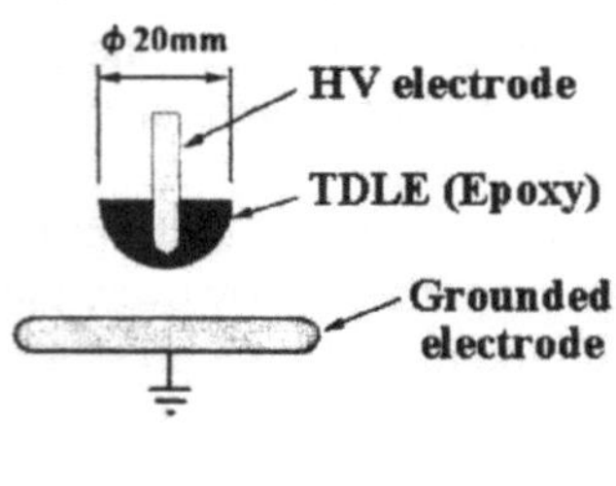

(b) Type2 TDLE of rod to plane electrode configuration.

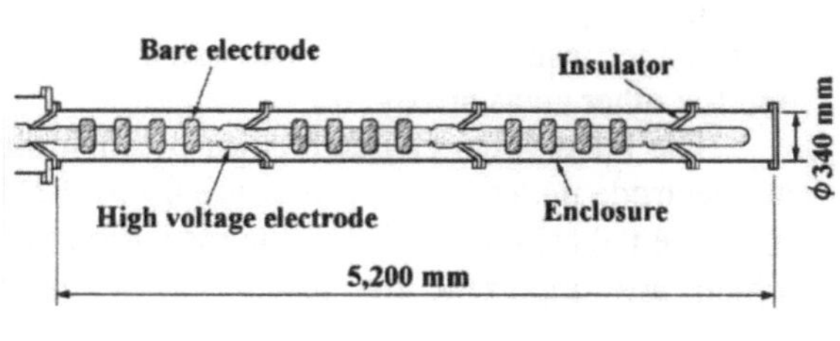

(c) Bare electrodes of coaxial electrode configuration using actual bus.

Figure 1. Experimental setups and test samples of TDLE and bare electrodes.

the cast epoxy of which the relative permittivity is 4.0 for Type 1 and Type 2.

In the case of Type 1, the outer diameter and the thickness of TDLE are 270mm and 30mm, respectively. The TDLE was set on the HV electrode of the bus. As to Type 2, the outer diameter is 20mm and the minimum epoxy thickness at the rounded tip of 4mm diameter HV embedded electrode was 2mm, 3.5mm and 5mm. The gap between TDLE surface and grounded electrode was varied between 3.6mm and 7.7mm.

Figure 1(c) shows coaxial bare electrode system using the same bus as Fig. 1(a) to compare the breakdown properties of TDLE with bare electrodes.[4,5] The bare electrodes tested were made of aluminum alloy whose diameter and width were 240mm and 100mm, respectively.

The surface roughness of TDLE and bare electrodes was ± 3μm. The numbers of TDLE in Fig.1(a) and of bare electrodes in Fig.1(c) were varied between one and twelve to investigate the area effect of breakdown. The distance between each test sample in the bus was set to 300mm for TDLE and 150mm for bare electrodes to avoid the interference with each other, respectively. The utilization factor of test samples of Figs.1(a) and (c) was around 0.6 whose value was equivalent to an actual bus. On the other hand, the utilization factor of Fig.1(b) was varied between 0.61 and 0.81.

We measured the breakdown voltages at negative LI voltages (1.2/50μs) by up and down method. SF_6 gas pressure was varied between 0.4 and 0.6MPa for TDLE and between 0.3 and 0.7MPa for bare electrodes.

2.2. Breakdown Properties of TDLE Compared with Bare Electrodes

It is known that the breakdown phenomena of SF_6 gas obey the Weibull distribution.[7] The cumulative breakdown probability F is expressed as follows.

$$F = 1 - \exp\left\{ -K \left(\frac{E - E_0}{E_d} \right)^m \right\} \tag{1}$$

Where E_0, E_d, m and K are location parameter (minimum breakdown field), scale parameter (theoretical breakdown field), shape parameter and a constant, respectively.

Figure 2 shows the example of Weibull plots of the breakdown field at Type 1 TDLE and bare electrodes. These data were obtained when the number of tested electrodes was four in the experimental setup in Figs.1 (a) and (c). The gas pressure was 0.4MPa. The data agree well with the Weibull distribution.

We obtained minimum breakdown field $E_{0.1\%}$ based on the Weibull distribution whose breakdown probability was 0.1%. The shape parameter m is different between TDLE and bare electrodes. The value m for TDLE is about two, while that for bare electrodes is about four. The fact shows that the deviation of TDLE is larger than that of bare electrodes.

The examples of the sequence of breakdown field of TDLE and bare

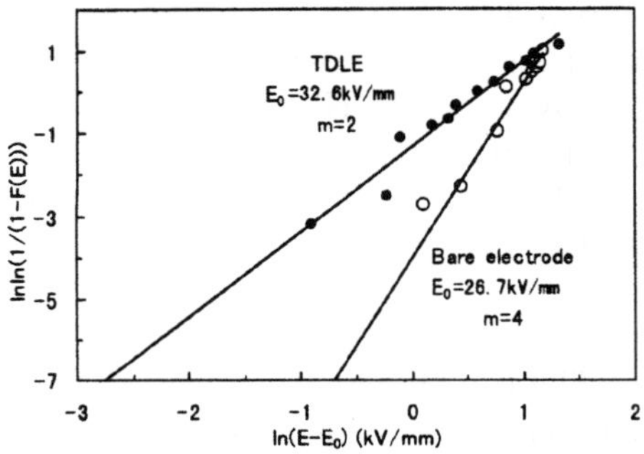

Figure 2. Weibull plots of breakdown field at 0.4MPa. Number of tested electrodes is four.

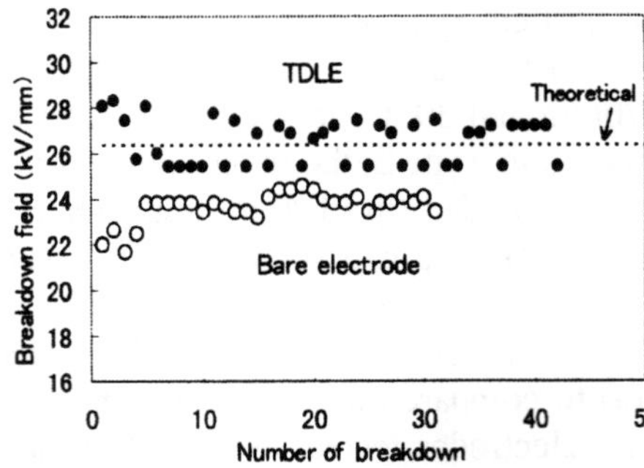

Figure 3. Sequence of breakdown field at 0.3MPa. Number of tested electrodes is one.

electrode are shown in Fig. 3. The bare electrode has a conditioning effect, showing the increase of breakdown field after several shots. The breakdown field never reaches the theoretical breakdown field even after conditioning.

On the other hand, the sequence for the TDLE shows the reverse pattern of the bare electrode. Several breakdowns at the beginning of voltage application occur at the higher field than the theoretical value. After several shots, the breakdown field becomes slightly lower and higher than the theoretical value. The phenomena might be influenced by the accumulated charges on the dielectric-layer surface caused by the last breakdown, since most of the low breakdown fields of TDLE occur after the breakdown at high electric field.

Figure 4 shows the minimum breakdown field $E_{0.1\%}$ as a function of effective electrode area. Here, the effective electrode area is defined as the area more than 95% of the maximum field of TDLE surface and bare electrodes. The values of $E_{0.1\%}$ decrease gradually and approach the constant value as increasing the effective electrode area for both TDLE and bare electrodes. However, note that the area effect of breakdown at TDLE is drastically reduced.

The pressure dependence of the minimum field $E_{0.1\%}$ (the asymptotic value

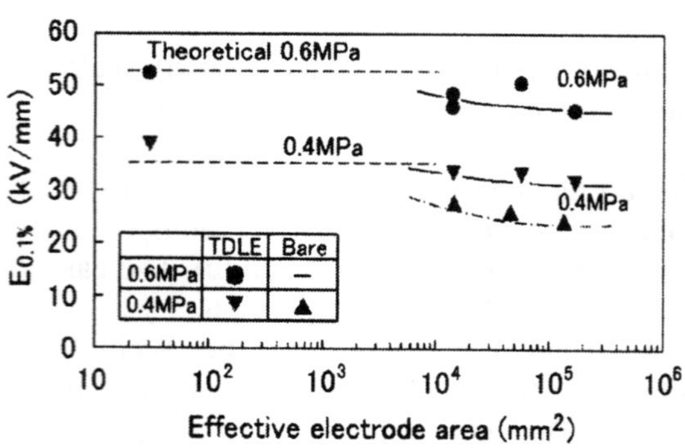

Figure 4. Minimum breakdown field $E_{0.1\%}$ as a function of effective electrode area for TDLE and bare electrodes at 0.4 and 0.6MPa.

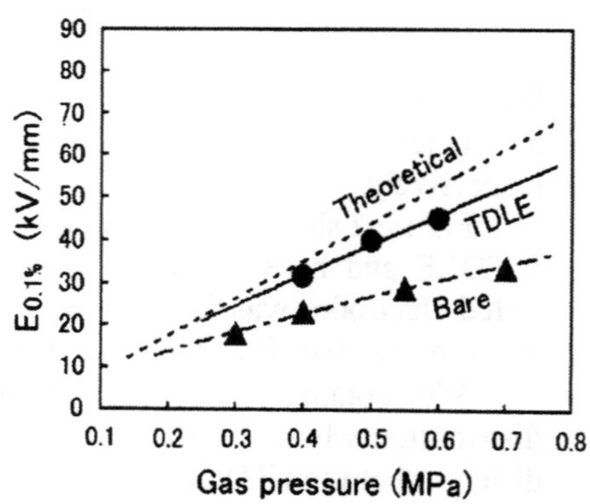

Figure 5. Dependence of minimum breakdown field $E_{0.1\%}$ on gas pressure for TDLE and bare electrodes when effective electrode area is infinitely large.

shown in Fig.4) is shown in Fig.5. It is remarkable that TDLE has higher pressure dependence of minimum breakdown field up to 0.6MPa than bare electrodes. The minimum breakdown field of TDLE reaches about 90% of the theoretical breakdown field of SF_6, and also the minimum breakdown field is 1.5 times higher than that of bare electrodes at 0.6MPa. Therefore, the TDLE has high breakdown field and are effective especially at high pressure region.

In general, area effect of breakdown is caused by the field concentration at micro protrusions on electrode surface. The protrusions are likely to initiate electrons due to field emission at negative polarity. A thin dielectric coating of several micrometers thickness on bare electrodes can inhibit the field emission from the protrusions. However, the thick dielectric-layer between several hundred micrometers and the order of millimeter would be needed to prevent the field concentration by covering the protrusions. The area effect of breakdown of TDLE seems to be remarkably improved, since TDLE has enough dielectric-layer thickness of 10mm and few harmful protrusions exist on the dielectric-layer surface compared with bare electrodes. Moreover, the protrusions on dielectric-layer surface cause less field concentration than those on bare electrodes. However, more detailed study is necessary to explain the above phenomena.

3. DEGRADATION AND PUNCTURE PROPERTIES OF TDLE DUE TO REPETITIVE DISCHARGES IN SF_6

3.1. Essential Properties in Applying TDLE to HV Equipment

TDLE has to maintain the dielectric performance without degradation of dielectric-layer surface even after repetitive breakdown in SF_6 if we apply TDLE as the substitutes of metallic electrodes for the HV equipment.

Another point is as follows. When breakdown in SF_6 occurs, the dielectric-layer surface of TDLE is grounded and the voltage at the moment of breakdown is applied between the embedded HV electrode and the dielectric-layer surface.

Figure 6 shows the results of electric filed analysis for Type 2 TDLE at the moment of breakdown. Since the field at the head of HV electrode is quite high, the breakdown in SF_6 would cause puncture of the dielectric-layer (epoxy) initiated from the head of the HV electrode, if the field were not fully reduced.

Therefore, the properties on surface degradation and dielectric-layer puncture by repetitive discharges in SF_6 are quite essential for TDLE application to the HV equipment.

To investigate the above properties, we basically used Type 2 TDLE for the measurements in the following sections 3.2 and 3.3 and applied negative LI voltages (1.2/50 μ s) until puncture of the dielectric-layer occurred.

3.2. Properties of Dielectric-layer Surface Degradation

Figure 7 shows the dependence of TDLE breakdown field in SF_6 on number of breakdown. Regarding the discharge process, the gas gap breakdown between

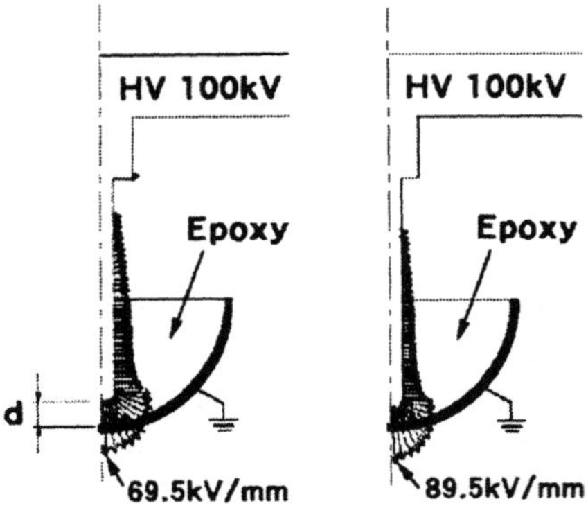

(a) d=3.5mm (b) d=2mm

Figure 6. Electric field distribution of embedded electrode inside dielectric-layer at the moment of breakdown.

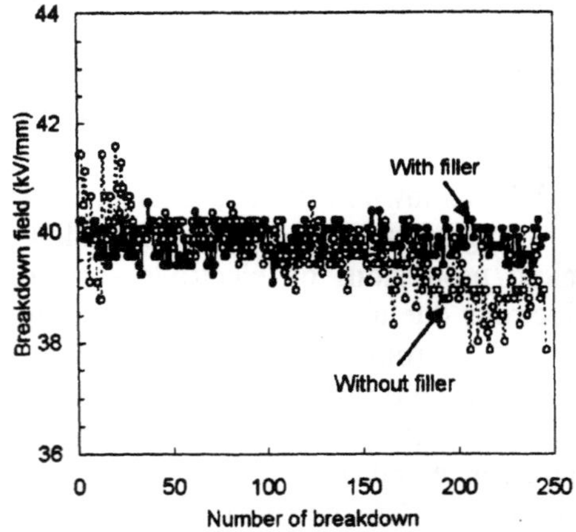

Figure 7. Dependence of TDLE breakdown field in SF_6 on number of breakdown. (Type 2 TDLE)

the dielectric-layer and the grounded electrode is followed by the creeping discharge along the dielectric-layer surface. The breakdown field decreases as increasing the number of breakdown for the epoxy samples without filler, while the field is constant in the case of the samples with silica filler.

The repetitive creeping discharges along the epoxy dielectric-layer surface might cause the degradation of epoxy and the decrease of breakdown voltages. However, the filler in epoxy prevents the degradation against the repetitive creeping discharges, since the filler has excellent performance against the tracking of epoxy surface. Therefore, the dielectric-layer surface degradation of TDLE can be prevented by selecting appropriate filler in the dielectric for the practical application to the HV equipment.

3.3. Properties of Dielectric-layer Puncture

Figure 8 shows the properties of dielectric-layer puncture dependent on number of breakdown in SF_6 and the influence of filler in epoxy. The vertical axis is the puncture field of epoxy defined as electric field of embedded electrode inside epoxy. The breakdown field decreases as increasing the number of breakdown for both cases with filler and without filler.

The relation between the puncture field E and the number of breakdown N is given by the following equation.

$$E \propto N^{-1/n} \tag{2}$$

Note that the values of n representing the gradient of the puncture properties are the same for both cases with filler and without filler. Therefore, filler in epoxy does not influence the puncture properties of epoxy.

Figure 9 shows the properties of dielectric-layer puncture dependent on the number of breakdown in SF_6 and the influence of TDLE configurations. The configurations used for the experiments were electrode samples of Type 1, Type

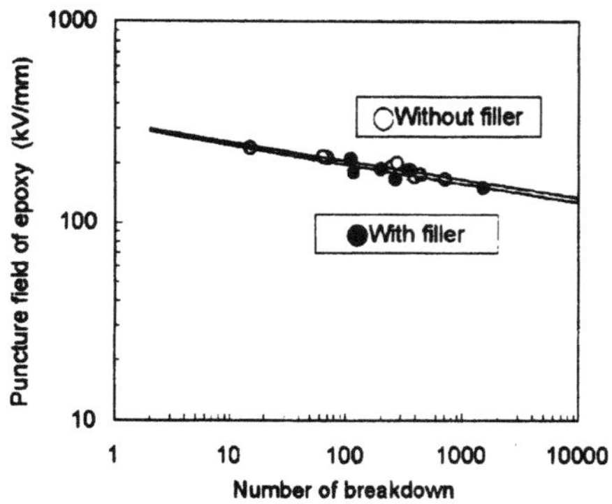

Figure 8. Properties of dielectric-layer puncture dependent on number of breakdown in SF_6 and influence of filler in epoxy. (Type 2 TDLE)

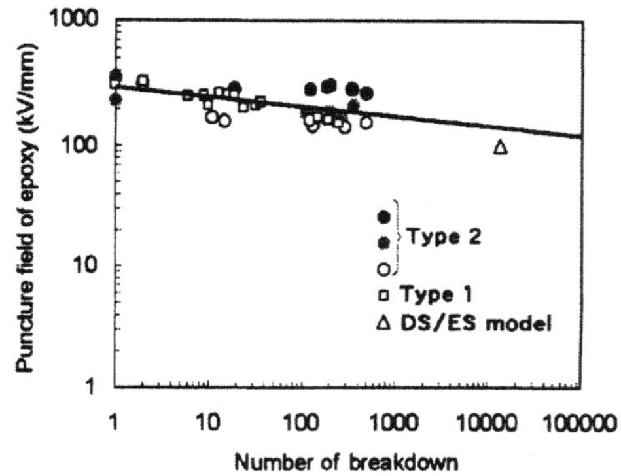

Figure 9. Properties of dielectric-layer puncture dependent on number of breakdown in SF_6 and influence of TDLE configurations.

2 and disconnecting switch (DS)/earthing switch (ES) model. The diameter of HV embedded electrode of Type 2 was 2, 4 and 6mm in this case. Here, DS/ES model corresponds to the similar configuration (TDLE-TDLE) shown in Fig.11, which will be explained later. Note that the electric field of embedded electrode inside the dielectric-layer was intentionally increased to facilitate the experiments, compared with the field of TDLE of practical application to DS/ES.

The puncture field decreases as increasing number of breakdown for all cases and the values of n in Eq.(2) is about 13. Puncture properties for all samples including actual DS/ES model have the same tendency, which indicates that the variation of TDLE configurations such as various electrode systems and sample diameters does not affect the puncture properties.

Therefore, the insulation design of TDLE as substitutes for metallic electrodes especially used in case of repetitive discharges can be conducted based on the results of Fig.9.

4. APPLICATION OF TDLE TO HV EQUIPMENT

4.1. Advantages of TDLE

The characteristics of TDLE are described in the previous chapters 2 and 3, and the advantages of TDLE are summarized as follows.

(1) TDLE has high minimum breakdown field of around 90% of theoretical breakdown field and is effective especially at high gas pressure region.

(2) TDLE can be applied as the substitutes of metallic electrodes by using appropriate filler in the solid dielectric to prevent the dielectric-layer surface degradation and also by adopting the insulation design criteria against the puncture of the dielectric-layer.

On the other hand, TDLE has another advantage of reducing maximum field strength of electrode systems.[2] Figure 10 shows the electric field distribution

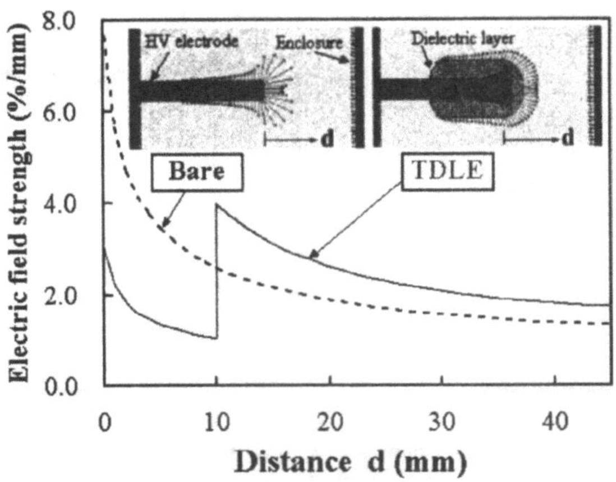

Figure 10. Electric field distribution between HV electrode and enclosure as the example of Type 1 coaxial electrode system.

between HV electrode and enclosure as the example of Type 1 coaxial electrode system. Note that "Bare" in the figure corresponds to the bare electrode without thick dielectric-layer in the case of Type 1 TDLE. The field distribution of the bare electrode is non-uniform, and the maximum field is quite high. However, the maximum field of TDLE appears on the dielectric-layer surface, and the maximum field is drastically reduced. Therefore, TDLE has also the following advantage.

(3) Reducing the maximum field of electrode system especially for electrode forming non-uniform field.

4.2. Application of TDLE to GIS

Considering the above three advantages of TDLE to improve a current SF_6 gas insulation system, we would like to introduce the application example and the trial design of TDLE to GIS.

Figure 11 shows the comparison between bare electrodes and TDLE application to 300kV DS and ES. The external view of TDLE as the substitutes of DS moving side and ES moving/stationary side shielding electrodes is also shown in Fig.12. The outer diameter of the TDLE is around 150mm.

The shielding electrodes with large curvature are basically used for a current design, since quasi-uniform field distribution is required to avoid local field concentration. If applying small curvature electrodes (small diameter) as shown in Fig.11, the distance between moving and stationary electrodes has to be elongated to 115%, compared with a current electrode of larger diameter.

However, TDLE enables to apply the embedded electrode with quite small curvature due to the effects of increasing breakdown field and of reducing maximum field of electrode system. The diameter of a current DS is downsized to 74% without elongating the distance between moving and stationary shielding

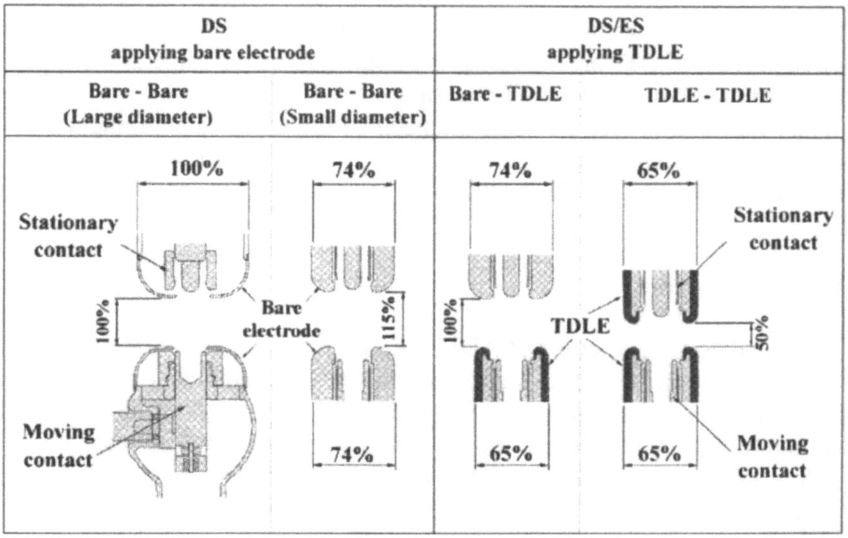

Figure 11. Application of TDLE to 300kV DS/ES.

electrodes, combining TDLE for moving side and bare electrode for stationary side electrodes. If applying TDLE for both moving and stationary side, the diameter and the distance between the electrodes are downsized up to 65% and 50%, respectively, although our application at the moment is limited to the combination of moving side TDLE and stationary side bare electrode for DS. Note that the same TDLE as DS can also be applied to ES.

Figure 13 shows a trial design applying TDLE to 145kV three phase enclosure type current transformer(CT). The TDLE outer diameter is not changed

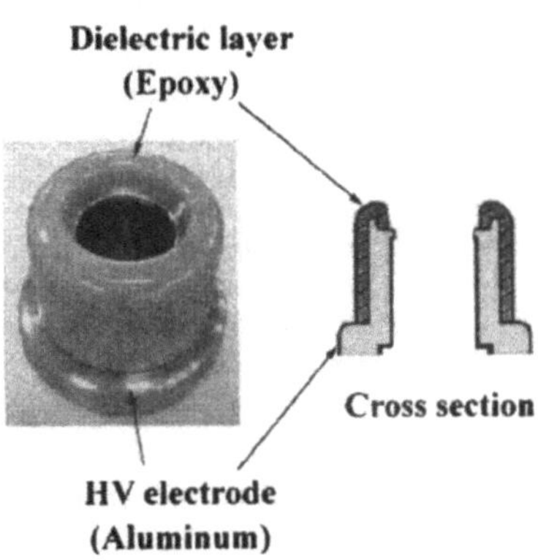

Figure 12. TDLE applied for DS moving side and ES moving/stationary side shielding electrodes.

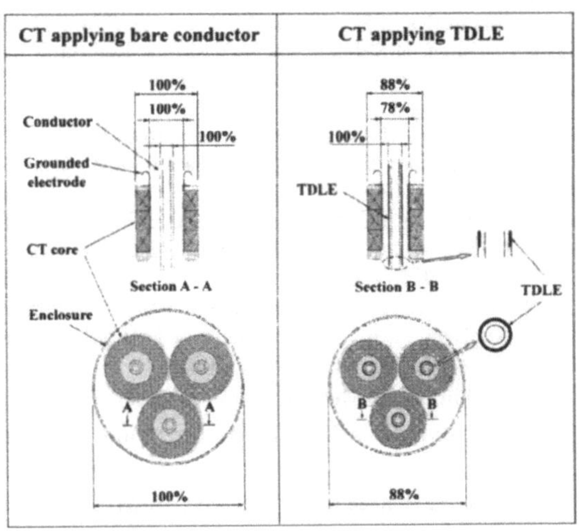

Figure 13. Trial design applying TDLE technique to 145kV CT.

from the bare conductor diameter to prevent a lowering of the current capacity at rated current. The thickness of dielectric-layer of the TDLE is 5mm in this case. The cross section of CT core by TDLE is designed to be the same as that by bare conductor to maintain the same CT accuracy.

The result shows that the inner diameter of the CT core applying TDLE is reduced to 78%, compared with a current CT using a bare conductor, which contributes to downsize the CT enclosure to 88%.

5. CONCLUSIONS

The breakdown properties of TDLE in SF_6 have been investigated and the practical application of TDLE techniques to the HV equipment has also been studied to downsize the equipment. Regarding the advantages of TDLE and the application to the HV equipment, the following points are summarized.

(1) TDLE has high minimum breakdown field of around 90% of theoretical breakdown field and are effective especially at high gas pressure region.

(2) TDLE can be applied as the substitutes of metallic electrodes by using appropriate fillers in the solid dielectric to prevent the dielectric-layer surface degradation and also by adopting the design criteria against the puncture of the dielectric-layer.

(3) TDLE can reduce the maximum field of electrode system especially for electrodes forming non-uniform field.

(4) TDLE is effective to downsize the component of GIS such as DS, ES and CT.

REFERENCES

1. M.Honda, H.Okubo, H.Aoyagi and A.Inui, Impulse Breakdown Characteristics of Coated Electrodes in SF_6 Gas, *IEEE Trans. on Power Delivery*, PWRD-2, No. 3, pp.699-708 (1987).
2. H.Goshima, H.Shinkai and M.Yashima, Lightning Impulse Breakdown Characteristics of High-pressure N_2 as an Alternative Insulation Gas to SF_6, *Gaseous Dielectrics IX*, pp. 359-364 (2001).
3. T. Yoshida, M. Miyagawa, I. Ohshima, N. Masaki and S. Yanabu, Increase of Breakdown Voltage due to Composite Insulation in SF_6 Gas, *T. IEE Japan*, Vol. 110-B, No.4, pp.348-356 (1990).
4. M.Yoshimura, K.Inami, H.Hama and H.Fujii, Area Effect of Breakdown at Lightning Impulse in N_2/SF_6 Mixtures, *XIII International Conference on Gas Discharges and their Applications*, pp.984-987 (2000).
5. H. Hama, K.Inami, M.Miyashita and M.Yoshimura, Dielectric Properties of Gas Insulated Bus Applying Low SF_6 Content and Highly Compressed N_2/SF_6 Gas Mixtures, *Gaseous Dielectrics IX*, pp. 487-496 (2001).
6. M.Yoshimura, H.Muto, K.Inami, Y.Ozaki, T.Ootsuka and H.Hama, Breakdown Characteristics of Epoxy Resin Insulator by Repetition Discharge in SF_6, *Proceedings of the 14th Annual Conference of Power & Energy Society, IEE of Japan, No.300* (2003).
7. T.Nitta, Y.Yamada and Y.Fujiwara, Area Effect of Electrical Breakdown in Compressed SF_6, *IEEE PES Summer Meeting*, PaperT73 449-6 (1973).

N_2/SF_6 GAS MIXTURES – LARGE SCALE APPLICATION IN GAS-INSULATED TRANSMISSION LINES (GIL)

Hermann Koch, Günther Schöffner *

1. INTRODUCTION

The application of N_2/SF_6 gas mixtures with SF_6-contents between 10 % and 20 % reduces the costs for the insulating gas remarkably. This is of great importance when GIL is to be used for long distance bulk power transmission. By this, also the acceptance for long distance applications of this technique is improved on the user side.

N_2/SF_6 gas mixtures may be applied in GIL because GIL is intended only for transmitting electrical power. No breaking, switching, or earthing function is used in a GIL system, where the arc extinguishing properties of SF_6 would be necessary. The only purpose of the gas mixtures in GIL is electrical insulation.

Already a low percentage of SF_6 improves the insulation capability of the gas mixture strongly. If only 10 % SF_6 is added about 60 % of the insulation capability of pure SF_6 is reached. With 20 % SF_6 this value increases to about 80 %. Using this gas mixture with low pressures below 0.7 MPa, the diameter for the enclosure pipes for the 400 kV GIL is 500 mm just like with 400 kV GIS.

The insulation performances of the gas mixtures have been tested systematically in real size laboratory test set-ups covering the whole range of SF_6 contents (0 - 100 %) and the gas pressure range from 1 - 10 bars. The tests have been carried out by the Siemens HV Lab and at the Institute of High Voltage Engineering of the Technical University of Munich.

* Hermann Koch, Günther Schöffner, Siemens AG, Erlangen, Germany D-91052

Gaseous Dielectrics X.
Edited by Christophorou *et al.*, Springer, New York, 2004.

Besides the systematic investigations of the gas mixture physics also a gas mixture handling system has been developed and successfully used with the first two GIL installations applying N_2/SF_6 gas mixtures. The gas mixture handling equipment consists of the mixing device, vacuum pumps, filling pumps and gas separation device for storing SF_6 and N_2 separately. For the storage of gas mixtures the SF_6 content is reduced to a value between 1 - 10 % to enable high pressure storage (~ 200 bars) in standard pressure vessels.

Finally, experiences with the application of N_2/SF_6 gas mixtures in the first projects and with long term gas tightness tests of conventional O-ring sealings have been gained.

2. GIL BASIC UNITS

A long distance GIL supplied by Siemens can be assembled by using only four different GIL units:

- straight unit
- angle unit
- disconnector unit
- compensator unit

The GIL basic units are shown in Figure 2 [1, 2, 3, 4, 5, 8, 9, 10].

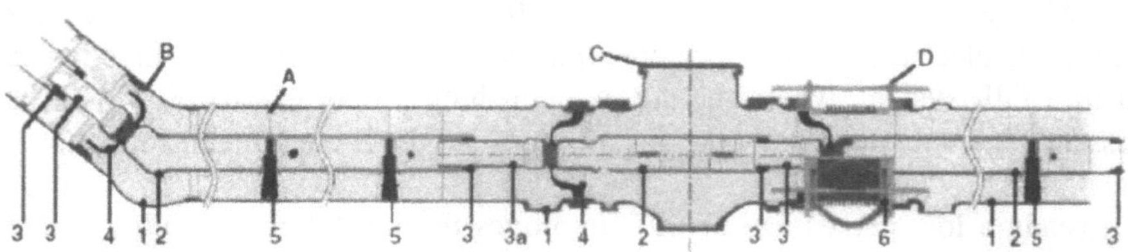

Figure 2. GIL basic units including straight, angle, disconnector and compensator unit.

The straight unit (A) consists of a conductor pipe coaxially arranged in the enclosure (1) by support insulators (5). Enclosure (1) and inner conductor (2) are made of aluminium. An angle unit (B) is used for directional changes. The conical insulator (4) fixes the inner conductor to the enclosure. In case of long distance GIL installations the gas compartments are separated by disconnecting units (C) approximately every 1...1.5 kilometers. At the disconnecting unit (C) the conical insulator (4) is of gas-tight type.

For outdoor and tunnel installations a compensator unit (D) is used to enable the thermal expansion of the enclosure. The thermal expansion of the conductor is compensated by sliding contact systems (3a, 3b).

Figure 2 shows flanges with O-ring sealings which are used to connect the disconnector and compensator units to the straight and angle units. The straight units are connected by welding. Typical lengths between disconnector and compensator units are several hundred metres up to 1,5 km. This means in average an effective O-ring sealing length of approx. 10 m for one gas compartment of 1 km length, or 250600 dm^3, or 250 m^3.

3. BASIC INFORMATION OF N_2/SF_6 GAS MIXTURES

N_2/SF_6 gas mixtures are applied in GIL to reduce the costs and to increase the environmental acceptance. In GIL no switching, earthing or breaking processes take place. The only purpose of the gas mixture is insulation. Therefore mixtures can be applied successfully. N_2 is the major gas component of the mixture because of its huge availability. It is inert, non-corrosive and of low cost. The handling of N_2 is simple and the handling equipment is widely available.

N_2 and SF_6 is the best combination for a mixture which is to be applied in GIL. Both gases are inert and non-corrosive and can be handled by the same gas handling equipment. Also many years of experience with N_2/SF_6 gas mixtures in low temperature ambient conditions are available world-wide. These experiences proof that the gas mixture has a long time stability when used in high voltage systems. Therefore, N_2/SF_6 is the best combination of a gas mixture for the use in GIL. Other mixtures have also been investigated several times but pointed out to have some decisive disadvantages compared to N_2/SF_6.

The most important properties of the mixtures [13] are shown in Figure 1.

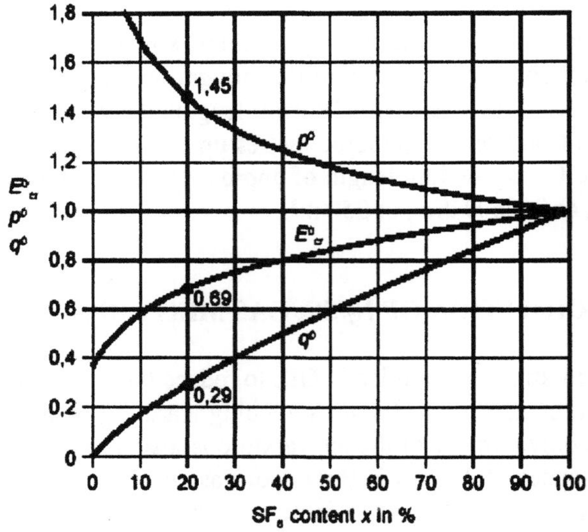

Figure 1. Normalised intrinsic dielectric strength E^0_{cr}, normalised required pressure p^0 for equal dielectric strength and resulting normalised quantity q^0 of SF_6 as a function of the SF_6 content x.

The graph shows three curves:

E^0_{cr} Normalised intrinsic dielectric strength
$E^0_{cr} = (E/p)_{cr} / (E/p)_{cr\ SF6}$

p^0 Normalised required pressure
$p^0 = 1 / E^0_{cr}$

q^0 Equal dielectric strength and resulting normalised quantity q^0 of SF_6
$q^0 = Xp^0$

For the SF_6 content x in % the three curves give information about the dielectric strength in a homogeneous field. The graph contents therefore information about one basic dielectric layout criteria.

The meaning of the graph shall be explained by an example of a mixture with an SF_6 content of 20 %. Taking x = 20 %, the normalised intrinsic dielectric strength E^0_{cr} is 0,69. That means that with an SF_6 content of 20 % a dielectric strength of approx. 70 % of that of pure SF_6 is reached. With this physical behaviour of the N_2/SF_6 gas mixture a great reduction effect of the use of SF_6 can be reached without reducing much the dielectric strength. This effect is related to the strong electronegativity of SF_6. The savings of SF_6 are directly shown by curve q^0, the equal dielectric strength and resulting normalised quantity of SF_6. The value of $q^0 = 0,29$ indicates that for equal dielectric strength in pure SF_6 the savings in the N_2/SF_6 gas mixture is 71 % of SF_6.

The curve of normalised required pressure p^0 shows that at an SF_6 content of 20 % the gas pressure is only 45 % higher than with pure SF_6 to reach the same dielectric strength.

The curves in Figure 1 are based on measured data which have been drawn from two scale one test set-ups in two laboratories in Munich and Berlin to get system independent data. Additionally the measured data have been compared to measured data of the test laboratory of EDF in France for further confirmation.

4. GAS TIGHTNESS OF GIL

The GIL of the second generation is designed for economic long distance bulk power transmission [11]. Therefore, a laying technology has been developed based on the pipeline laying technique. One basic element of this laying technique is the use of an automated orbital welding machine to produce high quality welds without failures. To prove this an automated orbital measuring system has been developed to control the quality of the weld. The quality criteria of the welds are the porosity and gas tightness. The porosity represents number and size of voids in the weld. If both values are below a distinct limit given by the welding standards for aluminium welds, the gas tightness will be guaranteed. The welded connections can therefore be seen as gas tight so that no gas refill is needed during the whole life time.

Nevertheless, at the ends of the GIL at the disconnecting units and at compensators flanges are needed. These flanges have sealings as used in GIS (O-rings). They have a limited gas tightness. For small gas compartments as used in GIS the tightness is indicated with values below 0,5 % per year. Experiences show that the realistic values of today's sealing technology is below 0,1 % per year.

In the following clause long-time measurements are explained to prove that this high gas tightness of the O-ring sealing can also be assumed for GIL filled with N_2/SF_6 gas mixtures [6, 7].

5. LONG DURATION TEST OF GAS TIGHTNESS

5.1. Test Set-up

The GIL unit having the most O-ring sealings is the disconnector unit as shown in Figure 3. The gas tightness of this unit was investigated during a long duration experiment.

Figure 3. Disconnecting unit test set-up

The volume of the disconnecting unit is 0,34 m^3 and has a total length of O-ring sealings of 6,7 m. That means that the O-ring sealing length related to the volume is relatively high (6,7 m/0,43 m^3) compared with the number of a one kilometer average gas compartment (10 m/250 m^3).

The pressure measurements were taken by a gauge of 0,1 % accuracy which was permanently connected to the gas compartment of the disconnecting unit. The pressure in the disconnecting unit is depending on the temperature. To correct the measured values to normalised values the temperatures of the wall of the pressure vessel and of the ambient air was also measured permanently. The used gas mixture had a volume content of 80 % N_2 and 20 % SF_6. The nominal filling pressure was 0,6 MPa.

5.2. Measuring Results

The gas density of the disconnecting unit was measured in time spans of some days. The total duration of the measurements was 2 years. The measuring results are shown in Figure 4.

Time	Ambient temperature	Gas pressure	Ambient air pressure
	°C	Bar	hPa
30.10.98 10:30	21,0	6,01	1000
06.11.98 14:20	20,0	5,98	1009
16.11.98 09:30	20,0	5,97	1012
23.11.98 07:30	16,0	5,88	1027
30.11.98 11:00	17,4	5,94	1025
07.12.98 08:35	17,8	5,92	1012
14.12.98 13:30	19,1	5,96	1021
22.12.98 07:30	18,5	5,93	1017
04.01.99 07:20	17,4	5,95	1008
13.01.99 07:30	15,2	5,92	1001
18.01.99 08:15	19,6	5,99	1018
25.01.99 07:30	19,8	6,00	1014
01.02.99 07:15	16,5	5,92	1027
08.02.99 07:30	18,2	5,97	993
17.12.99 08:30	19,8	6,01	989
22.02.99 07:40	21,2	6,05	983
01.03.99 07:10	20,6	6,02	1007
08.03.99 07:50	20,4	6,01	1005
15.03.99 07:55	20,0	5,98	1023
22.03.99 07:55	20,8	6,03	998

Figure 4. Measuring results of normalised temperature (T/T_0), pressure (p/p_0), and density (σ/σ_0)

The first value was measured on 30.10.1998 at an enclosure temperature of 17,4 °C, ambient air pressure 1025 hP and a gas pressure in the disconnecting unit of 0,514 MPa. The normalised pressure at $T_0 = 290{,}4$ K is $p_0 = 0{,}6965$ MPa.

The changes of the gas density during the measuring time were related to the filling pressure at the beginning indicated with the relative value 1. The normalised density has been calculated using the gas pressure and temperature. The accuracy of the measurement of the pressure had a maximum failure of $\pm$ 0,016 MPa and the temperature of $\pm$ 0,2 K. This equals a total measurement failure of $\pm$ 0,3 %.

In Figure 5 three curves are drawn for the normalised temperature (T/T_0), pressure (p/p_0), and density (σ/σ_0).

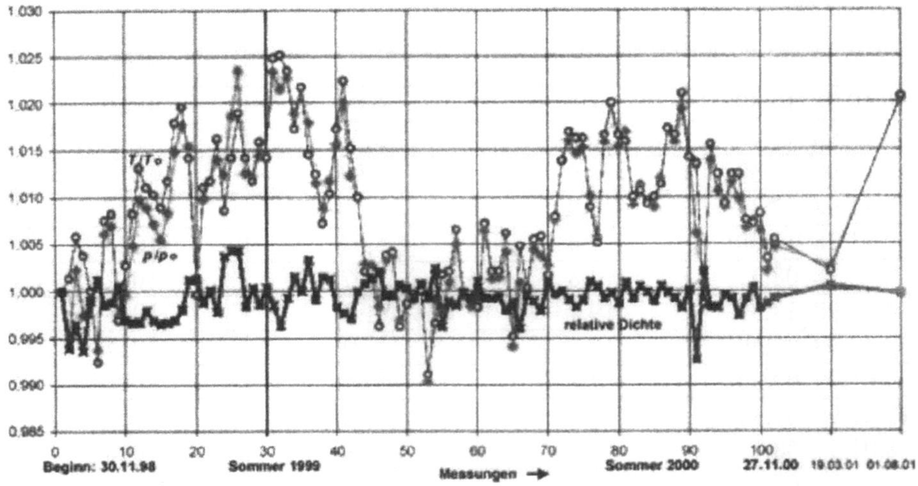

Figure 5. Normalised temperature (T/T_0), pressure (p/p_0), and density (σ/σ_0)

During the total time of the measurement no gas losses could be indicated which are higher than the maximum tolerance of $\pm$ 0,3 %.

This long duration gas tightness measurement showed that the O-ring sealing system as used today has the same high gas tightness for N_2/SF_6 gas mixtures as for pure SF_6. For a real directly buried GIL of 1000 m for example, this means a volume loss of approx 0.00025 %/a [12]. With an assumed filling pressure of 0.7 MPa this may result in a gas loss which causes a maximum pressure difference of only 0.016 MPa after 100 years of operation. This points out that a refilling will not be necessary.

6. CONCLUSION

The application of N_2/SF_6 mixtures in GIL resulsts in lower costs. Since the gas fulfills insulation purposes only the mixture can be applied successfully.

The long duration test over more than 2 years of gas tightness of O-ring sealings used in GIL showed that the high gas tightness as known for GIS can also be expected for GIL using N_2/SF_6 gas mixtures. Taking into account that a real GIL has relatively large gas compartments of 1 km length and that the large majority of the joints are welded and gas tight, it can be assumed that the gas tightness of GIL is that high that no refill is needed during the whole life time under normal operation.

7. REFERENCES

[1] Alter, Ammann, Boeck, Degen, Diessner, Koch, Renaud, Pöhler: N2/SF6 gas-insulated line of a new GIL generation in service, CIGRE Session 2002, Paris, 08/02

[2] Vich Piputvat,Wirach Rochanapithyakorn, Hillers T., Koch H., Pöhler S., Schoeffner G.: 550 kV Gas-Insulated Transmission Line for High Power Rating in Thailand, CIGRE Session 2004, Paris, to be published in 08/04

[3] Koch, H.: Experience with 2nd Generation Gas-Insulated Transmission Line GIL, JICABLE 03, Versailles 06/03

[4] Koch, H., Schoeffner, G.: Gas-Insulated Transmission Line - To Solve Transmission Tasks of the Future, IPEC Conference 2003, Singapore 05/03

[5] Koch, H.: Future needs of high power interconnections solved with Gas-Insulated Transmission Lines (GIL), PowerCon 2002 Conference, Kunming, 10/02

[6] Dießner, Finkel, Grund, Kynast: Dielectric properties of N2/SF6 mixtures for use in GIS or GIL, ISH London 99, paper 3.67, S. 18, 22.-27.08.1999

[7] Henningsen, Kaul, Koch, Schütte, Plath: Electrical and Mechanical Long-Time Behaviour of Gas-Insulated Transmission Lines, CIGRE 2000, Paris, 08/2000

[8] Schichler, U., Diessner, A., Gorablenkow, J.: Dielectric On-Site Testing of GIL, Proceedings of the 7th Int. Conference on Properties and Applications of Dielectric Materials, Nagoya, Japan, June 1-5, 2003

[9] Okubo, Hoshino, Takahashi, Hikita, Miyazaki: Insulation design and on-site testing method for a long distance, gas insulated transmission line (GIL), IEEE Electrical Insulation Magazine, 1998

[10] Okubo, Yoshida, Takahashi, Hoshino, Hikita, Miyazahki: Partial discharge measurement in a long distance SF6 gas insulated transmission line (GIL), IEEE Trans on PD, Vol. 13, 1998

[11] Schoeffner, G.; Koch, H.: New Possibilities of Power Grid Modernization with Gas Insulated Lines. Iasted PowerCon 2003: Blackout, New York, 2003

[12] Schoeffner, G.; Graf, R.: Suitability of N_2/SF_6 Gas Mixtures for the Application at Gas Insulated Transmission Lines. IEEE PowerTech, Bologna, Italy, 2003

[13] CIGRE TF D1.03.10: N_2/SF_6 Mixtures for Gas Insulated Systems, CIGRE Report, Paris, 2004.

EFFECT OF DIFFERENT DIELECTRIC COATINGS IN DIVERGENT FIELDS

Silvio Stangherlin, Christian Rein, Gerhard Salge, Friedrich Koenig*

1. ABSTRACT

Dielectric coatings are widely used in electric power equipment to increase the breakdown voltage and to reduce the insulation distance.

Two different coating materials, epoxy and polyamide, and two different coating thicknesses, 3 and 5 mm, were experimentally investigated in divergent field gaps (coated rod – bare plane arrangements). The experimental campaign was performed in air at 100 kPa.

Previous measurements referring to bare electrodes, under the same experimental conditions, have been used as reference for calculating the increase in the breakdown voltage $U_{50\%}$ as well as modification in the standard deviation σ.

The streamer inception model was also used to work out predictions of the breakdown voltage.

Discrepancies as well as similarities are put in evidence and discussed in the conclusions.

2. INTRODUCTION

It is well known that given a certain space and electrode shapes and dimensions, a considerable increase in the average breakdown voltage $U_{50\%}$ can be achieved by means of electrostatic shields[1-2], dielectric barriers[3], and/or dielectric coatings[4-5]. Advantages and drawbacks of each method have been identified in the past and are reported in the literature.

* Silvio Stangherlin, Gerhard Salge, and Friedrich Koenig, ABB Switzerland Ltd., Corporate Research, Segelhof 1, Baden-Daettwil, Switzerland CH–5405. Christian Rein, ABB AS, Power Technology Division, Amtm. Aallsgate 73, P.O. Box 108 Sentrum, Skien, Norway N-3713.

Gaseous Dielectrics X.
Edited by Christophorou *et al.*, Springer, New York, 2004.

Industrial knowledge and know-how about the use of thermosets and elastomers as dielectric coatings can be found in the literature as well as inside all OEMs (Original Equipment Manufacturers); while a certain gap still exists as far as thermoplastics are concerned.

The present contribution reports and compares the major engineering discharge parameters (the average breakdown voltage $U_{50\%}$ and the standard deviation σ) of epoxy and PA (PolyAmide) coatings in atmospheric air for the positive polarity, which is the weakest one. A coated rod – bare plane arrangement was used as benchmark. The gap distance and the coating thickness were varied.

3. EXPERIMENTAL SET-UP

The experimental setup was obtained from a previously reported test arrangement[6], using coated rods against a bare plate. It was intended to assess the dielectric behaviour of different coating materials and thicknesses in air in a divergent field.

The test voltage U (step voltage with ~300 ns rise time) was applied to the coated rod – bare plate electrode system. The coated rod consisted of a core and a solid coating, while the plate was made of aluminium with a 200 mm circular diameter and was very fine polished. The core was made of copper with a length of 110 mm and an 8 mm radius spherical tip at one end. The coated length of the core was 90 mm. Two different materials and two thicknesses were used to realise the solid coating. The coating materials were common cast epoxy ($\varepsilon_r = 4$) and a moulded composite thermoplast obtained by short glass fibres in a PA matrix ($\varepsilon_r = 3.6$). The thickness of the coatings was 3 and 5 mm. Four different sets of coated rods were prepared; each set consisted of 20 individual samples.

The maximum and the minimum divergences of the electric field k_{MAX} and k_{min} are respectively the maximum field in air E_{MAX} and the minimum field in air E_{min} over the average field in case of bare electrodes E_{avg}. They have been calculated using analytical formulas and are reported in Table 1.

Table 1. Field divergences k_{MAX} and k_{min} vs. gap distance d, coating material and thickness.

d	Bare rod		3 mm epoxy		5 mm epoxy		3 mm PA		5 mm PA	
[mm]	k_{MAX}	k_{min}	k_{MAX}	k_{min}	k_{MAX}	k_{min}	k_{MAX}	k_{min}	k_{MAX}	k_{min}
10	1.893	0.691	1.630	1.020	1.647	1.302	1.602	1.003	1.595	1.261
30	4.309	0.377	3.028	0.493	2.523	0.566	2.944	0.488	2.477	0.556
50	6.787	0.257	4.643	0.330	3.788	0.375	4.594	0.327	3.725	0.368
70	9.277	0.195	6.285	0.249	5.092	0.281	6.222	0.246	5.010	0.276
100	13.02	0.143	8.764	0.181	7.069	0.204	8.677	0.180	6.958	0.201

The experimental campaign was performed in atmospheric air. The test gap was housed in an aluminium vessel. The maximum working pressure that can be safely handled by the vessel is 1 MPa.

A LeCroy fast oscilloscope (2 GHz sampling rate) with high memory storage (50×10^3 samples, 8 bits resolution) acquired the gap voltage by a coaxial capacitive coupling to the HV busbar.

The negative polarity was applied to the plate while the rod was grounded.

4. EXPERIMENTAL RESULTS

The very popular "up and down" method[7] was applied to determine the average breakdown voltage $U_{50\%}$. Each experimental point comes from an "up and down" procedure formed by 20 valid shots. The step voltage U_{step} was initially set, the "up and down" procedure was performed and then the standard deviation of the breakdown voltage σ was estimated. The initial choice of U_{step} was then checked, as it has to be of the same order of magnitude of σ. Moreover, the voltage level of each breakdown was plotted vs. the shot number to highlight possible trends typical of electrode conditioning. 2 to 3 experimental points were acquired under the same conditions using different samples from the same set of coated rods.

The experimental error bars reported in all figures were estimated assuming a normal distribution of the population sampled and a 95% confidence interval.

Figure 1 compares the experimental results obtained with coated rods to both similar previous measurements[6] and calibrations obtained with bare rods.

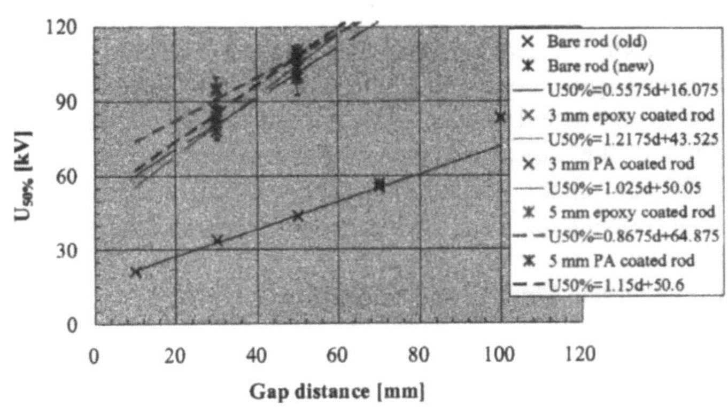

Figure 1. $U_{50\%}$ vs. gap distance d, coating material, and thickness in atmospheric air for the positive polarity. Comparison with previous measurements using an uncoated rod[6], used as reference and named "Bare rod (old)", and new measurements, used as calibrations and named "Bare rod (new)", are reported. Linear best fittings of all experimental results are also given.

The discharge always started from the tip region and propagated in both directions along the coating surface and across the gap. The coating was never punctured. Visual inspections revealed neither carbonisation nor discharge traces on the coating surface.

As expected, the benefit given by the use of coated rods is clearly visible in the picture and may be represented by the ratio $k = U_{coated} / U_{bare}$. Increasing the gap distance from 30 to 50 mm slightly reduces the benefit. Changing coating material from epoxy to PA produces no significant effect, while increasing the coating thickness from 3 to 5 mm gives a very limited benefit.

For practical applications, the benefit $k \sim 2 \div 2.5$ may be considered as independent from the gap distance as well as the coating material and thickness.

The statistical spread in the breakdown voltage, given by different samples from the same set of coated rods, increases when changing from epoxy to PA. This may be explained considering the different local surface charging and recombination processes as well as random local field enhancements produced by the short glass fibres sticking out of the PA surface.

5. COMPARISON WITH PREDICTIONS

Estimations of the streamer inception voltage U_i were performed. The electric field was calculated using a self-developed internal code based on finite element method (FEM). No shape modification was adopted in case of coated rods; while, according to previous papers[6, 8], the bare rod was simulated assuming the tip of the rod having a 100 µm protrusion with 30 µm tip radius. Notice that this is a pure numerical stratagem used to fit the experimental points. The dimensions of a real protrusion are more than one order of magnitude smaller.

The streamer inception voltage U_i was calculated considering the net ionisation coefficient λ/N of air from the literature[9]. The critical avalanche number for PA and epoxy coated rods was respectively increased to 62.55 and to 71.77 to fit the experimental results. This is another pure numerical stratagem as a physical value for the critical avalanche number is not far from 10.

The comparison among predictions, experimental results and the threshold for the streamer propagation from the literature[10] is reported herebelow.

5.1. Maximum Electric Field in Air – Rod Tip

On one hand, from the whole set of U_i the maximum electric field in air E_i^{MAX}, in the middle of the rod tip, was evaluated applying Eq.(1). k_{MAX} values are reported in Table 1.

$$E_i^{MAX} = k_{MAX} \frac{U_i}{d} \qquad (1)$$

On the other hand, the maximum electric field in air $E^{MAX}_{50\%}$ with equal probability for withstand or breakdown was then derived in a similar manner.

The whole set of E^{MAX}_{i} predictions is reported and compared to the $E^{MAX}_{50\%}$ experimental points in Figure 2.

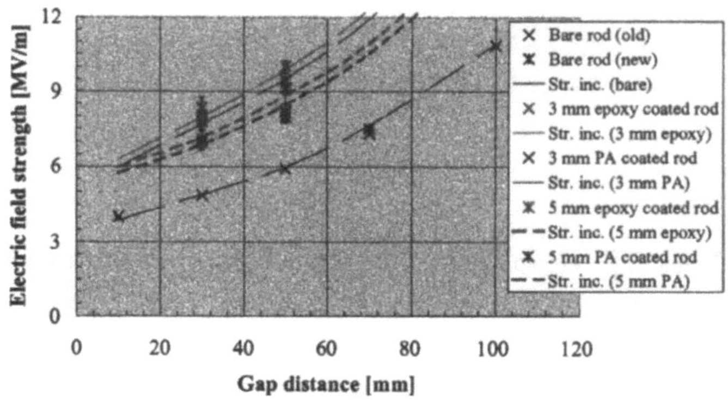

Figure 2. E^{MAX}_{i} and $E^{MAX}_{50\%}$ vs. gap distance, coating material, and coating thickness in atmospheric air for the positive polarity.

An extremely good agreement was found among predictions and measurements. The streamer inception model can predict the breakdown voltage.

5.2. Minimum Electric Field in Air – Plate Surface

Again, from the whole set of U_i the minimum electric field in air $E^{\min}_{i}$, in the middle of the plate surface was evaluated applying Eq.(2). Also $k_{\min}$ are reported in Table 1.

$$E^{\min}_{i} = k_{\min} \frac{U_i}{d} \tag{2}$$

The minimum electric field in air $E^{\min}_{50\%}$ with equal probability for withstand or breakdown was then derived in a similar manner.

The whole set of $E^{\min}_{i}$ predictions is reported and compared to the $E^{\min}_{50\%}$ experimental points in Figure 3. Comparison to the streamer propagation threshold from the literature[10] is also given.

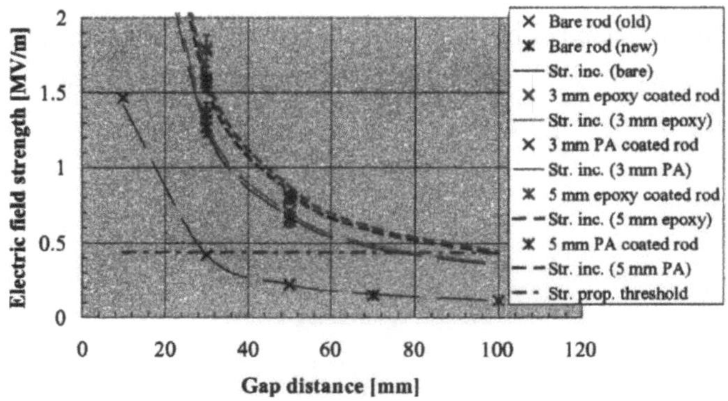

Figure 3. E_i^{min} and $E_{50\%}^{min}$ vs. gap distance, coating material, and coating thickness in atmospheric air for the positive polarity. The streamer propagation threshold[10] is also reported.

As expected, the extremely good agreement among predictions and measurements is confirmed.

At d = 30 mm for the bare rod the minimum electric field is 425 kV/m, quite close to 440 kV/m which is the streamer propagation threshold. For longer gaps, in spite of the fact that the minimum electric field is far below such a threshold, the streamer can still bridge the gap and run into breakdown. The field distortion in front of the streamer head can not be neglected, as it is usually done, and the real poissonian field must be considered. That applies also to coated rods, considering the translation of the curves. Nevertheless, the streamer inception model, with the modifications above described, may be used to obtain reasonable predictions of the breakdown voltage.

6. CONCLUSIONS

The effect of epoxy and PA coatings in atmospheric air for the positive polarity was experimentally investigated. A coated rod – bare plane arrangement was used as benchmark. The gap distance and the coating thickness were varied.

The discharge always started from the tip region and propagated in both directions along the coating surface and across the gap. The coating was never punctured. Visual inspections revealed neither carbonisation nor discharge traces on the coating surface.

The experimental campaign highlighted the benefit given by the use of coatings. Such a benefit was quantified and is: slightly reduced when increasing the gap distance, almost independent of the coating material, very limited when increasing the coating thickness. The statistical spread in the breakdown voltage is increased when changing from epoxy to PA.

Calculations of the streamer inception voltage U_i were also performed. A good numerical agreement with the physical reality was obtained considering either a certain electrode roughness (bare rods) or a bigger critical electron avalanche (coated rods). Such approach is a pure numerical stratagem and has no correspondence with the physical reality.

The streamer inception model is confirmed as a very simple, general and powerful tool to utilise when reasonable predictions of the breakdown voltage are required.

7. REFERENCES

1. H. Roser, Schrime zur Erhörung der Durchschlags-Spannung in Luft, *Elektrotechnische Zeitschrift* **B17**, 411-412 (1932).
2. H. El-Haéry, Ladungsverteilung auf Isolierschirmen im inhomogenen Feld bei Stossbeanspruchung, Ph. D. Thesis, University Stuttgart, (1981).
3. L. Ming, M. Leijon, T. Bengtsson and M. Darveniza, Barrier effects in a rod/rod air-gap under DC voltage, in: *Gaseous Dielectrics VII*, edited by L. G. Christophorou and D. R. James (Plenum Press, New York, 1994), 169-175.
4. G. Baldo and G. Pesavento, Impulse performance of air gaps in series with thick insulating layers, *5th International Symposium on High Voltage Engineering* **1**, 14.25 (1987).
5. L. Ming, U. Fromm, M. Leijon, D. Windmar, L. Walfridsson, A. Vlastos, M. Darveniza and J. Kucera, Insulation performance of covered rod/plane air-gap under lightning impulse voltage, *10th International Symposium on High Voltage Engineering* **3**, 235-238 (1997).
6. S. Stangherlin, G. Salge and F. Koenig, Measurements of discharges and their branching behaviour in atmospheric air, 2002 Conference on Electrical Insulation and Dielectric Phenomena, 782-785 (2002).
7. W. Hauschild and W. Mosch, *Statistical techniques for high voltage engineering* (Verlag Techniques, Berlin, 1984).
8. S. Stangherlin, G. Salge and F. Koenig, Dielectric behaviour of compressed N_2-O_2 gas mixtures in a slightly divergent field, 14th International Conference on Gas Discharges and their Applications, 239-242 (2002).
9. G. Friedrich, Ph. D. Thesis ETH N. 9992 (1992).
10. N. L. Allen and A. Ghaffar, The conditions required for the propagation of a cathode-directed positive streamer in air, *J. Phys. D: Appl. Phys.* **28**, 331-337 (1995).

Effect of the electrode coating and the field utilization factor on the insulation property of carbon dioxide gas under AC voltage

Satoshi Matsumoto, Nobuhiko Yokoyama, Kazuharu Kiyama, Shinya Ohtsuka and Masayuki Hikita

1. Introduction

Carbon dioxide gas (CO_2) is an inert gas and easily obtainable in a pure state. Recently, highly-pressurized CO_2 gas is focused on as one of the major candidates of SF_6 substitution gas for gas insulated equipments, and several researches on the insulation properties or the partial discharge characteristics of carbon dioxide gas have been carried out[1-3].

Although a large number of studies has been made on the effect of electrode coating as for the SF_6 gas insulation[4-15], little is known about those for the pressurized CO_2 gas. The purpose of this paper is to clarify the insulation properties of pure CO_2 gas under AC voltage.

Firstly, the sparkover voltages were measured using the bare electrodes and test results were evaluated with Paschen curve.

Secondary, comparison tests were conducted between bare electrodes and the coated electrodes at the gas pressure of 0.6MPa and the gap length ranging from 5mm to 15mm using sphere-to-sphere electrodes. The test results show that the first spark over voltage of the semi-sphere aluminum electrodes covered with the enamel resin is about 10% higher than those without the coating.

Lastly, the effect of field utilization factor (mean electric field divided by maximum electric field of the gap) is studied changing the electrode configurations. This paper discussed the relationship between electric field utilization factors of the electrodes and the spark over electric fields.

We come to the conclusion that electrode coating might serve as an important technique of improving insulating property of high-pressurized carbon dioxide gas as an insulation medium for gas insulated apparatus.

2. Experimental

Figure 1 shows the experimental setup for the spark over voltage measurement. Table 1 summarizes the electrode materials and geometries together with the equi-potential lines. The semi-sphere electrodes were made from aluminum and iron covered with or without insulated coating. The electrodes were shaped by lathe and polished with a cloth. The coating material is the enamel resin, and the thickness of the coating is about 30-40 micro meters. Needle to plane electrodes were also used for non-uniform electric field tests. Needle is made from the stainless steel. The electric field utilization factors were varied by changing the gap distance and electrode configuration.

Experiments were carried out in the 154kV GIS model tank. Figure 1 shows the setup of the

Kyushu Institute of Technology, Sensui-cho 1-1,Tobata-ku,Kitakyushu-shi, 804-8550 Japan

Gaseous Dielectrics X.
Edited by Christophorou *et al.*, Springer, New York, 2004.

electrodes. AC 60Hz voltage was applied to the gas gap by 250kV testing transformer. Test voltage is increased by step-up method keeping the duration time of 1 minute and step voltage of 5kV. The value of AC voltages was expressed by the peak values. Gas pressures were varied ranging from 0.2MPa to 0.6MPa at room temperature.

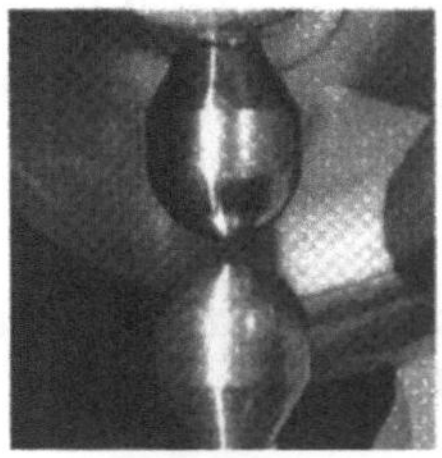

Fig. 1 Experimental setup using Semi-sphere electrodes
(Gap Length: 10mm, Radii of electrodes: 25mm/25mm)

3. Experimental Results

3.1 Paschen curve for Carbon Dioxide gas in quasi-uniform field

Table 2 shows the average values and percent standard deviations of the sparkover voltages for each electrode shown in table 1. Figure 2 shows the dependence of sparkover voltage on the whole range of the product of gap length and the gas pressure (Paschen curve) for CO_2 gas. The figure shows that the sparkover voltages increase in proportion to the product of the gap length and the pressure and our results agree with those reviewed by Z.Krasucki[16] under DC or AC voltage applications, with no apparent discontinuities present. However, the figure gives an indication of the magnitude of departure from Paschen's law observed at higher pressures.

3.2 Test with the coated electrodes

The electrodes covered with insulating material are expected to have high insulation properties as in SF_6 gas. Figure 3 shows the variations of sparkover voltages with aluminum electrodes covered with enamel coating (Electrode No.5 in table 1) and aluminum bare electrodes (Electrode No.2 in table 1) at the gas pressure of 0.6MPa. The figure shows that the first sparkover voltage for coated electrodes has the highest sparkover voltage about 10% higher compared with other ones. However sparkover voltages decrease as the trial numbers increased. And it has a large scatter as shown in figure 3 or no.5 in table 2. On the other hand, bare electrodes have the conditioning effect; sparkover voltage increases as the number of sparkover increased and approaches to a given voltage (conditioned voltage). The value of the first sparkover voltage for coated electrodes and the conditioned voltages for bare electrodes is almost the same.

Figure 4 shows the variations for iron electrodes covered with enamel coating. In this figure, reduction of sparkover voltage is little compared with those obtained by aluminum coated electrodes. Moreover iron bare-electrodes have high dielectric strength compared with aluminum electrode as shown in figure 4. It seems that the electrode material affects the sparkover characteristics, namely, the coating to aluminum electrode reduces the generation of the initial electron from an electrode surface and it may protect the start of streamer.

Figure 5 shows the marks of sparkover discharges. The photographs show that the coated film was damaged by discharges and the electrode surface is directly touching the gas.

Table 1 Electrode configurations and the field utilization factor

No.	Electrodes				Equi-potential lines (Left:HV, Right LV)
	Configuration	Material	Gap length	Field utilization factor η	
1	Semi-sphere to Semi-sphere (Bare electrodes, R25/R25)	Aluminum	5mm	0.906	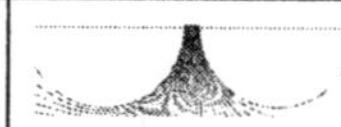
2			10mm	0.858	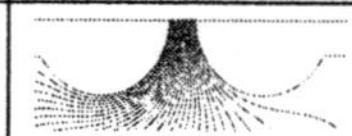
3			15mm	0.792	
4		Iron	10mm	0.858	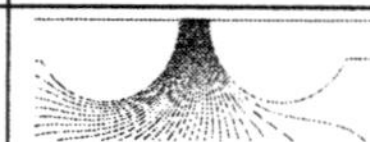
5	Semi-sphere to Semi-sphere (Coating, R25/R25)	Aluminum	10mm	0.855	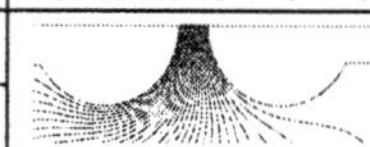
6		Iron	10mm	0.855	
7	Needle to plane	Stainless steel ϕ 1/ ϕ 40	10mm	0.067	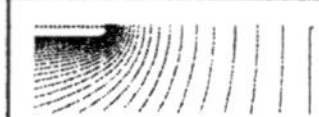
8		Stainless steel ϕ 2/ ϕ 40	2mm	0.394	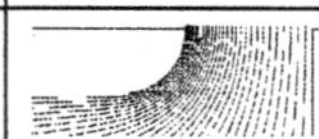
9		Stainless steel ϕ 2/ ϕ 40	10mm	0.122	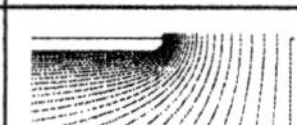

Table 2 Spark over voltages for the electrode without coating

Electrodes			CO_2 gas pressure					
			0.2MPa		0.4MPa		0.6MPa	
No.	D (mm)	η	V_{50} (kVp)	St.Dev (%)	V_{50} (kVp)	St.Dev (%)	V_{50} (kVp)	St.Dev (%)
1	5	0.906	28.2	2.9	49.4	1.3	68.2	1.8
2	10	0.858	47.9	3.2	85.4	1.0	117.1	6.6
3	15	0.792	70.1	1.5	125.2	1.6	165.4	7.1
4	10	0.855					124.8	2.6
5	10	0.855					102.5	13.9
6	10	0.855					120.3	4.6
7	10	0.067	24.6		38.4		43.0	
8	2	0.394					25.2	2.1
9	10	0.122	29.4	6.8	44.0	6.0	48.3	13.2

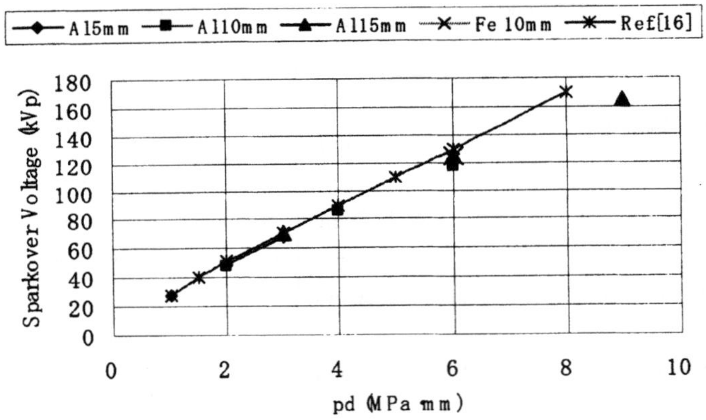

Fig.2 Paschen curve for CO_2 Gas
(Semi-sphere electrodes, R25mm/R25mm)

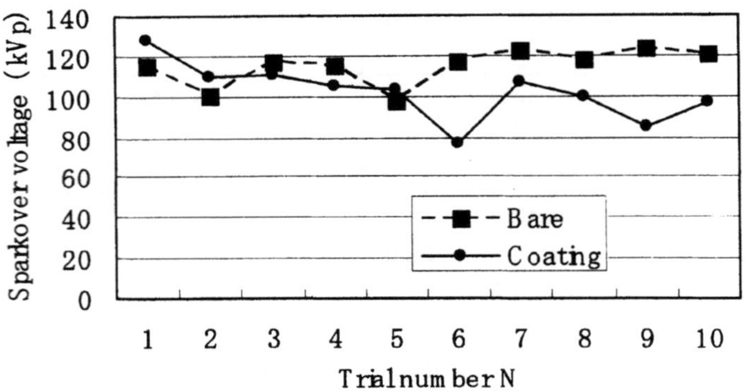

Fig.3 Variation of spark overvoltage
(Aluminum electrodes, gap length: 10mm, pressure: 0.6MPa)

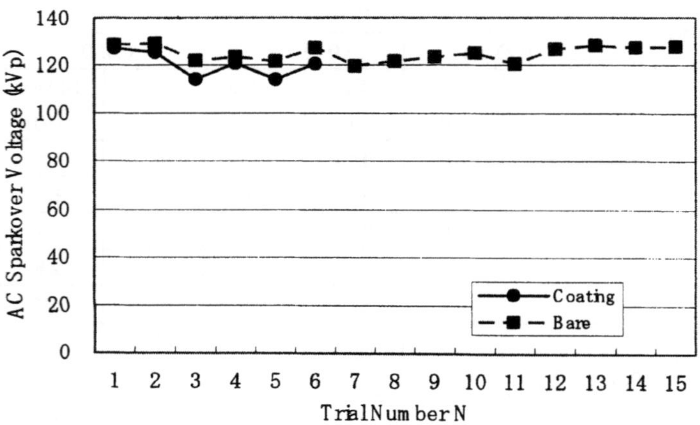

Fig.4 Variation of sparkover voltage (Iron electrodes, gap length:10mm, pressure:0.6MPa)

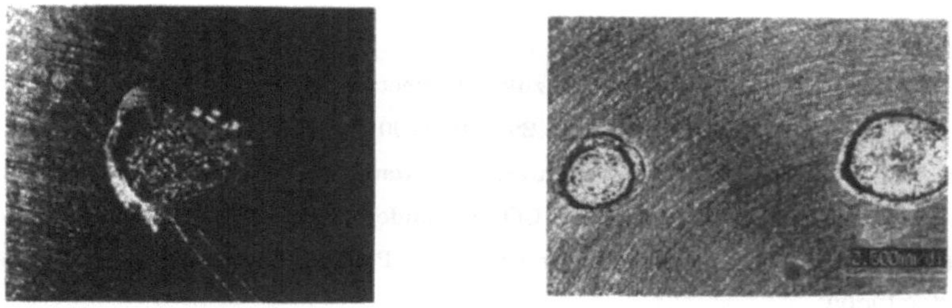

(a) Aluminum electrode (b) Iron electrode

Fig.5 Marks of sparkover discharge for enamel coating electrode

Figure 6 shows the relationship between field utilization factor and mean values of sparkover electric field. There is no corona discharge for the field utilization factor more than 0.792, but not less than the value of 0.122. Partial discharge inception voltages for the electrode of no.9 at the gas pressure of 0.2MPa, 0.4MPa and 0.6MPa are 25.5kVp, 31.9kVp and 38.9kVp respectively.

We have only limited information on the relationship between the field utilization factor and sparkover mechanism, we will continue the experiments.

Conclusions

We have carried out the experiments to clarify the spark over characteristics of CO_2 gas under AC voltage. The following results were obtained:

(1) The sparkover characteristics under quasi-uniform electric field using semi-sphere electrodes obeys the Paschen's law.

(2)The aluminum electrode covered with enamel coating has the higher sparkover voltage compared with one of the bare electrode about 10%. This means that the coating prevents a reduction of AC sparkover voltage.

(3)The relationship between the field utilization factor and spark over voltage was studied.

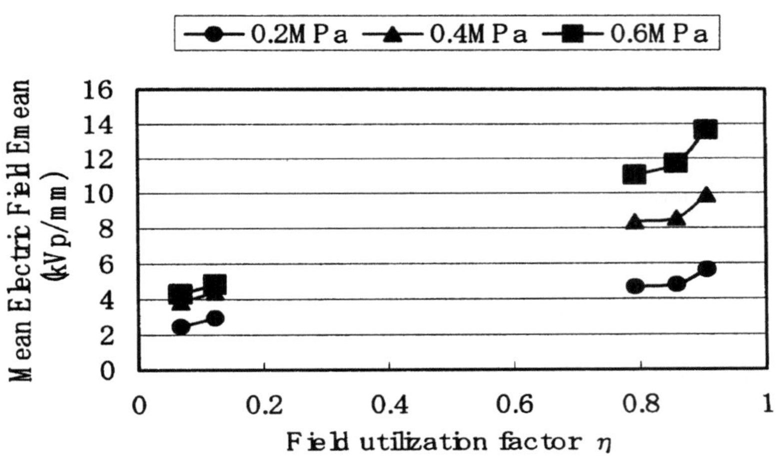

Fig.6 Field utilization factor vs. mean electric field at sparkover discharge

References

[1]M.Shiiki, M.Sato, M.Hanai and K.Suzuki: "Dielectric performance of CO_2 gas compared with N_2 gas", Gaseous Dielectrics IX, pp.295-300 (2001)

[2]S.Ohtsuka, M.Onomoto, T.Nishimura, S.Matsumoto, M.Shibuya and M.Hikita: "Insulation properties of High-pressure CO_2 gas under existence of a particle", Proceedings of the third IASTED International Conference on Power and Energy Systems, Sept.3-5, pp.721-726 (2003)

[3]S.Ohtsuka, K.Kiyama, M.Cho and M.Hikita:"Time-developing Partial Discharge Properties Leading to Sparkover in SF_6, N_2, CO_2 and their gas Mixture under Lightning Impulse Application", to be published in Gaseous Dielectrics X March 30 (2004)

[4]J.M.K.MacAlpine and A.H.Cookson:"Impulse Breakdown of Compressed Gases between Dielectric-Covered Electrodes", Proc. IEE Vol.117, No.3, pp.646-652 (1970)

[5]D.J.Chee-Hing and K.D.Srivastava:"Insulation Performance of Dielectric-Coated Electrodes in Sulfur Hexafluoride Gas", IEEE Trans. on Electrical Insulation, Vol.EI-10, No.4, pp.119-124 (1975)

[6]A.E.Vlastos and S.Rusck: "The Influence of a Thin Electrode Coating on the AC Breakdown of SF_6", Fourth International Conf. on Gas Discharge and their Applications, pp.59-62 (1976)

[7]E.Gockenbach: "Influence of Electrode Coating on the Breakdown Voltage of SF_6", IEE Conf., Publ.165, pp.181-184 (1978)

[8]V.N.Borin and I.M.Bortnik: "Dielectric Coating for Gas-Insulated High Voltage Equipment", Sixth International Conf. on Gas Discharges and their Applications", Part 1, pp.208-209 (1980)

[9]A.E.Vlastos and S.Rusck: "Influence of Coating Polished and Rough Electrode on the Impulse and AC Breakdown of SF_6", Second International Symposium on Gaseous Dielectrics, pp.210-214 (1980)

[10] A.E.Vlastos and S.Rusck: "Influence of Electrode Surface State on the Breakdown of Pressurized SF_6", Sixth International Conf. on Gas Discharge and their Applications, Part 1, pp.200-203 (1980)

[11]A.E.Vlastos, Chun YiYu and A. Salem: "Influence of Coating on the Impulse Breakdown of SF_6 for Electrodes with Different Surface Roughness", Seventh International Conf. on Gas Discharges and their Applications, pp.273-276 (1982)

[12] A.E.Vlastos and S.Rusck: "Influence of Electrode Coating on the Breakdown Voltages of SF_6", Third International Symposium on High Voltage Engineering, 32.08 (1979)

[13]A. Rein and J.Kulsetas: "Impulse Breakdown of SF_6/N_2 Insulation Influence of Electrode Covering Polarity Effect", Third International Symposium on Gaseous Dielectrics, pp.315-321 (1982)

[14]A.Inui, Y.Togawa, H.Okubo, M.Honda and S.Yanabu:"Effect of Electrode Coating on Impulse Breakdown in SF_6", 4th International Symposium on High Voltage Engineering, No.32.04 (1983)

[15]M.Honda, H.Okubo, A.Inui: "Impulse Breakdown Characteristics of Coated Electrodes in SF_6 Gas", IEEE Trans. PD Vol.PWRD-2, No.3, pp.699-707 (1987)

[16]Z.Krasucki:" Paschen Curve for Carbon Dioxide" Electra No.52, pp.75-78 (1974)

INSULATION DEGRADATION MECHANISM OF EPOXY SPACER SAMPLES FOR GIS

H.Okubo[1] S.Watanabe[2] N.Hayakawa[2] and T.Kumai[3]

1. INTRODUCTION

Solid insulators such as epoxy spacers in GIS have been exposed to various stresses during a long-term operation, which can deteriorate the insulation performance of GIS. Therefore, the insulation diagnosis techniques of GIS will play a key role for keeping electrical insulation performance of solid insulators and then for reliable operation of GIS.

From the above viewpoints, we have been investigating the partial discharge (PD) inception characteristics of epoxy spacer samples and the subsequent PD generation characteristics leading to breakdown (BD), using PD current pulse waveform analysis (PD-CPWA)[1] under ac voltage application. In this paper, we used two types of epoxy spacer samples; thermal degradation samples and transparent samples. We measured the PD inception and subsequent PD generation characteristics of thermal degradation samples for different thermal stress conditions. We also identified the PD location in the epoxy spacer samples using the transparent samples, and discussed the PD and BD mechanism.

2. EXPERIMENTAL SETUP

Figures 1 and 2 show epoxy spacer samples and the PD measuring system, respectively. The epoxy spacer sample was made of epoxy resin filled with alumina, which was the same material as that used for the actual GIS spacer. The glass transition temperature of the epoxy resin was 122°C. High voltage (HV) and grounded (GND) electrodes made of aluminum with the edge radius of 6mm were embedded in the epoxy resin. The embedded electrode surface

[1] Center for Integrated Research in Science and Engineering, Nagoya University, Nagoya, 464-8603, Japan
[2] Department of Electrical Engineering, Nagoya University, Nagoya, 464-8603, Japan
[3] Electrical Engineering Department, Chubu Electric Power Co., Inc., Nagoya, 461-8680, Japan

Gaseous Dielectrics X.
Edited by Christophorou *et al.*, Springer, New York, 2004.

(a) Normal sample (with filler) (b) Transparent sample (without filler)

Figure 1 Epoxy spacer sample.

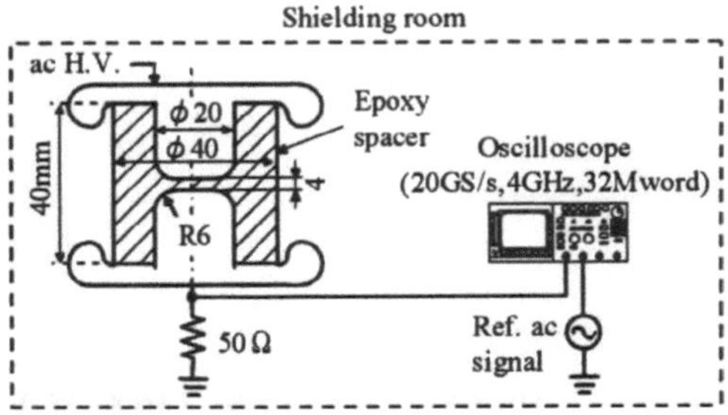

Figure 2 Experimental setup.

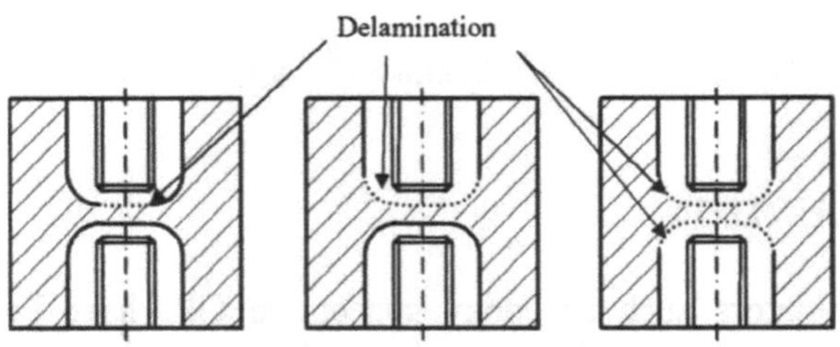

(a) Delamination A (b) Delamination B (c) Delamination C

Figure 3 Three different delamination defect for transparent sample.

was sandblasted for the better adhesion with the epoxy resin. The gap length between the both electrodes was 4mm.

The thermal stress conditions of thermal degradation samples and the fabrication conditions of transparent samples are as follows:

(a) Thermal degradation sample

The epoxy spacer sample described above has been verified to be PD free up to $154kV_{rms}$. Then, the thermal stress was applied to the virgin sample in an oven for different temperature and duration.

(b) Transparent sample

The transparent sample was fabricated without filler. Three types of the transparent samples with different delamination region were fabricated, as shown in Fig.3. The delamination was made on the electrode surface by mirror finish and a mold release agent before the casting process.

Each sample was placed in a 154kV experimental GIS filled with SF_6 gas at 0.4MPa. The ac 60Hz high voltage was applied to the sample. PD current pulse signals were detected by a digital oscilloscope (20GS/s, 4GHz, 32MWord) through 50Ω resistance as a detecting impedance[2]. The PD detection sensitivity is less than 0.1pC. Owing to the long memory of the oscilloscope together with PD-CPWA, all of the PD current pulse waveforms with nano-second time resolution can be recorded during the PD measurement from voltage application to BD. Moreover, in the case of the transparent samples, PD light emission images were observed by a digital camera with an image intensifier (I.I.).

3. PD CHARACTERISTICS OF EPOXY SPACER SAMPLES

3.1. Thermal Degradation Sample

3.1.1. PD Inception Characteristics

Figure 4 and Table 1 show the thermal stress conditions of the thermal degradation samples. The total number of thermal degradation samples was 29, where 16 types of thermal stress were applied. As the result of voltage application to each thermal degradation sample, PD signal was detected in 9 samples at the applied voltage lower than 154kV$_{rms}$. Figure 4 and Table 1 also show the PD inception voltage (PDIV). PDIV of each sample was 114.4-153.6kV$_{rms}$, which means that the electrical insulation performance was deteriorated by the thermal stress. Since the degree of thermal degradation may depend on the temperature and the heating duration, it is necessary to take into account their influence on the thermal degradation characteristics of epoxy spacer samples.

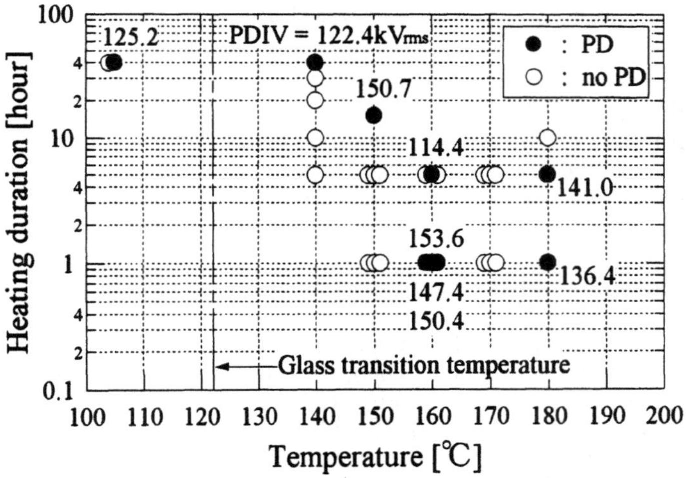

Figure 4 Thermal degradation conditions and PDIV.

Table 1 Thermal stress conditions and measured PD/BD characteristics of thermal degradation samples.

Temperature [°C]	Heating duration [hour]	PDIV [kV$_{rms}$]	PDIE$_{average}$ [kV$_{rms}$/mm]	Time duration to BD t_{BD}	Polarity of BD	Number of PD pulses till BD $N_{Positive}$	Number of PD pulses till BD $N_{Negative}$
105	40	125.2	31.3	2min 26sec	Negative	60	1
140	40	122.4	30.6	12min 22sec	Positive	12194	17336
150	15	150.7	37.7	33sec	Positive	697	737
160	1	153.6	38.4	1min 55sec	Negative	191	143
		150.4	37.6	1.7sec	Positive	90	38
		147.4	36.9	4.7sec	Positive	248	114
	5	114.4	28.6	> 3hours	-	-	-
180	1	136.4	34.1	10sec	Positive	437	382
	5	141.0	35.3	11min 22sec	Positive	3340	2026
Without heating		> 154.0	> 38.5	-	-	-	-

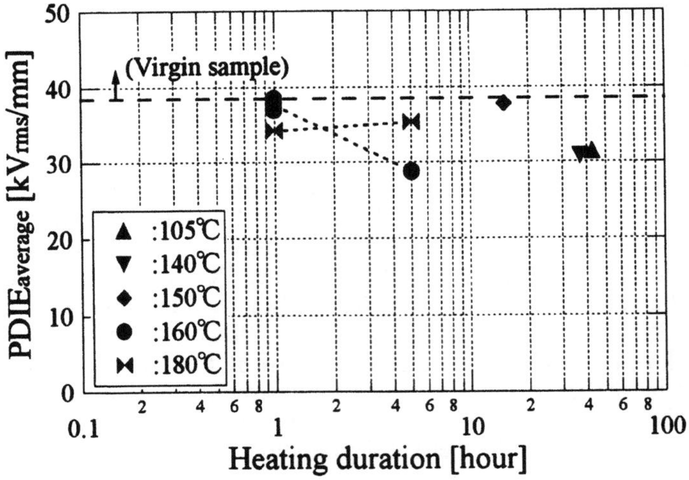

Figure 5 Relationship between $PDIE_{average}$ and heating duration.

Figure 5 shows the relationship between the average PD inception electric field ($PDIE_{average}$ = PDIV / 4mm) and heating duration. The influence of the heating duration on $PDIE_{average}$ seems to be greater than that of the temperature. The decrease of $PDIE_{average}$ under the longer heating duration may be attributed to the deterioration of insulation performance of epoxy resin by chemical reaction and so on.

3.1.2. PD Inception to BD Characteristics

After the PD inception of 9 samples in Table 1, the applied voltage was kept constant at PDIV of each sample. Then, 8 samples resulted in BD within 3 hours. Table 1 also shows the time duration (t_{BD}) to BD after the PD inception, the polarity of BD and the number of PD pulses. Figure 6 shows the time transition of PD pulse current from PD inception to BD. The horizontal time axis was normalized by t_{BD}. PD current and the number of PD pulses gradually increased with the elapse of time and drastically increased before BD. In addition, PD pulse generation was not continuous, but intermittent from PD inception to BD. Such tendencies in PD generation were similar in most samples, irrespective of the difference in PDIV and t_{BD}.

PD-CPWA gives us various information from the viewpoint of the PD current pulse waveform parameters. Figure 7 shows the time transition of the steepness di/dt at the wave front of each PD current pulse waveform. In the similar way as that in Fig.6, di/dt also gradually increased with the elapse of time and drastically increased before BD. Such a time transition of PD waveform parameters may be attributed to the change of physical properties of the epoxy resin under the thermal and electric field stresses. The insulation performance of the epoxy resin would be deteriorated due to electron impact, local temperature rise, chemical reaction and so on, where di/dt could be closely related to the electron avalanche. These effects could be accumulated during the long-term voltage application, and resulted in BD when a certain PD waveform parameter such as di/dt reached the critical level.

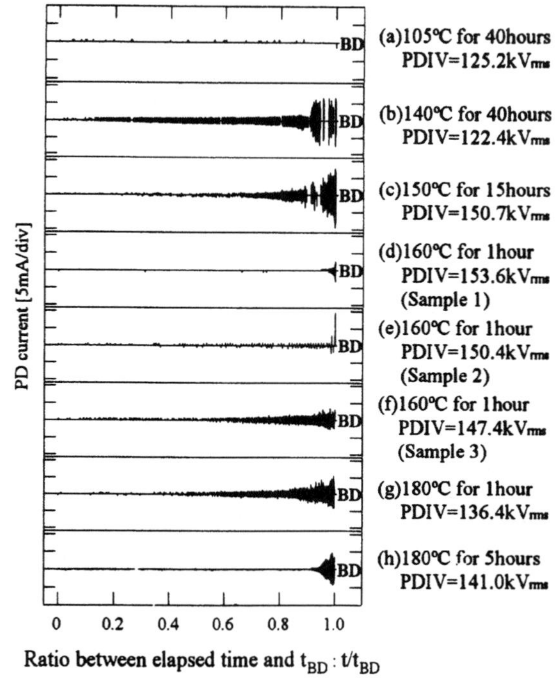

Figure 6 Time transition of PD pulse current from PD inception to BD.

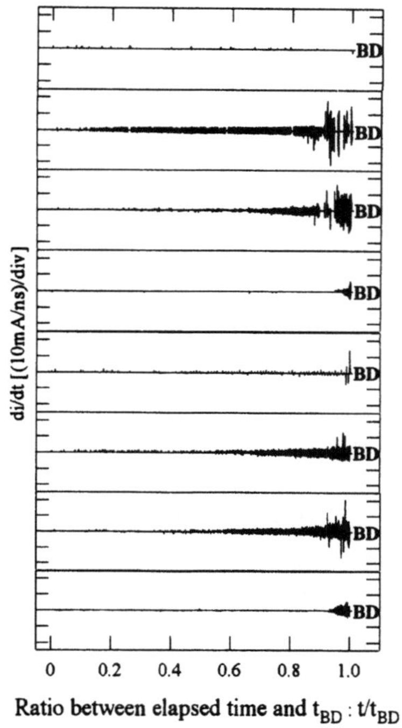

Figure 7 Time transition of di/dt of PD current pulse.

3.2. Transparent Sample

3.2.1. PD Generation Characteristics

Since PDIV of a transparent sample with the Delamination A was higher than $154kV_{rms}$, the applied voltage was kept constant at $154kV_{rms}$. As the result, PD was generated at 18min 34sec after the voltage application, and resulted in BD at 60sec after the PD inception. Figure 8 shows the time transition of PD current from PD inception to BD. PD pulse generation after the PD inception was not continuous but intermittent, and drastically increased before BD, which means that PD generation characteristics from PD inception to BD of the transparent sample reproduced fundamental PD generation characteristics of epoxy spacer samples in Fig.6.

Figures 9(a)-(c) show the PD light emission images of the transparent sample with the Delamination A, which correspond to time width in Fig.8(b), respectively. PD location from PD inception to BD was unique at the edge of the HV electrode, but not on the delamination region. In addition, PD light intensity drastically increased before BD.

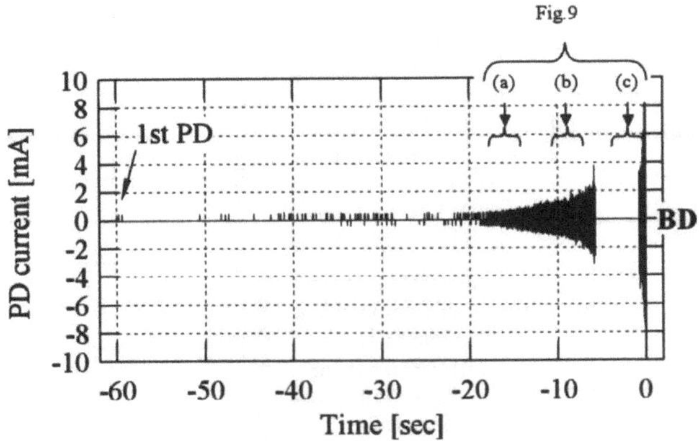

Figure 8 Sequential PD generation characteristics of transparent sample (Delamination A).

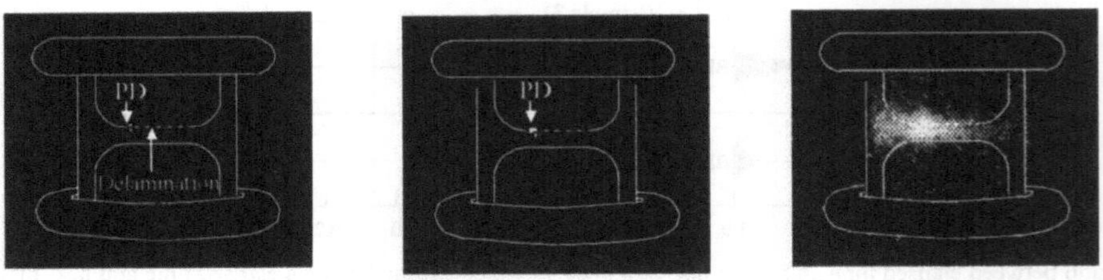

(a) 14sec to 17sec before BD (b) 8sec to 11sec before BD (c) 1sec to 4sec before BD

Figure 9 PD light emission images of transparent sample (Delamination A).

3.2.2. PD and BD Mechanism

Figure 10 shows the PD generation part and the electric field distribution of transparent samples. PD tends to generate not only around the high-stressed edge of HV and GND electrodes, but also on the low-stressed electrode surface without the delamination. After the experiment, some transparent samples were decomposed, and the BD trace was found to be unique.

The above results were discussed in terms of the interface condition between the embedded electrode and the epoxy resin. Figure 11 shows the interface condition between the embedded electrode and the epoxy resin observed by an optical microscope. As shown in Fig.11(b), the delamination thickness at the interface was less than 4μm. On the other hand, the electrode surface roughness was 3.3μm for mirror finish and 26.7μm for sandblast, respectively. These results tell us that PD generation would be attributed to the local high electric field at a microprotrusion on the sandblasted surface rather than the existence of the delamination in the test samples.

The thermal degradation sample led to BD in the previous section was also decomposed, and a unique BD trace was found just like the transparent sample. Therefore, PD in thermal degradation samples would also be generated at a unique weak point at the interface between the electrode and the epoxy resin, which could bring about the erosion of the epoxy resin. The erosion of the epoxy resin would deteriorate the insulation performance at the interface and activate the PD development. Such a positive feedback process between the PD generation and the epoxy erosion could result in the final BD.

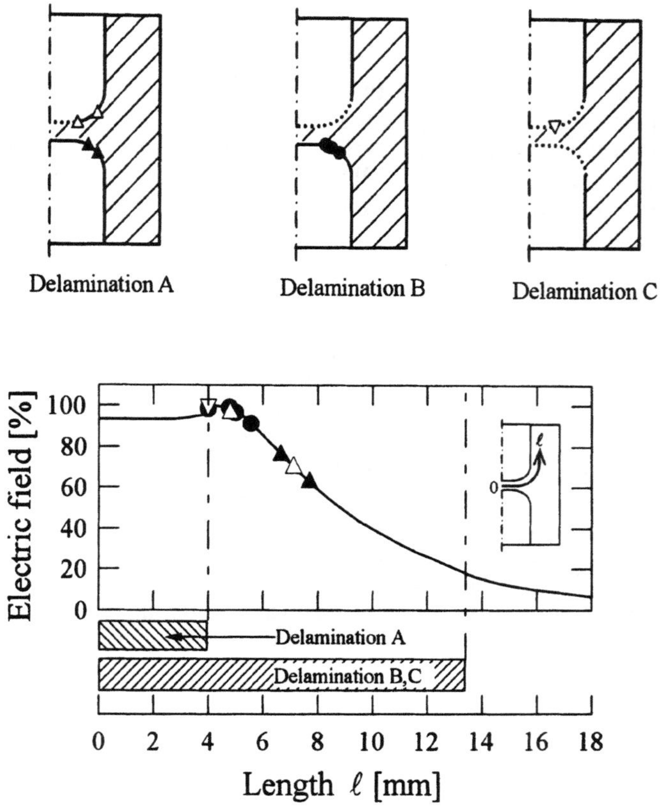

Figure 10 Electric field distribution and PD generation part.

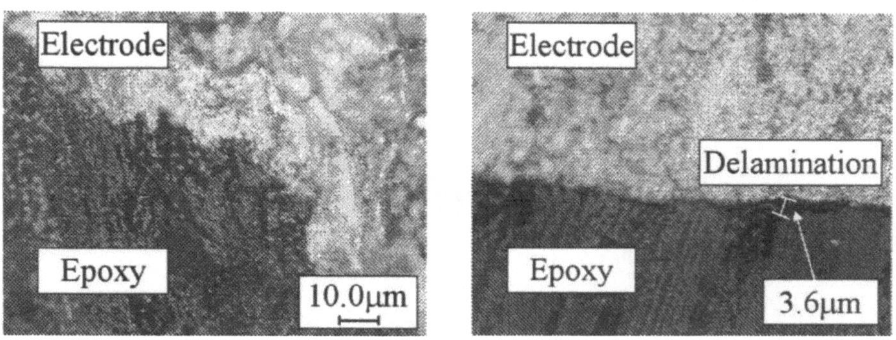

(a) Sandblasted region

(b) Delamination region

Figure 11 Interface condition between embedded electrode and epoxy resin.

4. CONCLUSIONS

In this paper, we fabricated two types of epoxy spacer samples; thermal degradation samples and transparent samples, in order to clarify the insulation degradation characteristics of solid insulators in GIS. PD inception characteristics, time transition of PD waveform parameters and the PD light emission images for thermal degradation samples and transparent samples revealed the followings:

1. PD inception voltage of the thermal degradation samples was lower than that of the virgin samples, which means that the electrical insulation performance was deteriorated by the thermal stress. The influence of the heating duration seems to be greater than that of the temperature.
2. PD pulse generation after the PD inception was not continuous but intermittent, and drastically increased before BD, in the thermal degradation samples as well as the transparent samples.
3. Time transition of the steepness di/dt at the wave front of each PD current pulse waveform reproduced that of the magnitude of PD current, which may be attributed to the change of physical properties of the epoxy resin under the thermal and electric field stresses.
4. PD location from PD inception to BD was unique, and finally resulted in BD by a positive feedback process between the PD generation and the epoxy erosion.

REFERENCES

1. H.Okubo et al.: "The Relationship Between Partial Discharge Current Pulse Waveforms and Physical Mechanisms", *IEEE Electrical Insulation Magazine*, Vol.18, No.3, pp.38-45 (2002).
2. H.Okubo et al.: "Investigation of Electrical Insulation Diagnosis of Solid Insulators in GIS by Partial Discharge Current Pulse Waveform Analysis (PD-CPWA)", *IEEE/PES T&D*, Yokohama, pp.925-930 (2002).

SECTION 9: GAS-INSULATED EQUIPMENT

INSTALLATION, MAINTENANCE, AND END OF LIFE OF A GAS-INSULATED SUBSTATION

Wolfgang Degen*

1. INTRODUCTION

The metal-enclosed gas-insulated switchgear is made as to ensure the best possible relationship between design, used materials, maintenance effort and maintenance intervals. The result is optimized utilization. The switchgear therefore offers a high degree of reliability and availability, even if the limits of wear (erosion, aging, corrosion etc.) are approached.

The reliability of the GIS is very important for the availability and the long service time of the switchgear. Therefore the complete process after manufacturing of correct installation, intelligent maintenance in combination with the adequate management strategy is the only possibility to achieve this aim with a minimum on effort and time out for the equipment.

Installation as final stage of the quality control before putting into service will be done according to the recommendations of the manufacturer, finished with tests where necessary to prove the correctness of the equipment.

Several options for maintenance can be chosen like scheduled maintenance, condition based maintenance, reliability centered maintenance or even a risk based corrective maintenance as correction after a failure occurs.

* Dipl. Ing. Wolfgang Degen, Siemens AG Germany, Power Transmission and Distribution, High Voltage Division, GIS Substations, PTD H2 TC QM, D-13623 Berlin

However the switchgear can only function reliably if it is correctly maintained. The purpose of inspection and maintenance therefore is:

- to determine to what extent certain parts heave been worn and to assess their status
- to ensure that parts still in good condition are kept in this state
- to replace certain parts in right time

Liberalization, deregulation and, in many cases, privatization has led to profound changes in the business environment of these industries. More and more it will be discussed the option of outsourcing of maintenance.

Maintenance is necessary but it has also to be considered, that about 30h percent of the failures in GIS are caused by human interference activities like maintenance. A reduction of human action thus contributes to the reliability of the equipment.[1, 2]

Furthermore inspections of GIS supplied in the early 1970's have shown that these are generally in a condition as new. Although these had been designed originally for shorter inspection and maintenance intervals, due to the lack of experience then, the experience gained during those inspections have backed the decisions to prolong the maintenance intervals.

While in the past most evaluations on economic benefits quite often end up with the capital cost of the investment only, today the overall life cycle cost consideration (LCC) including maintenance cost are regarded. Also environmental aspects have to be considered. Only this approach offers a fair comparison of solutions. Different maintenance strategies will be presented and its impact of the various influences on the assets will be defined.

Although almost 85% of the material of the equipment can be reused, special procedures how to handle e.g. used SF6 and contaminated materials must be clearly indicated.

2. INSTALLATION OF GAS INSULATED SUBSTATIONS

In the last years the development of gas insulated substations was mainly driven by the aim to reduce the use of material and costs by still extremely high reliability. Main steps of the development were as follows:

- progress of circuit-breaker technology, which allows to reduce the number of interrupter units despite increasing breaking capability
- progress of casting and machining technology of aluminum casted parts, which allow the use of minimized shapes and volumes

- use of computerized production and testing equipment with high quality standards
- design of integrated components with several functions like disconnectors and grounding switches within one gas compartment
- use of intelligent monitoring and diagnostic tools to postpone maintenance activities and to avoid unnecessary tasks

As a result very compact substations designs are on the market which allow the delivery of completely sealed and tested bay units up to 245 kV. The advantages of such small equipment are shipment of complete, factory assembled, pre-tested and pre-filled double-bays (245 kV) and triple-bays (145 kV) with shortest erection time on site.

The change to smaller and more compact substations seems to be stopped by the needs of the utilities for more convenient service and maintenance activities. It is understandable, that the accessibility is influenced by the degree of compactness of the design.

A further possibility of the space saving design and reduction of installation activities is the installation within a container for a mobile use on different places.

Testing on site of these new developments can be limited only to newly made connections or to components, which must be opened for further assembly. The target of mandatory and optional site tests is to prove that the GIS and its components are fulfilling the intended functions after transport and assembly at site. The procedures and conditions are defined by the relevant IEC standards with some additional requirements if specified by customers. Site tests are carried out on components, GIS parts of full GIS after site erection and are concentrated on assuring the quality of site erection activities and possible effects of the transport to site. Site tests are the final step of quality control before putting the equipment into service and are quite often used for gathering characteristics of the equipment for further comparison with results of the aged equipment.

3. AGE DISTRIBUTION OF HIGH VOLTAGE COMPONENTS AND GAS INSULATED SWITCHGEAR

The lifetime of switchgear depends mainly on the quality of the products but also on their individual service conditions and the maintenance strategy, which is a question of asset management and the economic situation of the user. Before talking about maintenance the age distribution of the gas insulated switchgear should be regarded.

The following graph shows the age distribution of the GIS respectively in Germany; however these figures are valid for other countries as well. The picture shows also the medium value and standard deviation for the estimated life time of gas insulated switchgear.[3]

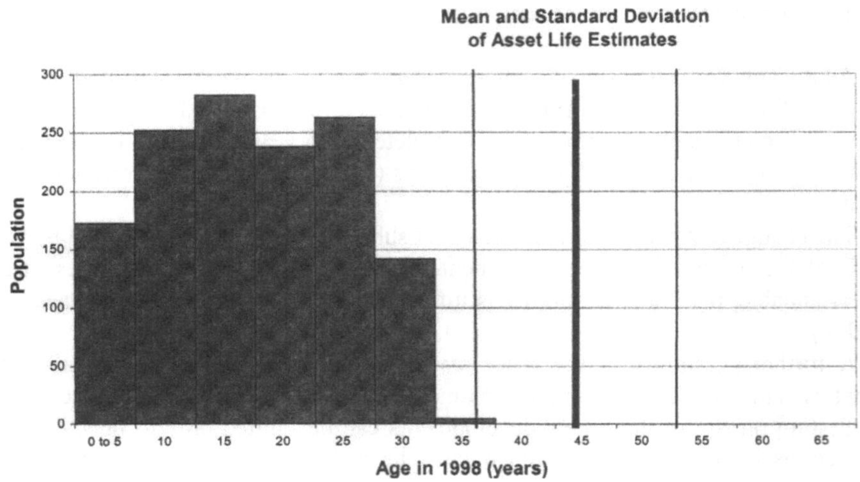

Figure 1 Age distribution of Gas Insulated Switchgear

It can be seen, that according to this study of CIGRE an average lifetime of 45 years can be expected. However it has to be considered, that this study is already more than five years old, which means that in less than 10 years the first GIS have to be replaced.

The ageing of the equipment will be influenced by the number of operations and the wear of parts, by electrical stresses and by temperature influence e.g. on sealing materials. The life expectancy can be enlarged by correct service conditions, doing maintenance if required and replacement of worn parts in time. The used components like insulating parts, contact systems and drive units indicate after more than already 35 years of practical experience a life time expectancy of almost 50 years.

But there are other factors which can influence the lifetime of a switchgear or even end the service period like

- reaching of the technical limits in a changing surrounding
- changing of the legal conditions

- changing of the environmental demands
- changing of the asset management strategy of the user

and last but not least

- lack of knowledge about the switchgear/missing of experienced personnel

Only the last point can be influenced by the manufacturer of the switchgear, all other aspects are out of control.

It is essential that already during the development of the switchgear as much as these aspects will be regarded. The high number of more than 10.000 operations also under service conditions, the environmental sustainability and the accomplishment of all existing standards and regulations at the time of development is part of it. Furthermore a responsible manufacturer will provide for the whole period of service life the necessary technical support as well as the delivery with spare parts just in case of a technical problem.

Concluding it has to be stated that every user of the switchgear has to decide by himself, if and when on which reasons he will end the service period of his equipment. The appropriate maintenance strategies, which have an important impact on this procedure, will be covered in the next chapter.

3. MAINTENANCE

Since the introduction of GIS into the market at 1968 progress has been made on the need of inspections and maintenance.

The effectiveness of maintenance depends mainly on the way manufacturers and users are working together.

Maintenance in general is per definition a required action on switchgear in order to sustain the availability of the equipment. The maintenance activity should

- replace specially stressed parts in time and prevent failures
- determine the wear of parts and assess their state
- ensure that parts are still in good shape and can be kept in place

The modern generations need less maintenance and allow longer service periods due to the excellent experiences of more than 35 years of service, as the following picture shows.

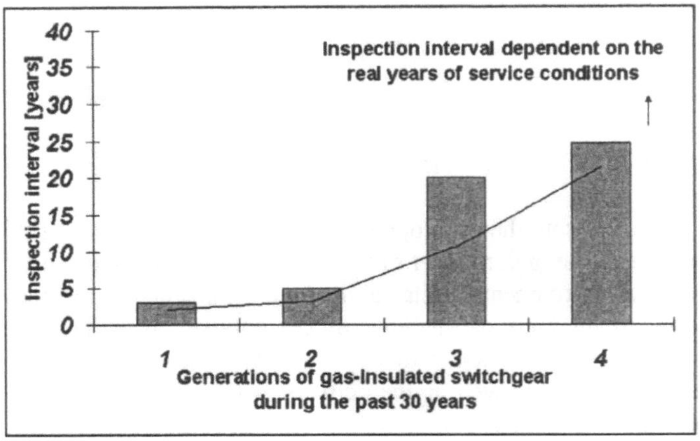

Figure 2 Development of maintenance intervals during the last 30 years

Maintenance can be done on basis of the manufacturers experience – time based maintenance TBM - as it is still done normally today, on basis of monitoring and the use of diagnostic systems - condition based maintenance CBM- and as reliability centered maintenance RCM (in some way a scheduled maintenance). As an effect of the deregulation another maintenance strategy, which can be called risk based maintenance, comes up. By all of these activities the maintenance costs should be minimized, the intervals between maintenance enlarged and the availability of the equipment increased.

The recent maintenance practices have been to give increased focus upon achieving greater efficiency. Business profitability, availability, reliability, revenue and customer satisfaction were all identified in the survey as key drivers. In the following the three most common possibilities should be presented in detail and commented:

3.1 Time Based Maintenance TBM

Today normally the time based maintenance activity is following a maintenance schedule according to the recommendations of the manufacturer. Following this approach the maintenance has to be done after reaching special limit values of

- number of operations
- interrupting current
- years of service (age)

due to the effects related to the service stresses. The limit values were specified by the manufacturer according to his experiences in development, manufacturing and service. All this is based on product reliability statistics of the manufacturer, which include the complete number of delivered units in service under all conditions, and will be adapted to changes if necessary.

The advantage of this method is a planned and organized procedure, harmonized for all elements of the equipment and substations and optimized for the effectiveness of maintenance effort. The disadvantage is that the activities are carried out independently if necessary or not. Failures might occur between the maintenance intervals. However the costs for labor and spare parts are predictable.

3.2 Condition Based Maintenance CBM

This kind of maintenance will be done on the basis of the results of diagnostic and monitoring devices for special parameters as there are e.g.

- drive characteristics/operating characteristics/timing
- density
- gas leakage/gas moisture/decomposition products
- partial discharges
- conductive particles
- breaking current
- local overheating
- wear/condition of contacts
- position of operating systems a.s.o.

The data gathered by sensors can be processed and used for the determination of trends, which indicate what activity might be necessary at which time.

Advantageous for this method is the knowledge about the actual state of the equipment, the prevention of predictable failures and the reduction of not necessary work. The disadvantage is that additional effort around the equipment (investment and labor) has to be done, that the overall reliability might be decreased and that possible actions cannot be planned. As the result of the diagnosis the maintenance of the circuit-breaker might be necessary at different times than that for disconnectors a.s.o.

3.3 Reliability Centered Maintenance RCM

RCM is a structured method of maintenance activity that establishes the maintenance requirements for the equipment based on the experiences and consequences of failure, as it is also done by the scheduled maintenance method. However the failure consequences are not determined by the manufacturer but by the (limited) experiences of the user with only his equipment in an operating system. The failure causes and failure modes for each component in this system have to be identified and than to be subjected to a failure mode effective analyses (FMEA).

The advantage of this method is expected to have cost savings in preventive maintenance and that the costs of implementation will provide a pay back within one year (according EPRI). However it is difficult to judge correctly the failures and to proceed with FMEA without the knowledge and experience of the manufacturer. Failure statistics require big numbers of units to be representative. If only small units will be considered, the conclusions might be misleading.

3.4 Corrective Maintenance CM

In case of corrective maintenance no preventive maintenance at all is carried out, maintenance is only performed when a failure occurs. This has an influence on the failure rate and consequently on the availability of the switchgear. Reducing the time based maintenance will as well result in a higher failure rate as the variation of the maintenance interval in case of TBM.

It has to be considered that this method has an impact on the overall life cycle cost, depending on the reliability of the equipment and the financial losses in case of outage.

In general it can be said, that the choice of the right maintenance method has a major impact on the economic situation of the equipment and even on decisions of its replacement. Therefore this issue is subjected to all kind of cost saving investigations and as a result the outsourcing of maintenance will become an important issue in the future.

3.5 Outsourcing of Maintenance

An "International Survey of Maintenance Trends and Policies" was published in September 2000. This reported the results of a questionnaire conducted during 1998 and was coordinated by a joint working group (WG 23/39) of CIGRE study committees B3 and 39 and reflected the recent, and ongoing, dramatic changes occurring in the maintenance policies of the electricity supply industry worldwide.[5]

For many utilities the tradition has been for internal staff to undertake all operation and maintenance activities - and to achieve this at the same high technical standard at all points in the transmission system. This situation has changed, and many utilities now report outsourcing of activities that are judged to be non longer core-business like tower painting and substation building repairs. Such activities are traditionally defined as non-core work while technical work, for example, primary plant or protection maintenance, is considered to be core work since it has direct consequences for the power supply. The typical level of outsourcing reported in the survey was that in 1998 utilities were reporting 30-35% of their operation and maintenance activity was being outsourced. In general this activity related to non-core activity. The exceptions were in some countries where there was not a strong technical capability and there was greater reliance upon external technical support for activities usually conceived to be "core". This result shows an increase compared to a survey in 1991, which reported an average of 26% outsourcing. More surprising is that 25% of respondents outsourced more than 50% of their maintenance, and of these 15% outsourced more than 90%. Asia and the Middle East had the highest average of 57% and North America the lowest at 5%. While this North American result has increased substantially since 1998, there remains evidence that many of the large U.S. utilities still have little exposure to outsourcing.

Without doubt this strategy includes a certain risk is for transmission utilities. Beside the operational risk there are other equally significant categories of risk associated with outsourcing which should be regularly reviewed to ensure that they are fully understood and managed. The risks include:

- Inadequate training and loss of technological skills base
- Business failure or pull out by contractor
- The level and availability of skilled resources, especially if the utility does not have its own field resources
- Domination of the utility by the contractor

Whilst the above risks are significant they are normal business risks that can be overcome provided that the market is sufficiently mature to allow choice of suppliers. Appropriate forms of contract and ensuring that an asset manager retains sufficient expertise and resources to adequately manage suppliers can also offset such risks. This latter capability is defined in the CIGRE Guidelines as a realisation capability.

The use of external service providers does enable risk to be shared. However, it is unreasonable to expect a contractor to carry large risks on behalf of a utility if the work to be undertaken is low cost and low margin. The allocation of risk is a key issue and it should be recognised that it is generally more practical to accept greater rather than less risk.

4. END OF LIFE OF THE GAS INSULATED SWITCHGEAR AND DISPOSAL

The average service life of gas insulated switchgear is nowadays between 20 to 30 years old. It has to be considered, that the average age of end of life is between 35 to 50 years and as a result of good experiences it can be estimated, that a service period of about 50 years can be economically completed.

The use and operation of switchgear needs qualified staff due to federal regulation not only for operation but also for disposal. Disposal of equipment will be done typically for HV equipment on site and mainly the by users themselves, with assistance of manufacturer only if requested. For MV a system has to be installed, until now however, there is no recycling association or network installed.

The end of life procedures of a gas insulated substation must be defined. For decommissioning in general the equipment will be disassembled down to its material components, which are typically metallic material – aluminum, copper, aluminum casting components, low voltage waste, hydraulic fluid and grease. Almost 90% of all materials can be reused. Each material has a waste processing number, which indicates what has to be done. Materials and equipment must be cleaned or disposed of so that the decomposition products they contain are not released. The materials have to be assorted on site before delivery to the waste collector. Switch dust, absorption agents and filters from gas compartments, respiratory equipment, cleaning and maintenance gear, disposable protective clothing and other materials which have been in contact with the decomposition products, must be deposited in sealed containers and disposed of in compliance with local regulations, stating refuse code ASN 59304 (materials contaminated with chemicals) as necessary.

The owner is always responsible for cleaning and disposal tasks. Due to the environmental impact of SF6 special procedures how to handle e.g. used SF6 and SF6 contaminated materials have to be followed. General considerations of recovering SF6 and a basic structure of the SF6 reclaiming from electrical power equipment were described in several papers of the CIGRE B3-02 working group.[6]

Before the gas is recovered from the electric power equipment it has to be assigned a gas category for reclaiming. This category is based on the severity of contamination with the levels non-arced, normally arced and heavily arced. Based on this category an adequate reclaimer and reclaiming procedure are chosen. Heavily arced gas is to be expected from equipment in which failures have occurred:

- Circuit breakers after interruption failure;
- Insulation compartments after internal arcing failure;
- Any kind of arcing failure.

The main elements of the gas reclaiming procedures are:

- Determination of the gas category;
- Selection of filters;
- Gas purification by filtering;
- Quality checks;
- Handling of gas that results non-reusable.

For final waste high voltage equipment manufacturers are offering their services to switchgear users in parallel to switchgear users (mainly utilities) who work under own responsibility. Individual and small installations especially in the medium voltage sector are not covered by existing de-commissioning companies besides the manufacturers.

Documentation about de-commissioning is indirectly documented in the overall SF6 inventory by the recycled and re-used SF6 quantities. For decommissioning especially the SF6 balance will be evaluated and reported by the acting parties.

5. REFERENCES

1. CIGRE Technical Brochure No. 83 Jansen, A. et al.: "Final Report of the second international enquiry on high voltage circuit-breaker failures and defects in service", CIGRE WG 13.06, 1994
2. CIGRE WG 23-10 Kopejtkova, D., et al.: Report on the second international survey on high voltage gas-insulated substations (GIS) service experience, CIGRE 23-102, 2000
3. CIGRE Report 176: "Ageing of the System - Impact on Planning" of the Working Group 37-27, December 2000
4. Balzer, G; et. al.: Life Cycle Assessment of Substations: A Procedure for an Optimized Asset Management. CIGRE 2002, Paris, report 13-304
5. CIGRE Survey "Maintenance Trends and Policies". Joint WG SC B3 and 39 (WG 23/39) September 2000
6. Guide "Practical SF6 Handling Instructions", Draft 3rd February 2004, of WG B3-02 TF1 (not yet published)

FUTURE TRENDS OF GAS INSULATED SUBSTATIONS

Chris Jones*
On Behalf of CIGRE Task Force of Working Group B3-02

1. INTRODUCTION

Gas Insulated Switchgear (GIS) technology has been undergoing continual development since its first introduction in the 1960's, and whilst some may consider this technology as mature, there is an ever increasing rate of change in GIS design, manufacture, test and operation technologies and practices. Some of the developments taking place are general to switchgear, some are specific to GIS and some of the general developments have particular implications on GIS.

The work of Task Force 04 of CIGRE WG B3-02 is to identify the trends which will determine how GIS will evolve in the future. Trends impacting on GIS design, manufacture, test, operation through to end of life are all within the scope of this work. The objective is to provide a framework that can be used to track the evolution of GIS, and thus enable users and suppliers to optimise the future utilisation of GIS.

This paper provides an introduction to this work and identifies some of the factors that will determine the GIS of the future, in this paper focusing particularly on issues surrounding the use of SF_6 gas.

2. ENVIRONMENT, BUSINESS AND TECHOLOGY DRIVERS

Primary environment, business and technology drivers provide opportunities and threats for the evolution of GIS. Some of the factors that will determine the GIS of the future include:

* VA TECH Transmission & Distribution, Hebburn, Tyne and Wear, NE31 1EJ. UK.
Email: chris.jones@vatechuk.com

Gaseous Dielectrics X.
Edited by Christophorou *et al.*, Springer, New York, 2004.

2.1 Cost

In the future the need to reduce initial capital cost will continue to be critical, however increasingly Life Cycle Cost (LCC)[1] evaluations will be used to assess the most appropriate solution for a particular system requirement.

The Overall LCC is the sum of the initial capital cost, cost of operation, maintenance, unavailability costs and even the cost of decommissioning. Increasingly, as such a cost analysis is used more and more, GIS will become the most cost effective solution. How these costs are evaluated will vary from utility to utility and also by application but will nevertheless become part of the evaluation criteria.

In recent years the capital cost differential between GIS and alternatives has been reduced. There are also increasingly more hybrid alternatives, where GIS technology is mixed with AIS arrangements, to evaluate. These trends are likely to continue but, Will such solutions evolve into full GIS solutions? Will outdoor GIS gain in popularity due to lower initial capital cost as the indoor/outdoor question is again raised? Will mixed technology solutions lead to more effective substation layouts, no longer being a copy of an AIS substation? These and other comparisons are key issues for the GIS of the future.

2.2 Effects of Utility Privatisation

The climate of change since the early 1990's encouraged the spread of privatisation of public utilities in many parts of the world. One effect of privatisation has been an emergence of Supply Industries with much greater commercial awareness compared to the older Public Utilities that tended to be engineering driven. As well as the impact this has had on utilities, with significant reductions in engineering manpower, there is a growing demand to drive equipment and systems harder to ensure that maximum return is achieved on the assets. These changes create significant challenges such as:

- The economic view is becoming more and more the major selection criteria for the user
- Reduced lead times are required in planning, design, manufacturing and commissioning
- Utilities are moving towards increased outsourcing to specialised companies for many services, some of which may have previously been considered as core within previous organisations.

These challenges are influencing the way that GIS is perceived by the user and also driving GIS suppliers to exploit and develop certain features of the technology that reflect the needs of these new customer organisations.

2.3 Environmental Awareness

Perhaps even more significant than the above changes is the ever-increasing public awareness of environmental issues that are having a growing influence on the development of transmission networks.

Concerns over the visual impact and potential electromagnetic effects of overhead lines are making it more difficult to obtain way-leaves for new line routes. This provides challenges to reduce the environmental impact of overhead lines by reducing line and tower profile, extending the rating of existing circuits or perhaps replacing them altogether with low impact buried solutions such as cable or gas insulated lines. The same constraints apply equally to substations.

In recent years, global climate change, especially the threat of global warming of the surface of the earth, has been focused and the effects of green house gases such as SF_6 have been argued[2]. After COP3 in Kyoto in 1997, every effort to reduce the emission reduction of SF_6 into the atmosphere is being taken in the various industries using SF_6, especially in the power industry, and the development of gas insulated equipment with less SF_6 gas, including SF_6 gas mixtures, has continued. Since an alternative gas has not been found yet in spite of great efforts in the concerned fields, SF_6 has to be used and handled with special attention. Therefore, measures are being taken to keep the SF_6 emission to a minimum so that we can make the best use of gas-insulated equipment. With the purpose of controlling emissions of SF_6, joint activities have been conducted in many countries. The studies show that SF_6 emission from modern gas insulated equipment with up to date technology is much less than 1% per year and that most of emissions are from improper gas handling. In this sense, improvement relating to sealing performance does not have the highest priority.

GIS provides inherent benefits when compared to conventional AIS with regard to environmental impact, generation of EMF and susceptibility to EMC and we can expect increased selection of GIS in the future for these benefits.

2.4 Security

There is now much more awareness of the vulnerability of transmission systems regarding terrorist attack. Reviews of existing substations may lead to changes in design and operational philosophies. For future substations the issue of security will become even more important than in the past. GIS solutions having some inherent advantages in relation to security.

2.5 Blurring of Traditional Technology Boundaries

The introduction of coordinated control techniques based on distributed system architecture and the rapidly increasing range of digital relays has permitted considerable simplification of control and protection schemes with increased flexibility.

Whilst the systems currently available "coordinate" the control and protection functions in the substation a greater degree of integration can be confidently expected to develop and the traditional separation of control, protection and monitoring functions will disappear as the boundaries of bay, substation and transmission network are broken down.

In the substation the use of parallel conventional wiring will be dramatically reduced as digitally based control and protection systems are introduced utilising serial optical fibre connections. This change will lead to a dramatic reduction in the time spent during installation of conventional multi-cores on a GIS substation.

One clear trend is the incorporation of devices and technologies into integrated units within the GIS substation environment. Due to the inherent "integrated" nature of GIS this trend is to be expected but is accelerating, particularly in the area of integration between primary and secondary technologies.

The integration of digital control into primary components also includes data and information transfer by optical serial links. Perhaps the most significant changes in the way that utilities operate GIS substations will arise from the combination of the traditionally separated control and protection systems.

However there is still considered to be a gap between the life expectancies of primary and secondary equipments.

2.6 Globalisation of Markets and Harmonisation of Standards

The current situation is that each user of GIS prepares its own GIS specifications based on its own company standards and interpretation of National/International standards. This results in a wide variety of approaches and possible areas for confusion between users and suppliers of GIS (particularly for the first relationships between the parties).

This position relies on considerable effort by utilities and suppliers to ensure a complete communication of requirements is achieved. There is scope for a standardised user specification to be developed which could allow considerable improvement to be achieved in the effectiveness of the client/customer interface and tender preparation/evaluation.

There are already international technical working groups established to consider the harmonisation of switchgear standards between IEC and IEEE. The international Standards like IEC and IEEE becoming more and more important as a basis for well established technical descriptions and requirements, it is believed that they will be for many the only basis for national requirements.

3. GIS FUTURE SOLUTIONS

In Section 2, some of the primary environmental, business and technology drivers for the electricity transmission business environment were discussed. The combination of these business drivers will influence the overall mix of GIS solutions that become available in the future. In this Section the specific GIS responses to these drivers are discussed to illustrate how GIS can be expected to evolve in the future.

3.1 Integration of Primary Functional Elements

As well as the integration of the Control, Protection and Monitoring functions there will also be a move towards the integration of primary functional elements within GIS. For some time it has been commonplace to combine disconnector and earth switch functions, a further step has been to integrate these with the circuit-breaking element.

Existing iron cored ring type current transformers and electromagnetic voltage transformers have been used since the introduction of GIS with little development. However it is now possible to use coreless transducers which can offer benefits in space, cost and more appropriate interfacing with developing secondary technologies. These transducers offer the possibility of being able to be integrated into the primary GIS hardware thus making the GIS layout more compact and provide more freedom for choosing the optimum location of the measuring transducer.

We can trace a continual development of GIS technology involving the integration of primary functions into single elements. The objectives of such developments have been to make the GIS:

- More compact
- Simpler, involving fewer components
- Increased performance and reliability
- Less costly

We can assume that such requirements will continue in the future, therefore the continued integration of the primary GIS elements will continue providing that the same objectives are achieved. Whilst we can see that the harmonisation of standards and merging of different Utilities and reduction in numbers of manufacturers might limit the variety of GIS available, it is also the case that innovation in how GIS primary elements are integrated will also provide for variety in terms of GIS design solutions.

3.2 SF_6

One issue generating considerable environmental interest in transmission systems is that of the insulating gas SF_6. The greenhouse gas effects of SF_6 are well-documented[2] and the electrical transmission industry has taken a proactive and responsible attitude in dealing with this issue. Guidelines on the handling and recycling of SF_6 have been produced[3], which together with tighter standards for equipment design should ensure a reduced loss of gas to the atmosphere from electrical applications. Also SF_6 inventory systems have been introduced to monitor SF_6 usage. Unfortunately other industries have not taken the same responsible attitude and the collective efforts of those in the electrical industry will be required to ensure the controlled use of SF_6 can continue. The detrimental environmental impact of a restriction in the use of SF_6 would be considerable. Future requirements for acceptable leakage rates are likely to be much less than the 0.5% per compartment per annum which is commonly accepted. Fortunately design and manufacturing technology is now available which enables such performance levels to be achieved. Sealed for Life designs that imply no maintenance inside the gas compartments will also become closer.

3.2.1 Ongoing SF_6 *emission reduction efforts*

Voluntary programs for the reduction of SF_6 emissions from the electric industry were initiated shortly after the high global warming potential of SF_6 had become known. These activities were organized by the electric industry (electro-technical committees, manufacturers and utilities), SF_6 producers and manufacturers of SF_6 handling and recycling equipment. They consisted in summary of:

- Issuing of recommendations for environmentally responsible SF_6 handling in the electric industry
- Improved equipment design to reduce the quantity of SF_6 required per performance and to reduce leakage
- Implementation of conservative SF_6 handling practice in test laboratories, factories and on erection sites.
- R&D activities to explore the feasibility of SF_6 substitution.
- Introduction of SF_6 inventories

With this background it is now worth considering in detail the possible impact on GIS equipment to the use of alternative gases. These options can basically be divided into two categories:

- SF_6 gas mixtures in order to reduce the total quantity of SF_6 gas used.
- Use of a completely new gas as an alternative

3.2.2 Gas Mixtures

The use of SF_6 gas mixtures was first used in cold environmental conditions to overcome problems with pure SF_6 gas liquifaction at the combination of high operating pressures and low operating temperatures. Using mixtures of SF_6 with either N_2 or CF_4 allowed equipment to operate at low temperatures without need for additional heating. This necessity was particularly the case for circuit breakers which have a relatively high operating SF_6 density. For GIS equipment other than circuit breakers the problem was often reduced due to the combined effect of the lower operating SF_6 density of GIS equipment and the fact that GIS is often installed inside a building (particularly in onerous environments).

For circuit breakers the relatively high pressure SF_6 is still needed in order that the current interrupting properties of the circuit breaker are maintained. More recently there has been an emergence of low percentage SF_6 gas mixtures where only dielectric characteristic need to be achieved as for GIS non-active parts and Gas Insulated Lines. Designs of GIS can be adapted relatively easily to achieve comparable performance as 100% GIS equipment, this is likely to involve an increase in overall gas pressure for the equipment. For active elements such as disconnectors and circuit breakers the required performance cannot be achieved with such low percentage SF_6 gas mixtures without significant modification and in the highest ratings will be impossible to achieve.

Another issue in the consideration of low percentage SF_6 gas mixtures is the gas handling. Due to the differences between the two gases (SF_6 and Nitrogen) it is not possible to use conventional SF_6 gas reclaimers with mixtures. Special SF_6/N_2 gas reclaimers have to be used making recycling of gas more difficult and costly. Nevertheless we now can foresee the application of GIS using low percentage SF_6/N_2 mixtures, particularly where long sections of GIS busbar or Gas Insulated Line are needed. However, for most applications mixtures are not environmentally or economically attractive in the short term, the ongoing work with CO_2 and other gas mixtures is awaited with interest.

3.2.3 Alternative Gases

As indicated in the above section it is possible to achieve good dielectric performance of GIS equipment using low percentage SF_6 gas mixtures. A next stage is obviously to consider using no SF_6 gas and 100% nitrogen. Performance levels can be achieved such that practical solutions are possible but with a dramatic de-rating of the GIS equipment. For example equipment which has a dielectric performance to IEC 60517 for a rated voltage of 170kV with SF_6 may be possible with adaptation to achieve a rating of 72,5kV using 100% N_2. This of course assumes that the circuit breaker performance is achieved using another gas (either CO_2 or perhaps a vacuum interrupter device).

Several decades of intense and comprehensive research have shown that a functionally equivalent substitute gas for SF_6 does not exist for physical reasons.

However it clearly points out that the number of dielectric gases envisaged, are restricted to a few gases (small atmospheric molecules, HFC's, PFC's and SF_6). However, all of them may not be usable and unfortunately, the higher the dielectric strength, the higher the greenhouse potential.

As a consequence, SF_6 substitution, if enforced, would become technically difficult, economically unacceptable and environmentally critical, particularly for high voltage transmission equipment. Nevertheless, researchers continue to explore new possible insulation medium combinations. Manufacturers are continuing research and development work to identify performance niches (mainly low voltage and low current) in which SF_6-free equipment might become feasible. This type of niche product could emerge as parallel advances are made in vacuum interrupter technology that is currently limited to a maximum voltage on a single unit of 72,5kV. Single devices that can economically operate at 145kV (or even higher) could considerably advance such developments.

Therefore a possible vision of the GIS of the future is GIS with vacuum interrupter circuit breaker and N_2 insulation for use up to 170kV in a size of equipment currently needed for 300kV equipment.

3.3 Sealed for Life designs

According to IEC 60694 'Gas and Vacuum Tightness' a sealed system is defined as one where no further gas or vacuum processing is required during its expected operating life where expected operating life has standard values of 20 and 30 years between maintenance intervals.

This implies that for the specified service life the sealing system for the SF_6 gas is designed and tested for a leakage rate assuring at least the minimum operating density for the whole life.

It also implies that a sealed system is operated without constant gas density supervision and requires no service access by gas connections.

For high voltage GIS these implications are feasible only for parts of the equipment requiring no maintenance during the specified life time. In practice this limits its application to static components. For today's switching equipment the maintenance cycles depend mostly on the use in service i.e. number of switching operations at load or no load (condition based maintenance).

Usually the service safety of high voltage GIS is too important to rely on sealing systems without gas density supervision.

The leakage of modern GIS is in the range of 0.2 % per year according to accessible service experience.

These facts and the relatively low cost of density supervision equipment allow the conclusion that sealed for life systems as defined above provide no significant advantage for the user of high voltage GIS and impose unnecessary risks in case of unnoticed leakage.

There appears to be no visible trend towards use of sealed for life GIS, based on the definition given, however the need to minimise leakage, maintenance and gas handling will dictate that the same features as needed for "sealed for life" equipment will be introduced into the GIS of the future.

3.4 Standardisation

There is a drive for more standard interfaces for cables, transformer bushings and perhaps even busbar systems for easier extensions of GIS. Standards for mixed technology and integrated function solutions will also continue to evolve.

3.5 Re-locatability of Equipment

Increased ability to re-locate "by easy dismantling" for system flexibility as well as moveable as a response to the demands of changing network conditions.

3.6 Urban Substations

More substations will be installed in cities so GIS will need to be integrated with urban environments. Part of this integration will consider security of the substation, therefore the inherent security offered by GIS in comparison with conventional substations will provide additional application opportunities.

3.7 Simplified Single Line Arrangements

Transmission substation plant is becoming increasingly more reliable. Increased reliability with commensurate reduction in maintenance requirements will lead to simplification of the electrical diagram of the substation. For example in AIS the practice of providing circuit breaker by-pass facilities will surely completely disappear. Whilst in GIS where additionally the primary insulation is protected from the polluting and aging effects of the atmosphere the use of simpler busbar systems forecast in successive CIGRE GIS Service Experience Review Reports[4] must inevitably take place. For utilities there may also be implications with regard to insurance requirements and premiums.

3.8 Increased Use of Condition Monitoring and Diagnostics

The economic and system security implications of taking equipment out of service for maintenance will dictate that maintenance is carried out on a predictive basis rather than on a simple time basis, as has normally been the case in the past. It is important that the predictive maintenance basis ensures that incipient problems are identified to avoid enforced outages whilst not requiring outages on plant which is performing satisfactorily. Effective application of predictive maintenance policies will yield significant cost benefits to the operating company. The tools used in the development of predictive maintenance solutions will consist of diagnostic techniques and condition monitoring.

Developments for the future will include the integration of all types of substation condition monitoring, making use of it's digital signal base, into the overall Substation Management System embodying all protection, control and monitoring systems. This will enable the user to optimise the operation and maintenance of his total network.

3.9 Functional/Standardised Specifications

The trend towards functional specifications is likely to continue. This allows suppliers to offer alternative solutions that may offer users particular advantages, however users need to be aware that such choices would require evaluation. Conversely standardised specifications may reduce engineering effort required within a user, but reduce choice.

4. CONCLUSION

In such a brief paper it has only been possible to identify a small selection of items that will continue to have an impact on the future development of GIS. The full work of this Task Force will be published as a CIGRE Brochure and will provide much more detail on this work. The Brochure will also be able to speculate as to the most likely combinations of possible future features that might appear in future GIS, thus giving the reader a view, based on the impressions of the Task Force, of the GIS of the future.

The resulting future vision will hopefully stimulate discussions on this subject One issue is clear that GIS has many further developments and changes to undergo in the future.

5. REFERENCES

1. CJ Jones et al, CIGRE 1996, paper 23-203, Guidelines for Monitoring, Control, and Supervision of GIS Incorporating Advanced Technologies.
2. G Mauthe et al, SF_6 and the Global Atmosphere, ELECTRA No 164 1996.
3. G Mauthe et al, SF_6 Recycling Guide, Re-use of SF_6Gas in Electric Power Equipment and Final Disposal, ELECTRA No 173 August 1997, pp 43-71.
4. D. Kopejtkova et al, CIGRE 1992, paper 23-101, A Twenty Five Year Review of Experience with SF_6 Gas Insulated Substations.

HYBRID DESIGNS FOR OPEN AIR SUBSTATIONS

Carl E Sölver*

1. INTRODUCTION

The transformation of the electric power industry that has taken place during the latest decade has had significant impact on the design of substations. There is strong focus on cost, both for new equipment and during the life cycle, as well as on availability and environmental properties. This is true both for new substations and for extension or refurbishment. The changing market offers the opportunity to change traditional practices, and to introduce new solutions. Many new switchgear devices have been developed based on either air insulated switchgear (AIS) or gas insulated switchgear (GIS) technology, or a combination of both. Such devices are used for refurbishment or upgrading of existing substations, and also for new substations.

The new switchgear devices typically lead to substantial space savings compared to traditional open air solutions, and also to higher overall availability of the substation. They normally require fewer foundations than traditional equipment, and have short and simple installation and commissioning procedures. Total life cycle cost and environmental impact is generally lower than for traditional equipment.

In the present paper, HV substations for sub-transmission and transmission voltages are considered.

2. CONVENTIONAL HV SUBSTATIONS

A typical feeder bay in a HV substation consists of a circuit breaker, one or more disconnectors, an earthing switch, as well as current and voltage transformers. There may be one, two, or in some cases even three busbars. In a conventional AIS substation the HV equipment is freestanding, and insulators of either porcelain or composite material is used. The circuit breakers normally apply SF_6 gas for insulation and arc quenching. In a conventional GIS substation the equipment, and also the busbars, are enclosed in earthed metal tanks with SF_6 gas.

* Carl E Sölver, ABB Power Technologies, SE-77180 Ludvika, Sweden

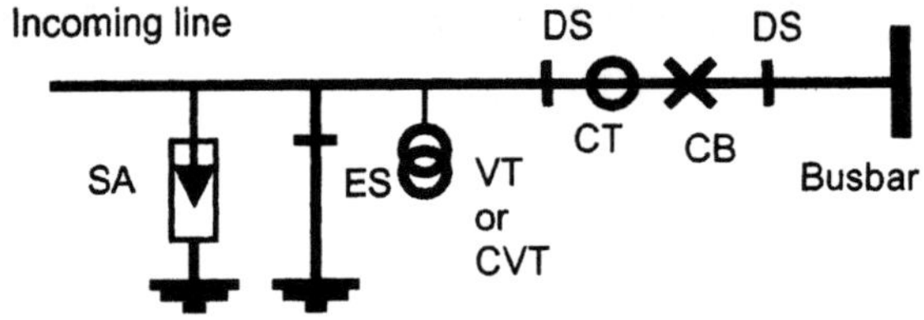

Figure 1. Conventional line bay in substation with single busbar.

One major function of conventional disconnectors is to isolate the circuit breakers when maintenance is to be performed. Historically, the circuit breakers, of either air-blast or oil type, required relatively frequent maintenance. Change to modern SF_6 circuit breakers has strongly reduced the maintenance demands (de-energization required typically only with 15 year intervals). Now there is a clear tendency that air insulated traditional disconnectors themselves require more maintenance than the circuit breakers. Thus, there are good incentives to move away from traditional disconnectors. By doing that, the total maintenance work is decreased, and as a result the overall availability of the substation will increase.

3. COMPACT AIS SUBSTATIONS

There are several ways to make a traditional substation more compact and also reliable. Traditional, freestanding equipment may be combined into groups or modules ("switching device combinations"). It is also possible to use new types of AIS equipment that combine the functions of two or more traditional apparatuses ("combined function switching devices"). One such device is the disconnecting circuit breaker.

A further step, that brings the AIS substation closer to a GIS hybrid, is to introduce metal enclosed, gas-insulated busbars.

3.1 Withdrawable Circuit Breakers

One way to eliminate conventional disconnectors is to place the circuit breaker on a movable trolley. The disconnecting function is then achieved by motion of the complete circuit-breaker unit. The circuit breaker can be moved to a first disconnected position, and also further on to a maintenance position. In the maintenance position, the circuit breaker may also be fully removed, and replaced in a simple manner with a spare unit in order to keep the downtime for maintenance short.

The withdrawable circuit breaker may easily be combined with current transformer, earthing switch, and even a busbar section, all placed on the same structure.

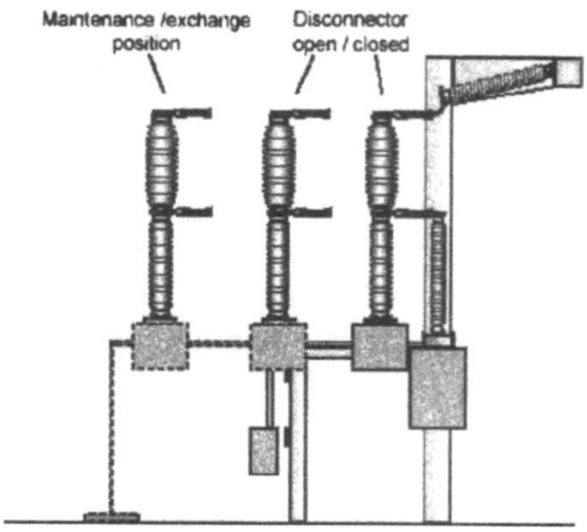

Figure 2. Withdrawable 145 kV circuit breaker with busbar section

An alternative way to achieve the disconnecting function is to rotate the circuit breaker poles around their axis [1]. For a single unit circuit breaker, the vertical breaking chambers are then equipped with horizontal disconnector arms, making contact e.g. with a current transformer on one side and a support insulator on the other side. In case of a T-shaped double unit circuit breaker, applied at higher rated voltage - 420 or 550 kV - the pole may also be rotated, and now the horizontal breaking units inherently serve as disconnector arms. In both cases the circuit breaker poles may easily be removed when in disconnected position, for maintenance or replacement with a spare pole.

Figure 3. Withdrawable 145 kV circuit breaker with integrated current transformer

A further step of integration is to combine a withdrawable SF_6 circuit breaker with an SF_6 insulated current transformer, as shown in Figure 3. The current transformer is vertically mounted, and supports the horizontal breaking chamber. In this way the same gas compartment is used for two functions. The disconnecting function is achieved by sideways motion of the circuit breaker poles, and the whole unit may be lowered down to ground level for maintenance or replacement. An earthing switch may be placed on the same structure.

Other equipment in the substation bay may also be integrated in similar ways, in order to make the substation more compact. One example is a line entrance module, consisting of a voltage transformer with integrated earthing switch. If applicable, surge arresters may also be placed on the same structure.

3.2 Disconnecting Circuit Breakers

In a disconnecting circuit breaker, the same contacts provide circuit breaker functions (make and break current) as well as disconnector functions (isolate parts of the network from each other). The concept of disconnecting circuit breakers for air-insulated substations has been discussed since the middle of the 1990's [2][3][4]. The disconnecting circuit breaker replaces the traditional combination of a circuit breaker and its associated conventional disconnectors. In principle, the device has to fulfill both applicable circuit breaker standards and disconnector standards. The long maintenance intervals of the disconnecting circuit breaker, compared with conventional open-air disconnectors, provide high availability also in combination with simplified substation configurations. In addition, the reduction in number of switching devices means that space requirements are reduced and that control systems can be simplified.

Figure 4. 420 kV disconnecting circuit breakers with two interrupters per pole.

A main advantage of the disconnecting circuit breaker compared to a conventional open-air disconnector is that the electrical contacts are enclosed in SF_6 gas, and thereby protected from the influence of ambient conditions, including the effects of pollution. The protected environment provides improved reliability and prolonged intervals between de-energization for maintenance.

The availability of simplified substations may be further improved if the disconnecting circuit breaker is equipped with manual disconnecting devices. Such devices are composed of standard substation hardware. They provide convenient and safe disconnection and reconnection in the rare case that need arises for maintenance or repair of the disconnecting circuit breaker itself. When the disconnecting circuit breaker is disconnected in this way, and proper safety distances are ensured, the other parts of the substation may be re-energized again.

Disconnecting circuit breakers may also be applied in GIS switchgear [5]. In this case the manual disconnecting devices are integrated into the GIS, in a separate gas chamber. Mechanical interlocking is applied in order to prevent mal-operations.

Disconnecting circuit breakers constitute a new class of equipment, and IEC (International Electrotechnical Commission) has concluded that there is a need for standardization of the requirements of such "combined function disconnecting circuit-breakers". The work on this standard is now well under way.

3.3 SF_6 Insulated Busbars

In a conventional AIS substation, the space beneath the busbars is normally poorly utilized. In cases where there are space constraints, it may then be beneficial to apply encapsulated, SF_6 insulated busbars instead of normal air-insulated busbars, although the cost will be higher [5]. Depending on the electrical scheme, the SF_6 busbar will be equipped with SF_6 disconnectors, and connected to AIS type equipment by means of SF_6 to air bushings. Substations with double busbar configuration may be arranged in a particularly compact way, using back-to-back arrangement of the bays. The SF_6 busbars also help to decrease the total height of the substation, since one level of conductors can be saved.

An SF_6 busbar will have low failure probability, but in the rare case of a failure there will be a relatively long outage time. A double busbar solution will lead to very high overall availability, but is expensive. A single busbar solution may well give adequate availability. It has also been proposed to combine a single SF_6 busbar with one spare AIS phase, to be used only in the event of a failure of a phase of the SF_6 busbar [5]. In this case the outage time may be considerably reduced.

4. HYBRID SOLUTIONS BASED ON GIS TECHNOLOGY

Hybrid gas insulated modules are based on standard metal enclosed gas insulated components, and are adapted for outdoor conditions. They are applied together with air-insulated busbars, and used when the full space saving of a traditional GIS substation is not required. Use of air insulated busbars gives a cost advantage compared to the corresponding traditional GIS solution.

For sub-transmission voltage levels, the hybrids may be either single-phase or three-phase enclosed, while single-phase enclosures will dominate at transmission voltages. In their simplest form the modules include a circuit breaker, a disconnector and earthing switch, and one or more instrument transformers, all in the same gas compartment. More complicated modules are also applied, in order to fulfill the specific demands of various substation configurations [6]. The hybrids may be combined with traditional instrument transformers, or make use of integrated, low level sensors for current and voltage measurements.

A typical, single-phase enclosed, hybrid solution for a substation with single busbar is shown in Figure 5. The unit contains circuit-breaker, disconnector and earthing switch, and current transformer. The disconnector and earthing switches make use of the same moving contacts and operating mechanism. Each pole has its own enclosure, and all the internal components are in the same SF_6 gas compartment. The current transformers are of ring core design and are placed on the bushings. There may be either one set of current transformers or two. Voltage transformers are not included.

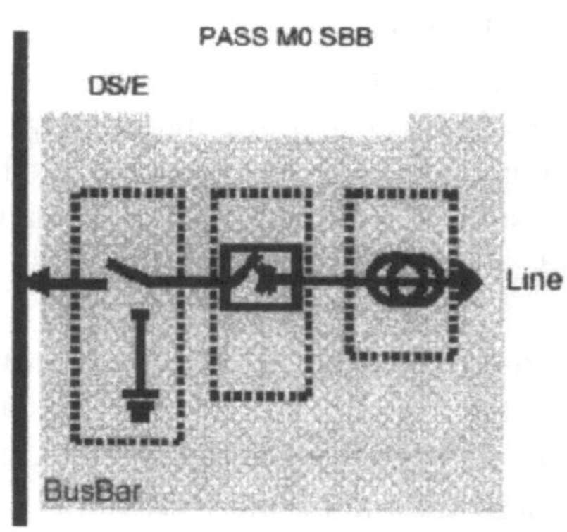

Figure 5. 145 kV GIS hybrid for single busbar substation, and corresponding single line diagram.

For a substation with double busbar, it is straightforward to add another disconnector and bushing to each of the poles, as illustrated in Figure 6. The same operating mechanism is used for the earthing switch and the two disconnectors.

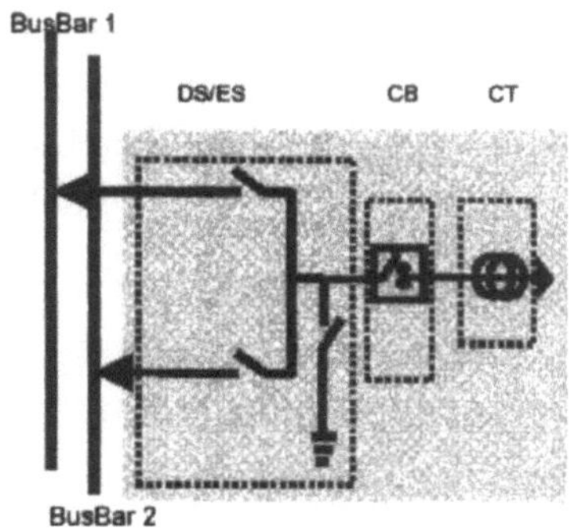

Figure 6. Single line diagram for GIS hybrid in double busbar substation.

The GIS hybrid solutions may also be specially adapted to other situations. One example is a hybrid intended for connection of a transformer directly to an overhead line. The hybrid is then adapted to contain a circuit breaker for the transformer, as well as disconnectors and earthing switches for sectionalizing of the line. See Figure 7.

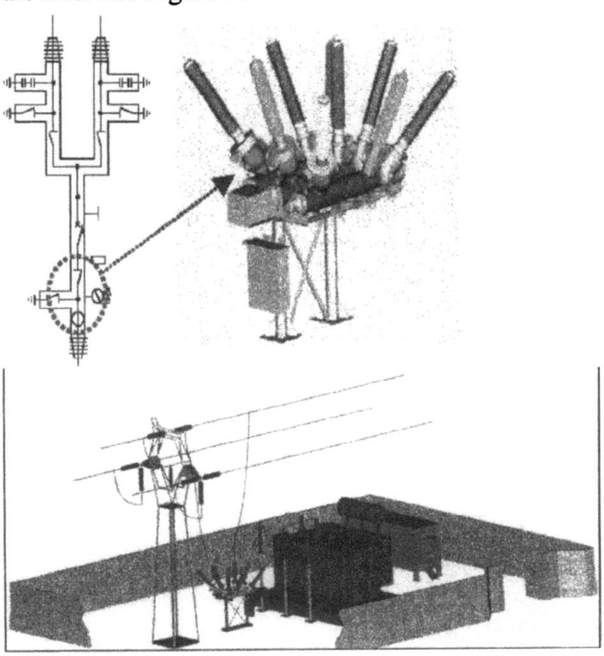

Figure 7. Connection of transformer to overhead line, using a GIS hybrid.

5. AVAILABILITY CONSIDERATIONS

A strong driving force behind the use of compact and hybrid substation solutions is availability considerations. Unavailability may be caused by maintenance of equipment in a substation as well as by failures. Maintenance is often the dominating reason for unavailability, and one of the most important factors is maintenance of conventional open-air disconnectors. Substation solutions where conventional open-air disconnectors are eliminated are therefore beneficial. In this case it is often possible to simplify the substation configuration, and still obtain an improved overall availability.

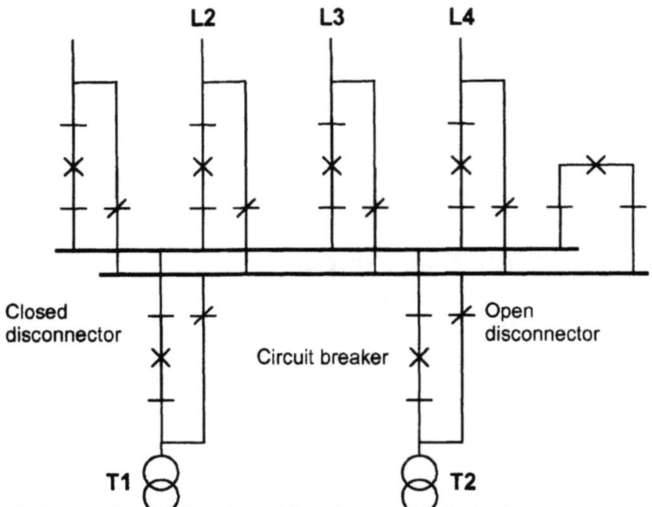

Figure 8. Conventional substation with main and transfer busbar

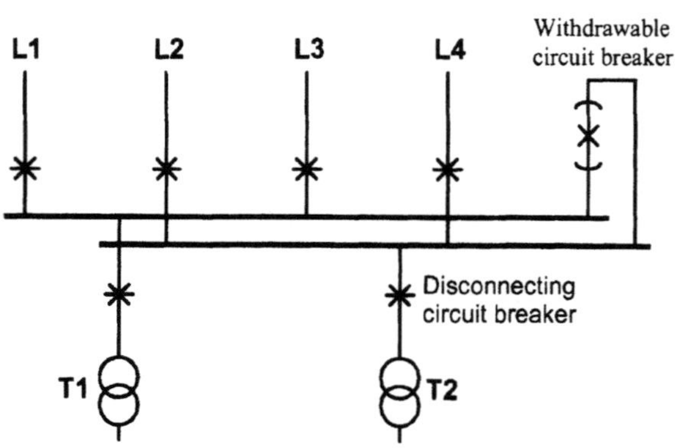

Figure 9. Compact substation with sectionalized single busbar and disconnecting circuit breakers

A simple example will serve to illustrate the improvement in availability that may be reached [4]. Figure 8 shows a traditional substation layout with four line bays and two transformer bays. There are two busbars (one main busbar and one transfer busbar) and conventional AIS equipment. Figure 9 shows a corresponding substation using disconnecting circuit breakers. In this case, there is only a single, sectionalized busbar. The circuit breaker used for sectionalizing of the busbar is of withdrawable type. Table 1 shows the data that have been used for availability calculations, in order to compare the properties of the two designs.

Table 1. Maintenance and failure data used for availability calculations.

Type of equipment	Maintenance		Failure	
	Interval, years	Duration, hours	Rate, 10^{-6}/hour	Repair time, hours
Circuit breaker, disconnecting circuit breaker	6	6	0,3	48
Disconnector	4	4	0,1	8
Withdrawable circuit breaker	6	6	0,3	24
Fixed contacts of withdrawable circuit breaker			0,01	8
Power transformer	4	8	1,0	48
Busbar, including instrument transformers			0,1	8

By use of the data given in Table 1, it is possible to calculate the unavailability of various parts of the substation. An important case is when both transformers are out of service simultaneously. In this case the substation is totally unable to transfer power between the two voltage levels. The calculated unavailability figures for this situation is:

Conventional substation (Figure 8) 8,3 hours per 100 years
Compact substation (Figure 9) 1,1 hours per 100 years

Similar calculations may be made for other compact and hybrid substations, and generally show considerable gains in availability when comparing to traditional solutions. Such improvements in availability should be considered as a factor in overall LCC calculations when different substation concepts are compared.

6. SPACE SAVINGS

Change from a conventional open-air substation to compact or hybrid design leads to considerable space savings. In an extreme example, illustrated in Figure 10, there is a space saving from 1000 m^2 to 120 m^2. In practical cases, the saving will typically be 50-60%, due to requirements for access roads, buildings etc.

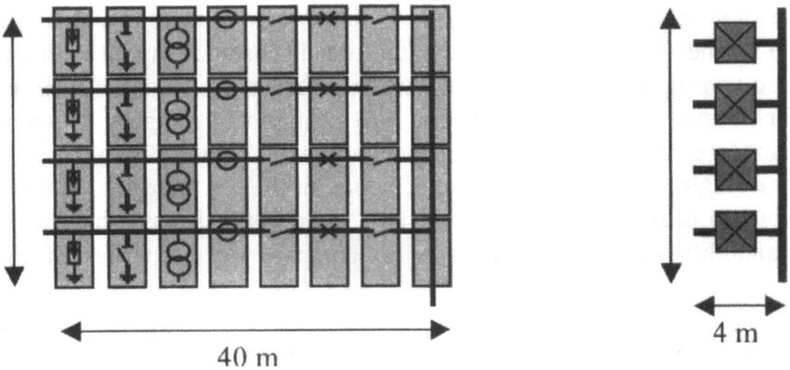

Figure 10. Single busbar 145 kV substation with four line bays. Theoretical space saving when changing from conventional to compacts AIS design.

7. CONCLUSIONS

There is an ongoing evolution of substation designs from traditional air-insulated types. One major alternative is to apply compact and integrated equipment based on normal air-insulated apparatus, while another alternative is to introduce switching units based on GIS technology.

Both alternatives generally lead to improvements in overall compactness, require less space, and have better visual impact. They also generally lead to higher availability compared to traditional solutions. In many cases it is also possible to simplify the busbar configuration.

8. REFERENCES

1. P. Roussel, L. Hossenlopp, and F. Gallon, Technical and economical evaluation of new air-insulated substation concepts, *Cigré* Report 23-205, Session 2002
2. B. Wahlström, Y. Aoshima, Y. Mino, C. Lajoie-Mazenc, D.R. Torgerson, and A.N. Zomers, The Future Substation: a reflective approach, Cigré Report 23-207, Session 1996
3. P. Norberg, M. Tapper, W. Lord, and A. Engqvist, The Future Substation - Reflection About Design, Cigré Report 23-105, Session 1998
4. C.E. Sölver, H.E. Olovsson, W. Lord, P. Norberg, and J. Lundquist, Innovative substations with high availability using switching modules and disconnecting circuit breakers, *Cigré* Report 23-102, Session 2000
5. H. Aeschbach, E. Mikes, P. Ponchon, and F. Gallon, Influence of electrical arrangement and rated voltages on substation space requirements and total costs for various gas- and hybrid-insulation solutions, *Cigré* Report 23-201, Session 2002
6. F. Leclerc, Y. Doin, A. Giboulet, and G. Triay, Hybrid solutions for high voltage substations, *Cigré* Report 23-202, Session 2002
7. A. Carvalho, P. Bosshart, U. Christiansen, G. Guerra, L. Tang, and H.E. Olovsson, Functional specification as driver for technical/economical optimisation of substation, *Cigré* Report 23-101, Session 2000

PROBABILITY OF PARTIAL DISCHARGE DETECTION IN AGED GIS DUE TO VOID IN EPOXY

A. Pharmatrisanti, Sander Meijer, Johan J. Smit[1]

1. INTRODUCTION

For more than 20 years, PLN, the electricity utility company in Indonesia, has been operating GIS substations at 150 and 500 kV. 30% of the operated GIS are already in service for at least 10 years and served a quite high load. Afterwards, these substations are categorized as aged GIS. During these years some unnecessary outages occurred without any early notification. In other words, no alarm was generated since no monitoring method was installed. Even though the "system" was claimed as the main cause, the "GIS equipment" was also playing an important part (29% at 150 kV and 46% at 500 kV)[4], see figure 1.

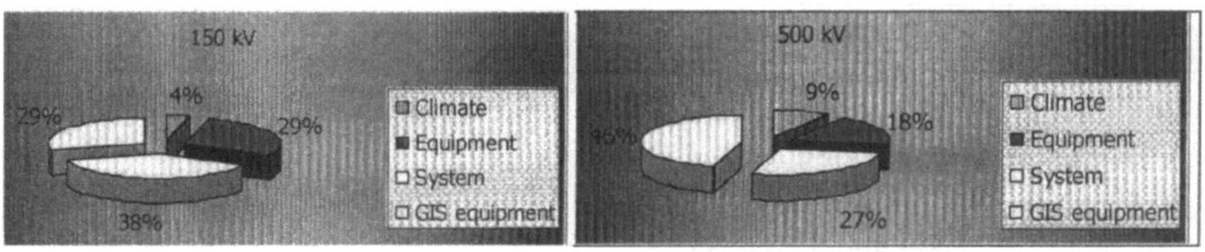

Figure 1: Outages differed by the source[4]

In order to raise the reliability, the company decided to apply a scheduled maintenance system. Therefore the investigation on the switching equipment condition, as the most important component of transmitting the electrical energy, should be conducted. To analyze the insulation condition of the equipment, partial discharge detection can be used. As the partial discharge activities excite ultra-high frequency (UHF) waves, the detection and recognition of partial discharge source can be used to observe the actual condition of the GIS. The VHF/UHF PD detection system has been used since 2002 and the investigation is focused on aged GIS[1].

To get the probability of PD due to void in epoxy, some newer GIS were also investigated in order to fulfill the requirement of random sample.

[1] A. Pharmatrisanti, PLN Engineering Center, Jakarta, Indonesia.
Sander Meijer and Johan J. Smit, Delft University of Technology, High Voltage Technology and Management, Delft, The Netherlands.

Gaseous Dielectrics X.
Edited by Christophorou *et al.*, Springer, New York, 2004.

2. THE MEASUREMENT OBJECTS

Operated GIS in Indonesia are categorized as follows:

1) Old with low load
2) Old with high load
3) New with low load
4) New with high load

Those GIS are operated either at 150 kV (HV = High Voltage) or 500 kV (EHV = Extra High Voltage). Furthermore, an additional categorization is based on the number of phases inside the enclosure, i.e.: one phase per enclosure (S = Single phase) and three phase per enclosure (T = Three phase). The measurement objects are 7 out of 63 GIS population in Indonesia (11% of population), which representing those categorized, see Table 1.

Table 1: Overview of the investigated installations.

		Age	
		Old	New
Load	Low	Kemang (HV-S)	Cipinang (HV-T)
	High	Mampang (HV-S) Karet (HV-S) Senayan (HV-S) Setiabudi (HV-S)	Kembangan (EHV-S)

Figure 2: Map of the measurement objects

Kembangan acts as one of Java Island 500 kV transmission network back bone, while the rest are 150 kV substations. Senayan and Kemang are considered as rather important due to their customers.

3. MEASURING SYSTEM

The measuring system consists of:

1) External coupler
2) Signal pre-amplifier 28 dB
3) 75 ohms coaxial cable
4) 1.8 GHz spectrum analyzer
5) Analyzer software

The measuring took at least a whole day for every GIS. The overall check up was done by measuring any PD activities in every phase in one bay. The acquisition time to get the PRPD is 13,020 ms. The power supply of the spectrum analyzer was taken from the output of the auxiliary transformer which has phase shift of 150° lagging to the high voltage side (refer to Table 2).

Table 2: Phase shift in spectrum analyzer power supply

Transformer voltage	Connection type	Remarks
High voltage – Medium voltage (150 kV to 20 kV)	YY	No phase shift
Medium voltage – Low voltage (20 kV to 380 V)	Dy5	150° lagging phase shift at low voltage

Therefore any phase resolved pattern gathered from the spectrum analyzer should be shifted 150° to the left.

4. MEASURING RESULTS

The predicted elementary results of the measurement are merely the source of the PD, i.e.: free moving particle, high voltage protrusions, low voltage protrusions, void inside the dielectric, floating electrode, and no PD at all. In fact some showed uncertain pattern, unspecified spectrum, and combination of two defect sources.

The stations in Kemang, Cipinang and Senayan were clean. The phase-resolved PD patterns obtained in the other stations are depicted in Figure 3.

Mampang	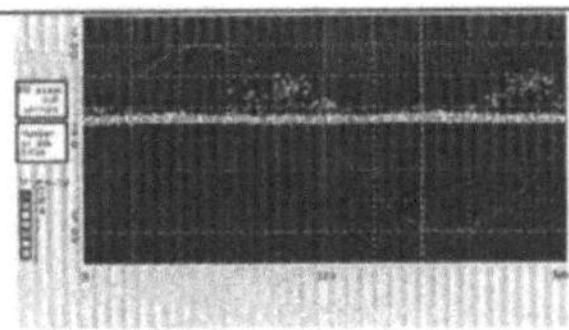 transformer bay 1 phase blue freq=22.5 MHz, void	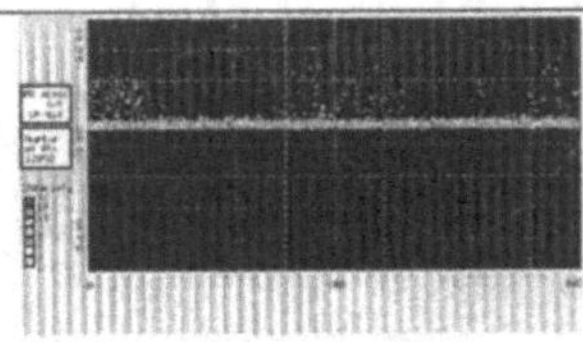 line bay 1 phase yellow freq=36 MHz, uncertain
Karet	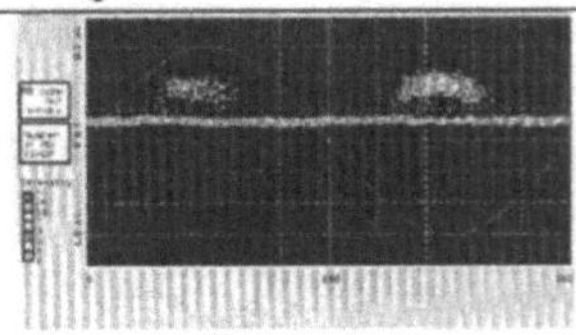 line bay 2 phase blue freq=14.5 MHz, floating electrode	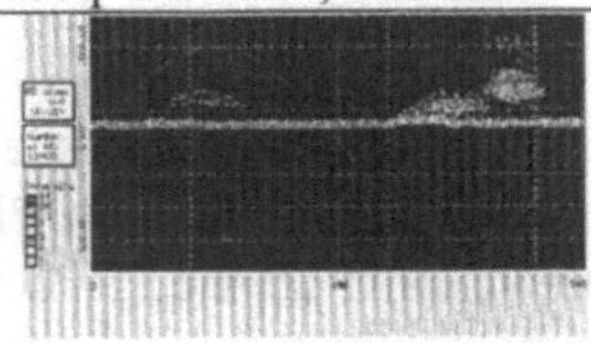 line bay 2, phase red freq=157.5 MHz, floating electrode and protrusions
	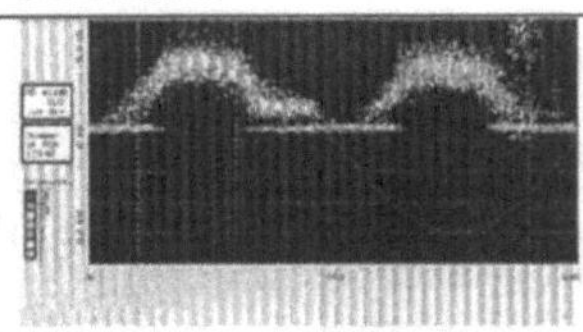 line bay 2 phase yellow freq=49.5 MHz, uncertain combination of two defects	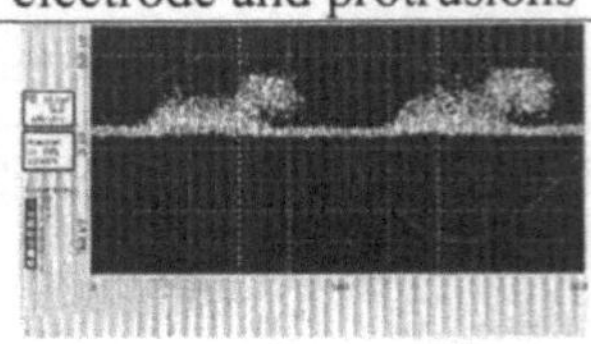 line bay 2 phase yellow freq=175.5 MHz, floating electrode and protrusions
	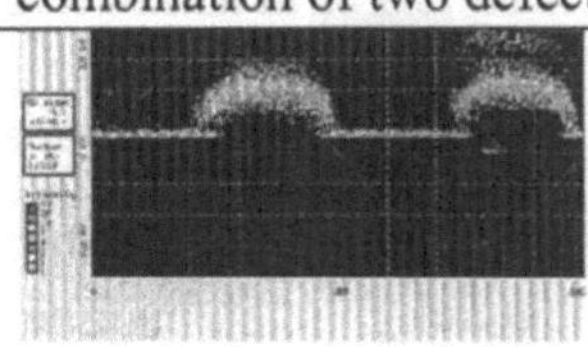 transformer bay 1 phase red freq=64.47 MHz, floating electrode	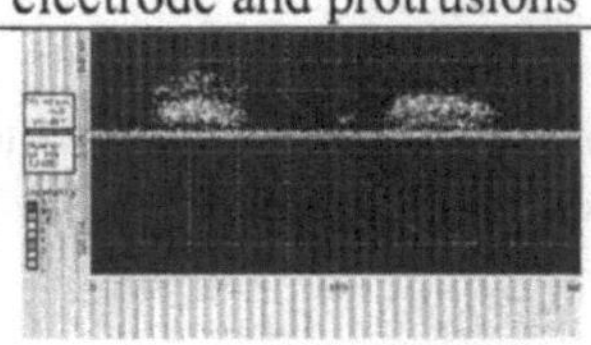 transformer bay 2 phase red freq=58 MHz, floating electrode
Setiabudi	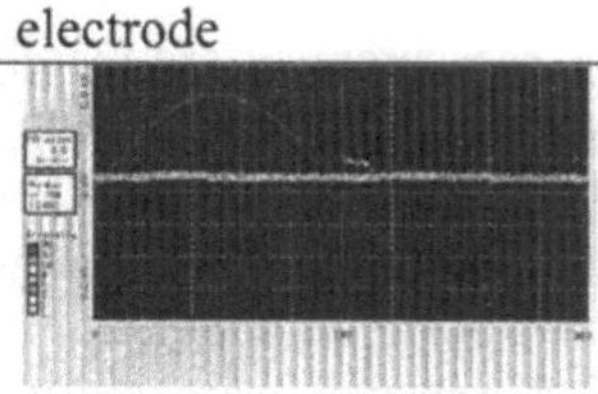 line bay 2 phase blue freq=13.5 MHz, protrusions	

Figure 3: Phase-resolved PD patterns obtained in three different GIS's.

By calculating the appearance of every source in the same bay at the same phase, the results can also be translated into statistical data as shown in Table 3. Note that the combination of two defects is calculated as two appearances, e.g.:

combination of floating electrode and protrusions gives one additional "frequency" each to floating electrode and protrusions.

Table 3: The results in statistical data

Elementary result	Frequency	Relatively Frequency
Free moving particle	0	0.000
Protrusions	3	0.200
Void inside dielectric	1	0.067
Floating electrode	4	0.267
No PD	3	0.200
*Unspecified	1	0.067
**Uncertain	3	0.200
	15	

Note that the VHF/UHF method has limitation in determining very small void. This means that during the measurement some small voids might be undetected by this method. It gives "zero" value of possibility.

In the case of large void, the method has quite high trust level. By assuming that the trust level of using the VHF/UHF method in determining the PD due to large void is 99%, the probability of PD occurrence due to large void inside dielectric which can be detected is able to calculate.

Determine that:

1) A = PD occurrence due to large void
2) B = positive detection

The statistical data can be expressed as follows:

1) $P(A)=0.067$
2) $P(B \mid A)=0.99$
3) $P(B \mid \text{non-}A)=0.467$ (PD due other defects, but positively detected)

To determine the probability of PD occurrence due to void inside dielectric which can be detected or $P(A \mid B)$, the Bayes method is used[3].

Table 4: Bayes method used to calculate P(A | B)

	A	Non-A	
B	**0.06633**	**0.467**	0.53333
Non-B	0.00067	0.466	0.46667
	0.067	0.933	1

Therefore the P(A | B) can be calculated with the following equation:

$$P(A|B) = \frac{P(AandB)}{P(B)} = \frac{0.06633}{0.53333} = 0.12437 \quad (1)$$

Thus the probability of PD occurrence due to large void inside the dielectric that can be positively detected is 0.12437 or 12.44%.

5. CONCLUSIONS

1) Even though the VHF/UHF method is a quite accurate "device" to detect PD occurrence only 12.44% of the positively detected GIS is "contaminated" by large void inside the dielectric.
2) In case of very small void, the VHF/UHF is unable to detect it.
3) High experience is still the main requirement to recognize the PD pattern to avoid any usage of ambiguous or uncertain data.

6. REFERENCES

1. A. Pharmatrisanti, S. Meijer, *VHF/UHF PD Detection on Aged GIS*, Proceedings of the 13th International Symposium on High voltage engineering, Delft, 2003
2. S. Meijer, *Partial Discharge Diagnosis of High-Voltage Gas-Insulated Systems*, 2001, Optima Grafische Communicatie, Rotterdam, pp. 87,98,99.
3. Ronald E. Walpole, et.al, *Probability and Statistics for Engineers and Scientists*, 7th ed., Prentice Hall, New Jersey, 2002
4. Internal communication and report of PT. PLN (Persero) P3B in the year of 2001-2002.

SECTION 10: ALTERNATIVE TECHNOLOGIES

THE POWER DELIVERY SYSTEM INFRASTRUCUTRE OF THE FUTURE[♦]

Clark W. Gellings[*]

1. INTRODUCTION

The electric infrastructure of the future will be integrated, self-healing, and electronically controlled to provide extraordinary resiliency and responsiveness. Transmission and distribution technologies will be designed with the flexibility to accommodate changing end-use needs. The will be fully capable of acting – in real time – on the billions of decisions made by consumers and their increasingly sophisticated microprocessor agents. The infrastructure will be always on, or "alive." It will be interconnected, interactive, and merged with communications in a complex network of real-time information and power exchange.

The blackouts 2003 in the Northeastern United States and Europe reminds us that electricity is indeed essential to our well-being. And it highlights one of the most fundamental of electric functions: getting electricity from the point of generation to the point of use. Power delivery has been part of the utility industry for so long that it is hard to imagine that this process has not already been optimized. However, the power delivery function is changing and growing more complex with the exciting requirements of the digital economy, the onset of competitive power markets, the implementation of modern and self-generation, and the saturation of existing transmission and distribution

[*] Clark W. Gellings, Vice President Power Delivery and Markets, Electric Power Research Institute (EPRI) 3412 Hillview Avenue, Palo Alto, CA. USA 94304

capacity. Without accelerated investment and careful policy analysis, the vulnerabilities already present in today's power system will continue to degrade.

2. TODAY'S ELECTRIC INFRASTRUCTURE

Simply stated, today's electricity infrastructure is inadequate to meet rising consumer needs and expectations.

An Aging Infrastructure

Today's aging electric power system is largely based on technology developed in the 1950s or before and installed over the last 30 – 50 years. As much as 25% today's infrastructure will need to be replaced or upgraded in the next 10 years. The first question which arises is what to replace or upgrade – the answer is not entirely clear.

Grid Constraints

During the last decade, total electricity demand in the U.S. rose by nearly 30% but the nation's transmission network grew by only 15%. Over the same period, expenditures by investor-owned utilities for distribution system construction fell by about 10% in real terms. The outlook for the next decade is even worse: Demand is expected to grow by 20% but planned transmission system growth is only 3.5%. The picture in any part of the world is not much different.

Reliability and Power Quality

The digital revolution rides on the back of new digital technologies that have broadly influenced improved energy efficiency, productivity, communications, automation, and other benefits. These technologies, while opening whole new horizons of commerce and innovation, also have exposed vulnerabilities in the traditional technologies, and methods used to interface electric power delivery with digital systems, processes and enterprises.

It is important to optimize the interface between electric power delivery and the digital economy. Optimizing this interface will require a strategy that must comprise all elements of the power delivery and end-use process – from the power plant, to the interconnecting systems, to the response of the digital systems, processes, and the enterprises themselves – and will be met with a combination of implementation techniques, new technologies, and new approaches to interfacing electricity supply with all forms of digital applications.

Infrastructure Security

The electricity sector needs to develop a comprehensive strategy to address security threats. Such a strategy must ensure the public that they are well protected, and free from threats posed by or to the electricity system. Technologies need to be developed to ensure rapid response from any threats which do materialize.

Open Markets

Open market access, the design and implementation of new market structures and greater choice among electricity providers brought by competition provides enormous opportunities for delivering on the promise of choice. The effects of the lag between demand and infrastructure investment are already being felt. In four of the last five years, the U.S. and Europe have faced serious reliability problems.

Efficiency

While end-use devices and appliances have improved substantially, there is still an opportunity to reduce demand by as much as 25 – 45%. While restructuring and the lack of cohesive national programs have not allowed the industry to coalesce around this opportunity, substantial opportunities exist to transform end-use energy – consuming devices and appliances.

The Vision

The envisioned power delivery system and electricity markets will enable achievement of the following goals:

- Physical and information assets that are protected from man-made and natural threats, and a power delivery infrastructure that can be quickly restored in the event of attack or a disruption: a "self-healing grid".
- Extremely reliable delivery of the high quality "digital-grade" power needed by a growing number of critical electricity end-uses
- Availability of a wide range of "always-on, price-smart" electricity-related consumer and business services, including low-cost, high value energy services, that stimulate the economy and offer consumers greater control over energy usage and expenses
- Minimized environmental and societal impact by improving use of the existing infrastructure; promoting development, implementation, and use

of energy efficient equipment and systems; and stimulating the development, implementation, and use of clean distributed energy resources and efficient combined heat and power technologies
- Improved productivity growth rates, increased economic growth rates, and decreased electricity intensity (ratio of electricity use to gross domestic product, GDP)

Barriers to Achieving this Vision

To achieve this vision of the power delivery system and electricity markets, accelerated public/private research, design, and development (RD&D), investment, and careful policy analysis are needed to overcome the following barriers and vulnerabilities:

- The existing power delivery infrastructure is vulnerable to human error, natural disasters, and intentional physical and cyber attack.
- Investment in expansion and maintenance of this infrastructure is lagging, while electricity demand grows and will continue to grow.
- This infrastructure is not being expanded or enhanced to meet the demands of wholesale competition in the electric power industry, and does not facilitate connectivity between consumers and markets.
- Under continued stress, the present infrastructure cannot support levels of power, security, quality, reliability, and availability (SQRA) needed for economic prosperity.
- The infrastructure does not adequately accommodate emerging beneficial technologies including distributed energy resources and energy storage, nor does it facilitate enormous business opportunities in retail electricity/information services.
- The present electric power delivery infrastructure was not designed to meet, and is unable to meet, the needs of a digital society – a society that relies on microprocessor-based devices in homes, offices, commercial buildings, industrial facilities, and vehicles.

Enabling Technologies

The new infrastructure will be designed with the consumer as its basis. Consequently consumers will have increased choice and greater control of energy use. New price-driven energy services subject to consumer choice will likely proliferate. Consumers will be able to tailor their electricity service to best suit individual preferences, including costs, environmental impacts, and levels of reliability and power quality.

A communications network will link consumers to the electricity marketplace on a "24/7" basis, with a consumer communications portal enabling interactive, two-way exchanges – including price signals that flow from service providers to consumers and their array of "smart" appliances and equipment (including self-generation), and consumption patterns and other information that stream back to service providers and others. Smart appliances and equipment will automatically adjust energy use in response to real-time and forecasted price signals and in combination with preset consumer preferences.

Due to advances in small. Efficient electric generators and storage technologies, consumers will become consumer/producers. Greater consumer choice and control – along with the influx of distributed energy resources – will force the redesign of the distribution system.

A major challenge in creating the transmission system of the future is that new line construction will continue to be severely constrained by environmental and siting issues. This means that meeting capacity growth will require more productive use of existing rights-of-way. A variety of new technologies will aid this effort.

Coupling information technology with power-electronic, power flow controllers will enable the transmission system of the future to be turned in real time to optimize utilization. Incorporating distributed intelligence into the system will enable the creation of a "self-healing" grid that is capable of automatically anticipating and responding to disturbances while continually optimizing its own performance.

EPRI has developed the following list of critical enabling technologies that are needed to move toward realizing the vision of the power delivery infrastructure and electricity markets:

- Automation: the heart of a "smart power delivery system"
- Communication architecture: the foundation of the power delivery system of the future
- Distributed energy resources and storage development and integration
- Power electronics-based controllers
- Power market tools
- Technology innovation in electricity use
- The Consumer Portal

These technologies are synergistic (i.e., they support realization of multiple aspects of the vision). Aspects of some of these enabling technologies are under development today. Each of these technologies calls for either continued emphasis or initiation of efforts soon in order to meet the energy needs of society in the next 20 years and beyond.

Automation: The Heart of a "Smart Power Delivery System".

Automation will play a key role in providing high levels of power SQRA throughout the electricity value chain of the near future. To a consumer, automation may mean receiving hourly electricity price signals, which can automatically adjust home thermostat settings via a smart consumer portal. To a distribution system operator, automation may mean automatic "islanding" of a distribution feeder with local distributed energy resources in an emergency. To a power system operator, automation means a self-healing, self-optimizing smart power delivery system that automatically anticipates and quickly responds to disturbances to minimize their impact, minimizing or eliminating power disruptions altogether. This smart power delivery system will also enable a revolution in consumer services via sophisticated retail markets. Through a two-way consumer portal that replaces today's electric meter, consumers will tie into this smart power delivery system. This will allow price signals, decisions, communications, and network intelligence to efficiently flow back and forth between consumer and service provider in real time. The resulting fully functioning retail marketplace will offer consumers a wide range of services, including premium power options, real-time power quality monitoring, home automation services, and much more.

Communication Architecture

To realize the vision of the smart power delivery system, standardized communications architecture must first be developed and overlaid on today's power delivery system. This "integrated energy and communications system architecture" (IECSA) will be an open standards-based systems architecture for a data communications and distributed computing infrastructure. Several technical elements will constitute this infrastructure including, but not limited to, data networking, communications over a wide variety of physical media, and embedded computing technologies. IECSA will enable the automated monitoring and control of power delivery systems in real time, support deployment of technologies that increase the control and capacity of power delivery systems, enhance the performance of end-use digital devices that consumers employ, and enable consumer connectivity, thereby revolutionizing the value of consumer services.

Distributed Energy Resources and Storage Development and Integration

Small power generation and storage devices distributed throughout – and seamlessly integrated with – the power delivery system ("distributed energy resources") and bulk storage technologies offer potential solutions to several challenges that the electric power industry currently faces. These challenges include the needs to strengthen the power delivery infrastructure, provide high quality power, facilitate provision of a range of services to consumers, and provide consumers lower cost, higher SQRA power. However, various impediments stand in the way of widespread realization of these benefits. A key challenge for distributed generation and storage technologies, for example, is to develop ways of seamlessly integrating these devices into the power delivery system, and then dispatching them so that they can contribute to overall reliability and power quality. Both distributed storage and bulk storage technologies address the inefficiencies inherent in the fact that, unlike other commodities, almost all electricity today must be used at the instant it is produced.

Power Electronics-Based Controllers

Power electronics-based controllers, based on solid-state devices, offer control of the power delivery system with the speed and accuracy of a microprocessor, but at a power level 500 million times higher. These controllers allow utilities and power system operators to direct power along specific corridors – meaning that the physical flow of power can be aligned with commercial power transactions. In many instances, power electronics-based controllers can increase power transfer capacity by up to 50 percent and, by eliminating power bottlenecks, extend the market reach of competitive power generation. On distribution systems, converter-based power electronics technology can also help solve power quality problems such as voltage sags, voltage flicker, and harmonics.

Power Market Tools

To accommodate changes in retail power markets worldwide, market-based mechanisms are needed that offer incentives to market participants in ways that benefit all stakeholders, facilitate efficient planning for expansion of the power delivery infrastructure, effectively allocate risk, and connect consumers to markets. For example, service providers need a new methodology for the design of retail service programs for electricity consumers. At the same time, consumers need help devising ways they can participate profitably in markets by providing dispatchable or curtailable electric loads, especially by providing reserves. And market participants critically need new ways to manage financial risk. To enable the efficient operation of both wholesale and retail markets, rapid, open access to data is essential. Hence, development of data and communications standards for emerging markets is needed. Further, to test the viability of various wholesale and retail power market design options before they are put into practice, power market simulation tools are needed to help stakeholders establish equitable power markets.

Technology Innovation in Electricity Use

Technology innovation in electricity use is a cornerstone of global economic progress. In the U.S., for example, the growth in GDP over the past 50 years has been accompanied by improvements in energy intensity and labor productivity. Improved energy-use efficiencies also provide environmental benefits. Development and adoption of technologies in the following areas are needed:

- Industrial electrotechnologies and motor systems
- Improvement in indoor air quality
- Advanced lighting
- Automated electronic equipment recycling processes

In addition, widespread use of electric transportation solutions – including hybrid and fuel cell vehicles – will reduce petroleum consumption, reduce the U.S. trade deficit, enhance U.S. GDP, reduce emissions, and provide other benefits.

The Consumer Portal

Once communications and electricity infrastructures are integrated – realizing the ability to connect electricity consumers more fully with electronic communications will depend on evolving a consumer portal to function as a "front door" to consumers and their intelligent equipment. The portal would sit between consumers "in-building" communications network and wide area "access" networks. The portal would enable two way, secure and managed communications between consumers equipment and energy service and/or communications entities. It would perform the work closely related to "routers" and "gateways" with added management features to enable energy industry networked applications including: expanded choice; real time pricing; detailed billing and consumption information; wide area communications and distributed computing. This could include – data management, and network access based on consumer systems consisting of in-building networks and networked equipment which integrate building energy management, distributed energy resources, and demand response capability with utility distribution operations.

3. CONCLUSION

The worldwide participation of energy companies, universities, government and regulatory agencies, technology companies, associations, public advocacy organizations, and other interested parties is necessary to refine this vision and evolve the needed technology. Only through collaboration can the resources and commitment be marshaled to reach these destinations.

♦ Based on a Briefing presented by Clark W. Gellings before the United States of America, Northeast-Midwest Congressional Coalition's September 8, 2003 Briefing: The Electrical Grid and The Blackout of 2003: Explanations and Solutions

NEW AND RENEWABLE ENERGY TECHNOLOGIES

Prospects for their deployment in small scale power production and distributed generation

Demosthenes P. Agoris*

1. INTRODUCTION

Renewable and new - environmental friendly - energy technologies are gaining acceptance as a safe way of maintaining and improving living standards without harming the environment. New and renewable energy technologies (NRETs) can be applied in the electricity, heat and transport markets; furthermore the new generation of fuel cells will power motor vehicles and will replace the conventional storage systems. But the main emphasis has been on electricity generation, because many of these technologies are best applied to produce electricity and that is the earliest market where restructuring provides unique occasion for their deployment. However, adapting to using NRETs widely requires significant changes in institutions and attitudes, by taking into account their most common typology that support their decentralized use. These technologies are small-scale, modular, and deployed close to the resource itself. A new philosophy of incorporating electric energy production in the power system is developed, the dispersed (or distributed) generation.

2. THE TECHNOLOGIES

Small scale and distributed generation technologies can be distinguished in two groups; technologies dedicated for renewable energy sources and those that exploit several types of fuels either with conventional, like gensets and gas turbines, or with new energy technologies like microturbines and fuel cells.

* University of Patras, Electrical & Computer Engineering Department, High Voltage Laboratory, Rio, Patras, Greece GR-26500

Gaseous Dielectrics X.
Edited by Christophorou *et al.*, Springer, New York, 2004.

Diesel and NG gensets that till today are used for emergency back-up can now be used as peaking load units, contributing to the network. Their power ranges from 50 kW to 5 MW and efficiency of the order of 30% to 40%. In several cases they are combined with heat or chill production, in combined heat and power (CHP) applications, and can be used even as base load units. Gas turbines, further to their application as peaking units by utilities, are commonly used in industrial applications for power in CHP plants. In case that electric power produced in a CHP plant is not to be consumed by the owner (industry, hospital, etc) he can feed it in the network, acting as a seller; otherwise, when power is not enough, the owner can buy from the grid, being a customer.

A technology that has recently been made commercially available is that of microturbines, where turbine, compressor and generator are on the same shaft that spins at very high speeds, that can be as high as 120.000 rpm for small units. The power produced is of high frequency and is rectified to 50/60Hz through power electronics. Their power ranges from 25 kW to 500 kW. The electrical efficiency is of the order of 25% and can be increased to an overall efficiency of 80% if the heat is also used. Microturbines can be operated on a variety of fuels, like natural gas, propane, diesel etc. With no or few modifications microturbines can be operated with low calorific value fuels like landfill gas or biogas.

The new promising technology for small scale generation is this of fuel cells. In principle, a fuel cell operates like a battery, producing energy in the form of electricity and heat as long as fuel is supplied. A fuel cell consists of two electrodes sandwiched around an electrolyte. Oxygen passes over one electrode and hydrogen over the other, generating electricity, water and heat. The ideal fuel for them is H_2 but some types can operate on gaseous hydrocarbons that are internally reformed into H_2. Fuel cells operate at temperatures that vary between 80-1000° C. Depending on the electrolyte used, fuel cells can be divided into four main types: Phosphoric acid fuel cells (PAFC), solid oxide fuel cells (SOFC), molten carbonate fuel cells (MCFC) and proton exchange membrane fuel cells (PEMPC). An overview of fuel cells is given in Table 1. Furthermore, alkaline fuel cells are used in space applications, requiring pure H_2. The combination of a fuel cell operated on hydrogen that is produced from renewable energy sources is a hybrid system that is environmentally absolutely friendly. In this case hydrogen is the energy carrier that is used for storage and generation of energy as well.[1,2,3]

3. RENEWABLE ENERGY TECHNOLOGIES - AN OVERVIEW

Renewable energy sources include solar energy, wind, wave, tidal, biomass, geothermal and hydro energy. Although it is convenient to talk about renewables as a single category, many different technologies and fuels are involved, all with very different typical characteristics. The various technologies for renewable energy exploitation for power generation are not uniformly mature or cost effective, and most of them still need further intense R&D efforts having a significant way to go before they are competitive with fossil technologies. Renewables can supply a significant portion of the world's energy needs, creating many benefits further to the environmental improvement, increasing fuel diversity, ameliorating national energy security and economic development since RES technologies can help create new jobs and income; this is sustainability. However, these benefits are often not reflected in the prices paid for energy from several renewable sources, placing them at a disadvantage when competing against fossil fuels and nuclear power technologies.

Table 1: Fuel cells characteristics for dispersed generation applications

Fuel Cells Overview				
	PAFC	*SOFC*	*MCFC*	*PEMFC*
Commercially Available	Yes	No	No	No
Size Range	100-200*kW*	1*kW* - 10*MW*	250*kW*-10*MW*	3-250*kW*
Fuel	Natural gas, landfill gas, digester gas, propane	Natural gas, hydrogen, landfill gas, fuel oil	Natural gas, hydrogen	Natural gas, hydrogen, propane, diesel
Efficiency	36-42%	45-60%	45-55%	30-40%
Environmental	Nearly zero emissions	Nearly zero emissions	Nearly zero emissions	Nearly zero emissions
Other Features	CHP (hot water)	CHP (hot water, LP/HP steam)	CHP(hot water LP/HP steam)	CHP (80°C water)
Commercial Status	Some commercially available	Likely commerciali-zation 2004	Likely commerciali-zation 2004	Likely commerciali-zation 2004

3.1. Wind turbines

Horizontal axis wind turbines dominate throughout the world, since they have proven more cost effective than the vertical axis machines. The WT components could be broken into 4 basic subsystems:

- a rotor, usually consisting of two or three blades, a hub through which the blades attach to the low speed drive shaft, and sometimes hydraulic or mechanically-driven linkage systems to pitch all or part of the blades
- a drive train, including a gearbox and generator, shafts and couplings, a nacelle cover for the entire drive train, a mechanical disk brake, and often a yaw system to keep the rotor orientated towards the wind
- a tower with foundation that supports the rotor and drive train
- controls, instrumentation, electrical equipment and cabling.

The technological evolution of wind turbines is fast (see Figure 1). On 1.1.2003 31.127 MW of wind power were operated in all over the world; 75% of them are in Europe and 15% in USA. The world market is increasing by 25% last year; commercial machines reaching 5 MW are installed while the installed price is dropping below 1000 €/kW. The technology is at a stage where it can deliver large-scale implementation reliably and at a price approaching that of conventional plants. The system cost effectiveness has improved by a factor of 3 over the last 10 years. Reliability is now very high with the machines available for generation for more than 96% of the time.

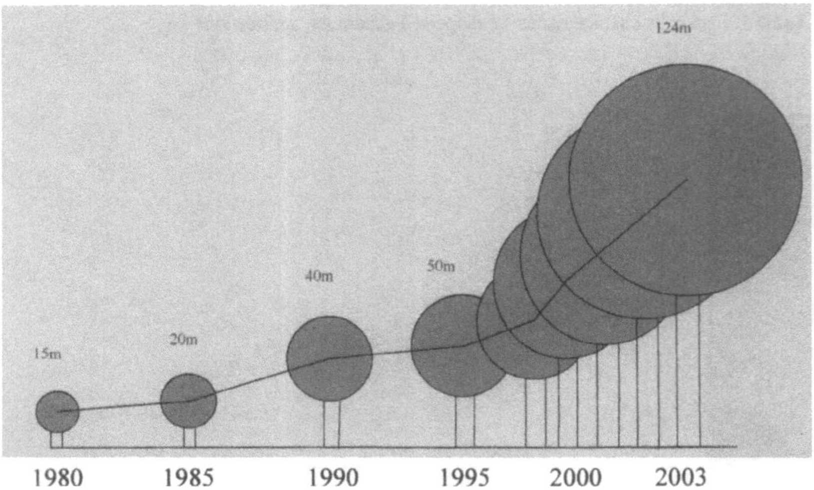

Figure 1: The evolution of wind turbines dimensions according to EWEA

3.2. Solar Power

The most common forms of solar cells are based on the photovoltaic (PV) effect. The power that is available from a PV device depends on the intensity and the wavelength of the sunlight and on the type of the material and its exposed area. The voltage of a PV cell depends on the design and materials of the cell, does not depend on its size, and remains fairly constant with changing light intensity. The current is almost directly proportional to light intensity and size. Inverters, batteries, battery charge controllers, etc. provide the interface for the connection with the grid for a specific application. Theoretical efficiencies of PV cells are 20-30%, practically they are much lower. The power output of a solar cell can be increased quite effectively by using a single or dual axis tracking mechanism to keep the PV panel perpendicular to the rays of the sun, or by using lenses or mirrors. Generally speaking, electricity generated from PVs cannot yet compete with efficient conventional central generating plants in terms of cost.

Silicon solar cells are made as mono-crystalline by using single crystal wafers, as polycrystalline wafers or as thin films. Polycrystalline wafers that are produced by casting are significantly cheaper than mono-crystalline cells, but less efficient. The most applicable technology of thin films is that of the amorphous silicon, while combinations of other materials (Cd, Cu, In, Ga, Se, etc) are also used lately. Among the advantages of thin film cells is the possibility for mass production and their ability to be incorporated on building structures.

Further to solar cells, configurations that concentrate sunlight for heating of a working fluid used for electricity generation in a thermodynamic cycle are the solar thermal power systems. Three are the most promising configurations. In all three cases the collected solar energy heats a working fluid, which then drives a thermal engine. The power-tower approach employs a field of tracking mirrors that reflect sunlight to a receiver a tower top. A second configuration employs a field of sunlight-tracking parabolic troughs that focus sunlight onto their axis. Both approaches use a turbine fed by all collectors in the range of 30 MW. Third approach employs parabolic dishes, either as single units or in fields tracking the sun. A receiver is placed at the dish focal point and each dish/engine unit (Stirling engine) is a self-contained power system of about 10-30 kW typically.[1,2]

3.3. Geothermal energy

Geothermal energy is the natural heat of the earth transferred by fluids that absorb the heat within crystal rocks. It is estimated that very large quantities of energy can be extracted from geothermal energy, which although is non-uniform, its distribution is globally widespread. In many parts of the world, geothermal electricity is competitive with conventional energy sources. A number of conversion technologies using the geothermal resource exist, namely:

- Dry steam power plants draw from underground resources of steam. Steam is piped directly to the power plant, where it is directed into the turbine.
- Flash steam power plants use geothermal reservoirs of water with temperatures greater than 180°C. The hot water flows up through wells under its own pressure. As it flows, the pressure decreases and some of the hot water boils into steam, which is separated and used to power a turbine. The leftover water and condensed steam are injected back into the reservoir.
- Binary cycle power plants operate on water at lower temperatures (100° - 180°C), and use the heat from the water to boil in a heat exchanger a working fluid, which then turns a turbine. The water is re-injected into the reservoir.[1,2]

3.4. Small-scale hydropower

A mature technology that can be well integrated at the distribution level, small-scale hydropower technology (installed capacity less than 10MW) is efficient and highly developed, with many schemes and suppliers existing in all over the world with an exploitable resource. The main parts of a small hydro plant include:

- a suitable rainfall catchment area, and a hydraulic head
- a means of transferring water from the intake to the turbine
- a turbine house for the generation equipment and the regulating valve
- a tailrace to return the water to its natural course
- a connection to the grid and control equipment

Projects are characterised by long life and low operating costs, with the bulk of their expenditure occurring during construction. The applicable size and type of the hydro turbine depends primarily on the head, the discharge and the runner speed. The range of suitable heads for various types of turbines according to Layman's is given in Table 2.[1,2]

Table 2: Range of heads of hydro turbines

Type of turbine	Range of head (m)
Kaplan	$2 < H < 40$
Francis	$10 < H < 350$
Pelton	$50 < H < 1300$
Turgo	$50 < H < 250$
Cross-flow	$3 < H < 250$

3.5. Biomass electricity

To generate electricity from biomass two systems of quite different character need to work together, namely a supply system that produces, collects and delivers the fuel, and a power station that generates the electricity. Four categories of fuel resources can be distinguished: wood, agricultural residues, energy crops, and wastes. Technologies can be categorised into three groups:

- Conventional steam cycle: Biomass is burned in excess of air to produce heat, which is in turn used to raise high-pressure steam in a boiler. Then the steam expands through a turbine, which drives an electrical generator.
- Gasification and other processes: They offer methods of power generation with higher efficiencies than combustion-based steam cycles, which are achieved by converting first the solid biomass to liquid or gaseous fuels and then burning these intermediates in engines or gas turbines.
- Co-firing with fossil fuels: This has been receiving increased attention recently. The concept is to fire a proportion of biomass with a fossil fuel in existing power plants.

A biomass power plant is constrained by the availability of the resource, the maximum being around 30…70MW$_e$. Plants using conventional steam cycles at such small scales have conversion efficiencies of only 25%. Gasification or pyrolysis can raise it to over 35% now, and 45% in the medium to long term. The capital costs per kW$_e$ produced are expected to be comparable with steam plants.

3.6. Wave energy and tidal energy

Wave power results from the harnessing of energy transmitted to waves by winds moving across the sea surface. Wave power generation devices fall into two general categories, fixed and floating. Fixed devices, mounted either to the seabed or the shore, have significant advantages over floating ones, particularly in the maintenance, but the number of sites available for such devices is limited. Oscillating water column and floating devices generate electricity through the harmonic motion of a floating part, are the most applicable methods for wave energy utilization. Much research occurs internationally to develop oscillating water and floating columns that require less stringent silting conditions.

Tidal energy exploits the natural rise and fall of coastal tidal waters caused principally by the interaction of the gravitational fields of the Sun and Moon. The modern version of tide mills is a semi-permeable barrage built across an estuary, allowing floodwaters to fill an impounded basin via a series of sluices. In future schemes the energy yield would be enhanced by pumping water into the estuary on the flood tide. A variant is tidal stream technology, aiming to exploit strong tidal currents found in shallow seas where natural constrictions exist. This technology is in its infancy. Devices similar to submerged wind turbines would be used to exploit the kinetic energy in tidal currents.[1,2]

4. POWER SYSTEMS & DISTRIBUTED GENERATION

Distributed resources introduce tremendous changes to the generation and distribution businesses, and to their management. Since many years, power systems are operated in a top-down fashion. Thus, large power generation plants, such as a nuclear plant or a lignite-fired plant, produced most of the power. This kind of generation is often located close to the required resources such as water for cooling purposes for nuclear plants or a lignite mine in the latter case. The power is then transmitted to large consumption centres over long distances and using high voltage lines. This very centralized structure is operated by a hierarchy of control centres, which facilitate the constant monitoring and control of the system, thus generation is instantly adjusted to the load of demand. However, the increasing reliability and decreasing production cost of renewable energy and emerging new technologies, environmental concerns, people reaction to new transmission lines, the deregulation of energy markets and the need for quality of service, security and diversification of energy supply, are contributing to the shift from centralised to de-centralised power generation including and small-scale cogeneration. Distributed generation has many benefits in terms of:

- reduced transmission and distribution losses
- avoidance of new electricity transmission infrastructure
- low investment cost and can be build incrementally to match demand
- potential to use locally available renewable energy sources
- local exploitation of the heat produced for the case of CHP applications
- shorter planning and erection phases
- mostly can be transported to other sites if need arises

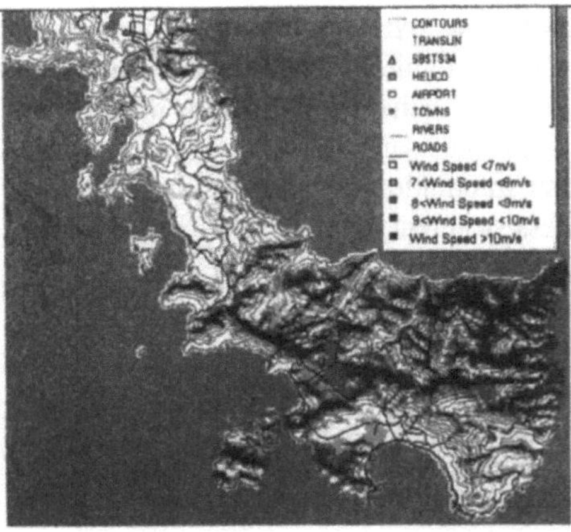

Computed values	Wind speed (m/sec)			
	>7	>8	>9	>10
Available Area (km^2)	69,46	44,42	26,42	15,75
Annual mean wind speed (m/sec)	9,00	9,80	10,70	11,50
Annually expected energy prod. (GWh)	8.015	5.752	3.768	2.401
Total installed capacity (MW)	2.364	1.518	911	548
Utilization Factor (%)	38,7	43,3	47,2	49,9
Electricity production cost (€/kWh)	0,042	0,038	0,035	0,033
Total installation cost (million €)	2686	1725	1035	623
Annual O&M cost (million €/year)	137,6	80,8	52,7	31,6
Discounted payback period (years)	5,10	3,70	3,00	2,60
Internal Rate of Return-IRR (%)	19,08	26,47	32,59	36,71

Figure 2: Wind penetration scenarios for the Greek island of Evia

There are also some potential drawbacks, including electrical quality issues in terms of voltage and frequency control, reactive power issues, high costs for some technologies, stochastic nature of some sources and the availability of grid in remote areas. However, the presence of a large number of small generation units on the distribution grid leads to technical challenges. Key to this strategy is distributed intelligence, a decentralized grid and smart use of Information and Communication Technologies (ICT). Indeed, ICT is going to be at the heart of the system architecture, as only ICT makes it possible to exchange immensely increased amounts of data, to control an increasing numbers of components, and to make it possible for new business actors to smoothly interact and go into markets, whilst maintaining the overall safety and security of the system.[3]

A high penetration of DER would also mean significant developments in gas infrastructures (including storage) to supply fuel cells, micro-turbines and Stirling engines. This will cause a high interdependency between gas, electricity and communication infrastructures and will affect other fuel supply structures as well. It is important to account and make sure that energy and intelligence are being distributed with an adequate level of reliability and security.[3]

Decision-support software tools are necessary for evaluating the penetration of dispersed generation from RES in transmission and distribution networks. Such a tool that has been developed is based on a Geographical Information System (GIS) and is taking as input the physical potential of renewables (e.g. wind, hydro, etc) and other parameters related to existing infrastructures, land use and availability, H.V. network, energy technology characteristics, as well as costs and schemes for financial support. A module of the tool digitizes the electrical network combining electrical topology information, geo-information for the distribution network and other geographical information of the area of interest. These data are stored in a geo-database and constitute the basic input for the two major applications of the tool, the technical and economical study of RES projects. The output of the system includes calculations of the available potential of RES, expected energy production and the implications of RES penetration under different scenarios. Steady state network analysis is performed for each scenario. Transient analysis can be also performed via interface with the EMTP-ATP. Finally, the tool calculates economic indicators for the evaluation of the economic potential of RES in given geographical areas. Some indicative results for the windiest area of Greece in Southern Evia are illustrated in Figure 2.[4]

5. PROSPECTS AND TRENDS

Worldwide, renewable sources of energy are estimated to meet 15-20% of final energy demand – predominantly from hydro, biomass and geothermal. Studies by the International Energy Agency (IEA) have shown that promotion programmes over the last decade, including R&D, demonstration and market stimulation, have typically reduced the capital cost of NRETs by about half. A further 50% reduction is projected for the next decade, although substantial variations around this average are expected. Several of studies have been published for the development of the potential global market for RES, carried out by reputable organisations, such as UN, IEA, WEC and Shell International. Modelling of all these studies is based on two main types of scenarios, those driven by economic growth and those driven by environmental concerns. In most of these studies the total contribution made by RES is aggregated (Table 3).[1,2,5,6]

During 1995 RES fulfilled just 6% of primary energy requirements in the EU, but a EC White Paper and Action Plan published in 1997 proposes the ambitious target of doubling this to 12% by 2010 (22.1% for RES-Electric), thereby saving over 400 million tonnes a year of CO_2 emissions. The EU White Paper set out specific targets for renewable energy technologies. The full list is given in Table 4. However, the overall contribution to the EU electricity market still remains small, around 3% when excluding large hydro.[1,2]

Table 3: Future prospects according to international modelling studies

Study	Date	Renewables contribution
IEA	2010	From 19 to 21% of the predicted electricity production in 2010 (excluding uses of biomass)
WEC	2050	Over 25% and up to 45% of the predicted global electricity output (between 21 and 30% of the expected energy demand)
UN	2050	Current contribution of 20% will rise to 60% of electricity demand and 40% of the fuel market
Shell International	2060	Up to 40% of total energy supply

Table 4: Targets set by the European Commission White Paper

Energy Source	Contribution in 1995	Contribution in 2010
Wind	2.5GW	40GW
Hydro - Large	82.5GW	91GW
Hydro - Small	9.5GW	14GW
PV	$0.03GW_p$	$3GW_p$
Solar Thermal Collectors	6.5 million m^2	100 million m^2
Biomass	44.8Mtoe	135Mtoe
Geothermal Electricity/Heat	0.5GW/1.3GW	1GW/5GW

6. DISCUSSION AND CONCLUSIONS

As a conclusion, it is generally anticipated that NRETs will be increasingly economically competitive with fossil and nuclear plant and some, such as wind, fully competitive within a time horizon of 10 years. Furthermore, it is possible to attempt to draw out those technologies that will make the most significant contribution at various time frames. These are:

Short term (available now): Small hydro, wind power, energy from wastes, biomass residues, active solar, stand alone PV and high enthalpy geothermal.

Medium term (commercial in the next 10 to 15 years): large scale PV, microturbines, energy crops, fuel cells, and geothermal aquifers.

Long term (may be commercial within 25 years): onshore and offshore wave, photoconversion, and solar thermal electricity.

The concept of distributed generation offers the basis for the incorporation of NRETs in the new market of electricity. Meanwhile, in autonomous isolated and relatively small power systems the penetration of renewables may reach very high levels, even up to 100% of demanded power for short time-periods as in the case of the Greek island Kythnos. Anyway, stable, favourable framework conditions, reflecting the environmental and social benefits of NRETs seem to be established for this stage that will support the prospects for their deployment.

7. REFERENCES

1. Smit J., Pryor B., Wallis D., Hjortsberg A., Price A., Davidson B., Agoris D., Stephen R., Boltza A., Povh D., Wahlstrom B., Philippe G., Baehr R., Janssen A., Robert A., Larsson S., Gonzalo F., Ziegler G., Gockenbach E., Muhr M., Lauby M., *Emerging technologies and material challenges-Panel on Power Systems 2020*, 39th CIGRE Session (Paris, 2002)
2. Agoris D., Malamatenios H., *An overview for renewable energy technologies and their market* CIGRE SC15/D1-JAG 02 TC (Paris, 2002)
3. S. Tselepis, S. Tentzerakis, N. Lymberopoulos, P. Chaviaropoulos, D. Agoris, *Distributed Energy Resources*, CIGRE WG:D1-18 (Turku, 2003)
4. D. Agoris, K. Tigas, V. Kilias, J. Kabouris, A. Koronides, J. Vlachos, M. Psalidas, E. Pyrgioti, *Decision – support information system for evaluating the penetration of dispersed renewable energy sources generation in transmission and distribution networks*, 40th CIGRE Session, paper C6-104 (Paris, 2004)
5. IEA, *Renewables for Power Generation: Towards a Better Environment* (London, 1997)
6. WEC, *Energy for Tomorrow's World* (London, 1994)

SUPERCONDUCTORS

An Emerging Power Technology[†]

Jack W. Ekin[*]

Superconductors offer effectively zero resistance to the flow of electricity and thus have the potential to significantly reduce Joule energy losses in the distribution and utilization of electric power, as well as the size and weight of power generators and motors. Although superconductivity was first discovered in 1911, temperatures near absolute zero were required for their operation, and so large-scale application of superconductivity was limited until the discovery in 1986 of a new class of oxide superconductors that operate at much higher (although still cryogenic) temperatures. Recent, remarkable progress has been made in the fabrication of practical conductors using the new oxide superconductors. Some of the properties of superconductors (both low- and high-temperature types) and the progress that has been made in overcoming the impediments to their commercialization are reviewed. Also discussed are ongoing demonstration projects for transmission lines, motors, generators, synchronous condensers, and transformers that are currently being constructed by teams of industry and government researchers to implement and test this new technology in commercial settings.

1. INTRODUCTION

Superconductors are materials that offer zero resistance to dc current flow if cooled to low enough temperatures and operated in low enough magnetic fields. What makes them practical is their ability to carry very large current densities, $\geq 10^5$ A/cm^2, with little or no power dissipation. As such, they are potentially

[*] National Institute of Standards and Technology, Boulder, Colorado 80305.

Gaseous Dielectrics X.
Edited by Christophorou *et al.*, Springer, New York, 2004.

useful in a wide range of power applications, including generators, motors, and power transmission lines.

Their ability to carry high current densities offers the prospect of increasing current levels by 5 to 7 times in existing utility corridors without raising the voltage. In urban settings this can be done by threading them through existing underground conduits without digging up streets or acquiring expensive property. The prospect is tantalizing. Superconducting power cables also offer a very environmentally friendly avenue for increasing power, without higher voltages, without hazardous oil, and without electromagnetic interference. In sensitive environmental cases, they offer an acceptable solution to increased energy capacity. Superconductors also provide a way to fabricate motors and generators that are smaller, lighter, and more energy efficient.

Here we consider superconductor performance limits and the tremendous advances that have occurred recently in their ability to operate at higher temperatures and magnetic fields. We also look at the potential for producing practical conductors and cables, and we briefly review the specific applications currently under development.

2. SUPERCONDUCTIVITY—WHAT ARE ITS LIMITS?

For practical applications, superconducting materials must have a critical current density $>10^5$ *A/cm*^{2} *under the conditions of temperature, magnetic field, and strain that they will see in service.* Otherwise they do not offer sufficient advantage over ordinary copper wire. (More specifically, the engineering requirement is that the *overall* composite wire, including other components, must have current densities $>3\times10^4$ A/cm^2 for rotating machinery designs and $>1\times10^4$ A/cm^2 for transmission lines).

Unfortunately, this requirement precluded the use of superconductors in practical devices for nearly half a century following their discovery in 1911.[1] These early superconductors operated only at very low temperatures (just a few degrees above absolute zero), and, to make matters worse, researchers soon learned that magnetic fields of only hundredths of a tesla destroyed their superconductivity. (The fields could arise either from self-fields generated in transmission lines or fields generated by winding them into magnet coils; either way, they were useless in practical applications.)

2.1. Discovery of Type-II Superconductivity

Then, in 1961, the superconducting compound Nb_3Sn was discovered to have a critical field over 20 T,[2] and suddenly the path was open for the production of high-field magnets. Paradoxically, these high-field superconductors were impure,

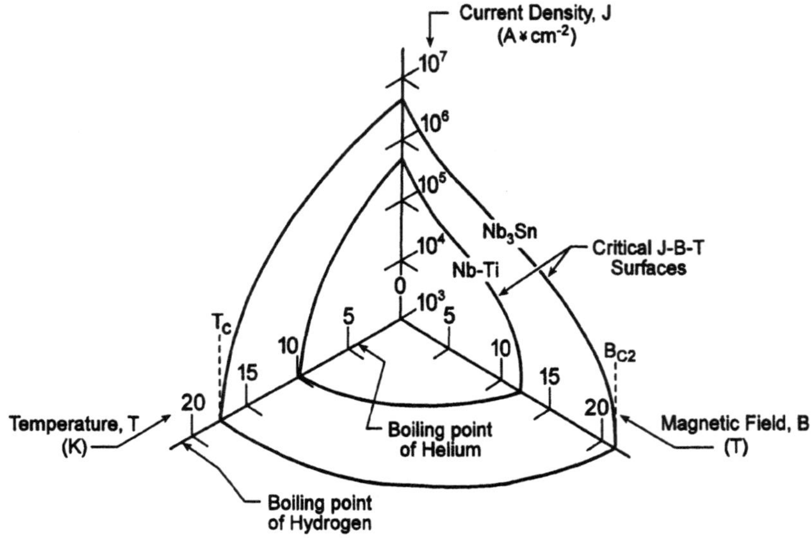

Figure 1. Critical-current density J_c of two commercial low-temperature superconductors as a function of magnetic field *B* and temperature *T*. In the region between the origin and the critical *J–B–T* surface, the material is superconducting; above the surface the superconductor reverts to the normal resistive state.

unlike the early pure elements that had been studied. The scattering of electrons in these materials caused by impurities resulted in a new type of superconductivity, called Type-II superconductivity, wherein magnetic fields could penetrate the material as small, localized *magnetic vortices* that did not destroy superconductivity in the bulk of the material. (A good introduction to Type-II superconductivity is given in the book by Rose-Innes and Rhoderick.[3])

Figure 1 shows the key parameter, the critical-current density J_c (vertical axis) as a function of magnetic field *B* and temperature *T*. The data from each conductor form a *critical surface*: below the surface, the material is superconducting; above the surface, the superconductor reverts to the normal (resistive) state.

Since, to be practical for large-scale applications, a superconductor typically should have a J_c greater than about 10^5 A/cm^2, we see from Fig. 1 that Nb_3Sn can be used in liquid helium (which boils at 4.2 K) in magnetic fields up to about 14 T to 16 T. Nb_3Sn is a brittle superconducting compound, however, and several decades of experience were needed to learn how to handle this delicate material. In the meantime, Type-II superconductivity was discovered in the *ductile* alloy Nb–Ti. As shown in Fig. 1, this material can be used in liquid helium at magnetic fields up to about 7 T to 8 T. Although more limited in magnetic field than Nb_3Sn, Nb–Ti is much easier to handle and so has been the workhorse superconducting material until the present time, principally for magnetic-resonance-imaging (MRI) systems.

2.2. Discovery of High-T_c Superconductors

The problem still remained that superconductors operated at only very low temperatures. Then in 1986, superconductivity was discovered in a very unlikely class of materials—ceramics.[4] With these materials emerged the possibility of applications at higher temperatures utilizing liquid-nitrogen cooling, which boils at 77 K. They could also operate at magnetic fields much higher than those possible with low-T_c superconductors.

However, just after their discovery, we discovered that the critical-current density of high-T_c superconductors (HTS) was limited to very low values by weak links at grain boundaries.[5] The weak links were a result of the charge carrier density being much less in high-T_c than in low-T_c superconductors as well as from the loss of coherence between the superconducting pairs of electrons as they cross very small crystalline imperfections at grain boundaries. In one of the most promising high-T_c superconductors, $YBa_2Cu_3O_7$ (YBCO), grain boundaries

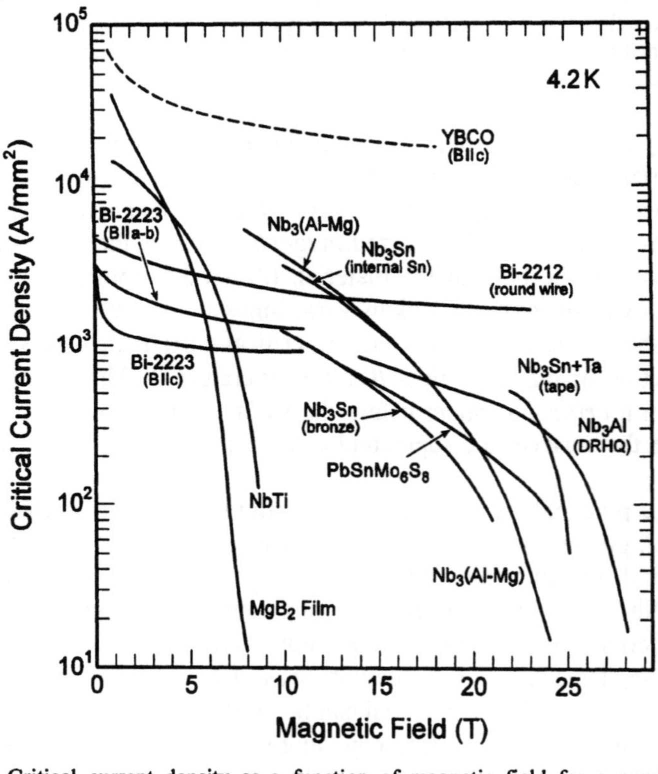

Figure 2. Critical current density as a function of magnetic field for a number of high-field superconductors at *liquid-helium* temperature (from a compilation by Lee[8]). Data are referenced to the superconducting-material area alone, as opposed to the entire cross section of the wire, which contains structural and stabilizer materials needed for the practical use of superconductors.

are misaligned by more than about four degrees form weak links.[6,7] This need for that are misaligned by more than about four degrees form weak links.[6,7] This need for highly aligned grains constitutes the main challenge of fabricating practical high-J_c HTS conductors. (In contrast, low-T_c superconductors do not lose superconducting coherence across grain boundaries and, in fact, the grain boundaries play the important beneficial role of immobilizing the magnetic vortices.)

Figures 2 and 3 show the current record holders for high critical-current density in both high-T_c and low-T_c superconductors: Fig. 2 at the boiling temperature of liquid helium (4.2 K) and Fig. 3 at the boiling point of liquid nitrogen (77 K). Notice that as the temperature is raised from 4.2 K to 77 K, all but YBCO and $Bi_2Sr_2Ca_2Cu_3O_{10}$ (Bi-2223) are eliminated from the plot. So these are the two materials currently being used to develop superconducting transmission lines that operate at liquid-nitrogen temperature.

Our work at NIST has focused on yet another limitation to the performance of high-T_c superconductors—their mechanical tolerances. At first it was feared

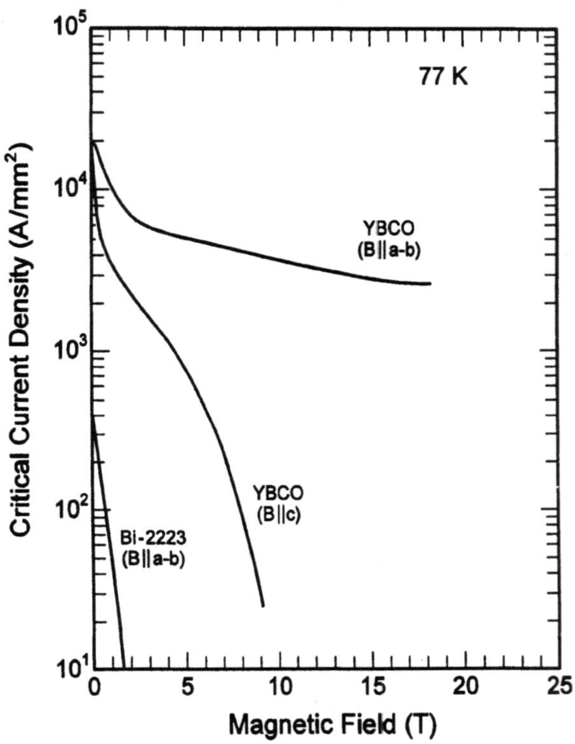

Figure 3. Critical-current density as a function of magnetic field for high-T_c superconductors at *liquid-nitrogen* temperature (from a compilation by Lee[8]). The critical-current density J_c is referenced to the superconductor area alone.

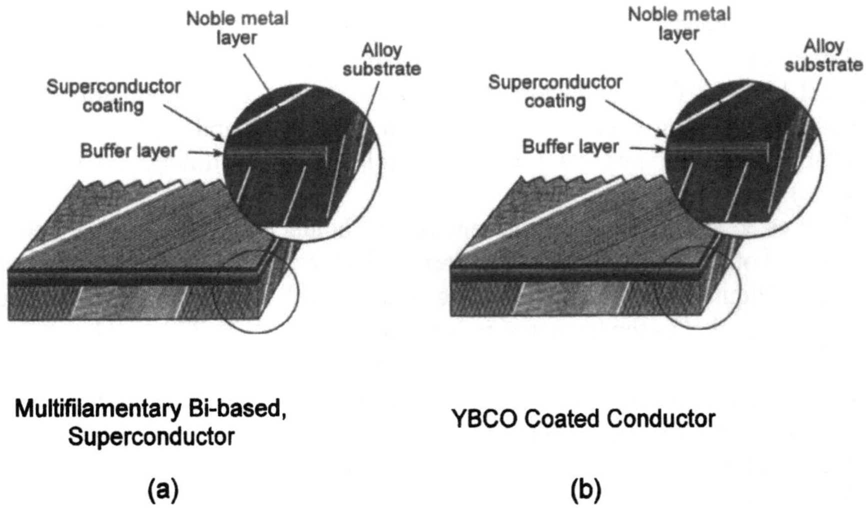

Figure 4. Cross-sectional view of practical high-T_c superconducting composites: (a) multifilamentary $(Bi,Pb)_2Sr_2Ca_2Cu_3O_{10-x}$ (Bi-2223) silver-alloy-sheathed high-T_c superconductor tape; (b) coated thick-film $YBa_2Cu_3O_7$ (YBCO) superconductor tape (from American Superconductor).

that the brittle superconducting component in these conductors would be susceptible to breakage at very small strain levels where the substrates of these composites yield, about 0.2 %. Very recently we discovered that nature has been kind to us and, in fact, YBCO performs well at strains far above the yield strain of the substrates.[9] This opens the door for potentially replacing the lower current density Bi-2223 materials, currently being used in high-T_c demonstration projects, with YBCO conductors, which can operate at much higher current densities (see Fig. 3).

Figure 4 shows cross-sectional views of practical Bi-2223 and YBCO conductors. More information about the fabrication of practical high-T_c superconductors can be found in the *Handbook of Superconducting Materials*.[10]

3. SUPERCONDUCTING POWER DEMONSTRATION PROJECTS

Annual production of energy by electrical utilities is about 3×10^{12} kWh in the United States. Energy losses associated with power transformation, transmission, and distribution amount to about 7 %. These losses would be cut approximately in half with the full implementation of superconducting technology, including transformers, power-quality devices, fault-current limiters, and underground transmission cables.

3.1. High-T_c Superconductor Power Cables

As of the year 2000, there were about 8000 km of high-voltage (> 69 kV) underground cable in the United States, generally in urban areas. Power utility companies could increase the current-carrying capacity of these underground cables by 5 to 7 times by retrofitting copper cables with superconducting cables through existing underground utility conduits. Thus, HTS power cables will most likely find their first application in these regions of highly priced real estate. HTS power cables would also increase system efficiency.

Table 1. High-T_c Superconducting Power Cable Demonstration Projects

Location / Partners	Type of Cable	Specification	Comments
Completed cable projects			
Carrollton, Georgia Southwire Cable Co.	3-Coaxial (separate cryostats) Cold dielectric Above ground	30 m long 1250 A 12.4 kV	First HTS cable demonstration project >26 000 h of fault-free operation
Copenhagen NKT	3-Separate cables	30 m long 2000 A 30 kV	First HTS cable to supply 50 000 customers
Tokyo Sumitomo Electric Tokyo Electric Power Cent.Res.Inst.of El.Power	3-Core cable (in one cryostat)	100 m long 1000 A class 66 kV	First test of three-core cable in one cryostat
Cable projects under construction in the United States			
Columbus, Ohio Southwire Cable Co. American Electric Power American Superconductor	Triaxial cable (in one cryostat) Cold dielectric Underground Includes splice and multiple 90° bends	300 m long 3000 A 13.2 kV	First three-phase coaxial cable Scheduled to be energized mid-2005
Albany, New York SuperPower Niagara Mohawk Power Sumitomo Electric BOC	3-Coaxial cables (in one cryostat) YBCO section 30 m long Underground Underground joint	350 m long 800 A 34.5 kV	Installation between two major urban substations Scheduled to be energized mid-2006
Long Island, New York American Superconductor Long Island Power Auth. Nexans Air Liquide	3-Coaxial cables (in separate cryostats)	600 m long 2400 A 138 kV	First transmission-level (138 kV) HTS cable Scheduled to be energized end 2005

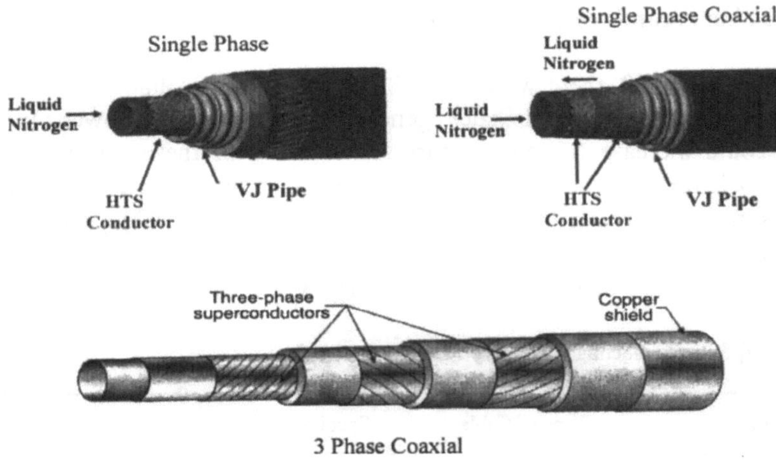

Figure 5. Three types of HTS power cables being developed and tested: single phase, single-phase coaxial, and 3-phase coaxial (or triaxial). All are cooled by flowing liquid nitrogen and thermally insulated with vacuum-jacketed (VJ) pipe (from Southwire).

Several superconducting power-cable projects have been completed, and three more are under construction in the United States. One of the completed projects was the first to supply electricity to 50 000 customers; another project has logged more than 26 000 h of fault-free operation (20 000 h unattended) while powering three manufacturing plants. A summary of details is given in Table 1.

Figure 5 shows three types of HTS power cables that are in use or planned for installation. To facilitate fabrication and maintenance of the vacuum jacket that thermally insulates these superconducting cables, all projects currently under construction in the United States employ dielectric materials operating at cryogenic temperatures inside the jacket.

Several new projects are planned throughout the world. The ones in the United States include the first three-phase triaxial cable, the first installation between two major urban substations, and the first transmission-level (138 kV) HTS cable.

3.2. High-*Tc* Superconductor Motors, Generators, and Transformers

3.2.1. Motors

Electric motors consume over 60 % of the total industrial demand for power.

Superconducting motors would be less than half the weight and size of conventional motors. For example, a 25 MW superconducting motor would weigh between 45 Mg and 63 Mg (compared with 122 Mg for a conventional motor), and its size would be between 2 m and 3 m in diameter (compared with a diameter of 4.5 m for a conventional motor). Substantial ac synchronous motors have been demonstrated by Siemens (400 kW), Rockwell (750 kW) and American Superconductor (3.7 MW). Of particularly high impact are motors for ship propulsion, where size and weight are a premium. A 5 MW high torque, low-rpm motor has been demonstrated by American Superconductor and Alstom, and a 36.5 MW motor is under development.

3.2.2. Generators

Superconducting generators offer even greater reductions in weight and size. For example, a 26 MW generator would weigh between 9 Mg and 18 Mg (compared with 54 Mg for a conventional generator), and its size would be between 0.8 m and 1.8 m in diameter (compared with a diameter of 3 m for a conventional generator). Superconducting generators also offer the benefit of a 0.35 % to 0.55 % efficiency gain. A 1.5 MVA demonstration machine is currently being built by General Electric and a 100 MVA HTS demonstration generator is scheduled for 2005.

3.2.3. Transformers

Superconductors also offer the potential for constructing smaller, environmentally friendly transformers. A conventional 30 MVA transformer weighs about 44 Mg and contains about 23 kL of oil. The HTS transformers would weigh from 22 Mg to 15 Mg and contain no oil. The HTS transformers also offer inherent overload protection without loss of equipment life, a smaller "footprint," elimination of expensive and environmentally risky oils, and potential for indoor siting without unnecessary hazard. A 5/10 MVA superconducting transformer is being build by Waukesha Electric Systems and SuperPower.

3.2.4. Other Power Projects

Other superconducting demonstration projects currently being built with HTS wire include fault-current limiters (General Atomic, with Southern California Edison Power), energy-storage flywheels with superconducting bearings (Boeing), superconducting magnetic energy storage (SMES) devices for uninterruptible power systems (Japanese Super-ACE program), magnetic separation for ore refinement and chemical processing (DuPont), and smaller, more accessible magnetic resonance imaging (MRI) systems. However, the application nearest to commercialization is the dynamic synchronous condenser, a rotating machine without prime mover that can inject inductive and capacitive out-of-phase power for reactive compensation and grid stabilization. The HTS design is compact, producing many times the VARs of a conventional machine with high efficiency and subcycle response times. An 8 MVAR unit by American Superconductor is under beta-test in the Tennessee-Valley-Authority power grid.

3.3. Future Applications

From 1970 to 2000, superconductivity slowly entered the market place in applications where performance could not be matched by existing products (research magnets, MRI, and high-energy physics accelerator magnets). Refrigeration requirements were a major impediment. From 2004 to 2010, HTS applications will enter niche markets where cost is not as important as size, weight, effectiveness, and ancillary benefits. This will be a period where cryogenic engineering and system reliability will improve significantly, along with ever lower cost wire, setting the stage for applications in large markets where cost as well as size, weight, efficiency, and performance are important.

4. REFERENCES

1. H. Kamerlingh Onnes, The resistance of pure mercury at helium temperature. *Leiden Commun.* No. 120b (1911).
2. J. E. Kunzler, Superconductivity in Nb_3Sn at high current in a magnetic field of 88 kilogauss, *Phys. Rev. Lett.* **6**, 89–91 (1961).
3. A. C. Rose-Innes and E. H. Rhoderick, *Introduction to Superconductivity* (Pergamon Press, 1976).
4. J. G. Bednorz and K. A. Mueller, Possible high-T_c superconductivity in the Ba–La–Cu–O system, *Z. Phys.* **B64**, 189–193 (1986).
5. J. W. Ekin, A, I. Braginski, A. J. Panson, M. A. Janocko, D. W. Caponne II, N. J. Zaluzec, B. Flandermeyer, O. F. de Lima, M. Hong, J. Kwo, and S. H. Liou, Evidence for weak link and anisotropy limitations on the transport critical current in bulk polycrystalline $Y_1Ba_2Cu_3O_x$, *J. Appl. Phys.* **62**, 4821–4828 (1987).
6. D. Dimos, P. Chaudhari, J. Mannhart, and F. K. Legoues, Orientation dependence of grain-boundary critical currents in $YBa_2Cu_3O_{7-\delta}$ bicrystals," *Phys. Rev. Lett.* **61**, 219–222 (1988).
7. D. Dimos, P. Chaudhari, and J. Mannhart, "Superconducting transport properties of grain boundaries in $YBa_2Cu_3O_7$ bicrystals, *Phys. Rev.* **B 41**, 4038–4049 (1990).
8. P. J. Lee, Univ. of Wisc.–Madison, http://www.asc.wisc.edu/plot/plot.htm (2002).
9. N. Cheggour, J. W. Ekin, C. C. Clickner, D. T. Verebelyi, C. L. H. Thieme, R. Feenstra, and A. Goyal, Reversible axial-strain effect and extended strain limits in Y–Ba–Cu–O coatings on deformation-textured substrates, *Appl. Phys. Lett.* **83**, 4223-4225 (2003).
10. D. Cardwell and D. S. Ginley, eds. *Handbook of Superconducting Materials,* Inst. of Physics Pub. Inc. (2003).

INSULATION SYSTEMS FOR SUPERCONDUCTING POWER DEVICES: NORMAL OPERATION VERSUS ABNORMAL CONDITIONS

J. Gerhold and C. Sumereder[1]

1. INTRODUCTION

Application of superconductivity in electrical power systems has been a fascinating option since the early sixties. The high current capacity of superconductors was claimed in an euphoric manner to open a way for improved devices with huge power ratings; prototypes for transmission lines with cold dielectric, exciter rotors for generators, Superconducting Magnetic Energy Storage (SMES) as well as model magnets for future fusion reactors, have been successfully built and operated[1]. Liquid or Supercritical Helium (LHe, SHe), respectively, was indispensable to cool the classical metallic superconductors with very low critical temperature, and cold helium performed also well as an insulating fluid up to more than 500 kV test voltages, notwithstanding many Cassandra Cries based on the known dielectric weakness of helium gas at ambient conditions. There is now undoubted some confidence on reliable voltage insulation at low temperatures.

However, economics with huge units came out disappointing. The discovery of superconductors with high critical temperature (HTS), which may be cooled with cheap liquid nitrogen (LN2) instead of helium, has raised again the interest in power applications, preferably for the medium power range. Some devices, e.g. field windings in generators, SMES and fusion magnets of course, still rely onto the use of LTS in order to produce high magnetic fields in the multitesla range. Ac apparatus, e.g. transmission lines, transformers, and Fault Current Limiters (FCL) in contrast produce moderate magnetic fields well below the Tesla limit and thus can use HTS with benefit. In fact, offering simultaneously an improved equipment performance, e.g. negligible thermal aging and drop of en-

[1] Graz University of Technology, Kopernikusgasse 24, A-8010 Graz, AUSTRIA

Gaseous Dielectrics X.
Edited by Christophorou *et al.*, Springer, New York, 2004.

vironmental impairment, may yield attractive economics even for medium sized units[1, 2]. The reader may be referred to the Handbook of Applied Superconductivity where a survey of the state of the art can be found[3].

Abnormal conditions have to be taken into account apart from normal operation. Ac equipment may be stressed by overvoltages, and adequate cooling may be disturbed in case of overload or fault conditions; this can impair the dielectric insulation seriously. Quenching of magnets is a very unique inherent phenomenon of superconductivity[4]. All these abnormal conditions have to be included into the respective high voltage design and testing scheme[5, 6].

2. VOLTAGE STRESS AND BASIC INSULATION CONFIGURATIONS

Prospective near - term power applications of superconductivity can be seen in the medium power range only; nuclear fusion being a distant future option. The running voltages may range from less than 1 kV up to about 110 kV. Figure 1 left sketches a very summarizing survey[2-5], whereas the right figures indicates usual test voltages of equipment for utility use [6].

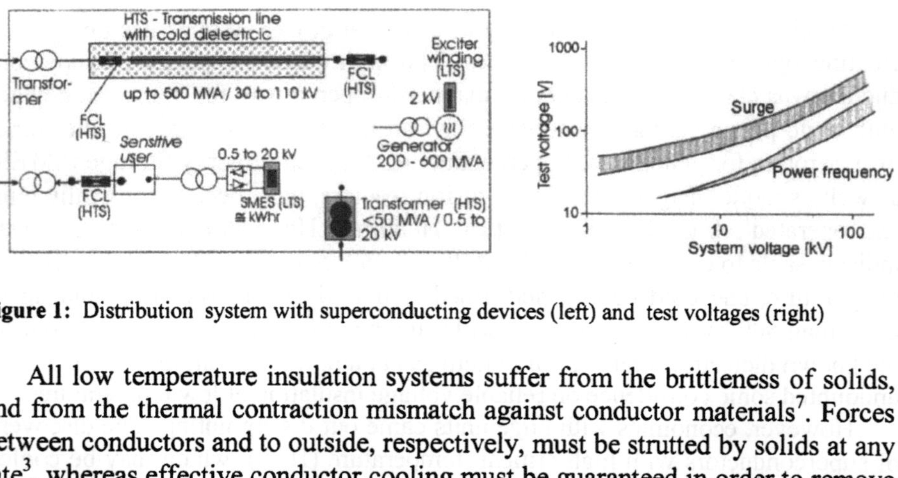

Figure 1: Distribution system with superconducting devices (left) and test voltages (right)

All low temperature insulation systems suffer from the brittleness of solids, and from the thermal contraction mismatch against conductor materials[7]. Forces between conductors and to outside, respectively, must be strutted by solids at any rate[3], whereas effective conductor cooling must be guaranteed in order to remove ac - losses from HTS; losses also arise in LTS dc magnets when varying the current[3]. Indirect conductor cooling is not matching the needs in power applications. Typical arrangements for directly cooled conductors are illustrated in figure 2. The main features are listed below[5]:

i) solid insulating spacers embedded in a cooling fluid are used in stiff fluid transparent structures, e.g. in SMES magnets. Voltage insulation in the interspace between solid spacers must be guaranteed by the fluid, which plays an active role[4]. Conducting surfaces may eventually be covered by some thin insulating sheets. Very stiff compact impregnated coils cooled by SHe seem to be adequate for the excitation windings of generators with quick response[8]. Transformer and

FCL windings, respectively, may be built up by shunted HTS ribbons applied as a single layer onto a solid insulator and are wetted by the coolant[9].

ii) fluid - impregnated wrapped insulation is a typical arrangement for the cores of transmission lines with cold dielectric. This insulation can be assembled in very long units, and can be bent[3]. The fluid again plays an active role[9].

iii) very stiff solid insulation in an evacuated environment might be preferred in internally forced - flow cooled large magnet structures with SHE, but assembly can become very sophisticated; delicate connections must not be embedded within a bulk solid. Voltage breaks for coolant fed in/out must be provided, but the fluid is not used as an active insulating medium. Stressed surfaces of solids in vacuum, and stressed open vacuum space can often not completely disregarded in a complex system with many conductor joints, diagnostic feedthroughs and hydraulic connections[10].

iv) superconductors must be fed with current and voltage. Leads to outside have to be provided. These leads should guarantee a safe insulation transition from the cryogenic region into ambient air over an extended temperature span. Adequate bushings are state of the art[11].

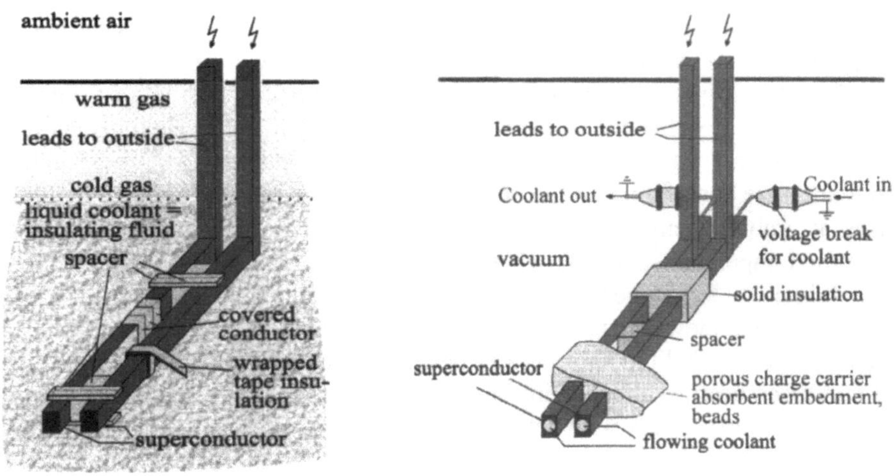

Figure 2: Basic cryogenic insulation configurations; left...embedded in coolant fluid right...in vacuum with internally forced-flow cooled conductors

3. DIELECTRIC PERFORMANCE OF INSULANTS

Table 1 gives some physical data for nitrogen and helium[3]. The liquefied gases can easily be vaporized. The low permittivity of the fluids (≈ 1 for cold gases; 1.05 for LHe, 1.44 for LN2)[7] may yield local stress enhancement within butt gaps and fluid filled voids in solids with higher permittivity[11]. Hence these regions may be weak points in a system. Partial discharges can arise which affect the useful lifetime; this has to be circumvented at least for normal operation. The fluids may be assumed to be self – healing but the solids are not. Other weak points are spacers where flashover may occur considerably below the strength of the fluid. Bubbles in a bulk liquid may also act as weak points[4, 7, 11].

Table 1: Physical data of cryogenic fluids

Gas	Normal boiling point *nbp* [K]	Mass Density ρ_0 at ambient conditions 300K / 100 kPa [kg/m³]	Tripel Point [K/kPa]	Critical Point [K / kPa / kg m⁻³]	Latent Heat at *nbp* [kJ/kg]	Relative Vapor Density at *nbp*	Relative Liquid Density at *nbp*	Specific Heat of gas at *nbp* [kJ / K kg]
^4He	4.21	0.162	--	5.25 / 226 / 70	20.9	103	700	5.2
N2	77.3	1.138	63.14 / 12.5	126.1 / 3350 / 314	200	4.0	646	1.04

3.1. Gas Discharges

The general similarity law states breakdown in gases being controlled by the density number times spacing product, $N_v \cdot a$; the gas temperature having no effect by itself [12]. However, the actual discharge buildup with time can in fact be quite different. It has been found for instance that statistical time lags of impulse breakdown are significantly enhanced in cold nitrogen gas[13]. Helium gas discharges with fast acting automatic current limitation just right of the Paschen minimum have been found on the other hand to develop via very sharp multiple pulses at ambient temperature, whereas a smooth transition has been seen at cryogenic temperature[14]. The Paschen curves for nitrogen and helium in figure 3 are well known. The somewhat unstable semi-vacuum range ends with high vacuum where breakdown does not depend on the actual gas[5]. The former CIGRE Paschen curve for helium showed a rather large scattering range; no single–valued curve has been established in those days[12]. A more specified Paschen curve has been verified recently with help of sophisticated UHV techniques[14].

The uniform field breakdown strength values of nitrogen and helium vapor at the respective *nbp* come close together; this is caused by the very different grade of densification[15] – see table 1. The strength values are similar to that of very slightly pressurized (0.14 MPa) warm SF6 gas.

Whenever the dielectric stress in a gas exceeds an order of 10 kV/mm, extrinsic effects arise and breakdown remains below the Paschen limit[7, 12]. The transition to high density or liquid state is illustrated in figure 4 for three gases; the characteristics are normalized to the density number at the critical point, N_{vcrit}. Extrinsic affected breakdown comes into the picture more and more and causes significant statistical scattering. The two – parameter Weibull distribution function is commonly used in high voltage design as an engineering tool[16, 17]; the inset in figure 4 illustrates the Weibull shape parameter decrease with rising density number for helium found with IEC 156 electrodes at 0.5 mm spacing.

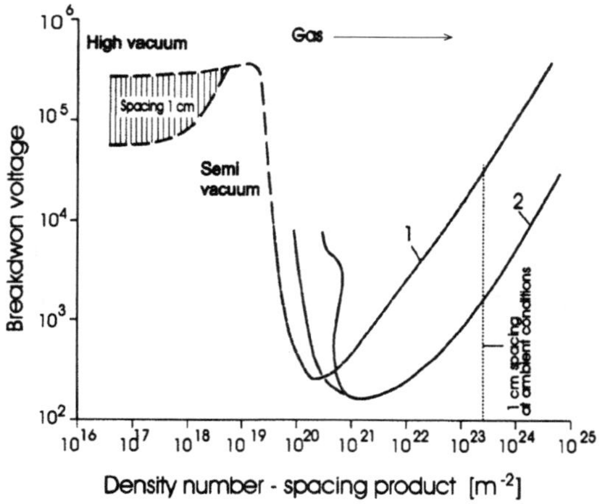

Figure 3: Paschen curves for nitrogen and helium
1...nitrogen 2...helium

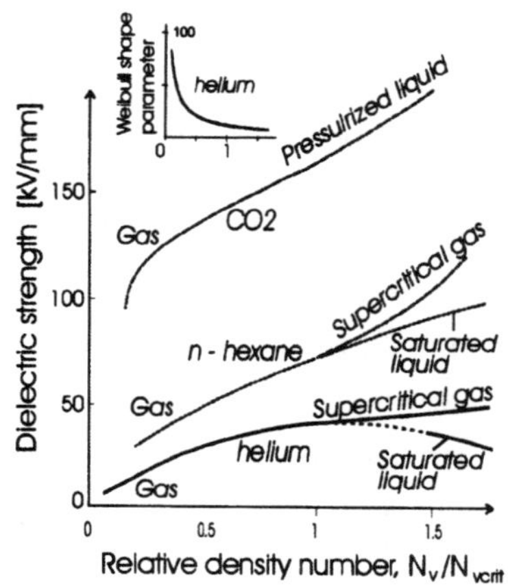

Figure 4: Transition from gas to high density state

Cold helium data found with 60mm Ø plane electrodes at 1mm have been published elsewhere[18]. Data for nitrogen are scarce but a similar tendency may be anticipated.

Figure 5 may give some impression on nonuniform - field corona onset and breakdown at cryogenic temperatures in nitrogen[19, 20]; the opposite point polarities yield higher voltage levels. The inherent complexity of low temperature discharges where ion diffusion is very weak may be illustrated for instance by the maximum breakdown voltage of 17.5 kV with 1 mm spacing which is known to occur at 8 K/0.8 MPa in helium[21]. However, information for cryogenic temperatures is quite far from being complete. Data for corona extinction are almost none[11]. Strongly nonuniform field configurations should be limited to low total voltages with preference.

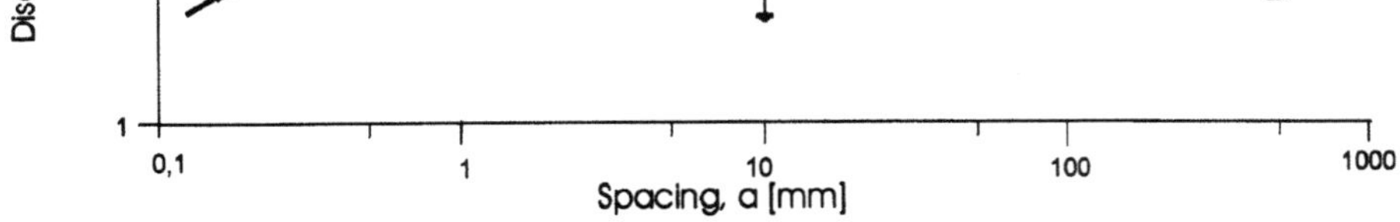

Figure 5: Nonuniform field breakdown and corona onset in nitrogen and helium vapor; uniform field breakdown characteristics are included for comparison
——— N2 - - - - Helium

3.2 Liquid Breakdown

Breakdown in liquids is almost fully extrinsic dominated; the actual liquid being just one of many parameters[7, 16]. Non-boling nitrogen and helium, respectively, show a very reasonable dielectric strength which is quite similar to that of oil in a uniform field; as a thumbs rule[7], near uniform field medium - term mean breakdown voltages V_b may be written as $V_b = C \cdot a^{0.8}$; C being between 20 to 30 kV mm$^{-0.8}$. Size effects where the breakdown strength decreases with the highly stressed liquid volume v almost with $v^{-0.1}$ as well as statistical time effects where the strength decreases with stressing time t as $t^{-1/n}$, have been discussed in detail; n may range from 19 to 27 in case of LN2, and about 15 in case of LHe [7, 11, 16, 22]. The experimental data range is still limited. However, no true liquid aging must be anticipated.

The impact of external heat input into LN2 is illustrated in figure 6. The dc or ac strength decrease down to the vapor level is drastic at normal pressure. However, elevated pressures are very prospective to increase the liquid strength; the situation with LHe is quite similar[11, 16]. Transmission lines for instance use subcooled LN2 under high pressure; no bubbles may arise within the fluid[9]. A distinct breakdown delay has been found in case of pulse voltage stress[23] - see figure 7; similar characteristics are also known for LHe.

Point to plane breakdown voltages on the other hand may be predicted with $7 \cdot a^{0.4}$ for LHe, and $18 \cdot a^{0.26}$ for LN2[7]. The superiority of the latter is obvious. However, corona occurs only somewhat below the breakdown limit[17]. Hence the liquids are quite prospective insulants at low to medium voltages[16, 24]; again, high voltages are difficult to control.

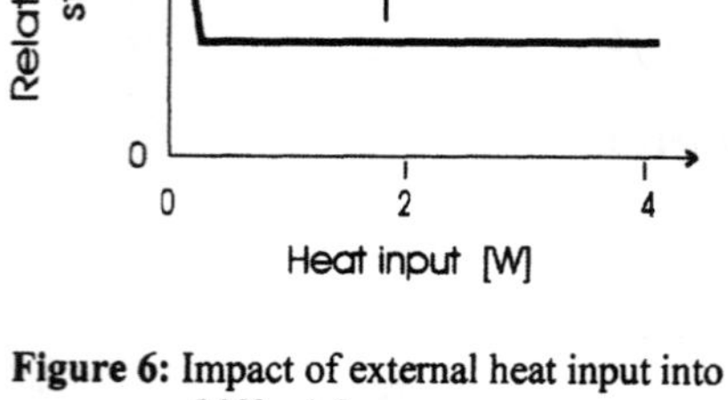

Figure 6: Impact of external heat input into LN2; 1.2 mm gap
1...ambient pressure 2...0.5 MPa

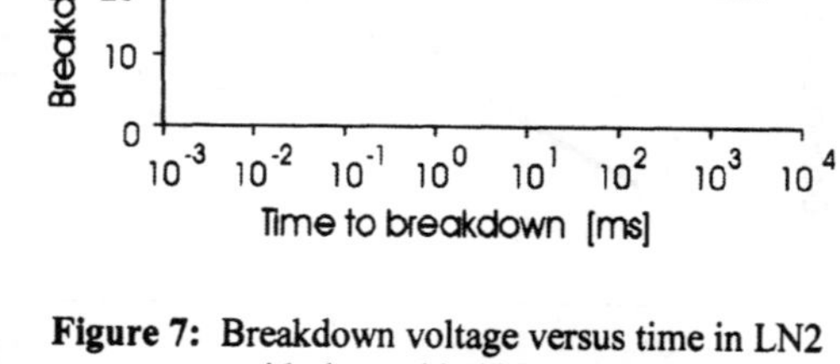

Figure 7: Breakdown voltage versus time in LN2 with thermal bubbles
1...envelope of minima; 2...0.2/1 ms pulse; 3...5/20 ms pulse; 4...115/415 ms pulse

4. ABNORMAL OPERATION

The definite design method should be applied for cryogenic insulation systems where an exchange of damaged parts in case of local failures may be extremely difficult[5]. Figure 8 left sketches the design philosophy. Hence, the basic experimental data which may end usually with 1% failure probability must be extrapolated by means of statistics down to an extremely low level, and overvoltages as well as test voltages must remain well below this withstand boundary. Figure 8 right illustrates an extrapolation of breakdown data in LN2; the large size data for instance could be extremely well fitted to a two-parameter Weibull distribution[16].

However, abnormal operation may not be limited to usual overvoltages. Running for instance an HTS transformer under overload conditions where the excess current produces heat in the silver - shunted HTS yields bubbles, without rising the temperature; thermal solid insulation aging thus can be disregarded completely[2]. A design relying onto cold nitrogen vapor instead of liquid may then be adequate. A Weibull distribution for nitrogen vapor where scattering is minute has been also included into figure 8. Again, the almost no-breakdown probability limit should not interfere with any actual overvoltages.

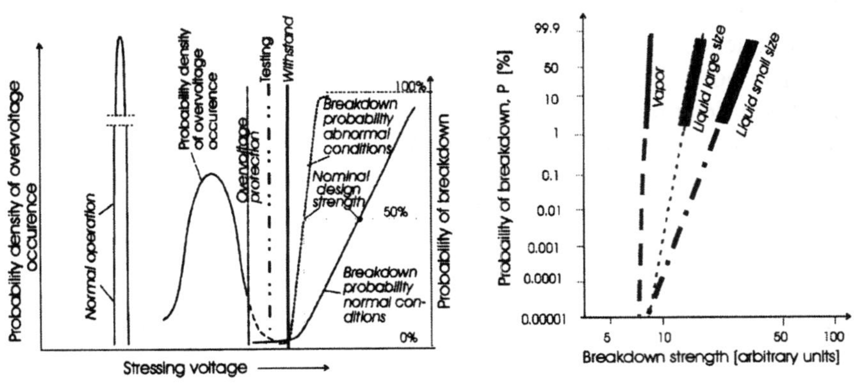

Figure 8: Scheme of definite design method (left); extrapolated Weibull plots for nitrogen (right)

In case of an FCL, normal insulation against the grounded vessel may rely onto LN2. Activation where very large amounts of heat are produced because the HTS goes normal very fast[2, 9] yields quite different conditions- see figure 9: a warm gas layer may adhere close to the quenched HTS, the rest of the gap being liquid or cold gas during the first voltage spike. Impeding the vapor from leaving the vessel may yield an increased pressure after some delay; the application of solid barriers between neighboring HTS coils may also bring benefits[17]. Design models using different routes –for instance with help of the characteristics shown in figure 6 and 7, respectively, have been sketched but still need further development[2].

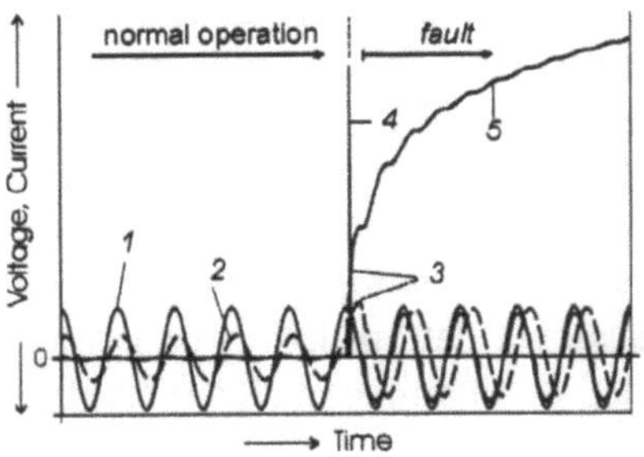

Figure 9a: FCL in the peak limiting mode 1...feeding voltage 2...normal operating current 3...limited current 4...voltage peak across FCL 5...HTS temperature

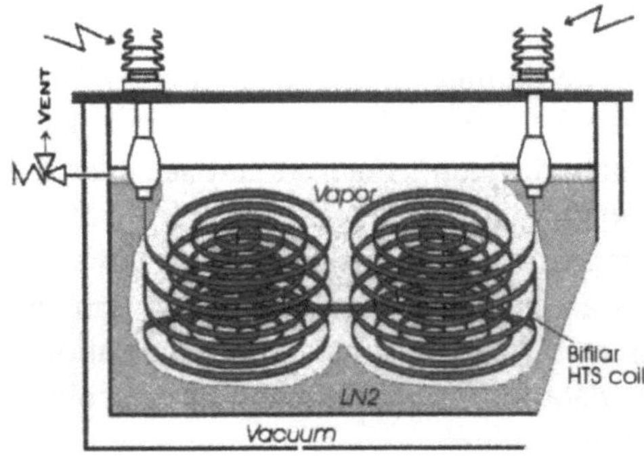

Figure 9b: Activated FCL with bifilar coils

Another popular failure which needs careful attention is a magnet quench. The quenched LTS heats up unless the current is forced to decrease; the large stored magnetic energy requiring an external dump equipment. The dump scheme is sketched in figure 10. The hot spot temperature should not exceed 100 K in order to not risk uncontrolled thermomechanical stress within the winding; this condition determines the time constant which may range from seconds to minutes[4, 5]. The internal voltage distribution cannot be determined from the terminal voltage, and the position of the quench can be somewhere[5].

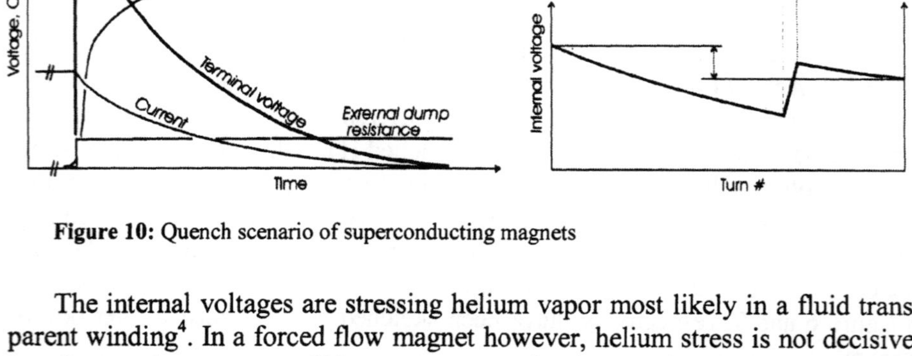

Figure 10: Quench scenario of superconducting magnets

The internal voltages are stressing helium vapor most likely in a fluid transparent winding[4]. In a forced flow magnet however, helium stress is not decisive, but abnormal vacuum conditions may occur simultaneously. Heating up of cold surfaces where frozen gases having leaked in from atmosphere previously, and even more dangerous helium leaking via heated joints into the vacuum space can become very dangerous: the Paschen curve is crossed from high vacuum via semi-vacuum to the Paschen minimum conditions[10]; the withstand voltage decreasing then significantly [5]. Large discharge paths have to be evaded. Sophisticated means, e.g. careful control of electrical potential and a Paschen – minimum related design are required[10]; electron absorbing insulating surfaces in critical regions or beads may also be helpful[2].

4. EXPERIENCE SURVEY

There are only few but nevertheless significant experiences with cryogenic insulation systems. Failures are best to learn from. The former LTS cables for instance performed very satisfying in the wrapped tape insulation but failed in the terminations at elevated voltage tests, for various reasons[3]. This illustrates clearly that the design of gas insulated transition regions was not fully adequate. Care must thus be recommended for new HTS cable design. Unsatisfying performance in LTS magnets also has been reported, especially in case of vacuum leaks[2].

However, real long term experience is not yet available. Extrapolation to 30 years power system lifetimes are based on various model experiments[2, 9, 25, 26].

5. SUMMARY

Cryogenic fluids can be characterized by a complete lack of hazard risk thanks to the chemical inactivity, a lack of possible environment pollution even in case of malfunction, a lack of thermal aging with option of overload capacity, and no disposal costs after end of life. The sound dielectrical performance of the liquids has been proven for many years; vacuum also has shown a quite reasonable characteristics but careful control is essential. Abnormal conditions however may require an alternative design based onto vapor and gas instead of liquids. Vacuum leaks in forced - flow cooled magnets need special care in order to prevent long distance discharges.

6. REFERENCES

1. W. G. Garlick, Power system applications of high temperature superconductors, *Cryogenics* **37**, 649-652 (1997)
2. Gerhold, High voltage in superconducting power equipment – prospects and limits, *Adv. Cryog. Eng.* **46**, 1-20 (2000)
3. *Handbook of Applied Superconductivity,* edited by B. Seeber (IOP Publishing, Bristol, 1998)
4. T. Satow, Specifications and performance experience of pool boiling cooled magnets, *Cryogenics* **38**, 1145-153 (1998)
5. M. Hara and J. Gerhold, Electrical insulation specification and design method for supercon-

ducting power equipment, *Cryogenics* **38**, 1053-1061 (1998)

6. S. W. Schwenterly, High-voltage testing of superconducting power apparatus, *Cryogenics* **38**, 1115-11122 (1998)
7. J. Gerhold, Properties of cryogenic insulants, *Cryogenics* **38**, 1063-1083 (1998)
8. K. Miyaike, T. Kitajima and T. Itoh, Development of 70 MW class superconducting generator with quick response excitation, *Cryogenics* **42**, 183-189 (2002)
9. CIGRE Task Force 15.00.01, Electrical insulation in HTS power cables, fault-current limiters and transformers, *ELECTRA* No. **186**, 11-29 (1999)
10. K. Bauer, S. Fink, G. Friesinger, A. Ulbricht and F. Wüchner, The electrical insulation of a forced flow cooled superconducting (sc) magnet, *Cryogenics* **38**, 1123-1134 (1998)
11. M. Hara and H. Okubo, Electrical insulation characteristics of superconducting power apparatus, *Cryogenics* **38**, 1115-11122 (1998)
12. CIGRE Working Group 15.03, Breakdown of gases in uniform fields, Paris, France (1977)
13. H. Fujita, T. Kuono, Y. Noguchi and S. Ueguri, Breakdown voltages of gaseous N_2 and air from normal to cryogenic temperatures, *Cryogenics* **18**, 195-200 (1978)
14. J.Gerhold, M.Hubmann, E.Telser, Near Paschen – minimum discharge phenomena in helium gas, *Proc. 12th Intern. Conf. on Gas Discharges and their Applications* Vol. I, Greifswald, Germany, 252-255 (1997)
15. J. Gerhold, Dielectric strength of gaseous and liquid insulants at low temperatures, *CIGRE-Symposium*, Vienna, paper 100-01 (1987)
16. J. Gerhold, Potential of cryogenic liquids for future power equipment insulation in the medium high voltage range, *IEEE Transactions on Dielectrics and Electrical Insulation* **9**, 878-890 (2002)
17. C. Sumereder, Dielektrische Untersuchungen an Tieftemperatursystemen, *Doctorate Thesis, Technical University Graz* (2003)
18. R.J. Meats, Pressurized-helium breakdown at very low temperatures, *Proc. IEE* **119**, 760-766 (1972)
19. M. Hara, J. Suehiro and H. Matsumoto, Breakdown characteristics of cryogen gaseous nitrogen and estimation of its electrical insulation properties, *IEEE Transactions on Electrical Insulation* **24**, 609-617 (1989)
20. H. Toyota, S. Zama, Y. Akkamine, S. Matsuoka and K. Hidaka, Gaseous electrical discharge characteristics in air and nitrogen at cryogenic temperature, *IEEE Transactions on Dielectrics and Electrical Insulation* **9**, 891-898 (2002)
21. M. Hara, Electrical insulations in superconducting apparatus, *Cryog. Eng. Japan* **24**, 72-81 (1989); see also reference 7
22. H. Okubo, H. Goshima, N. Hayakawa and M. Nikita, High voltage insulation of superconducting power apparatus, *Proc. 9th ISH, Graz, Austria*, 9007/1-9007/12 (1995)
23. M. Hara, H. Koishihara and K. Saita, Breakdown behavior of cryogenic liquids in the presence of thermal bubbles under ramped voltage, *IEEE Transactions on Electrical Insulation* **26**, 685-691 (1991)
24. R. Shively, Submerged cryogenic motor materials development, *IEEE Electrical Insulation Magazine* 19, 7-11 (2003)
25. A. Bulinski and J. Densley, High voltage insulation for power cables utilizing high temperature superconductivity, *IEEE Electrical Insulation Magazine* **15**, 14-22 (1999)
26. H. Okubo, M. Hazeyama, N. Hayakawa, S. Honjo and T. Masuda, V-t characteristics of partial discharge inception in liquid nitrogen/PPLP® composite insulation system for HTS cable, , *IEEE Transactions on Electrical Insulation* **9**, 945-951 (2002)

DIELECTRIC AND LIFE CYCLE BEHAVIOUR OF A CRYOGENIC INSULATION SYSTEM

C. Sumereder, J. Gerhold, M. Muhr*

1. INTRODUCTION

Since the discovery of high temperature superconductors (HTS) in 1986 several applications for power equipment were developed. The critical temperature of the used ceramic HTS (up to 110 K) lays over the boiling point of liquid nitrogen (77 K), which is used for cooling down to the operating temperature and as insulating medium. Because of this different specifications compared to conventional insulation systems the dielectric and life cycle behaviour of a cryogenic insulation system was analysed [1].

In the test arrangement a cryogenic insulation system was constructed, where a combination of solid, liquid and gaseous materials were used. As representative for gaseous and liquid insulation medium Nitrogen was applied and for solid a polymer material. The solid medium has to show excellent electrical, mechanical and chemical properties at cryogenic temperatures. For this test arrangement PET was used because it has the same structure as Mylar, which is an often used polymer for cryogenic insulation systems.

In the first test series the breakdown voltage was measured and analysed with statistical methods [2]. To minimize the standard deviation of the breakdown voltages the N_2 gas was filled in the system between the high voltage electrode and the solid insulation medium, which was placed within the grounded electrode. The test voltage was an AC step voltage with 5 kV steps up to the breakdown.

* C. Sumereder, M. Muhr, Graz University of Technology, Austria, Institute of High Voltage Engineering and System Management, Inffeldgasse 18, A-8010 Graz, surname@hspt.tu-graz.ac.at, J. Gerhold, Graz University of Technology, Austria, Institute of Electrical Machines and Devices, Steyrergasse 21, A-8010 Graz, gerhold@ema.tu-graz.ac.at

Gaseous Dielectrics X.
Edited by Christophorou *et al.*, Springer, New York, 2004.

The statistic analysis of the measuring results showed a two parametric weibull distribution over this test arrangement.

In the second test series a life time curve of this insulation system should be determined. For this reason the thermal long time stability was very important. This demand could be fulfilled by measuring in a thermal insulated chamber with constant climate. The life time curve should be constructed by applying characteristic loads on the system.

2. LIFE TIME OF INSULATION SYSTEMS

The life time of insulation systems is dependant of load, which can be described by its quality and quantity. The quality can be expressed by the aging factors due to electrical, mechanical, thermal, chemical or their individual combination as multi stress load. The stress quantity can be described as critical magnitude for electrical field, mechanical stress, temperature etc. The aging behaviour of technical equipment can be characterized in the bath tube curve which is illustrated in **Figure 1.**

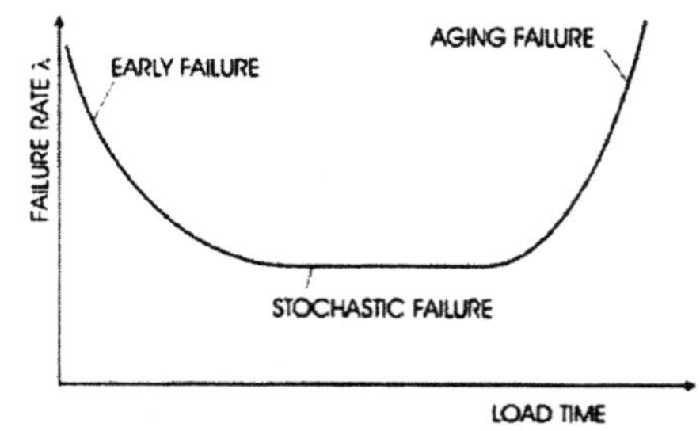

Figure 1. Bath tube curve of technical equipment

Looking at the aging behaviour of insulation systems the early failures and stochastic failures are not of interest. These failure types can be easily excluded taking a look to the Weibullexponent of the breakdown distribution. An exponent of 1 describes stochastic and an exponent smaller than 1 the early failures.

The aging behaviour can generally be expressed in the form of the exponential equation, which is linear in doubled logarithmic description. For electrical stress the aging behaviour can be described in the mathematic form of equation (1), with E... electric field, c_D ... constant, t_D ... load time, r ... life time exponent.

$$E = c_D * t_D^{-1/r} \qquad (1)$$

3. TEST VESSEL AND MEASURING CIRCUIT

The life time tests were done in a cryogenic test vessel with cylinder concentric electrodes. The test object was a polymer insulating material, PET. PET has a very similar structure to Mylar, which is an often used insulation material for cryogenic insulation systems. The test vessel was installed in a laboratory where constant thermal and humid conditions were guaranteed, that there was no influence to the quality of LN_2. This was very important because former investigations showed a big influence to breakdown behaviour in dependence of ice particles in LN_2.

The test vessel and test circuit is shown in **Figure 2.** The test voltage was generated with an AC transformer and applied to the centre electrode. The dielectrical behaviour was observed with partial discharge measuring system, the applied voltage and load time was recorded with PC system. Total three measuring series with over 90 test objects were done.

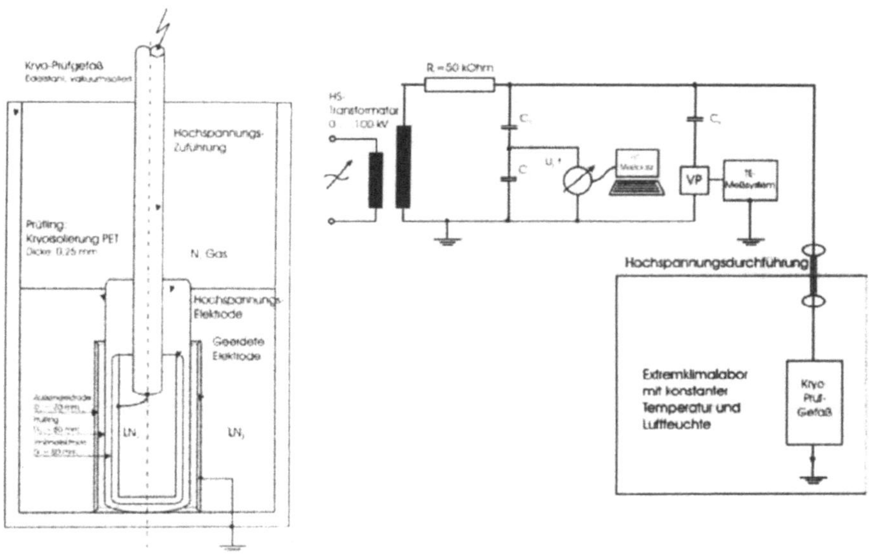

Figure 2. Cryogenic test vessel with electrical insulation system in cylinder concentric arrangement and measuring circuit for life time tests

4. RESULTS

Starting with voltage rise test the kind of breakdown distribution and the load parameter for constant voltage test could be determined. Evaluating the measuring results with the two parametric Weibull distribution the exponents were greater than 1, which confirmed the correct selection of the test voltages. The life time curve for the polymer insulation system could be constructed from the Weibull diagram. In **Figure 3** the evaluated test results and the life time behaviour is shown. [3]

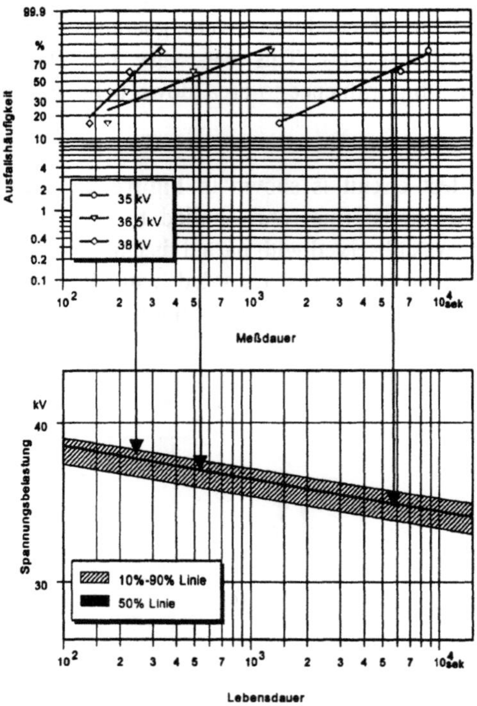

Figure 3. Test results of breakdown tests and life time curve of a cryogenic polymer insulation system.

The parameters of the life time curve were determined and can be expressed as equation (2):

$$U = 43{,}0 * t_D^{-1/40} \quad (2)$$

5. CONCLUSIONS

Comparing the life time exponent of this system with known values of PPLP, which is used as LN_2 impregnated insulant for cryogenic cables, the exponent is in the expected area. Life time exponents of cryogenic systems are higher because thermal aging can be excluded [4]. This means an important advantage to conventional insulation systems, because the life time exponent is more than doubled. The higher exponent of impregnated systems can be explained because of the trend to self healing effect. In the case of partial discharge the liquid medium fills the void [5]. The self healing effect can be influenced by increasing the pressure of the insulating liquid.

6. REFERENCES

[1] J. Gerhold: Potential of Cryogenic Liquids for Future Power Equipment Insulation in the Medium High Voltage Range, IEEE Transactions on Dielectrics and Electrical Insulation, Vol. 9, No. 6, December 2002

[2] L. A. Dissado, J. C. Fothergill, S. V. Wolfe, R. M. Hill: Weibull Statistics in Dielectric Breakdown Theoretical Basis, Applications and Implications, IEEE Transactions on Electrical Insulation Vol. EI-19 No. 3, June 1984, p. 227

[3] C. Sumereder "Dielektrische Untersuchungen an Tieftemperaturisolationssystemen", Doctoral Thesis, TU-Graz 2003

[4] G.C. Montanari: Absolute vs. Comparative End-of-Life Age, IEEE Transactions on Dielectrics and Electrical Insulation, Vol. 4, No. 2, April 1997

[5] R. Shively: Submerged Cryogenic Motor Materials Development, IEEE Electrical Insulation Magazine, Volume 19, Issue 3, May-June 2003, p. 7-11

GRID, SECURITY AND OPEN MARKET

George C. Contaxis*

1. INTRODUCTION

Power systems have been traditionally developed so that they guarantee three necessary fundamental conditions for the indispensable for the survival of modern society public good: uninterruptible service, sufficiency of supply, and affordable cost of electric energy. Among national grids of electric energy interconnections have been developed in order to provide mutual assistance to the interconnected networks but also to allow the trade of electric energy.

In the past, the Electricity Supply Industry (ESI) was dominated by vertically integrated electricity companies (production, transmission, distribution) operating at National or Regional level in a monopolistic status. In recent years restructuring of ESI takes place. Electricity now is considered as a "commodity", which is traded in a market. In this paper security issues related to the transmission before and after deregulation in ESI are presented. The contribution of deregulation to security is discussed and preventive measures are presented.

2. SECURITY ISSUES BEFORE DEREGULATION

Transmission system is the set of power devices that accommodate the transfer of bulk quantities of electric power from the generation sites to the loads. Due to the temporal and spatial distribution of both the generating units and the loads, transmission systems play a key-role in security of supply and influence significantly the Electricity Supply Industry (ESI). It comprises of Over-Head transmission Lines (OHL), underground or submarine cables, substations (i.e. subsystems equipped with transformers and autotransformers that allow the power transfer between systems operating at different voltage levels) and reactive power compensation devices (capacities and reactors). The main characteristics of transmission systems (compared to the distribution systems) is that they operate at high voltage levels (over 150 kV up to 1000kV) and their structure

* George C. Contaxis, National Technical University of Athens, Greece

Gaseous Dielectrics X.
Edited by Christophorou *et al.*, Springer, New York, 2004.

involves closed loops that allow higher reliability in transferring the electric power to the loads from the generators.

The operational requirements set by the transmission system are:

- **Security:** a transmission system is subject to random events (lightning, short-circuits, accidents etc) that may cause the loss of transmission lines or cables. In case of such contingencies the remaining system must be able to operate safely and it should not be led to major disturbances (i.e. brown-outs or even complete black-outs).
- **Smooth operation:** the electric power characteristics (voltage, frequency, power quality, etc.) should be kept within predefined limits.
- **Reliability:** the system should be able to provide the customers with electricity continuously even if some devices are out of operation due to outages or scheduled maintenance. Any interruptions cause inconvenience to the customers and curtailed revenues for the producers. Furthermore, restoration of service is always costly.
- **Minimum environmental impact**
- **Minimum cost:** Transmission services should be provided at minimum cost

The continuous increase of electricity demand requires the expansion and reinforcement of the existing transmission systems in order to accommodate the increasing transmission needs. Expansion of transmission systems includes the installation of new devices (mainly OHL and cables) or the reinforcement of existing transmission corridors and the upgrade of existing substations.

In monopolistic regimes, transmission expansion is a major issue and it is performed in long-term time horizon following the Least Cost Planning principles. Since transmission system influences (sometimes dramatically) the operation and the economics of the rest ESI subsystems (mainly the generators but the distribution as well), the transmission planning optimization is closely related to generation expansion planning. It is quite often that optimization of transmission expansion is performed jointly with the generation expansion.

Interconnections between European networks have been developed since 1950, under the aegis of the UCTE, aiming at the mutual assistance and electricity trading between the European countries.

3. SECURITY ISSUES AFTER DEREGULATION

Different market structures and network open access schemes have been proposed in order to materialize the restructuring of ESI. In the transmission level, new issues are under consideration such as access rules, transmission rights, transmission pricing, payment mechanisms etc. For the above issues there are not yet clear answers and practices that can be applied universally.

Under the new environment transmission is a common carrier that serves all the market participants and it is considered as a natural monopoly. Since transmission system operation may influence the power transfers from the generators to the loads and consequently the operation of the markets, its operation has to be transparent and non-discriminatory.

To guarantee transparency and avoid any discrimination, the European model requires the designation of an independent national Transmission System Operator (TSO) whose main tasks are: the operation, maintenance and expansion of the transmission network (without necessarily owning the network), cooperation and exchange of information with operators of interconnected systems ensuring a reliable and secure operation of the network, dispatching loads to the generating units, management of interconnections, provision of ancillary services and clearance of the market in a transparent and unbiased manner. The above tasks are performed under specific legislation (Grid Codes).

Furthermore, the association ETSO (European Association of Transmission System Operators) was set up. The objective of ETSO is the establishment of closer cooperation between national network operators regarding the trade of electricity, while ensuring the quality and reliability of the electricity supplied, as well as the development of a system for exchange of information on cross-boarder electricity flows that will enable early detection of possible network overloads. The national network operators manage not only the flows of electricity but also the transmission infrastructure. In most EU countries (except Greece and Italy) the network operators are at the same time transmission asset owners, which theoretically facilitates the expansion of the networks.

The sufficiency of infrastructure is one of the main tasks of the national operators. The operator is responsible for the expansion of the transmission networks and the investment costs are financed by transmission charges. In March 2002 EC approved the objective of a level of electricity interconnection equivalent to at least 10% of installed generation capacity for all Member States.

In deregulated electricity markets, since the behavior of the market participants is in general unknown, the transmission system optimum expansion is performed independently of the expansion planning of the rest ESI subsystems, under high uncertainties (concerning the new generation sites and future generators, power exchanges, etc). It should be noticed that:

- Investments in transmission systems are usually not profitable; they require high capital cost while the pay-back period is usually significant (up to 40 years)
- The realization of new transmission projects requires considerably long time (possibly several years). This is mainly due to the licensing procedures and the expropriations needed, but also due to the increasing protest by the public to the erection of the projects (especially in developed countries).

As a consequence of the electricity industry deregulation, regional electricity markets are developing and the bulk transmission systems are being subjected to flows in magnitudes and directions that were not contemplated when they were designed and for which there is minimal operating experience. In some areas, transmission systems are reaching their limits as the systems are subjected to new loading patterns resulting from increased electricity transfers and customer demand increases. If national regulatory frameworks regarding investments in the generation sector are not clarified, more countries will rely on imports instead of their own infrastructure, resulting in the concentration of generation in a few areas and in long-distance bulk energy transits, which would be detrimental to the reliability of the European grids.

4. BLACKOUTS AND DEREGULATION

Although there is a long term experience related to the design and operation of power systems and although power systems utilize modern control centers equipped with high tech communication and computer facilities, unforeseen situations lead to interruptions in the supply of electric energy, which in certain cases are widespread and long-lasting. Recently, such blackouts have appeared in the USA, Canada and Europe:

- **June 26, 2003**: Widespread interruption of electricity took place in Italy, during the peak hours, due to excessive increase in demand in conjunction with the reduction of the generating capacity of large units, caused by the deficit of available hydro for cooling purposes.
- **August 14, 2003**: Total interruption of electricity took place in all northwestern regions of USA and Canada. The region includes cities such as New York, Detroit, Boston, Cleveland, Ottawa. There were indications that the worst blackout in American history was caused by fault in three transmission lines; others concluded that it was caused by cascading events.
- **August 28, 2003**: For five hours a blackout affected one fifth of London and this blackout was characterized as the worst blackout of the last decade. The blackout was attributed to erroneous selection of a fuse.
- **September 23, 2003**: A blackout, the worst recorded in Sweden since 1983, in South Sweden and Eastern Denmark affected 5 million inhabitants.
- **September 28, 2003**: A general blackout affected almost all Italy. The blackout was attributed to faults in the grids of France and Switzerland.

Most of the above blackouts are related to the transmission system and they gave the opportunity to opponents of electricity market deregulation to strength their argument that deregulation can not guarantee the supply of electric energy as efficiently as a public electricity company. They argue that the main cause of the problem is the lack of interest in the issues of security of supply and reliability of power systems. However, the easy conclusion that liberalization is the main cause of blackouts is rather misleading. Referring, for example, to the Italian blackout, the current structure of the grid is the result of decisions taken under the previous state monopoly. In other cases (e.g. California), the blackouts were due to the imperfect implementation of the liberalization in the corresponding electricity markets.

Electricity grids are highly interconnected and interdependent. What happens in one point of the network affects areas very far away. Given this complexity, the electricity system requires carefully designed and consistent rules regarding the use of the transmission grid.

5. PREVENTIVE MEASURES

The negative consequences of blackouts indicate the need for preventive measures. These measures include construction of new power plants, construction of new interconnections with neighboring power systems, improvement of the information exchange between network operators, implementation of demand-side management techniques.

The continuous increase in electricity demand has to be balanced by a corresponding increase of the installed capacity. Unfortunately, despite the liberalization of the energy sector in most of the EU countries, there is lack of interest in investing in new capacity. Consequently, new regulatory frameworks and initiatives for independent generators are needed in order to enable the construction of new generation capacity. Moreover, priority has to be given to projects concerning cogeneration units and renewable energy sources.

An additional measure is the construction of new interconnections between the neighboring systems. Towards this effect, the European Council approved the target of 10% electricity interconnection capacity for Member States by 2005.

Besides the construction of new transmission lines, a very important issue is the improvement of the information exchange between network operators. The correct and prompt information enables potential network overloads to be detected in advance and appropriate measures to be taken in order to avoid incidents on the network. Consequently, the development of a system for the exchange of information between the network operators is a crucial issue, which should be addressed in the near future.

In addition to the above, the reinforcement of the reliability of power systems is a very significant factor for the prevention of blackouts. An adequate amount of spinning reserve should be guaranteed, so that the system is capable of handling contingencies. Furthermore, the promotion of decentralized energy production is also required, in order to minimize stability risks and thus to insure security of supply. The reliability of the power systems can be further improved by using voltage stabilizers and controllers such as static VAR compensators

(SVC), flexible AC transmission systems (FACTS), and novel methods of isolating parts of the grid to restrain instabilities.

The problems caused by the continuous increase of demand for electricity can be also confronted by implementing energy saving measures. Consequently, initiatives on saving off-peak electricity by demand management must be adopted, in order to improve the end-use efficiency.

In addition to the previous measures, the attraction of better skilled engineers of power systems should be taken into account. This goal can be achieved by improving the salaries of such engineers, an issue which should be addressed by the corresponding authorities and power industries. In any case, this expense is justified by considering the huge economic cost of a potential blackout.

The above measures can particularly assist in decreasing the probability of future blackouts in power systems. However, it is obvious that the successful implementation of these measures is strongly related with the immediate modernization of the regulatory framework of the liberalized energy markets.

6. CONCLUSIONS

In this paper security issues related to the transmission system before and after the restructuring of the electricity market were discussed. Emphasis was given to the contribution of the grid to blackouts, as well as the impact of the liberalization of the electricity market on power supply security. The conclusions of this paper can be summarized as follows:

- Recent blackouts cannot be attributed to the liberalization of the electricity markets.
- Regulatory frameworks regarding investments in the generation and transmission sectors should be clarified.
- The reliable operation of a power system should not be based on the existence of interconnections, but on sufficient domestic generating capacity.
- Despite the scheduled transmission expansion plans (TSO and ETSO), the construction of the necessary new transmission lines is quite difficult, due to difficulty in financing the projects, as well as public protest.
- Preventive measures are necessary in order to decrease the probability of blackout occurrence.

7. REFERENCES

1. Mrs de Palacio, Information Memo to the Commission, 'Initial assessment of the power cut in the United States and Canada on 14th August 2003. Comparison of the American and European Systems', August 2003
2. UCTE, 'Interim Report of the Investigation Committee on the 28 September 2003 Blackout in Italy', October 2003

SECTION 11: INDUSTRIAL OUTLOOK: REDUCTION IN SF_6 USE AND EMISSIONS

EPA's SF_6 EMISSION REDUCTION PARTNERSHIP: MAXIMIZING THE BENEFITS OF SF_6 EMISSION REDUCTIONS FOR ELECTRIC UTILITIES

Jerome Blackman[1] and Ravi Kantamaneni[2]

1. INTRODUCTION

First manufactured in 1902, sulfur hexafluoride (SF_6) is a fluorinated compound with an extremely stable molecular structure. This structure makes it far superior to oil or air in its insulation properties, dielectric strength, and arc quenching abilities. Electric equipment manufacturers and utilities therefore rely heavily on SF_6 in transmission systems. In 2001, electric utilities and electrical equipment manufacturers (OEMs), taken together, purchased over 80 percent of all SF_6 produced worldwide (1). Under ideal conditions, SF_6 would remain contained within transmission equipment. In reality, however, SF_6 is inadvertently emitted into the atmosphere as leaks develop during various stages of the equipment's lifecycle. SF_6 can also be accidentally released during equipment installation, servicing, or de-commissioning.

SF_6 has been identified by the Intergovernmental Panel on Climate Change (IPCC) as a highly potent greenhouse gas that contributes to climate change. According to the IPCC (2), SF_6 is 22,200 times more effective at trapping infrared radiation than an equivalent amount of carbon dioxide (CO_2) over a 100-year period. Additionally, it has an atmospheric life of 3,200 years; its accumulation in the atmosphere is virtually irreversible. Measurements taken from 1978 through 1996 indicate SF_6 concentrations are increasing in the atmosphere at a rate of 7 percent per year (3). In 2002, SF_6 emissions from the U.S. electric power industry totaled 589 metric tons. From a greenhouse gas

[1] U.S. Environmental Protection Agency, Climate Protection Partnerships Division, 1200 Pennsylvania Avenue, N.W, Washington, DC 20001, USA.

[2] ICF Consulting, 1725 Eye St., Suite 1000, Washington, DC 20006, USA.

Gaseous Dielectrics X.
Edited by Christophorou *et al.*, Springer, New York, 2004.

perspective, that equates to 14.1 million metric tons of CO_2, or 5 percent of total CO_2 and non-CO_2 greenhouse emissions from U.S. industrial processes (4). Although SF_6 is emitted in smaller quantities than other greenhouse gases, it has a significant long-term impact on global climate change. Consequently, the need to minimize SF_6 releases and avoid its accumulation in the atmosphere is important.

2. SF_6 EMISSIONS REDUCTION PARTNERSHIP FOR ELECTRIC POWER SYSTEMS – SF_6 Emission Reduction Activities

The U.S. Environmental Protection Agency (EPA) launched the SF_6 Emission Reduction Partnership with the electric power industry in 1999. Currently, more than 70 companies – representing all segments of the U.S. electric utility industry, have voluntarily committed to identify, implement, and report the results of their emission reduction activities. The Partnership provides a forum for the industry and EPA to share information of ways to reduce SF_6 emissions to technically and economically feasible levels.

The efficiency of SF_6 management can be improved through cost-effective operational improvements and equipment upgrades. These opportunities fall into four categories: SF_6 inventory tracking systems, SF_6 recycling, improved leak detection technology, and SF_6 management training programs to improve operation and maintenance practices.

2.1 SF_6 Inventory Tracking Systems

Each year, SF_6 Partners prepare and submit reports to EPA that detail their annual SF_6 emissions and emission reduction achievements. SF_6 emissions estimates are prepared using a mass-balance approach to systematically track and account for all company uses of SF_6 during the reporting year. SF_6 use accounted for in this approach includes: cylinders and equipment purchased during the reporting year, SF_6 sent off-site to destruction facilities or for recycling, SF_6 returned to suppliers, and SF_6 sold to other entities. Using this method, estimates are developed based on the assumption that any SF_6 gas that cannot be accounted for is emitted into the atmosphere.

SF_6 Partners are asked to use a standard protocol to weigh, leak check and inventory their SF_6 gas cylinders. This process enables companies to track gas purchases and the rate of gas use, thereby, ensuring SF_6 inventories are not overstocked. Since SF_6 cylinders are typically rented in the U.S., this approach also enables a utility to reduce their annual cylinder rental fee charge. Another important aspect of inventory tracking involves verification of the remaining gas or "heel" in returned cylinders. The term "heel" is used to describe the amount of gas that normally remains in the cylinder after use. It can account for as much as 10 percent of the original cylinder gas mass (e.g., for a standard 52 kilograms (kg) SF_6 cylinder, the remaining unused gas could be roughly 5 kg). In the past, utilities typically returned cylinders to the vendor without accounting for the heel; consequently, paying for the full reported mass of the cylinder. By weighing cylinders before return, many SF_6 Partner utilities have been able to receive credit for the remaining portion of gas. The SF_6 inventory tracking system also facilitates identification of specific SF_6-containing equipment requiring frequent refilling. Since increased SF_6 consumption (loss) may indicate mechanical or structural problems, leading to potential equipment failure, SF_6 Partners have used their tracking systems to schedule maintenance and equipment replacement activities before problems occur, thus saving money and reducing unplanned system disruptions.

2.2 SF_6 Recycling

In the past during equipment servicing or replacement activities, SF_6 gas was typically vented to atmosphere. Now, however, most companies use SF_6 gas recycling carts to remove, store, clean, and re-fill the SF_6 gas to the gas-insulated equipment. Most recycling systems provide automatic gas purification/cleansing during removal and re-filling operations. Thus, recycling their SF_6 gas has enabled Partners to reduce their SF_6 gas purchase requirements. Additionally, many gas carts enable the equipment gas compartment to be flushed and evacuated before re-filling to ensure the removal of moisture, which reduces the likelihood of corrosion and thus increases electrical equipment lifetime.

2.3 Improved Leak Detection Technology

Utilities traditionally relied on soap and water solutions or "gas sniffers" to locate SF_6 leaks. This process involves de-energizing the equipment, to enable the leak inspector to get close to the equipment in order to either spray liquid soap on it, or use gas “sniffer” detection devices to detect the presence of SF_6. Both techniques are labor intensive, often requiring extensive climbing and reaching by workers, which is time-consuming, costly, and potentially hazardous. New laser leak detection systems exploit the strong infrared absorption characteristics of SF_6. Not only do these systems enable the operator to identify leaks as small as 1 kilogram per year at distances greater than 30 meters in “real-time” (6), but they provide the ability to conduct inspection on “live” equipment. Consequently, Partners are detecting minor leaks without taking equipment out of service or incurring large downtime costs, and benefit from a significant reduction in the time required to detect leaks.

2.4 SF_6 Management Training Programs

Critical to any successful emission reduction strategy is training to improve employee SF_6 handling. Effective safety and maintenance procedures, leak detection procedures, gas cart operation, gas quality testing, and arc by-product detection are all dependent on worker expertise and diligence. By implementing additional training sessions that enhance employee understanding of SF_6 gas issues, Partners have increased employee efficiency in monitoring SF_6 loss. They have also become proactive in using monitoring data to prioritize and schedule equipment maintenance and replacement.

3. PARTNER ACCOMPLISHMENTS

Since the SF_6 Emissions Reduction Partnership's inception in 1999 through 2002, cumulative reductions of nearly 132,000 kg of SF_6 (or over 3 million metric tonnes of carbon dioxide equivalent ($MMTCO_2e$)) have been achieved. This translates into an overall 11 percent reduction of SF_6 emissions from 1999-baseline levels. Table 1 provides a summary of U.S. industry SF_6 emission reductions between 1999 and 2002. Additionally, during this timeframe, the average Partnership SF_6 emission rate (or loss rate), which is defined as the total Partnership emissions divided by the total nameplate capacity of SF_6-containing equipment held by Partners, has decreased from 17 percent in 1999 to 11 percent in 2002. This accomplishment illustrates the increasing efficiency of Partner operations through the reduction in the quantity of SF_6 gas lost from equipment leakage or gas lost (or vented) during general cylinder handling and equipment maintenance activities.

Table 1. Summary of SF_6 emissions reductions achieved by Partners between 1999 and 2002.

Reporting Year	1999[a]	2000	2001	2002
SF_6 Nameplate Capacity (lbs. – millions)	3.4	3.8	3.9	4.0
Total SF6 Emissions (lbs. - thousands)	590	580	550	480
Total SF_6 Emissions ($MMTCO_2e$)	15.77	15.18	14.90	14.08
Emission Reduction from Baseline	—	4%	6%	11%

[a]Baseline Year

With the cost of SF_6 ranging from US$13 to US$18 per kilogram, SF_6 emissions reductions through 2002 equate to a financial benefit ranging between US$1.8 to US$2.4 million dollars (approximately €1.4 to €1.9 million). The environmental benefit of this reduction is equivalent to eliminating the emissions from over 590,000 cars or planting more than 10 million trees. (7)

3.1 One US Utilities' Success Story

The following information was offered by one of EPA's SF_6 Partners. This company has a service territory of 70,000 square miles and operates transmission equipment bearing over 200,000 pounds (nameplate capacity) of SF_6. This utility's effort to reduce SF_6 emissions resulted in discovering that significantly more equipment was leaking than previously thought. The use of advanced leak detection technology paid for itself though the cost savings gained from reduced leaks. This company implemented policies and procedures that initially cost $100,000 but resulted in savings in avoided gas purchases of $400,000; yielding a net savings of $300,000 (€237,300). The company now also recycles 90% of its SF_6 gas from decommissioned equipment.

Using a 1998 emissions baseline, this company reported a 50 percent reduction in annual SF_6 emissions in 2002. This success is based on a two-pronged approach aimed at enhancing both the company's SF_6 management system and facility-level maintenance procedures. The company developed a new SF_6 handling protocol and implemented training programs to educate employees on better leak detection practices, and SF_6 handling procedures, such as those required when evacuating SF_6 from circuit breakers, or transferring SF_6 from cylinders. This training has helped staff identify significant leaks on equipment, which was previously considered gas-tight.

With respect to the company's SF_6 management system, new measures were implemented to control purchases of SF_6 by improving inventory control and by reducing cylinder rental charges for late cylinder returns. The SF_6 tracking system has also provided a means to track equipment re-filling operations and coordinate leak detection activities more efficiently.

4. CONCLUSION

SF_6 is the most potent greenhouse gas known. The electric power industry is the largest user and emitter of this gas. Emissions will continue to be scrutinized as greenhouse gas mitigation options are explored. While there is currently no "silver bullet" replacement for this chemical, information from the SF_6 Emission Reduction Partnership shows that companies can achieve significant SF_6 emission reductions now, using cost-effective technologies and practices that provide additional financial benefits.

5. REFERENCES

1. Smythe, K., Production and Distribution of SF_6 by End-Use Application. International Conference on SF_6 and the Environment: Emission Reduction Strategies. San Diego, CA, November 21-22, 2002.

2. IPCC, Climate Change 2001: The Scientific Basis. Intergovernmental Panel on Climate Change, Cambridge University Press. Cambridge, U.K., 2001.
3. Maiss, M., Brenninkmeijer, C., Atmospheric SF_6, Trends, Sources and Prospects, Environmental Science & Technology, **32**:3077-3086, 1998.
4. U.S. EPA, Inventory of U.S. Greenhouse Gas Emissions and Sinks: 1990-2002, U.S. Environmental Protection Agency, Office of Atmospheric Programs, 2003. Draft.
5. van der Zel, L., EPRI SF_6 Research: Past Highlights and Future Plans. International Conference on SF_6 and the Environment: Emission Reduction Strategies. San Diego, CA, November 21-22, 2002.
6. FHWA, 2002. Federal Highway Administration, Highway Statistics, 2000
7. American Forest. 2002. Climate Change Calculator. Available at *http://www.americanforests.org/resources/ccc/index.php*
8. U.S. EPA, Company Profile – U.S. Environmental Protection Agency, Office of Atmospheric Programs, 2004. Draft.

ENVIRONMENTAL COMPATIBLE USE OF SF_6

Silvio Stangherlin*

1. ABSTRACT

SF_6 is a strongly electronegative gas (i.e.: it tends to attract free electrons). It has a unique combination of physical properties: high dielectric strength (about 3 times air), high thermal interruption capabilities (about 10 times air), high heat transfer performance (about twice air).

For that reason, since early 1960's, SF_6 has been successfully used by the Electric Industry in power equipment for the HV transmission and MV distribution of electricity (gas insulated substations, ring main units, circuit breakers, transformers, cables, etc.).

The Kyoto Protocol, in 1995, listed SF_6 as a potent and strongly persistent greenhouse gas, affecting the global warming of the planet.

In spite of many attempts, a green SF_6 substitute with reasonable dielectric and / or interruption capabilities, suitable for industrial applications, is still unknown. As a consequence of that, an environmental compatible use of the gas is fundamental to minimise the environmental impact of HV and MV electric power equipment.

SF_6 must be handled in a close cycle, to avoid any deliberate release in the environment. Among all voluntary initiatives, gas recovering and recycling have top importance. IEC Standards in force are going to be modified accordingly.

Voluntary agreements among manufacturers, users and authorities have been signed in some European Countries and Japan with the aim of controlling the use and emissions of SF_6 from the electric power equipment.

In general, state-of-the-art technologies and procedures shall be applied to minimize SF_6 emissions during development, manufacturing, installation, operation and maintenance of electric power equipment.

---

* Silvio Stangherlin, ABB Switzerland Ltd., Corporate Research, Segelhof 1, Baden – Daettwil, Switzerland CH – 5405.

2. INTRODUCTION

Sulphur hexafluoride, SF_6, is a synthetic molecule obtained by 6 atoms of fluorine gathered around a centrally located atom of sulphur. The chemical bond between fluorine and sulphur is one of the most stable existing atomic bonds (the most stable is carbon-fluorine): six of them grant the molecule very high chemical and thermal stability.

SF_6 is strongly electronegative (i.e.: it tends to attract free electrons). It has a unique combination of physical properties: high dielectric strength (about 3 times air), high thermal interruption capabilities (about 10 times air), and high specific heat transfer ratio (about twice air).

Pure SF_6 is an odourless, tasteless, non-toxic, non-corrosive, non-flammable and chemically inert gas ~5 times heavier than air under normal conditions. Its compatibility with materials used in electric constructions is similar to nitrogen, up to temperatures of ~180 °C.

First SF_6 investigations arise to the first half of the XX century[1]. Since early 1960's, it has been successfully used by the Electric Industry in power equipment for the HV transmission and MV distribution of electricity (gas insulated substations, ring main units, circuit breakers, transformers, cables, etc.). Other non-electric industrial applications include metallurgy, electronics, scientific equipment, ocular surgery, military, etc.

In the middle of the last decade, the Kyoto Protocol[2] listed it as a potent and strongly persistent greenhouse gas, affecting the global warming of the planet.

A systematic search for environmental friendly alternatives to SF_6 started and is still ongoing within the International Scientific Community. For the time being, no green alternative gas exists with adequate dielectric and/or interruption capabilities. Less performing substitutes have been identified and all of them require a complete redesign of the power electric equipment. The operating pressure and the insulating distance must be increased to comply with IEC Standards in force[3-7]. The additional cost, weight and dimensions result in an increase of the overall environmental impact.

Therefore, SF_6 must be used in a closed cycle, any deliberate release in the environment must be avoided, the SF_6 concentration in the atmosphere is monitored and the Electric Industry has commissioned itself to reduce the emission of "electric" SF_6 down to the minimum functional level for the equipment.

State-of-the-art electric power equipment is designed for low emissions. This has been recently recognised in the new revision of the IEC Standards 61276-200[6] and 61276-203[7] for MV and HV equipment, respectively.

The focus has moved now on SF_6 handling. New purity levels have been proposed by the International Council on Large Electric Systems[8] (CIGRE) and are going to be introduced in the 2nd edition of the IEC Standards[9-10] on SF_6. Reuse and recycling of SF_6 are going to become an industrial reality. Discussion is still ongoing at CIGRE concerning best practices to implement in appropriate SF_6 handling procedures. The latter should be adopted in a specific IEC mandatory standard, which is still missing today.

The following paragraphs describe the activity of CIGRE on SF_6 as well as its influence on IEC Standards. Then a case study using the Life Cycle Assessment (LCA) methodology is mentioned in which the proper use of SF_6 minimises the environmental impact in order to carry electricity to a rural area. Last, but not least, conclusions are drawn at the end.

3. ACTIVITY OF CIGRE ON SF_6

The mission of the so-called CIGRE SF_6 task force (Working Group B3-02 – on Gas Insulated Substations – Task Force 01 – on SF_6) is to promote "responsible environmental care" by assessing the environment relevance of SF_6, identifying best practices for SF_6 handling and recycling, and spreading out technical information and knowledge on SF_6.

Many articles and brochures have been published in the past regarding SF_6 and its use in electric power equipment[8, 11-15]. The detailed list of topics, which cover the most important aspects of the SF_6 technology, is summarised below:

- Proper handling of the gas and its by-products[11];
- Impact of the gas on the global atmosphere[12];
- First guide for recycling the gas[13] and its subsequent consolidation in a new revision[8];
- Mixing SF_6 with other gases to minimise the amount of SF_6 used[14];
- The current status of the SF_6 technology[15].

A template for voluntary agreements among SF_6 producers, original manufacturers of electric power equipment utilising SF_6, electric associations and utilities, industrial users and, possibly, Public Authorities and/or Environmental Bodies on the use of SF_6 and on measures for reducing the SF_6 emissions has been prepared and will be published as an IEC Guide in the near future.

During the regular quarterly meetings, the following items are discussed:

- Practical SF_6 handling instructions: a standard guide to compile customised instructions for end-users;
- Tightness of electric power equipment: sensitivity and limits vs. measurement technique, what can be experimentally measured and what can be guaranteed and can be matter for contractual obligations;
- Residual life concepts applied to SF_6 electric power equipment: what are the factors limiting the life duration.

Technical brochures and articles have been accordingly planned and will be published in the next future.

The next paragraphs report the purity requirements for SF_6 reuse recently recommended by CIGRE together with some theoretical considerations concerning best practices for SF_6 handling. Author's opinion regarding both air and SF_6 residual pressures in gas compartments is also expressed.

3.1. Purity requirements for SF_6 reuse

When SF_6 is used in electric power equipment, four different kinds of contaminants are present[8]. They are:

- Inert gases, e.g. CF_4, air, CO_2;
- Reactive gases that are toxic and/or corrosive, e.g. SO_2F_2, SO_2, HF, SOF_2, SOF_4, SF_4, WF_6;
- Moisture (water vapour);
- Condensed contaminants, e.g. oil, dust, particles.

For each contaminant, the choice of the maximum contamination level for SF_6 reuse is suggested considering:

- The maximum tolerable level in equipment that is the minimum between the functional deterioration limit and health risk limit;
- The sensitivity of the measurements to perform on-site utilising commercially available portable equipment;
- Functional over-specifications have not been considered as they would lead to an increase of the cost of the gas handling and thus discourage the reuse.

3.1.1. Inert Gases

Up to 3% vol. inert gases can be tolerated in equipment. This is the limit for reuse.

3.1.2. Reactive Gases

50 ppmv total concentration of reactive gases*, which is the cumulative sum of the concentration of each reactive gas, can be tolerated in equipment. A factor 2 as safety margin has been adopted.

3.1.3. Moisture

When dealing with moisture, three levels require to be considered:

* Commercially available reaction tubes sensitive to SO_2+SOF_2 are recommended for on-site measurements. As the typical stoichiometric ratio between SO_2+SOF_2 and HF is 1:2÷1:4, 50 ppmv total reactive gases approximately corresponds to a tube reading of 12 ppmv and to 25 ppmv HF concentration.

- The condensation limit of the liquid phase, corresponding to a 0 °C dew point or a 611 Pa partial water vapour pressure;
- The limit imposed by IEC 60694, which contains a safety margin, corresponding to a –5 °C dew point or a 400 Pa partial water vapour pressure.
- The typical moisture level in equipment, which is 100÷300 Pa. This level is mainly established as equilibrium of sorption-desorption processes at the inner polymeric surfaces of the equipment and the adsorbers.

150 Pa partial water vapour pressure, corresponding to –15 °C dew point is the limit providing adequate safety margins that are 2.7 and 4 in relation to IEC 60694 and the condensation of the liquid phase, respectively.

Figure 1 shows the above mentioned moisture levels expressed in ppmv vs. the nominal SF_6 filling pressure.

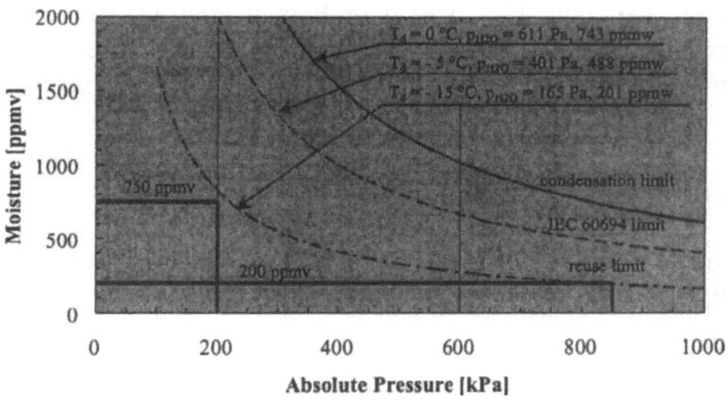

Figure 1. Moisture levels expressed in ppmv as a function of the nominal SF_6 filling pressure.

The definition of individual ppmv levels for reuse as a function of the nominal SF_6 filling pressure is clearly cumbersome. Therefore the suggested practical approach splits the whole range of nominal SF_6 filling pressures down into two ranges, as indicated by the step curve:

- The complete range of pressures, below 850 kPa, covers both HV and MV insulation systems as well as all circuit breakers. The required general purpose reclaiming equipment has to compress SF_6 to liquefaction.
- The low range of pressures, below 200 kPa, is typical of MV insulation systems. Low cost reclaiming devices are available and can be utilised, as there is no need to compress SF_6 to liquefaction.

The reuse criteria for both pressure ranges are chosen such that the limit curve corresponding to a water vapour pressure of 150 Pa is practically fulfilled at the highest nominal SF_6 filling pressure in the range, namely, at 200 kPa for the low range of pressures and at 850 kPa for the complete range of pressures.

3.1.4. Condensed Contaminants

No limit is set for solid contaminants. The use of dust filters with a 1µm pore size is required instead, as that is enough to remove dust and particles.

If all pieces of equipment in contact with SF_6 are oil-free, then it is not required to check for oil content. By the way, an oil trap reduces the oil content down to less than 1 mg/m^3, which is more than enough for practical SF_6 reuse.

3.2. Theoretical considerations on SF_6 handling

The next paragraphs present some theoretical considerations concerning best practices for SF_6 handling. The focus is on both air and SF_6 residual pressure in gas compartments vs. SF_6 dilution and handling losses, respectively. Author's opinion, as this is still under discussion within the CIGRE SF_6 task force, is also reported.

3.2.1. Air residual pressure vs. SF_6 dilution

When considering SF_6 dilution due to the air residual content in the gas compartment, the following aspects require to be considered:

- Inert gases can be separated compressing SF_6 down to liquefaction. By the way, this requires more expensive gas handling equipment;
- The new concept of "technical grade SF_6" supplied, which is going to be introduced by the 2nd edition of IEC 60376[10], allows up to 1% vol. air and 0.4% vol. CF_4, that is 1.4% vol. for the sum of both inert gases;
- The purity requirements for SF_6 reuse and recycling recommended by CIGRE, that are going to be introduced by the 2nd edition of IEC 60480[9], allow up to 3% vol. for the sum of both air and CF_4;
- Each time a gas compartment is evacuated down to the air residual pressure p_{air} and filled with SF_6 up to the nominal SF_6 filling pressure p_{SF6}*, the gas is diluted by a factor $1-p_{air}/p_{SF6}$;

Considering starting from "technical grade SF_6", then the maximum number of handling operations (evacuation + filling) before reaching the SF_6 reuse limit and avoiding SF_6 liquefaction becomes a function of p_{air} and p_{SF6}. This has been evaluated and reported in Table 1.

* Typical nominal SF_6 filling pressures are: 100÷150 kPa for MV insulation, ~300 kPa for MV breakers, ~500 kPa for HV insulation, and ~700 kPa for HV breakers.

Table 1. Maximum number of handling operations vs. air residual pressure p_{air} and nominal SF_6 filling pressure p_{SF6}.

p_{air} [Pa]	p_{SF6} [kPa]				
	100	150	300	500	700
100	16	24	49	81	114
300	5	8	16	27	38
500	3	4	9	16	22
1000	1	2	4	8	11

State-of-the-art MV equipment have a typical nominal SF_6 filling pressure not greater than 300 kPa, they are sealed pressure systems (commercially denominated as "sealed for life"), and require no SF_6 handling on-site. Therefore there is no practical need to reduce p_{air} in order to grant a high number of handling operations in the low-end part of the p_{SF6} range.

On the contrary, the following considerations applies to HV equipment, for which the nominal SF_6 filling pressure is typically greater than 300 kPa:

- The air residual pressure shall be checked once evacuation is completed and immediately before commencing of the SF_6 filling operation;
- Vacuum* shall be applied for at least 1 hour;
- The vacuum pump shall be detached from the gas compartment so that the reading is independent on the location of the pressure gauge;
- A reasonable number of handling operations shall be granted without liquefying SF_6 to separate it from inert gases.

Author's opinion, as this is still under discussion within the CIGRE SF_6 task force, is to recommend no more than 300÷400 Pa as air residual pressure. That is also in line with IEC 60694 requirement on –5 °C dew point in equipment, which corresponds to 401 Pa moisture partial pressure.

3.2.2. SF_6 residual pressure vs. SF_6 handling losses

The SF_6 handling losses per each handling operation can be easily evaluated, as it is the ratio between the SF_6 residual pressure p_r and the nominal SF_6 filling pressure p_{SF6}. They are reported in Table 2.

* Within this context, the word "vacuum" is used to indicate an air residual pressure in the gas compartment lower than 300 Pa.

Table 2. SF_6 handling losses in % per handling operation vs. SF_6 residual pressure p_r and nominal SF_6 filling pressure p_{SF6}.

p_r	p_{SF6} [kPa]				
[Pa]	100	150	300	500	700
300	0.30	0.20	0.10	0.06	0.04
500	0.50	0.33	0.17	0.10	0.07
1000	1.00	0.67	0.33	0.20	0.14
3000	3.00	2.00	1.00	0.60	0.43
5000	5.00	3.33	1.67	1.00	0.71
10000	10.0	6.67	3.33	2.00	1.43
30000	30.0	20.0	10.0	6.00	4.29

Again, as no SF_6 handling on-site is required for MV equipment, the following considerations apply to HV equipment, for which the nominal SF_6 filling pressure is typically greater than 300 kPa:

- The SF_6 residual pressure shall be checked once the gas recovery is completed and immediately before flushing the compartment with air;
- A reasonable target to achieve is both handling and leakage losses of the same order of magnitude;
- An average time between two consecutive handling operations of 5÷7 years has been assumed, as this value is representative of the reality*.

Author's opinion, as this is still under discussion within the CIGRE SF_6 task force, is to recommend no more than ~5 kPa SF_6 residual pressure. Such a figure shall be accordingly reduced in case SF_6 is handled more frequently.

In practice, for an efficient and fast SF_6 recovering, the compressor requires to be realised connecting two compression stages in series:

- The main compression stage, usually realised by a piston type compressor, which operates between a gas inlet pressure ~100 kPa (typically >50 kPa) and the pressure in the gas storage container;
- The auxiliary compression stage, only when required, operates between the pressure in the gas compartment and the pressure at the inlet of the main compressor. The auxiliary compressor enables reaching a SF_6 residual pressure ~5 kPa. A vacuum pump is needed for further lowering of the SF_6 residual pressures down to less than 100 Pa.

* Actually, state-of-the-art HV equipment require to be inspected for maintenance after 20÷25 years and then reaches the end-of-life after 40 years. Therefore there is no real need to handle SF_6 more than twice in 40 years.

4. INFLUENCE ON IEC STANDARDS

IEC Standards on SF_6 are revised/amended by the Technical Committee 10 – Fluids for electrotechnical applications – Maintenance Team 26 (TC 10 MT 26). Two standards are in force at the moment and both of them are at an advanced stage of the revision process. In particular:

- IEC 60480 2nd edition[9] reached the Final Draft of International Standard (FDIS) state. It concerns SF_6 removed from power electric equipment and also defines purity requirements for reuse of reclaimed SF_6, improperly called "used SF_6".
- IEC 60376 2nd edition[10] is now a Committee Draft for Voting (CDV). It concerns SF_6 supplied and introduces the concept of "technical grade SF_6".

Table 3 shows purity requirements among "technical grade SF_6", "used SF_6" and the CIGRE proposal using the same units for direct comparison.

Table 3. Direct comparison among "technical grade SF_6" (IEC CDV 60376), "used SF_6" (IEC FDIS 60480), and the CIGRE proposal[8]. The same units are used.

Contaminant	Maximum acceptable levels				
	CDV IEC 60376	FDIS IEC 60480		CIGRE proposal	
	All pressures	Rated absolute pressure < 200 kPa	Rated absolute pressure > 200 kPa	Rated absolute pressure < 200 kPa	Rated absolute pressure > 200 kPa
Air	1% vol	not given	not given	not given	not given
CF_4	0.4% vol	not given	not given	not given	not given
Air and/or CF_4	not given	3% vol.	3% vol.	3% vol	3% vol
Moisture	200 ppmv	750 ppmv	200 ppmv	750 ppmv	200 ppmv
Acidity in HF	6 ppmv	25 ppmv	25 ppmv	25 ppmv	25 ppmv
Mineral oil*	10 ppmw	10 ppmw	10 ppmw	use oil trap	use oil trap

With the only exception of oil, which according to CIGRE can be reduced far below the IEC limit just using an oil trap if required, IEC 60480 2nd edition states the maximum acceptable contamination levels proposed by CIGRE. Therefore SF_6 reuse and recycling is allowed with concrete facts and not only in theory.

"Technical grade SF_6", which is intended as the supplied SF_6 pure enough for the Electric Industry, is derived from "used SF_6" keeping the lowest moisture content, reducing the amount of inert gases by a factor 2, and the acidity by a

* If all pieces of equipment in contact with SF_6 are oil-free, then it is not required to check for oil content. Under these circumstances, the oil trap is not required as well.

factor 4. This resulted from a three decades long experience dealing with SF_6 gained by the Electric Industry, which is applied to cut down the environmental impact of the SF_6 technology itself.

Both IEC Standards 60480 and 60376 are not the only documents issued by IEC about SF_6, as IEC Technical Report 61634[16] on use and handling is also available. As all technical reports issued by IEC, it reports the state-of-the-art (dated 1995) together with recommendations. TC 10 MT 26 recently has suggested IEC to revise the document before adopting it as effective standard.

5. ENVIRONMENTAL IMPACT OF THE SF_6 TECHNOLOGY

The environmental parameters defined by the Kyoto Protocol[2] (e.g. Ozone Depletion Potential, Global Warming Potential, Atmospheric Life Time) are not adequate to measure the environmental impact of electric power equipment based on SF_6 technology. The latter must be evaluated from the global point of view, accounting for all components as well as the system design solutions enabled by the technology used. The appropriate procedure is called Life Cycle Assessment (LCA) and is defined in ISO Standard 14040[17].

Two LCA case studies were carried on for a regional power supply (HV grid)[18] and for the power distribution[19] (MV grid). The conclusion was that the use of SF_6 technology in electric power equipment leads to considerable environmental benefit over the use of SF_6-free technology. The superior electric properties of SF_6, state-of-the-art electric power equipment with low leakage rate, and the implementation of best practices on SF_6 handling and recycling enable the synergetic effect of saving land occupation, materials, and energy losses. The impact on the environment is therefore minimised.

This example clearly shows that an environmental compatible use of SF_6 can be realised and shall be practically encouraged.

6. CONCLUSIONS

SF_6 is a synthetic gas used by the electric industry in electric power equipment for three decades. On one hand, it has superior dielectric and interruptive performance. On the other hand, it is one of the most potent and persistent greenhouse gases.

As, for the time being, no alternative gas exists with adequate characteristics, it must be used in a closed cycle and the Electric Industry has commissioned itself to reduce the emission of "electric" SF_6 down to the minimum functional level for the equipment.

State-of-the-art electric power equipment is designed for low emissions. The focus has moved now on SF_6 handling. According to recommendations from

CIGRE, the IEC Standards in force are going to be modified to allow SF_6 reuse and recycling on industrial scale.

Discussion at CIGRE is now on best practices to implement in appropriate SF_6 handling procedures. Such a critical topic should be internationally regulated, while an appropriate mandatory standard is still missing for the time being.

An LCA case study was mentioned in which the SF_6 technology minimises the environmental impact if state-of-the-art electric power equipment is used and best practices on SF_6 handling and recycling are implemented. Therefore a proper use of the SF_6 technology gives environmental benefit. That is a practical example of environmental compatible use of SF_6, which shall be encouraged.

7. REFERENCES

1. H. C. Pollock and F. S. Cooper, The Effect of Pressure on the Positive Point-to-Plane Discharge in N_2, O_2, CO_2, SO_2, SF_6, CCl_2F_2, Ar, He, and H_2, *Physical Review*, **56**, 170-175 (1939).
2. IPCC (Intergovernmental Panel on Climate Change), Climate Change 1995: *The Science of Climate Change*, edited by J. T. Houghton et al. (Cambridge University Press, Cambridge, 1996).
3. IEC Standard 60694, Common specifications for high-voltage switchgear and controlgear standards, 2.2 edition 2002.
4. IEC Standard 61276-100, High-voltage switchgear and controlgear – Part 100: High-voltage alternating-current circuit-breakers, 1st edition 2001 and amendment 1 2002.
5. IEC Standard 61276-102, High-voltage switchgear and controlgear – Part 102: High-voltage alternating current disconnectors and earthing switches, 1st edition 2001.
6. IEC Standard 61276-200, A.C. metal-enclosed switchgear and controlgear for rated voltages above 1 kV and up to and including 52 kV, 1st edition 2003.
7. IEC Standard 61276-203, Gas-insulated metal-enclosed switchgear for rated voltages above 52 kV, 1st edition 2003.
8. P. Glaubitz, S. Stangherlin, J. L. Bessede, W. Degen, Y. Fushimi, J. Henriot, A. Holm, G. Mauthe, M. Meguro, L. Niemeyer, P. O'Connell, M. Pittroff, P. Prieur, L. Roethlisberger, P. Sieber, and S. Theoleyre, SF_6 Recycling Guide (Revision 2003), CIGRE Brochure N. 234, 2003.
9. IEC FDIS 60480, Guide to the Checking and Treatment of SF_6 taken from Electrical Equipment and Specification for its Re-use, 2nd edition 2004.
10. IEC CDV 60376, Specification of technical grade SF_6 for use in electrical equipment, 2nd edition 2004.
11. G. Mauthe, K. Pettersson, D. Gleeson, D. König, J. Lewis, T. Molony, P. O'Connell, A. Porter, and L. Niemeyer, Handling of SF_6 and its Decomposition Products in Gas Insulated Switchgear (GIS), *Electra* **136**, 69-89 and **137**, 81-108 (1991).
12. G. Mauthe, L. Niemeyer, B. M. Pryor, R. Probst, H. Bräutigam, P. O'Connell, K. Pettersson, H. D. Morrison, J. Poblotzki, and D. König, SF_6 and the Global Atmosphere, *Electra* **164**, 120-131 (1996).
13. G. Mauthe, B. M. Pryor, L. Niemeyer, R. Probst, J. Poblotzki, H. D. Morrison, P. Bolin, P. O'Connell, and J. Henriot, SF_6 Recycling Guide, CIGRE Brochure N. 117, 1997.
14. L. Niemeyer, H. D. Morrison, G. Mauthe, B. M. Pryor, H. Knobloch, J. Henriot, M. Pittroff, R. Probst, J. Poblotzki, B. Mazzoleni, J. Castonguay, W. Boeck, Y. Murayama, and P. Bolin, Guide for SF_6 Gas Mixtures, CIGRE Brochure N. 163, 2000.
15. P. O'Connell, F. Heil, J. Henriot, G. Mauthe, H. Morrison, L. Niemeyer, M. Pittroff, R. Probst, and J. P. Taillebois, SF_6 in the Electric Industry, Status 2000, *Electra* **200**, 16-25 (2002).

16. IEC Technical Report 2 61634, 1st edition 1995 "High-voltage switchgear and controlgear – Use and handling of sulphur hexafluoride (SF_6) in high-voltage switchgear and controlgear".
17. ISO Standard 14040, Environmental management – Life cycle assessment – Principles and framework, 1997.
18. Project Group ABB, PreussenElektra, RWE, Siemens, and Solvay, Electricity supply using SF_6 technology, summary given in: B. Zahn and E. Ruess, Economical and ecological system comparison for the electricity supply of an urban area, CIGRE SC23.99 (COLL) IWD, Zurich, 1999
19. Solvay Management Support: SF6-GIS-Technologie in der Energieverteilung – Mittelspannung. Life Cycle Assessment study commissioned by ABB, Areva T&D, EnBW Regional, e.on Hanse, RWE, Siemens, and Solvay Fluor und Derivate. Solvay: Hannover/Germany, 2003 (in German, abstract and summary available in English)

INDUSTRIAL OUTLOOK: HOW TO REDUCE SF_6 USE AND EMISSION – VARIOUS AGGRESSIVE APPROACHES TO REALIZE LESS SF_6 ENVIRONMENT

T. Kawamura, M. Meguro, H. Hama and T. Yamagiwa *

1. INTRODUCTION

In Japan, gas insulated equipment has been applied widely in various voltage classes because the equipment can meet the social demands that the huge electric power should be supplied with high reliability to the metropolitan and industrial districts of comparatively restricted areas where very severe environmental conditions are required. The reason for the wide application is due to the superior characteristics of the gas insulated equipment such as realization of downsizing substation, increasing reliability and safety by applying the metal enclosing construction. For these reason, gas insulated equipment, such as circuit breaker (GCB), gas insulated switchgear (GIS) and gas insulated transformer (GIT) have been developed and widely applied.

Since SF_6 gas was identified as a greenhouse gas in recent years, electric power companies and equipment manufacturers in Japan have been studying on countermeasures for SF_6 to limit the release into atmosphere. A joint study under the Electric Technology Research Association (ETRA) was organized and the result was published in 1998. Based on this report, another joint study consisting of the representatives of electric power companies, equipment manufacturers and gas producers was set up in order to promote the recovery and reuse of SF_6 in electric power industries in 1999.

This joint study was carried out to promote the voluntary action for SF_6 emission reduction from gas insulated equipment in order to minimize the SF_6 emission from the equipment. The effect of this voluntary action has been

* T. Kawamura, Shibaura Institute of Technology, Tokyo, Japan 108-8548. M.Meguro, Toshiba Corp., Tokyo, Japan 105-8001. H.Hama, TMT&D Corp., Hyogo, Japan 661-8661. T.Yamagiwa, Japan AE Power Systems Corp., Ibaraki, Japan 316-8501.

Gaseous Dielectrics X.
Edited by Christophorou *et al.*, Springer, New York, 2004.

admitted.

Research and development have been carried out extensively to minimize the SF_6 usage and emission to the atmosphere. For this purpose, improvement of insulation characteristics of the equipment, application of the procedures by the advanced maintenance techniques, fundamental investigations and their practical applications of the alternative gases have been carried out.

This paper describes the intention and the recent achievements [1-5] obtained by the efforts that Japanese power utilities and manufacturers have intensively carried out for the reduction of SF_6 emission from the gas insulated equipment.

2. RECENT ACTIVITIES FOR SF_6 EMISSION REDUCTION AND CURRENT SITUATION IN JAPAN

2.1. Joint Study & Voluntary Action Plans

Since 1996, Japan has intensively been working for the emission reduction of SF_6 for the power industry. Joint study for technical standards for recycling and handling of SF_6 was conducted from 1996 to 1998 by the representatives of academy, utilities, equipment manufacturers and gas producers in Japan under the coordination of ETRA (Chair: Professor Takuma, Kyoto University). By this study, the concerned people in electric power industry in Japan shared the following information, which became the basis of every activity towards SF_6 emission reduction.[1-4]

(1) SF_6 balance in the power industry in Japan as well as in the world
(2) Stage of SF_6 emission (When and How SF_6 emission happens)
(3) Leakage rate of SF_6 from the GCBs under the commercial operation
(4) Basis for SF_6 quality criteria
(5) Basis for the recovery guideline
(6) Basis of SF_6 recycling scheme

In 1998, Federation of Electric Power Companies (FEPC) and Japan Electric Manufacturers' Association (JEMA) announced the voluntary action plans for SF_6 emission reduction from electrical equipment so as to minimize the SF_6 emission from the equipment in electric power industry in accordance with the criteria, basis and common understanding of the joint study, which are as follows.

(1) Suppression of SF_6 emission at all the stage from the development to the disposal of gas insulated equipment
(2) Establishment and promotion of recycling system
(3) Brush up the current SF_6 inventory system
(4) Development of new technology to minimize the usage of SF_6 in equipment
(5) Development of high-performance SF_6 recovery equipment (reclaimer)[5]

2.2. Follow-up Activities

Another joint study among the representatives of the electric power companies, the equipment manufacturers and gas producer was set up to promote

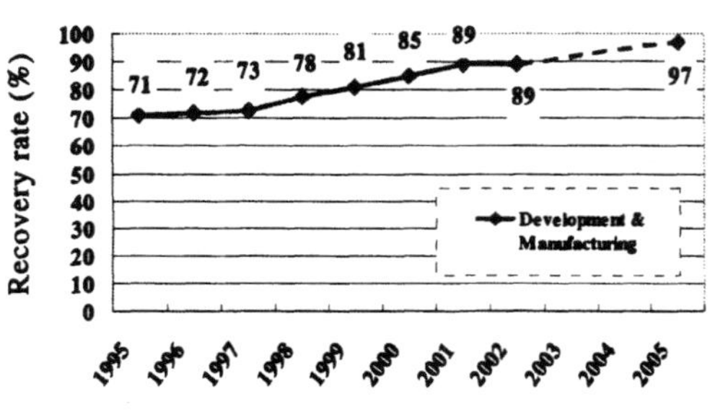

Figure 1. Recovery rate on development & manufacturing stage.

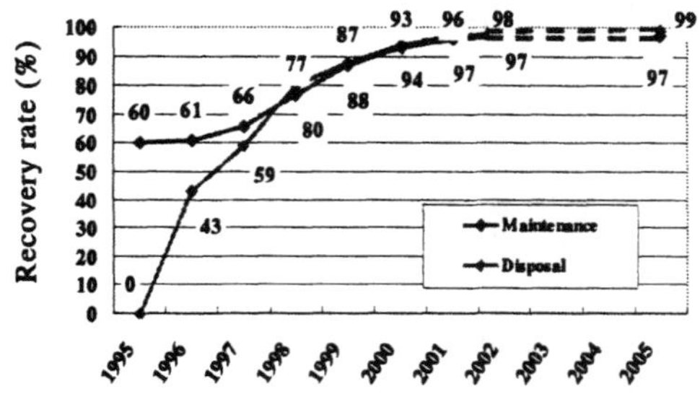

Figure 2. Recovery rate on maintenance & at disposal stage.

the recovery and reuse of SF_6 in electric power industry in Japan in October 1999. This joint study worked as the follow-up activity to let the voluntary action plans realistic. Main topics were as follows.

(1) Efficient use of SF_6 recovery equipment among the electric power industry
(2) Brush up the existing inventory system
(3) Standardization of procedure and common understanding for SF_6 handling

By those activities, the procedure for SF_6 handling and the measuring equipment were standardized and the consensus for the handling of recycled gas was established as well.

2.3. Current Status of SF_6 Emission from Electric Power Industry

As a consequence of those activities, the recovery rate on development & manufacturing stage (by equipment manufacturers), on maintenance stage and at equipment disposal (by utilities), are all improved as shown on Fig.1 and Fig. 2.

SF_6 emission from the electric power industry in Japan at the year of 2002 was 65 tons as shown in Fig.3, while the amount of SF_6 emission from

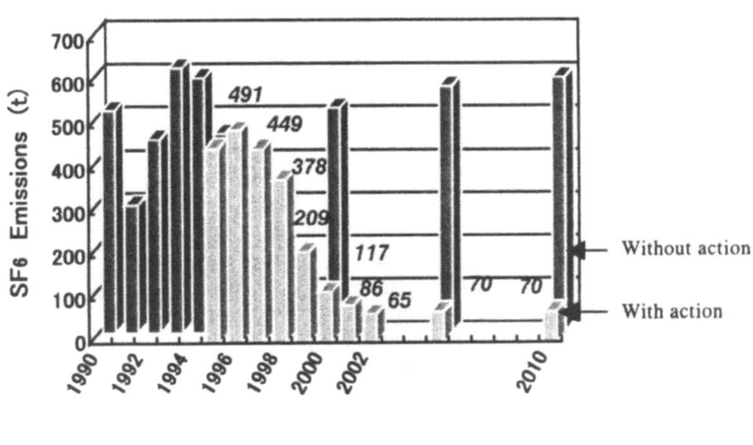

Figure 3. SF_6 emission from electric power industry in Japan.

1990 to 1995 was 450 tons/year and that expected to be more than 500 tons without any proper action.

3. TECHNICAL RESEARCH AND DEVELOPMENT FOR LESS SF_6 USAGE AND PERSPECTIVE OF EMISSION FOR THE FUTURE

In Japan, research and development have been continued from the view point of minimizing SF_6 usage and emission to the atmosphere as follows; (1) improvement of insulation system of gas insulated equipment to reduce SF_6 usage, (2) minimization of emission from the on-site equipment based on condition based maintenance (CBM) techniques, and (3) research on alternative gases and realization of the prototypes.

3.1. Improvement of SF_6 Gas Insulation System

The application of thick dielectric-layered electrode (TDLE) [6,7] is proposed as one of solutions to reduce SF_6 usage by improving the insulation system. TDLE has high breakdown field of around 90% of theoretical field, and is effective especially at high gas pressure region as shown in Fig.4.

Figure 5 shows a trial design applying TDLEs to 145kV three phase enclosure type current transformer (CT).[7] The TDLE outer diameter is not changed from the bare conductor diameter to prevent a lowering of the current capacity at rated current. The thickness of the dielectric-layer of TDLE is 5mm in this case. The cross section of CT core by TDLE is designed to be the same as that by bare conductor to maintain the same CT accuracy. When applying TDLE, the enclosure diameter is reduced to 88%, compared with a current CT using a bare conductor. The result shows that TDLE contributes to reduce SF_6 usage by downsizing the equipment.

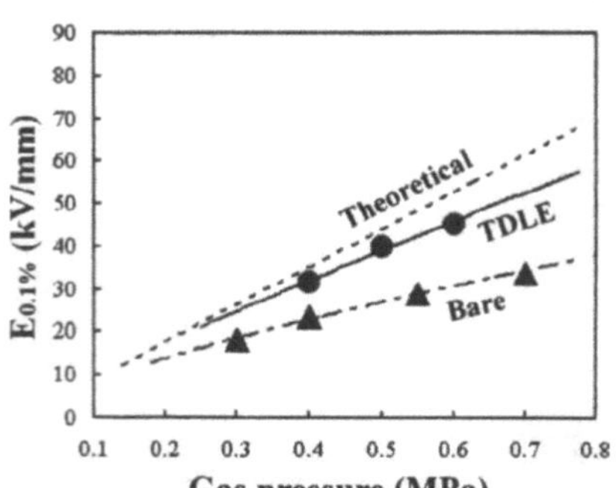

Figure 4. Dependence of minimum breakdown electric fields on gas pressure for TDLE and bare electrodes when effective electrode area is infinitely large. The broken line (Theoretical) shows theoretical breakdown fields of SF_6 gas.

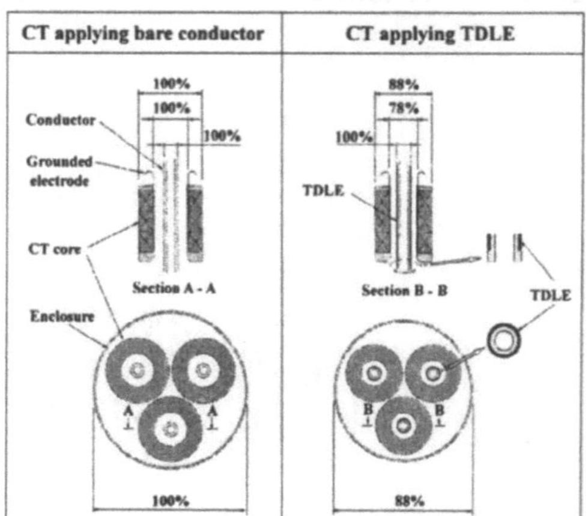

Figure 5. Trial design applying TDLE technique to 145kV gas insulated three phase enclosure type CT.

3.2. SF_6 Emission Reduction Based on CBM Techniques

3.2.1. Minimization of Open Inspection by External Diagnostics

Figure 6 shows a concept of CBM. In general, conventional monitoring techniques for gas insulated equipment diagnose the equipment as faulty condition if measured data exceeded a criteria. However, in order to realize CBM, it becomes important to use the change of measured data of the equipment to presume the future performance as shown in Fig.6. Thus, when the tendencies are observed that could lead to failures together with the use of test apparatus designed for external diagnosis, the necessity for open inspection of equipment is avoided and the amount of emission gas is consequently reduced.

As the examples of the CBM which could be applied to sealed type equipment, external diagnostics using X-ray image processing system and hybrid sensor detecting slow gas leak are described below.

(1) X-ray Image Processing Diagnostic Method

An imperfect contact sometimes generates Joule's heat at its contact surface. In the worst case, the contact starts melting by the heat and breakdown follows. Therefore, it is important to check the contact condition early before breakdown. Opening or overhauling inspection is the basic method for checking contact conditions. These inspections required gas-handling, outages of GIS and high cost. It is widely noted overhauling sometimes leads to the equipment failure, and therefore it is desirable to perform external diagnosis without opening the sealed vessels. One diagnostic tool for this purpose is mobile X-ray image processing. As a method for external diagnosis for power transmission and distribution equipment, X-ray projection techniques have been applied. A new method[8] was developed which featured highly accurate projection and detailed observation of internal condition of GIS. The system is shown in Fig.7.

(2) External Diagnostics for GCB

A new method[8] to diagnose the wear of contacts, the wipe dimension of contacts and the tightening conditions of bolts has been established by taking photos of conditions of the contact part with X-ray and processing those images.

The above mentioned diagnostic apparatus is composed of a compact movable X-ray generator and a detector. This diagnosis can be done in a shorter

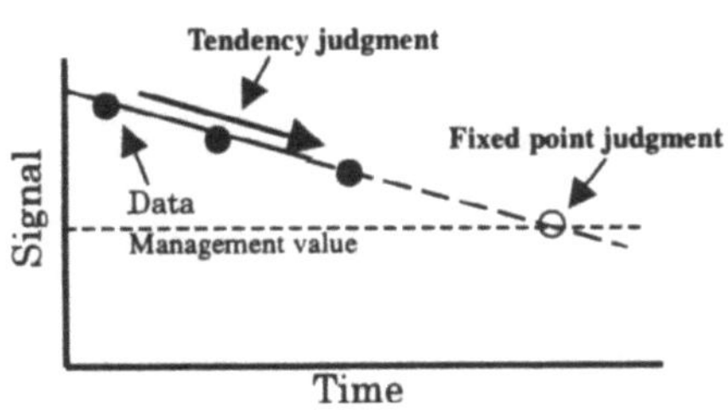

Figure 6. Concept of CBM.

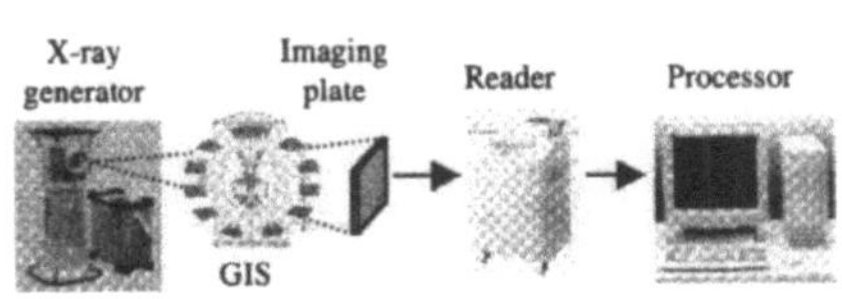

Figure 7. X-ray image processing system.

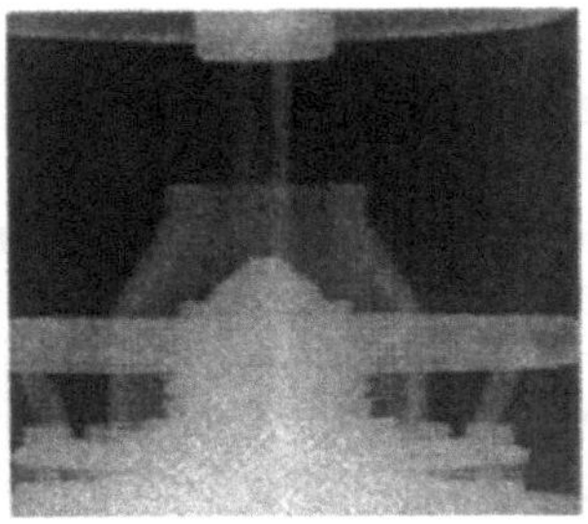

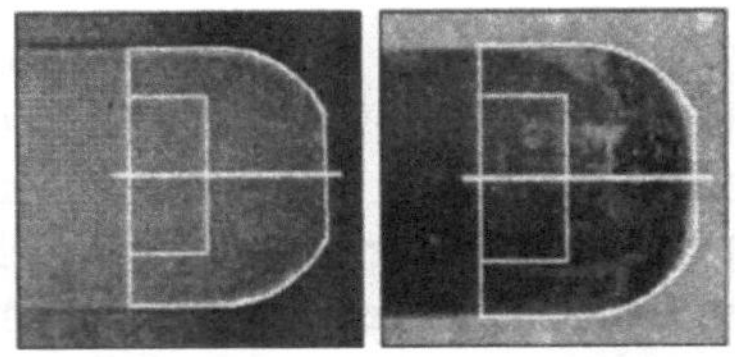

Figure 8. X-ray image of opening position of GCB.

Figure 9. Comparison between X-ray image and actual condition of stationary contact.

time by using the apparatus; the detector stores the photo images, the photo-image reader processes the images, and dimension of the object is measured with the dedicated software.

Contact wear of GCB was measured using above X-ray techniques. Figure 8 shows X-ray image of the GCB open position. The result was the same as the dimension actually measured on the disassembled GCB as shown in Fig.9.

This diagnosis can be used as an alternative measure of overhaul of GCB.

3.2.2. Monitoring of SF_6 Slow Leak by Hybrid Sensor

A state-of-the-art sensor[9] to monitor SF_6 condition of GIS/GCB is also effective to decrease SF_6 emission. The sensor is a hybrid type and has three high-resolution sensors decomposed gas sensor, gas pressure sensor and temperature sensor integrated in one package as shown in Fig.10. The hybrid sensor has four different functions, that is, monitoring of decomposed gas and gas slow leak, fault location and also gas pressure gauge.

Applying both the gas pressure sensor (high resolution of 20Pa and stability of 0.04%/year) and the temperature sensor to 550kV GIS, the stability of the sensors in detecting gas density has been monitored in the field. Figure 11 shows the results during six months. The variation of gas density for compartment of

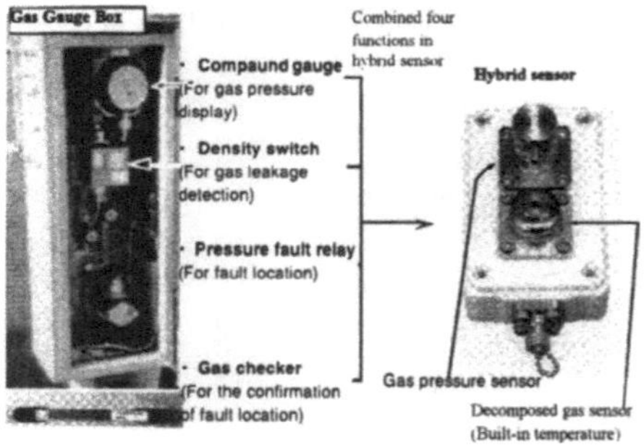

Figure 10. Construction of hybrid sensor.

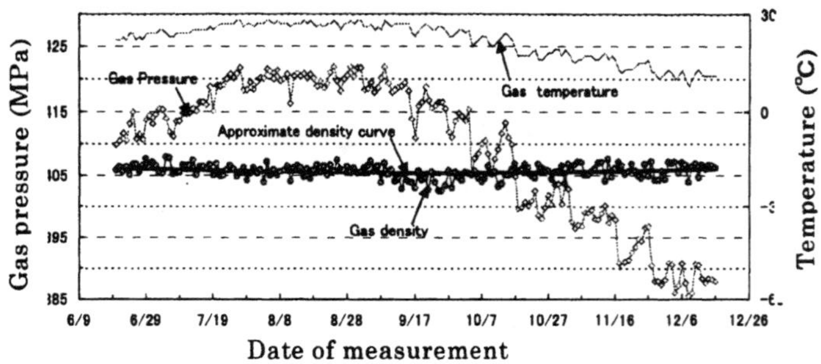

Figure 11. Stability of hybrid sensor in detecting gas density in the field during six months.

0.4MPa.gauge is around 0.004MPa which leads to the error of nearly 1%/6months(=2%/year). However, the approximate density curve calculated statistically using data during six months (thick solid curve) shows much smaller variation of 0.0007MPa which corresponds to the stability of 0.18%/6months. The fact shows that the sensors could detect the slow leak of 1%/year (0.5%/6months) within six months, whereas the conventional sensor for a gas density needs at least ten years to detect the same gas leak of 1%/year. Therefore, the hybrid sensor is effective to prevent SF_6 emission to the atmosphere during operation. In near future, the sensor could be expected to detect the slow leak of 0.5%/year standardized in IEC 60517.

3.3. Research and Development on Alternative Gases

3.3.1. Recent Research and Development in Japan

Table 1 shows research and development trend on alternative gases up to 2000 and from 2000 to 2003 in Japan. Up to 2000, N_2/SF_6 gas mixtures were mainly studied, since the mixtures were the most promising substitutes for SF_6 for the equipment due to high insulation performance and reasonable cost.

Since the basic properties of N_2/SF_6 gas mixtures and its application to the equipment have been fully studied, the research and development have been focused on the substitutes and the equipment without SF_6 after 2000.

As the recent topics in Japan, the prototypes of vacuum circuit breaker (VCB) and of GCB applying CO_2 are introduced as follows.

3.3.2. Prototype of 145kV VCB

For voltages over 72kV, GCB has advantages to other type circuit breakers in downsizing and cost reduction due to its high interrupting performance of SF_6 gas. However, CO_2 or other environmentally benign gases might be the candidates for the quenching gas as an alternative to SF_6 gas.

For middle-voltage class (3.8kV-38kV) and up to 84kV, VCB is extending their use. Two breaks porcelain type 168kV VCB has already been in service

Table 1. Research and development trend on alternative gases up to 2000 and from 2000 to 2003 in Japan.

	Categories	Contents	Gases	
			up to 2000	2000 - 2003
Insulation/Interruption/Thermal properties	Insulation properties	Basic physics	N_2/SF_6, $C\text{-}C_4F_8/N_2$ or CO_2, $N_2/SF_6/CO_2$, N_2/O_2, Air, N_2	N_2/C_2F_6, N_2/C_3F_8, CO_2/C_2F_6, CO_2/C_3F_8
		Uniform field	N_2/SF_6	Air, N_2, CO_2, C_3F_8, $CF_4/N_2/C_3F_8$, $c\text{-}C_4F_8$
		Non-uniform field	N_2/SF_6, $C\text{-}C_4F_8/N_2$ or CO_2, $N_2/SF_6/CO_2$, N_2/O_2, Air, N_2	N_2/SF_6, CO_2, N_2, N_2/CO_2, N_2 or CO_2/C_3F_8, N_2 or CO_2/C_2F_6
		Practical equipment test	N_2/SF_6	CO_2, (Vacuum)
	Interruption capability	GCB	N_2/SF_6	CO_2
		VCB	(Vacuum)	(Vacuum)
		DS	N_2/SF_6	N_2/SF_6
	Current capacity	Temperature rise of conductor and enclosure	N_2/SF_6	N_2
Equipment	Equipment design	GIL/GIB	N_2/SF_6	N_2/SF_6, N_2
		GIS	N_2/SF_6	Silicon oil
		Transformer	N_2/SF_6	-
	Prototype	GCB	-	CO_2
		VCB	(Vacuum)	(Vacuum)
	Recycle	Pressure swing adsorption	N_2/SF_6	-
		Polymer membrane	N_2/SF_6	N_2/SF_6

since 1979 and the current interrupting capacity has been improved up to 40kA[10]. However, its cost and size were less competitive than those of GCB applying the one break system. In order to substitute the GCB, one break VCB over 84kV class has been strongly expected.

Figure 12 shows the photo of a 145kV prototype vacuum interrupter. The rated current is 2000A. Techniques on analysis, experiments and designing were applied to achieve the optimal design for the interrupter[11], which enabled us to realize simple structure and low cost. Moreover, the reduced bottle size was almost equivalent to a traditional type 72/84kV vacuum interrupter. The interrupter withstood the lightning impulse withstand voltage of 650kV, which corresponds to the requirement by IEC standard. The interrupter was installed into the porcelain type VCB and synthetic short circuit tests were carried out using a Wile circuit. The symmetrical current of 40kA was successfully interrupted. And also, "Duty 4" and "Duty 5", required by IEC specification,

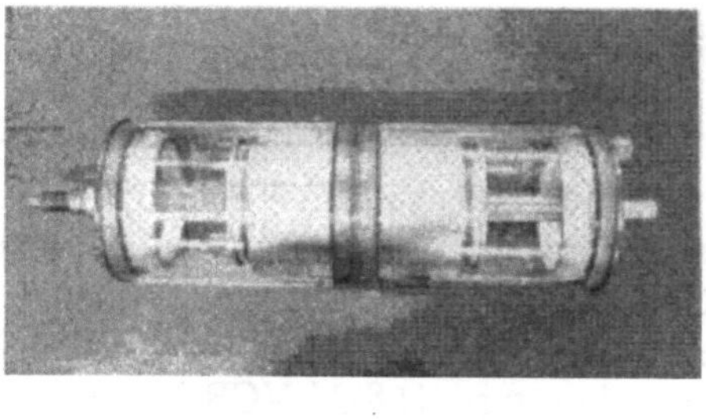

Figure 12. Prototype of 145kV vacuum interrupter.

were cleared satisfactorily.

Therefore, the VCB of 145kV / 40kA class would contribute to reduce SF_6 usage of the gas insulated equipment in the near future.

3.3.3. Prototype of 72kV CO_2-GCB

A 72 kV-31.5 kA class CO_2 gas circuit breaker (CO_2-GCB) model, which does not contain SF_6 gas at all, is designed and produced, adopting CO_2 gas as an alternative arc quenching and insulating medium.[12] Some effective technologies for current interruption by CO_2 were applied as described blow:

(1) A Hybrid-puffer™ type interrupter specially designed for CO_2 gas as shown in Fig.13 was adopted. A Hybrid-puffer™ is one of the techniques enhancing the blasting pressure by utilizing arc energy effectively during a current interruption. This technique is more effective for CO_2 than SF_6, because gas density and heat capacity of CO_2 are smaller than SF_6 and, in addition, arc voltage in CO_2 is higher than SF_6.

(2) An ablation element made of appropriate polymer material was installed in the puffer cylinder as one of breakthrough technologies improving interruption performance of the CO_2-GCB. Both the blasting pressure and the arc quenching performance of the blasting gas itself can be enhanced due to the generation of the ablated gas, including H or O elements, which have relatively satisfactory arc quenching capability.

Current interruption and dielectric tests of the CO_2-GCB model were carried out according to the standard of The Japanese Electrotechnical Committee JEC-2300-1998. As a result, the CO_2-GCB model showed satisfactory performance for major test-duties; namely, capacitive current switching, short-line fault 90%, terminal fault 100%(symmetrical and asymmetrical) interruption, power-frequency and lightning impulse insulation. These test results shows that the CO_2-GCB model achieved practical levels of performance, which is first completely SF_6-free and Freon-free GCB in the world.

Furthermore, a life cycle assessment was carried out to evaluate the environmental contribution of the CO_2-GCB model. As a result, it was noted that

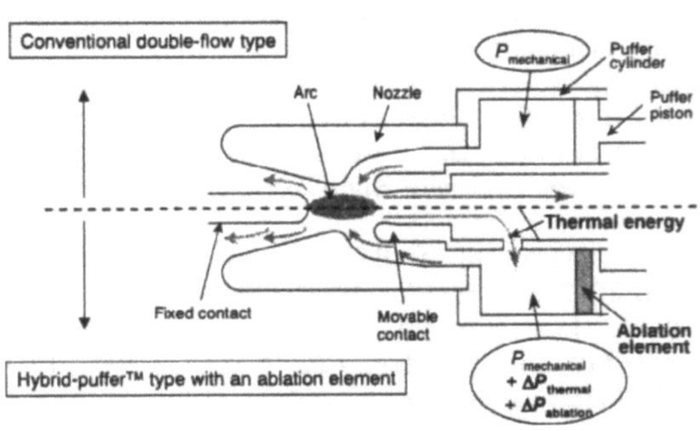

Figure 13. Schematic diagram of Hybrid-puffer™ type interrupter with an ablation element adopted for the CO_2-GCB model, compared to a double-flow type interrupter as a typical one.

the CO_2-GCB model could reduce the global warming impact by about 45% compared to the latest SF_6 GCB in the same rating for 40 years operation including two maintenance opportunities.

4. FUTURE PERSPECTIVE

In Japan, the use of gas insulated equipment is essential for the realization of huge electric power with high reliability to the metropolitan and industrial districts of comparatively restricted and small areas. It is also very important to meet the severe environmental and safety condition required in these areas.

As a consequence of the voluntary action by the electric power companies, equipment manufacturers and gas producers, SF_6 emission from the electric power industry remarkably decreased to 65 tons in the year of 2002 as compared with more than 500 tons without any specific action.

Several technical research and development for less SF_6 usage resulting SF_6 emission reduction have been carried out extensively. Recent results are presented and their effects are discussed.

Stimulated researches and developments are being carried out in Japan in order to keep SF_6 emission minimum.

REFERENCES

1. M.Meguro and S.Maruyama, SF_6 emission reduction from gas insulated electrical equipment in Japan; *USEPA conference at San Diego*, November (2000).
2. H.Yasutake and M.Meguro, SF_6 emission reduction from gas insulated electrical equipment in Japan; *USEPA conference at San Diego*, November (2002).
3. T.Kawamura et al., SF_6 gas handling in Japan, focused on emission reduction from gas insulated electrical equipment, *Gaseous Dielectrics IX*, pp. 575-584 (2001).
4. S.Maruyama and M.Meguro, SF_6 emission reduction from gas insulated electrical equipment in Japan, *International workshop on CIGRE WG23-02 Japan meeting*, April (2001).
5. T.Yamagiwa, M.Meguro et al., SF_6 gas emission reduction strategy in Japan, *CIGRE SC23 Venezuela colloquium*, October (2001).
6. H.Goshima et al., Lightning Impulse Breakdown Characteristics of High-pressure N_2 as an Alternative Insulation Gas to SF_6, *Gaseous Dielectrics IX*, pp. 359-364 (2001).
7. H.Hama et al., Improvement of SF_6 Gas Insulation System by Applying Thick Dielectric-layered Electrode and Its Application to High Voltage Equipment, *Gaseous Dielectrics X* (2004).
8. K.Kawakita et al., Life Management of GIS Based on Deterioration Diagnosis of Aged GIS", *IEEE/PES T&D 2002 Asia Pacific*, OR-18 (2002).
9. F.Endo, H.Hama et al., Innovation of Gas Insulation Monitoring Techniques and Application to Remote Monitoring System, *CIGRE Session 2002*, 15-103 (2002).
10. T. Takahashi et al., 168kV class vacuum circuit-breakers, *Meiden Rev.*, No.56, pp.9-26, (1979).
11. H.Saitoh et al., "Research and Development on 145kV / 40kA One Break Vacuum Circuit Breaker", *IEEE/PES T&D 2002 Asia Pacific*, PN-13 (2002).
12. T.Uchii et al., Investigations on SF_6-free Gas Circuit Breaker Adopting CO_2 Gas as an Alternative Arc Quenching and Insulating Medium, *Gaseous Dielectrics X* (2004).

APPENDIX A : BANQUET TALK

THE OLYMPICS – THEN AND NOW
A VIEW FROM ATHENS

P. Kitromilides*

Today the Olympics is a gigantic commercial and advertising enterprise, as witnessed by the latest games in Atlanta and Sidney and as Athens is discovering this year. It was not like that at all in the original incarnation of the games in ancient Greece and it was not meant to be this way by those who dreamt of the revival of the games in the eighteenth century and by those who worked with dedication and vision to bring about this revival in the late nineteenth century. In turning our thinking to the games in this Olympic year I should like to briefly survey the story of the original inception of the games, of its growth and significance and of the subsequent vicissitudes of the idea until its revival in Athens in 1896.

The Olympic games as an athletic contest grew out of a religious cult connected with the Doric expansion into the Peloponnese some time after the year 1000 BC. The Dorian tribes brought from the North the worship of Olympian Zeus and his cult was superimposed on earlier forms of religious worship in the sacred sanctuary on the river Alphios in the region of Elis in Western Peloponnese. The new cult gave to the old religious centre the name Olympia. This became the foremost centre of worship of the king of the ancient gods on whose cult focused the religion of the Greeks. To honour their gods at their annual religious festivals the ancient Greeks organized various sporting events and other kinds of competition in music, poetry and recitation, whereby the worshipers dedicated themselves, through the exhibition of their prowess and skills, to the god. For all these forms of competition they used the term agon. It is important to retain a sense of this original character of the games as an active expression of piety and their standing as part of the ritual of religious worship in order to grasp their broader significance in the Greek world and the moral authoriry ascribed to them.

* National Research Foundation of Greece

Gaseous Dielectrics X.
Edited by Christophorou *et al.*, Springer, New York, 2004.

The actual historical origins of the Olympics are lost in the mist of time. But the conventional date marking their beginning as an institution of some significance is the year 776 B.C., when the first victor was recorded. On that occasion the contest or agon was simply a running race following the sacrifices to the god. Both sacrifices and running formed integral parts of the ritual of worship. The fact, however, that the victor's name was recorded and thus transmitted to posterity suggests that some significance was already accorded to the sporting part of the ritual. The first Olympic winner was a local youth, Coroebus of Elis. This too suggests that originally the games were of local rather than panhellenic significance, a local occasion of communal worship of the king of Gods. From 776 B. C. the games took place every four years in an uninterruptad sequence until the year 394 A. D. when they were suspended by the Eastern Roman Emperor Theodosius. As a religious festival celebrating shared values in Greek culture the Olympic games grew in prestige and authority to such an extent that the four-year intervals of their organization became accepted throughout the Greek world as the basis of chronology and all other events were dated according to the Olympiads.

The games at Olympia took place every fourth year during the first full moon after the summer solstice. This means that the games were organized some time in June. Responsibility for the games rested with the city state of Elis in the Western Peloponnese, whose territorial jurisdiction included the sacred grove and the attached religious and athletic buildings at Olympia. Every fifth year, that is the year following the latest Olympics the citizens of Elis sent out an embassy to the other Greek cities announcing the convocation of the next games and proclaiming a sacred truce between warring cities for the period of the games. The armistice was necessary to allow pilgrims and athletes safe passage to Olympia. Ten months before the games the judges of the games, called Hellanodikai, were chosen. The judges, originally two, later nine and finally twelve, exercised strict control over the games. Upon their selection they were trained in their duties and then they took an oath to show total impartiality in exercising them.

Participation in the games was open to all free citizens of Greek blood from all over the Greek world. Although originally participants were mostly of Peloponnessian origin and the games were dominated by Sparta, gradually athletes from further afield, including the islands and Greek colonies in Sicily, South Italy and Asia Minor, were attracted. Barbarians –meaning people of foreign tongue– and slaves were allowed to watch but not to participate in the contests. Married women were barred from watching the games and this rule was strictly enforced. The only exception recorded was that of the Rhodian lady Kallipateira, who disguised herself as a man in order to watch her son take part in the games but was caught and would be condemned to death under the laws governing the games. She was pardoned, however, as the daughter of the famous Olympic winner Diagoras, and also as the sister, mother and aunt of Olympic victors. The incident known from ancient sources has been immortalized by the Modern Greek poet Lorenzo Mavillis in a moving poem. Married women were excluded but unmarried virgins were admitted to the stadium according to the ancient historian of the games Pausanias. It is also mentioned that while the Olympics were in progress special races for girls were held nearby as part of the festival of goddess Hera, called the Heraia. This too reflects the influence of Sparta on the games and her concern for the physical training of women.

Coroebus's victory in 776 BC, marks the beginning of the history of the Olympic games. Sports competitions were popular in archaic Greece and this led to the gradual amplification of the types of athetic contest included in the games. Thus whereas the sacred rites remained unchanged the sporting component of the worship of the god expanded and grew over time in to a major athletic event. The motivation for this stemmed from the broader cultural environment of archaic Greece and had to do with a general tendency emerging after the period called the «dark century» to overcome anonymity and obscurity and to affirm personal identity through some form of distinction, be it in exceptional physical performance, in music or in poety. For the rest of the eighth century new contests were added to the games: in 724 BC at the fourteenth games a two-lap race, diaulos, was instituted in addition to the original one-length race, stadion. At the fifteenth games in 720 BC an endurance race, dolichos, appeared. In 708 appeared the most famous of the contests, the five-disciphine pentathlon, comprising running, jumping, discus-throwing, javelin-throwing and wrestling. Boxing did not make its appearance until the twenty-third games in 688 BC., while four-horse chariot-racing appeared in 680 at the twenty-fifth games. By the middle of the seventh century the evolution of the Olympic programme had been completed with the appearance of horse-back riding and the pankration, a combination of boxing and wrestling, at the thirty-third games in 648 BC. Other forms of contests were included later on, but as variations and innovations of the basic disciplines institutionalized during the first century and a half in the history of the Olympics.

The enrichment of the programme also involved the extension of the length of the Olympics. From an one-day religious rite originally, it became a two-day affair by 680 and later on it was extended to four or five days. As time went by the sporting contests completely overshadowed the sacral element and the Olympics became emblematic of ancient Greek athletic values. The gradual dislocation of the religious element was contingent upon the growth of the civic culture of the Greek city-states and it was reflected in the whole history of the stadium and of the entire site at Olympia. While originally the stadium was part of the sanctuary in the sacred grove, gradually the stadium expanded in dimensions to accommodate the requirements of new contests and eventually it was separated from the sanctuary by the building of the stoa of Echo.

Great honours were bestowed upon victors in the Olympic contests. Upon winning a contest the victor was handed a palm-tree branch and at the conclusion of the games a wreath made of branches of a wild olive tree, the much-vied for kotinos, was placed upon their heads in a ceremony in the temple of Zeus. The wild olive tree branches came from Zeus's sacred tree in the sanctuary and they had to be cut by a youth whose both parents had to be alive: the games symbolized a celebration of life over death. A herald announced the names of the victors and their cities of origin. The cities were equally hailed as victors in the games along their athletes. Some cities distinguished themselves exceptionally in the games. After Sparta's prominence in the early history of the games other cities became famous for the victories of their athletes: Croton in Southern Italy rejoiced in the glories of its son Milon, the most famous athlete of the ancient world, known as the "King of Wrestlers". He first won the Olympic kotinos in 540 BC. at the age of fourteen and he gained the Olympic wreath seven more times, and he also scored victories seven times in the Pythian games at Delphi, ten times in the Isthmian games in Corinth and nine times in the Nemean games. His last recorded victory was in 516 BC. Other cities which repeatedly distinguished themselves in the Olympics through the prowess of their athletes included Athens, Miletus in Asia Miror, the island of Naxos and Ialysos in Rhodes, famous for the great boxer Diagoras.

The conclusion of the games was marked by the signing of hymns exhalting the victors and by sumptuous symposia organized by the city of Elis in the sacred grove in honour of the winning athletes and of the theoroi, the official representatives of their cities. Upon their return to their home city-states the victors were greeted with great expressions of joy. Many cities demolished a part of their fortifications for the Olympic victor to enter, claiming that with athletes such as these walls were unnecessary for their defense. The prestige of the games and the glory associated with victory in them were reflected in the exquisite Olympian Odes composed by the greatest lyric poet of antiquity, Pindar, in celebration of some of the best known victors. In all fourteen of Pindar's Olympian Odes survive and bring to us through the centuries something of the emotive climate of the games.

The religious origin of the games was practically forgotten in the ancient world, or at least overshadowed by their cultural and political symbolism in the classical, Hellenistic and Roman periods. It was recalled, however, when the Empire of New Rome decided to stamp out the remnants of the old pagan religion from the new Christian state. So in 394 A. D. the emperor Theodosius closed the sanctuary of Zeus and abolished the games. Between 776 BC. and 394 AD 293 Olympiads had taken place in ancient Olympia. The 294the Olympiad did not take place until after one thousand five hundred and two years and this happened in Athens and not in Olympia. If the Olympics had not been discontinued in the year 394 the games in 1896 would have been the 668th Olympiad. 2672 years intervened between 776 BC and the year 1896, but the spirit had lived on and brought the Olympics back to the country, if not the original site of their inception. How this remarkable revival came about is the second part of my story.

It did not of course come out of the blue. There had been a considerable history of reawakened interest in the Olympics in the period of classical revival in the age of the Enlightenment. Awareness of the symbolic significance of the Olympics in European culture is reflected in Shakerpeare's references to the «Olympian Games» in Henry VI and in Troilus and Cressida. John Milton also recalled «the Olympian Games» in Paradise Lost in 1667. But it was in the eighteenth century that the Olympics became a source of inspiration in European literary and musical life. Voltaire referred with admiration to the «jeux olympiques», the Italian playwright Pietro Metastasio wrote a play entitled «L'Olimpiade», a «Teatro Olimpico» was established in the northern Italian city of Vicenza, at least two musical compositions included the Olympics in their titles. Finally in 1788 the abbé Jean-Jacques Barthélemy published his great work Voyage of Young Anarcharsis in Greece. This became a great medium of popularizing knowledge about the ancient world among educated Europeans on the eve of the French Revolution. The book included an extensive survey of the site of ancient Olympia and detailed descriptions of the games, drawing on the accounts of Pausanias. Thus the Olympics returned to the foreground of European public opinion in an age of revolutionary changes.

These European sources soon became manuals of self-knowledge for the Modern Greeks. They were discovered among others by the great forerunner of the Greek revolution Rhigas Velestinlis, a great patriot and visionary who visualized the liberation of Greece from Ottoman rule at the time of the French Revolution. Rhigas translated Metastasio's play and Barthélemy's work and thus became the first Modern Greek author to write on the Olympics. It is significant that Metastasio's play includes an ode to liberty, which reveals Rhigas's deeper motivation in translating the work. Discussion about the prospects of reviving the Olympics went on in the patriotic clubs that were set up in the Ionian Islands, when these islands were taken over by French revolutionary troops in 1798. This was the cultural prehistory of the inception of the idea of reviving the Olympics as part of the return of Greece to the society of free nations. The eventual realization of this idea, however, was achieved thanks to the enthusiasm and dedication of another great visionary, the distinguished 19th-century writer Dimitrios Vikelas.

A diaspora man, who combined a genuine cosmopolitan outlook with sincere Greek patriotism, Vikelas had visited Olympia in April 1884 and he discovered there not only the irresistible beauty of the landscape but the quintessence of the spirit of ancient Greece. «Here antiquity presents itself more vivid and more magnificent than even on the Acropolis itself». His commentary on the treasures of classical sculpture unearthed by German and Greek archeological excavations in the sanctuary and the stadium is quite revealing of the admiration he felt for the place, its legacy and its symbolism . His evocative description concludes with a wish to see a museum erected at the site of Olympia to house and protect and show to the world the masterpieces of ancient art that conveyed to posterity Olympia's highest ideals. This visit which turned into a pilgrimage was in a way Vikelas's spiritual preparation for the role he was called to play in 1894 when he was named president of the international congress convened at the Sorbonne in Paris for the revival of the Olympic Games. This idea had been voiced in 1892 by another noble idealist, the baron Pierre de Coubertin who dedicated his life to the Olympic ideal. Vikelas as a literary personality of international renown used his prestige at the Paris congress to secure a unanimous decision of the national delegations for the organization of the first modern games in Athens. The idea to have the new games return to Olympia was excluded on account of the still primitive conditions prevailing at the site and of essentially non-existent transport and communication facilities.

This is how the modern Olympics began. To relate the story of the modern sequel to the games would require another and quite different narrative. I would, therefore, like to conclude with some reflections on the idea of the Olympic Games that might be of some relevance to today's world. When the Olympics were revived in April 1896 in Athens the expectation was widely shared that this international athletic event would mark two things: first it was hoped, by people like Vikelas and other leaders of Greek public opinion who enthusiastically supported his efforts to bring the Olympics back to Greece, that that huge, by contemporary standards, athletic event would seal Greece's recognition as a developed liberal European state and as an equal participant in international life. No one expressed the hopes associated with the revived Olympics better than the leading poet of the age, Costis Palamas in the «Olympic Hymn» he composed for the occasion, a poem that has been since the first modern Olympics established as the official anthem of the games. In it Palamas evokes the spirit of antiquity as the authentic source of beauty and truth, to be learnt by the moderns through the exertions of the games. The second expectation associated with the revival of the games in the 1890s was a broader hope, repeatedly articulated by the baron de Coubertin throughout his life: it was hoped that the Olympics by coming back to life would bring into the modern world something of the nobility of spirit marking the ancient games and especially that these athletic encounters would promote international understanding and solidarity and thus contribute to peace and justice in the world.

In view of what has come to pass in the world in the course of the twentieth century, it may sound a little too optimistic and even naive to recall these hopes more than a century after the revival of the games. As it turned out the actual history of the modern Olympics was at many points a far cry from the hopes voiced by de Coubertin, Vikelas and Palamas.

Despite the high idealism and moving sentiments elicited by the actual transaction of the Athens Olympics in 1896, when peasant boys ran the Marathon and working sailors swam in the Saronic gulf to Olympic victory, the subsequent history of the games diverged radically from these noble beginnings: in the interwar period the games were used by the fascist regimes to promote their totalitarian ideology and during the Cold War the noble spirit of emulation in the service of a shared code of humanist values was overshadowed by the irreconcilable competition of hostile political systems, which through the games attempted to score ideological victories against the lethal enemy.

The commercialization of the games in our time has turned most athletes and the huge complexes of interests behind them into agents of materialist glamour, totally oblivious to the authentic Olympic values and their religious and moral origin. Hence the skepticism with which the games in their contemporary gigantic guise are viewed in many quarters. I must admit that to a considerable extent I share this skepticism. As we are moving ahead in this Olympic year, however, when the games are returning to a metropolitan Athens incredibly different from the modest city of the first modern Olympics, I find my thinking quite often turning to those original ancient ideals of serene beauty and balance, civic spirit and frugality, sacred truce and reconciliation, which inspired the ancient games and wish that the hugely different modern games, by evoking the Olympic spirit at least serve in reminding us of the continuing significance and relevance of these ancient values.

The Hermes of Praxiteles, c. 340 BC, found in the Temple of Hera.

PARTICIPANTS

PARTICIPANTS

Adamiak, Kazimierz
Dept. of Electr. and Comp. Eng.
Univ. of Western Ontario
London, Ontario, Canada N6A 5B9
kadamiak@eng.uwo.ca

Agoris, Demosthenes
Electrical & Computer Engineering Department
University of Patras, Greece
dagoris@ee.upatras.gr
dimos@agoris.gr

Alexandri, Ioanna
PPC of Greece
70 Ag. Annis, 12241 Athens, Greece

Alexopoulou, Yota
Substation Subdivision of Power Section, PPC
70 Ag. Annis 70, 12241 Athens, Greece
galexop@dsm.dei.gr

Andreopoulos, Andreas
School of Chemical Engineering
National Technical University of Athens
9 Iroon Polytechniou Str.
Zografou 157 73, Athens, Greece
andreo@chemeng.ntua.gr

Antonopoulos, Grigoris
PPC of Greece
30 Halkokondyli Str.
Athens 10432, Greece

Castle, Peter
Dept. of Electr. and Comp. Eng.
Univ. of Western Ontario
London, Ontario, Canada N6A 5B9
pcastle@eng.uwo.ca

Chevrier, Pierre
Sti VATECH TD SA - C1 Plant - 1,
rue de la Niva BP 178 38004 GRENOBLE
CEDEX 1 – FRANCE
pierre.chevrier@vatech-td.fr

Christophorou, Loucas
Academy of Athens,
Panepistimiou 28, 106 79 Athens, Greece
lgchrist@otenet.gr

Contaxis, George
School of Electrical and Computer Engineering
National Technical University of Athens
9 Iroon Polytechniou Str.
Zografou 157 73, Athens, Greece
gcont@softlab.ece.ntua.gr

Daskalakis, John
Hellenic Transmission System Operator
11 Anphitheas Ave
171 22 Nea Smyrni, Athens, Greece
idaskalakis@desmie.gr

De Urquijo, Jaime
Centro de Ciencias Fisicas
UNAM, P. O. Box 48-3
62251 Cuernavaca, Mor., México
jdu@fis.unam.mx

Degen, Wolfgang
Siemens PTD H2 TC QM
Berlin, Germany
wolfgang.degen@siemens.com

Dervos, Constantine
School of Electrical and Computer Engineering
National Technical University of Athens
9 Iroon Polytechniou Str.
Zografou 157 73, Athens, Greece
cdervos@central.ntua.gr

Diamantis, Constantine
72 Athinon Ave
Neo Phaliro, Athens 18547, Greece
tsksd@otenet.gr

Dratsas, Athanasios
72 Athinon Ave
Neo Phaliro
Athens 18547, Greece
tsksd@otenet.gr

Drossos, Nikolas
72 Athinon Ave
Neo Phaliro, Athens 18547, Greece
tsksd@otenet.gr

Dujko, Sasha
Institute of Physics, P. O. Box 68
11080 Zemun, Serbia and Montenegro
sasha@phy.bg.ac.yu

Economou, Demetre
Plasma Processing Laboratory
Department of Chemical Engineering
University of Houston
Houston, TX 77204-4004, USA
economou@uh.edu

Ekin, Jack
National Institute of Standards and Technology
Boulder, CO 80305, USA
jackekin@earthlink.net

Elborki, Magdi
Dept. of Electrical & Electronic Engineering
UMIST, Manchester, UK
m.elborki@stud.umist.ac.uk

Forys, Mieczyslav
University of Podlasie
Chemistry Department
Siedlce, Poland
forys@ap.siedlce.pl

Fréchette, Michel
Institut de Recherche d'Hydro-Québec
Québec, Canada
frechett@ireq.ca

Gellings, Clark
Power Delivery and Markets
Electric Power Research Institute
3412 Hillview Ave
Palo Alto, CA
USA 94304
cgelling@epri.com

Georgantzis, George
PPC, Halkokondyli and Patission
Athens, Greece
g-georgantzis.dksd@ppc.gr

Gerhold, Jurgen
Technical University Graz
Institute of Electrical Machines and Devices
Steyrergasse 21, A-8010 Graz, Austria
gerhold@ema.tu-graz.ac.at

Gleizes, Alain
Centre de Physique des Plasmas et de leurs
Applications de Toulouse
UMR CNRS 5002, Université Paul Sabatier
118 route de Narbonne
F31062 Toulouse cedex 4, France
gleizes@pop.cpat.ups-tlse.fr

Goshima, Hisashi
Electrical Insulation Department, CRIEPI
2-11-1 Iwado-kita, Komae-shi
Tokyo 201-8511, Japan
goshima@criepi.denken.or.jp

Hadjiantoniou, Antony
Greek Atomic Energy Commission
Aghia Paraskevi, Athens, Greece

Hadjiantoniou, Dimitris
Academy of Athens, Soranou Ephesiou Str. 4
Athens 115 27, Greece
dimitris_h_1999@yahoo.co.uk

Hama, Hiroyuki
Strategic Planning Department
High-Voltage Switchgears Division
8-1-1 Tsukaguchi-Honmachi,
Amagasaki, Hyogo 661-8661, Japan
hiroyuki.hama@tmt-d.co.jp

Hayakawa, Naoki
Nagoya University, Furo-cho, Chikusa-ku
Nagoya, 464-8603 Japan
nhayakaw@nuee.nagoya-u.ac.jp

Hemmert, David
Pulsed Power and Power Electronics Center
Dept. of Electr. and Comp. Engineering
Texas Tech University
Lubbock, TX 79409, USA
druidelf@angelfire.com

Herńandez-Ávila Jose Luis
Universidad Autonoma Metropolitana
CBI-Energia
02220 Mexico
hajl@correo.azc.uam.mx

Hikita, Masayuki
Kyushu Institute of Technology
Dept. of Electrical Engineering
1-1 Sensui-cho, Tobata, Kitakyushu,
Fukuoka, Japan
hikita@ele.kyutech.ac.jp

Hotop, Hartmut
Fachbereich Physik
Universität Kaiserslautern
D-67653 Kaiserslautern, Germany
hotop@physik.uni-kl.de

Illenberger, Eugen
Institut fuer Chemie - Physikalische und
Theoretische Chemie
Freie Universitaet Berlin, Takustrasse 3
D-14195 Berlin, Germany
iln@chemie.fu-berlin.de

Izcara, Jesus
Ormazabal y Cia
Barrio Gezala 2
48140 IGORRE(Vizcaya), Spain
jih@ormazabal.com

Jannick, Peter
Solvay Fluor und Derivate GmbH und Co. KG
– Technical Service SF_6
Hans-Böckler-Allee 20
30173 Hannover, Germany
Peter.Jannick@solvay.com

Karayiannopoulos, Constantine
School of Electrical and Computer Engineering
National Technical University of Athens
9 Iroon Polytechniou Str.
Zografou 157 73, Athens, Greece
ckarag@central.ntua.gr

Karsten, Juhre
SIEMENS, Germany
karsten.juhre@siemens.com

Kawamura, Tatsuo
Strategic Planning Department
High-Voltage Switchgears Division
8-1-1 Tsukaguchi-Honmachi,
Amagasaki, Hyogo 661-8661, Japan
tatsuo@tkg.att.ne.jp

Kim, Yong-Ki
National Institute of Standards and Technology
Gaithersburg
Maryland 20899-8422, U.S.A.
kim@nist.gov

Kimura, Mineo
Graduate School of Science and Engineering
Yamaguchi University
Ube, Yamaguchi 755-8611, Japan
mineo@po.cc.yamaguchi-u.ac.jp

Kitromilides, P.
National Research Institute
Vassileos Constantinou 48
Athens 116 35, Greece
kne@eie.gr

Knobloch, Hartmut
Siemens AG, Berlin, Germany
hartmut.knobloch@siemens.com

Korasli, Celal
Unersity of Bahrain
Department of Electrical Engineering
P. O. Box 32038
Isa Town, Kingdom of Bahrain
korasli@eng.uob.bh

Koshino, Naoto
Nagoya University
Integrated Building 7F 701
Furo-cho, Chikusa-ku
Nagoya, 464-8603, Japan
koshino@okubo.nuee.nagoya-u.ac.jp

Kuffel, Edmund
Department of Electrical and Computer
Engineering, University of Manitoba
Winnipeg R3T 5V6, Canada
kuffel@cc.umanitoba.ca

Lazouras, George
Enginnering of Substation Subdivision of
Regional Section for Maintenance of North
Greece, PPC
POB 10757, 54110, Greece

Makochekanwa Casten
Faculty of Engineering, Yamaguchi University
Ube, 755-8611, Japan
B1273@yamaguchi-u.ac.jp

Matsumoto, Satoshi
Kyushu Institute of Technology
Department of Electrical Engineering
1-1 Sensui-cho, Tobata, Kitakyushu
Fukuoka, Japan
matsumoto@ele.kyutech.ac.jp

Mayhew, Chris
Molecular Physics Group
School of Physics and Astronomy
University of Birmingham
Edgbaston, Birmingham, B15 2TT, UK
C.MAYHEW@bham.ac.uk

Meijer, Sander
Delft University of Technology
Faculty of Electrical Engineering, Mathematics and Computer Science
High Voltage Technology and Management
P. O. Box 5031
2600 GA Delft, The Netherlands
S.Meijer@ewi.tudelft.nl

Mergos, John
School of Electrical and Computer Engineering
National Technical University of Athens
9 Iroon Polytechniou Str.
Zografou 157 73, Athens, Greece
jamergos@mail.ntua.gr

Mikropoulos, Panteleimon
Aristotle University of Thessaloniki
Department of Electrical & Computer Engineering
High Voltage Laboratory, Building D
Egnatia Str., Thessaloniki 541 24, Greece
pnm@auth.gr

Okubo, Hitoshi
Center of Integrated Research in Science and Engineering
Nagoya University
Nagoya, 464-8603 Japan
okubo@nuee.nagoya-u.ac.jp

Olthoff, James
National Institute of Standards and Technology
100 Bureau Dr., Mail Stop 8110
Gaithersburg, MD 20899-8110, USA
james.olthoff@nist.gov

Ohtsuka, Shiwya
Kyushu Institute of Technology
Department of Electrical Engineering
1-1 Sensui-cho, Tobata, Kitakyushu
Fukuoka, Japan
ohtsuka@ele.kyutech.ac.jp

Papagiannakopoulos, Panos
National Research Institute
Vassileos Constantinou 48
Athens 116 35
Greece
panosp@eie.gr

Renza, Berti
CESI,
Via Rubalttino 54
20235 Milano, Italy
BERTI@cesi.it

Schmidt, Martin
Institut für Niedertemperatur Plasmaphysik,
17489 Greifswald, Germany
schmidtm@inp-greifswald.de

Schöffner, Günther
Siemens AG, Erlangen, Germany D-91052
guenther.schoeffner@siemens.com

Schramm, Heinz H.
Siemens AG, Berlin, Germany
heinzh.schramm@berlin.de

Skalny, Jan
Department of Plasma Physics
Comenius University, Mlynska dolina F-2
84248 Bratislava, Slovakia
and
Department of Physics and Astronomy
Centre of Molecular and Optical Sciences
The Open University
Walton Hall, Milton Keynes MK7 6AA, UK
Jan.Skalny@fmph.uniba.sk

Skarby, Per
ABB, Zurich, Switzerland
per.skarby@ch.abb.com

Sölver, Carl-Einar
ABB Power Technologies
PTHV/HV/MT, SE-771 80
Ludvika, Sweden
carl-ejnar.solver@se.abb.com

Stoffels, Eva
Department of Biomedical Engineering
Eindhoven University of Technology
P. O. Box 513, 5600 MB Eindhoven
The Netherlands
e.stoffels.adamowicz@tue.nl

Stangherlin, Silvio
ABB Switzerland Ltd., Corporate Research
Segelhof 1, CH – 5405 Baden – Daettwil – Switzerland
silvio.stangherlin@ch.abb.com

Stassinopoulos, Costantine
Aristotle University of Thessaloniki
Department of Electrical & Computer Engineering
High Voltage Laboratory, Building D
Egnatia Str., Thessaloniki 541 24, Greece
costis@stassinopoulos.com / cas@eng.auth.gr

Stavropoulos, Demetre
PPC, 30 Halkokondyli Str.
Athens 104 32, Greece
d-stavropoulos.dksd@ppc.gr

Szamrej, Iwona
University of Podlasie
Chemistry Department
Siedlce, Poland
iwona@ap.siedlce.pl

Takuma, Tadasu
2-11-1 Iwado-kita, Komae-shi
Tokyo 201-8511, Japan
takuma@criepi.denken.or.jp

Trapezontas, Nikolas
72 Athinon Ave, Neo Phaliro
Athens 18547, Greece
tsksd@otenet.gr

Tsekouras, Nikolas
Transmission System Department
70 Ag. Annis Str., 12241 Athens, Greece
ntsek@tee.gr

Tsiroulis, Konstantine
Transmission Planning and Performance Management, PPC
89 Dyrrachiou Str., Athens 10443, Greece
ktsiroulis@dei.gr

Uchii, Toshiyuki
TMT&D Corporation
Kawasaki, Japan 210-0862
toshiyuki.uchii@tmt-d.co.jp

Vassiliou, Panayota
School of Chemical Engineering
National Technical University of Athens
9 Iroon Polytechniou Str., Zografou 157 80
Athens, Greece
pvas@chemeng.ntua.gr

Wagner, Hans-Erich
Ernst-Moritz-Arndt University of Greifswald
17489 Greifswald, Germany
wagner@physik.uni-kl.de

Weltmann, Klaus-Dieter
Ernst-Moritz-Arndt University of Greifswald
17489 Greifswald, Germany

Yamagiwa, Tokio
Japan AE Power Systems Corporation
1-1, Kokubu-cho, 1-chome
Hitachi-shi Ibaraki-ken
316-8501, Japan
yamagiwa-tokio@mb.jaeps.com

Yanabu, Satoru
Department of Electrical Engineering
Denki University
101-8457 2-2 Kandanishiki-cho
Chiyoda-ku, Tokyo, Japan
yanabu@e.dendai.ac.jp

PHOTOGRAPHS OF PARTICIPANTS

1.	Martin Schmidt	9.	Hiroyuki Hama	17.	Peter Castle
2.	Hans-Erich Wagner	10.	Edmond Kuffel	18.	Eugen Illenberger
3.	Peter Jannick	11.	Hitoshi Okubo	19.	Hartmut Hotop
4.	Harmut Knobloch	12.	Iwona Szamrej	20.	M. Kimura
5.	K.O.Weltman	13.	Michel Frechette	21.	John Mergos
6.	George Lazouras	14.	Eva Stoffels	22.	Sato. Yanabu
7.	J. Gerhold	15.	Wolfgang Degen	23.	Kaz Adamiak
8.	T. Yamagiwa	16.	Loucas G. Christophorou	24.	J.D. Skalny

1. Heinz Shramm
2. Panayota Vassiliou
3. Toshiyaki Uchii
4. Per Skarby
5. Sasa Dujko
6. Demetri Ekonomou
7. C. Tsiroulis
8. Sander Meijer
9. Yiota Alexopoulou
10. A. Hatziantoniou
11. D. Hatziantoniou
12. J. de Urquijo-Carmona
13. Karsten Juhre
14. Mieczyslaw Forys
15. Casten Makochekanwa
16. Yong-Ki Kim
17. M. Hikita
18. S. Ohtsuka
19. H. Goshima
20. Alain Gleizes
21. M. Elborki
22. Jack Ekin
23. Silvio Stangherlin
24. Satoshi Matsumoto
25. Naoki Hayakawa
26. David Hemmert

1.	Chris Mayhew	6.	C. A. Stassinopoulos
2.	James Olthoff	7.	Tatsuo Kawamura
3.	Naoto Koshino	8.	Celal Korasli
4.	C. G. Yakinthos	9.	J.L. Hernandez-Avila
5.	P. N. Mikropoulos	10.	T. Takuma

From left: George Contaxis, Carl-Einar Solver and Constantine T. Dervos

AUTHOR INDEX

AUTHOR INDEX

SUBJECT INDEX

SUBJECT INDEX

MIX
Papier aus verantwortungsvollen Quellen
Paper from responsible sources
FSC® C105338
FSC
www.fsc.org

If you have any concerns about our products,
you can contact us on
ProductSafety@springernature.com

In case Publisher is established outside the EU,
the EU authorized representative is:
Springer Nature Customer Service Center GmbH
Europaplatz 3, 69115 Heidelberg, Germany

Printed by Libri Plureos GmbH
in Hamburg, Germany

MIX
Papier aus verantwortungsvollen Quellen
Paper from responsible sources
FSC® C105338
www.fsc.org

If you have any concerns about our products,
you can contact us on
ProductSafety@springernature.com

In case Publisher is established outside the EU,
the EU authorized representative is:
Springer Nature Customer Service Center GmbH
Europaplatz 3, 69115 Heidelberg, Germany

Printed by Libri Plureos GmbH
in Hamburg, Germany